QUANTUM OPTOMECHANICS AND NANOMECHANICS

Lecture Notes of the Les Houches Summer School:

Volume 105, 3–28 August 2015

Quantum Optomechanics and Nanomechanics

Edited by

Pierre-Francois Cohadon, Jack Harris,
Florian Marquardt, Leticia F. Cugliandolo

OXFORD

UNIVERSITY PRESS

OXFORD
UNIVERSITY PRESS

Great Clarendon Street, Oxford, OX2 6DP,
United Kingdom

Oxford University Press is a department of the University of Oxford.
It furthers the University's objective of excellence in research, scholarship,
and education by publishing worldwide. Oxford is a registered trade mark of
Oxford University Press in the UK and in certain other countries

© Oxford University Press 2020

The moral rights of the authors have been asserted

First Edition published in 2020
Impression: 1

Published in the United States of America by Oxford University Press
198 Madison Avenue, New York, NY 10016, United States of America

British Library Cataloguing in Publication Data
Data available

Library of Congress Control Number: 2019947647

ISBN 978–0–19–882814–3

DOI: 10.1093/oso/9780198828143.003.0001

Printed and bound by
CPI Group (UK) Ltd, Croydon, CR0 4YY

École de Physique des Houches

Service inter-universitaire commun
à l'Université Joseph Fourier de Grenoble
et à l'Institut National Polytechnique de Grenoble

Subventionné par l'Université Joseph Fourier de Grenoble,
le Centre National de la Recherche Scientifique,
le Commissariat à l'Énergie Atomique

Directeur:
Leticia F. Cugliandolo, Sorbonne Universités, Université Pierre et Marie Curie, Laboratoire de Physique Théorique et Hautes Energies, Paris, France

Directeurs scientifiques de la session:
Pierre-Francois Cohadon, Laboratoire Kastler Brossel, École Normale Supérieure, Paris, France

Jack Harris, Department of Physics, Department of Applied Physics, and the Yale Quantum Institute, Yale University, New Haven, CT, USA

Florian Marquardt, Max Planck Institute for the Science of Light and Friedrich-Alexander Universitat Erlangen-Nurnberg, Erlangen, Germany

Leticia F. Cugliandolo, Sorbonne Universités, Université Pierre et Marie Curie, Laboratoire de Physique Théorique et Hautes Energies, Paris, France

Previous Sessions

Publishers

- — Session VIII: Dunod, Wiley, Methuen
- — Sessions IX and X: Herman, Wiley
- — Session XI: Gordon and Breach, Presses Universitaires
- — Sessions XII–XXV: Gordon and Breach
- — Sessions XXVI–LXVIII: North Holland
- — Session LXIX–LXXVIII: EDP Sciences, Springer
- — Session LXXIX–LXXXVIII: Elsevier
- — Session LXXXIX– : Oxford University Press

Preface

The Les Houches Summer School 2015 covered the emerging fields of cavity optomechanics and quantum nanomechanics. Optomechanics is flourishing (with the field growing from only very few groups 20 years ago to a dozen by 2006 and presently more than 80 worldwide), and its concepts and techniques are now applied to a wide range of fields.

Although it has its earliest roots in Einstein's 1909 derivation of the photon momentum and in the famous debates between Einstein and Bohr in the 1920s, modern quantum optomechanics was born in the late 70s in the framework of gravitational-wave interferometry. The initial focus of these debates was on the quantum limits of displacement measurements. Carlton Caves, Vladimir Braginsky, and others realized that the sensitivity of the anticipated large-scale gravitational-wave interferometers (GWI) was fundamentally limited by the quantum fluctuations of the measurement laser beam. The combination of detector shot-noise and radiation-pressure fluctuations lead to a Standard Quantum Limit (SQL)—the smallest observable displacement using coherent laser light. After tremendous experimental progress, the sensitivity of the upcoming next generation of GWI will effectively be limited by quantum noise.

In this way, quantum-optomechanical effects will directly affect the operation of what is arguably the world's most impressive precision experiment. The Laser Interferometer Gravitational-Wave Observatory (LIGO) has captured the spotlight in science due to its spectacular detection of gravitational waves in September 2015. Several of this school's speakers are part of the LIGO-Virgo Collaboration, and so they received the exciting news only a few weeks after returning from Les Houches, even though the subsequent detailed analysis would mean that 'the rest of us' only learned about it in February 2016. In these lecture notes, the relation of optomechanics to gravitational wave detection is explored in depth by David Blair, who also describes new experimental techniques that exploit optomechanical effects and may eventually be applied to LIGO. The basic theory of measurements in the gravitational wave context was discussed at the school by Yanbei Chen.

However, optomechanics has gained a life of its own with a focus on the quantum aspects of moving mirrors. Laser light can be used to cool mechanical resonators well below the temperature of their environment. After proof-of-principle demonstrations of this cooling in 2006, a number of systems were used as the field gradually merged with its condensed matter cousin (nanomechanical systems) to try to reach the mechanical quantum ground state, eventually demonstrated in 2010 by pure cryogenic techniques and just one year later by a combination of cryogenic and radiation-pressure cooling. Similar experiments now operate with mechanical elements cooled by either optical or microwave resonators.

The lecture notes of Antoine Heidmann and Pierre-François Cohadon provide an overview of these historical development of optomechanics, as well as giving a general introduction to the elementary physical concepts of the field. While radiation pressure is the 'standard' light force that comes to mind first, in micro- and nanomechanical systems many other forces can be present and should be taken into account. This is discussed in the notes by Ivan Favero. The basic quantum theory of cavity optomechanics is nicely summarized in the notes by Aashish Clerk, who then adds a detailed discussion of the quantum theory of measurement. One of the curiosities of quantum optomechanics is that the effective temperatures are very different for the optical photons (cold) and the phonons (warm). This could be used to even build a small optomechanical heat engine, an example of the active field of quantum thermodynamics, as explained by Pierre Meystre.

The concepts of optomechanics have also been extended to include hybrid quantum systems, where optomechanical interactions may be exploited to convert quantum information from the GHz domain in the solid state to flying qubits (photons), or cold trapped atoms. Pulsed schemes inspired by quantum information can be envisioned to manipulate quantum mechanical states. An additional promising goal is to test fundamental questions of quantum physics in optomechanical setups.

Coherent wavelength conversion from microwaves to optics may become one of the most promising applications of quantum optomechanics, and this is described in the lectures by Konrad Lehnert, who introduces microwave circuit optomechanics and its applications.

Andrew Cleland explains how one may couple superconducting qubits to both electromagnetic cavities (photons) and piezoelectric resonators (phonons). The qubit-phonon coupling is a perfect example of a hybrid quantum system involving mechanical motion. Other two-level systems are being considered for different applications. One outstanding controllable solid-state two-level system lives in diamond: the NV centre. Recently, it has been understood that NVs can be made to couple to mechanical motion, and that is described in the lectures by Ania Bleszynski Jayich.

One ever-present challenge in opto- and nanomechanics is to reduce the friction. A good part of the mechanical damping is often due to emission of sound waves into the chip to which the resonator is attached. This can be completely avoided by levitating the mechanical object. One example of this is atomic clouds, trapped in a laser wave. They can be coupled to another, more mesoscopic mechanical resonator via the light field, and this is the approach that Philipp Treutlein describes. However, instead of atoms, it is also possible to levitate micron-scale dielectric spheres. The hope is that such experiments can lead one to novel insights into fundamental decoherence processes and to searches for deviations from conventional quantum theory. This is a topic that Oriol Romero-Isart explores.

The School has brought together experts and students from different communities: gravitational-wave detection, quantum optics, atomic physics and condensed matter, for a global view of wide-ranging subject matter. As organizers, we were delighted to see this resulting in very lively discussions, which of course were mostly thanks to the students.

They were eager to exploit the opportunity to pose questions directly to the expert speakers and also to interact with them during the breaks and hikes.

In addition to the lectures, shorter research seminar talks were given by a variety of international experts (Eva Weig, Klemens Hammerer, Ivan Favero, Peter Rakich, Eugene Polzik, Tobias Kippenberg, Olivier Arcizet, Serge Reynaud, Jess Riedel and Nikolai Kiesel).

As with all such endeavours, collecting the lecture notes and creating this book was a lengthy process. In the end, some of the lecturers could not provide notes (Nergis Mavalvala and Johannes Fink). However, and fortunately, we were able to record a good fraction of the lectures on video, and they are hosted online. Please find the link at https://houches.univ-grenoble-alpes.fr/the-school/online-lectures/.

We would like to thank all the lecturers as well as the research seminar speakers for their contributions in making this school a success. We also acknowledge support from the EU-funded Marie Curie Initial Training Network cQOM and the Franco-German University. Special thanks go to Mrs Gesine Murphy for helping out with editing the LaTeX manuscript. We hope that you will enjoy these lecture notes as much as we enjoyed the lectures in August of 2015, in the inspiring atmosphere of Les Houches, in the French Alps next to Mont Blanc.

Pierre-François Cohadon
Jack Harris
Florian Marquardt

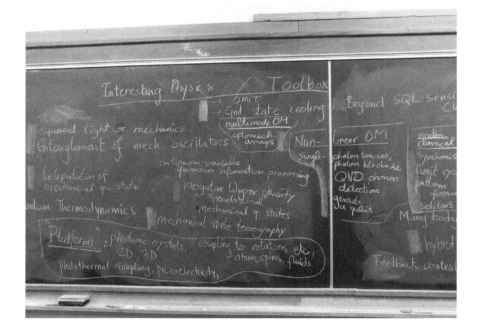

Contents

List of Participants

Organizers

COHADON Pierre-François
Laboratoire Kastler Brossel, Ecole Normale Supérieure, Paris, France

HARRIS Jack
Yale University, New Haven, USA

MARQUARDT Florian
Max Planck Institute for the Science of Light and University
Erlangen-Nürnberg, Germany

Lecturers

BLAIR David
University of Western Australia, Perth, Australia

CHEN Yanbei
California Institute of Technology, Pasadena, USA

CLELAND Andrew
University of Chicago, USA

CLERK Aashish
Institute of Molecular Engineering, University of Chicago, Illinois, USA,
and McGill University, Montréal, Canada

HEIDMANN Antoine
Laboratoire Kastler Brossel, Sorbonne Université, Paris, France

JAYICH Ania
University of California, Santa Barbara, USA

LEHNERT Konrad
University of Colorado, Boulder, USA

MAVALVALA Nergis
MIT, Cambridge, USA

MEYSTRE Pierre
University of Arizona, Tucson, USA

ROMERO-ISART Oriol
IQOQI, Innsbruck, Austria

TREUTLEIN Philipp
University of Basel, Switzerland

Seminar Speakers

ARCIZET Oliver
Institut Néel, CNRS, Grenoble, France

FAVERO Ivan
Université Paris-Diderot, Paris, France

FINK Johannes
California Institute of Technology, Pasadena, USA

HILD Stefan
University of Glasgow, UK

HAMMERER Klemens
Leibniz University, Hannover, Germany

KIPPENBERG Tobias
EPFL, Lausanne, Switzerland

KIESEL Nikolay
VCQ, Vienna, Austria

POLZIK Eugene
Niels Bohr Institute, Copenhagen University, Denmark

RAKICH Peter
Yale University, New Haven, USA

REYNAUD Serge
Laboratoire Kastler Brossel, Paris, France

RIEDEL Jess
Perimeter Institute, Canada

WEIG Eva
University of Konstanz, Germany

Students

AMITAI Ehud
University of Basel, Switzerland

ARRANGOIZ Patricio
Stanford University, USA

ASJAD Muhammad
University of Camerino, Italy

BARFUSS Arne
University of Basel, Switzerland

BRUCH Anton
Freie University, Berlin, Germany

BRUNELLI Matteo
Queen's University, Belfast, UK

CERNOTIK Ondrej
Leibniz University, Hannover, Germany

COHEN Martijn
University of Delft, Netherlands

CRIPE Jonathan
Louisiana State Unviersity, Baton Rouge, USA

DOMINGUEZ GOMEZ Diana Melisa
University of Antioquia, Medellin, Colombia

EDBLOM Christin
Chalmers University, Göteborg, Sweden

ELOUARD Cyril
Institut Néel, CNRS, Grenoble, France

FEDOROV Sergey
MIPT, Moscow, Russia

FIGUEIREDO ROQUE Thales
University of Campinas, Brazil

FOGLIANO Francesco
Pisa University, Pisa, Italy

GARCES MALONDA Rafael
Universidad de Valencia, Burjassot, Spain

GERBER Justin
University of California, Berkeley, USA

GUCCIONE Giovanni
Australian National University, Acton, Australia

HAMOUNI Mehdi
Université Paris Diderot, Paris, France

HELOU Bassam
California Institute of Technology, Pasadena, USA

HO Melvyn
University of Basel, Switzerland

KAIKKONEN Jukka-Pekka
Aalto University, Finland

KASHKANOVA Anna
Yale University, New Haven, USA

KATAOKA Yuu
Tokyo Institute of Technology, Japan

KOROBKO Mikhail
University of Hamburg, Germany

KRALJ Nenad
University of Camerino, Italy

LEMONDE Marc-Antoine
McGill University, Montreal, Canada

LUNA Jose Fernando
University of California, Santa Barbara, USA

MAILLET Olivier
Institut Néel, CNRS, Grenoble, France

MARKOVIC Danijela
ENS, Paris, France

MERCIER DE LEPINAY Laure
Institut Néel, CNRS, Grenoble, France

MESTRES JUNQUE Pau
ICFO, Castelldefels, Spain

MILBURN Thomas
Vienna University of Technology, Vienna, Austria

MOLLER Christoffer
Niels Bohr Institute, Copenhagen University, Denmark

NOGUCHI Atsushi
University of Tokyo, Japan

NOURY Adrien
ICFO, Castelldefels, Spain

PAGE Michael
University of Western Australia, Crawley, Australia

PANG Belinda
California Institute of Technology, Pasadena, USA

PEREIRA MACHADO Joao
University of Delft, Netherlands

PFEIFER Hannes
Max Planck Institute for the Science of Light, Erlangen, Germany

PINO Herman
IQOQI, Innsbruck, Austria

RAKHUBOVSKY Andrey
Palacky University, Olomouc, Czech Republic

ROULET Alexandre
University of Singapore, Singapore

SAFIRA Arthur
Harvard University, Cambridge, USA

SHKARIN Alexey
Yale University, New Haven, USA

SINGH Swati
Harvard University, Cambridge, USA

SINHA Kanupriya
University of Maryland, College Park, USA

SONG Taegeun
ICTP, Trieste, Italy

STEINMEYER Daniel
Leibniz University, Hannover, Germany

SUNDARESAN Neereja
Princeton University, USA

WANG Tian
California Institute of Technology, Ithaca, USA

WELKER Gesa
University of Leiden, Netherlands

1

Early History and Fundamentals of Optomechanics

Antoine Heidmann and Pierre-Francois Cohadon

Laboratoire Kastler Brossel, Sorbonne Université, CNRS, ENS-PSL Research University, Collège de France, Paris, France

Antoine Heidmann

Antoine Heidmann and Pierre-Francois Cohadon, *Early History and Fundamentals of Optomechanics* In: *Quantum Optomechanics and Nanomechanics.* Edited by: Pierre-Francois Cohadon, Jack Harris, Florian Marquardt, Leticia F. Cugliandolo, Oxford University Press (2020). © Oxford University Press. DOI: 10.1093/oso/9780198828143.003.0001

Chapter Contents

1.1 Introduction

Optomechanics is an emerging research field that deals with the interaction between a laser beam and a mechanical resonator through radiation pressure. It was born in the context of gravitational-wave detection and still has many applications in sensing, but it also addresses some fundamental aspects of quantum mechanics. In its simplest and historical form, optomechanics amounts to two complementary coupling effects: mechanical motion changes the path followed by light, but light (through radiation pressure) can drive the mechanical resonator into motion as well.

Optomechanics allows one to control resonator motion by laser cooling down to the quantum ground state, or to control light by using back-action in optical measurements and in quantum optics. Modern optomechanics takes advantage of dedicated mechanical resonators with low mass and high mechanical quality factor, inserted in very sensitive optical interferometers based on high-finesse optical cavities. Its main applications are optomechanical sensors to detect tiny mechanical motions and weak forces, cold damping and laser cooling, and quantum optics.

The objectives of this course are to provide a brief account of the history of the field, together with its fundamentals. We will in particular review both classical and quantum aspects of optomechanics, together with its applications to high-sensitivity measurements and to control or cool mechanical resonators down to their ground state, with possible applications for tests of quantum theory or for quantum information.

1.2 Optomechanics at the Classical Level

1.2.1 A Brief History of Radiation Pressure

In 1619, Johannes Kepler was the first to postulate the existence of a force which pushes one of the tails of comets away from the Sun. In *De Cometis Libelli Tres*, he wrote: 'The direct rays of the Sun strike upon the comet, penetrate its substance, draw away with them a portion of this matter, and issue thence to form the track of light we call the tail.'

Maxwell's theory of electromagnetism indeed stipulates that electromagnetic waves carry energy and momentum described by the Poynting vector $\vec{\Pi} = \vec{E} \wedge \vec{B}/\mu_0$. This implies a momentum transfer $\vec{P}_{abs} = \vec{\Pi}/c$ for an absorbing media and $\vec{P}_{rad} = 2\vec{\Pi}/c$ for

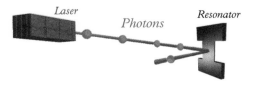

Fig. 1.1 *A light beam reflected on a moving mirror induces a radiation pressure proportional to the number of photons per second impinging the mirror, whereas the phase of the reflected field is changed by the mirror motion.*

a perfect reflector. For a propagating light beam, Π is the incident intensity (normalized as an energy flux) that can be written in terms of photons as $\Pi = \hbar\omega_0 \times I(t)$ where ω_0 is the light frequency and $I(t)$ the photon flux per second. Radiation pressure now appears as $P_{rad} = 2\hbar k \times I(t)$ where $k = 2\pi/\lambda = \omega_0/c$ is the wavevector (λ is the wavelength). This leads to the well-known interpretation of radiation pressure as the momentum transfer $2\hbar k$ for every incident photon, multiplied by the number of photons per second.

Radiation pressure is a tiny effect: for the light from the Sun at the surface of the Earth, the power flux of 1 kW/m^2 corresponds to a pressure of only 3 μPa, or a 10-pN force over a disc of 1-mm radius. Even for a full 1-W continuous-wave laser, the total force only amounts to a few nN.

The first experimental observations of radiation pressure were made in the early 1900s independently by Lebedev (Lebedev, 1901) and Nichols and Hull (Nichols and Hull, 1903). Both were based on improved versions of the Crookes radiometer (Crookes, 1874), which is a mill with vanes having black and white sides (and accordingly different absorption and reflection from the faces), leading to a net effect of mill rotation when irradiated by light. But as usual radiometers are dominated by thermal effects in the residual gas around the vanes, they turn in the opposite direction from what one would expect from radiation pressure (Crawford, 1985)! Improved versions with silver coating on the vanes, a proper choice of gas pressure to reduce thermal effects, and a transient exposure to light led to the demonstration of radiation pressure with a 1% accuracy, which was considered at the time as a demonstration of Maxwell's equations.

Among the many consequences of radiation pressure, let's note the astronomical implications of solar radiation pressure, which induces orbital perturbations (of dust, asteroids, spacecrafts...) but also leads to possible spacecraft propulsion using solar sails. As an example, the Ikaros mission took advantage of a 14×14 m^2 sail with embedded LCD panels of variable reflectance to control the trajectory, and managed to make a 6-month journey to Venus. On a very different scale, another well-known example is the laser cooling of atoms for which the 1997 Nobel Prize was awarded to S. Chu, C. Cohen-Tannoudji and W. Phillips.

Other consequences of radiation pressure are optomechanical effects induced by light on mechanical resonators. These effects have been first studied in the classical domain, with the demonstration of modification by light of the stiffness and damping of a mechanical resonator (as soon as in the 1970s), or the observation of light-induced bistability (in the 1980s).

1.2.2 Classical Description of Optomechanical Coupling

We will first derive the fundamental equations that govern an optomechanical system, excluding quantum noises. The simplest classical description for such a system consists of a lossless single-ended cavity with a movable mirror (Fig. 1.2). Mirror motion is given by the time-dependent position $x(t)$, whereas the intracavity optical field is described by the complex amplitude $\alpha(t)$ which is the classical counterpart of the annihilation operator $\hat{a}(t)$. Superscripts *in* and *out* are used for the incident and reflected fields (in the case of a lossless cavity, we have a single input–output port located at the input mirror).

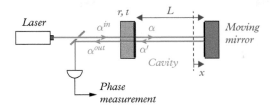

Fig. 1.2 *The simplest optomechanical system: a single-ended lossless cavity with one movable mirror. A light beam (incident field amplitude α^{in}) is sent into the cavity. The phase of the reflected field α^{out} monitors the mirror motion.*

The intensity is normalized as a number of photons per second: $I(t) = |\alpha(t)|^2$, and the radiation pressure experted by the intracavity field on the mirror is as previously given by $F_{\mathrm{rad}}(t) = 2\hbar k \times I(t)$.

Mechanical response of the moving mirror: optical spring and damping effects.

The mirror response to a force is described in the following as a harmonic oscillator of mass m, resonance frequency Ω_m (we will actually only refer to *angular* frequencies), mechanical damping Γ (or, equivalently, mechanical quality factor $Q = \Omega_m/\Gamma$). Linear response theory gives the Fourier transform $x[\Omega]$ of the mirror position as a function of the external force F applied on the mirror:

$$x[\Omega] = \chi[\Omega]F[\Omega], \tag{1.1}$$

$$\text{with } \chi[\Omega] = 1/m\left(\Omega_m^2 - \Omega^2 - i\Gamma\Omega\right), \tag{1.2}$$

where the Fourier transform is defined as $x[\Omega] = \int e^{i\Omega t}x(t)$.

Mirror motion changes the cavity length and then the detuning of the cavity with respect to the incident field. The intracavity intensity and the radiation pressure therefore become sensitive to the mirror motion, which can change both the stiffness and the damping of the mirror. Such effects were first observed by Braginsky in 1970 using a torsional oscillator at the end of a microwave resonator. Performing an optical measurement of the oscillations, he demonstrated stiffness and damping changes under microwave radiation (Braginsky, Manukin and Tikhonov, 1970). Related effects were also observed in the 1990s by M. Tobar and D. Blair within the context of resonant gravitational-wave detectors, using a 10-cm long high-Q (10^7) sapphire bar with a parametric coupling between the acoustic (50 kHz) and the microwave (10 GHz) modes in the crystal (Locke, Tobar, Ivanov and Blair, 1998).

Cavity dynamics: optomechanical bistability

The fundamental equations of the fields in presence of a movable mirror are deduced from the input–output relations of the fields at the coupling mirror, and the field

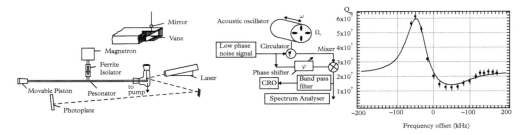

Fig. 1.3 *First experiments of dynamical radiation-pressure effects in a detuned cavity: radiation pressure appears as a viscous force which changes the damping of the resonator, both in a torsional oscillator inserted in a microwave cavity (left, (Braginsky, Manukin and Tikhonov, 1970)) and in a high-Q sapphire bar (centre, (Locke, Tobar, Ivanov and Blair, 1998)). The curve on the right shows the resulting quality factor change of the bar as a function of cavity detuning. Figure adapted with permission from references cited.*

propagation in the cavity. For example, the output field is given by $\alpha^{out} = -r\alpha^{in} + t\alpha'$ where r and t are the reflection and transmission amplitude coefficients of the coupling mirror, and one gets for the intracavity propagation $\alpha' = \alpha e^{i\psi}$ where $\psi = \psi_0 + 2kx$ is the position-dependent cavity detuning.

For a high-finesse and lossless cavity, reflection is close to 1: $r = 1 - \gamma$ and $t = \sqrt{2\gamma}$ where $\gamma \ll 1$ is the cavity damping. This yields:

$$\tau \frac{d\alpha}{dt}(t) = (-\gamma + i\psi(t))\alpha(t) + \sqrt{2\gamma}\alpha^{in}(t), \tag{1.3}$$

$$\alpha^{out}(t) = \sqrt{2\gamma}\alpha(t) - \alpha^{in}(t), \tag{1.4}$$

where $\tau = 2L/c$ is the cavity round-trip time.

Equation (1.3) is the usual equation for a Fabry–Perot cavity except that the detuning $\psi(t)$ now depends on the position $x(t)$ of the moving mirror. Equation (1.4) gives the reflected field as a result of the interference between the incident field directly reflected onto the coupling mirror, and the transmitted intracavity field. These equations may be rewritten in a more usual form for optomechanics, introducing the cavity damping rate $\kappa = 2\gamma/\tau$ and normalizing the intracavity field as a number of photons in the cavity rather than a flux: $\tilde{\alpha} = \sqrt{\tau}\alpha$. One gets:

$$\dot{\tilde{\alpha}}(t) = \left(-\frac{\kappa}{2} + i\Delta + iGx(t)\right)\tilde{\alpha}(t) + \sqrt{\kappa}\alpha^{in}(t), \tag{1.5}$$

$$\alpha^{out}(t) = \sqrt{\kappa}\tilde{\alpha}(t) - \alpha^{in}(t), \tag{1.6}$$

$$F_{rad}(t) = \hbar G|\tilde{\alpha}(t)|^2, \tag{1.7}$$

where $\Delta = \omega_0 - \omega_{cav} = \psi_0/\tau$ is the detuning between the laser optical frequency and the cavity resonance, and G represents the *strength of the optomechanical coupling*. It corresponds to the optical frequency shift per unit displacement and may depend on the precise geometry of the system, but it takes a simple expression in the case of our monodimensional model:

$$G = -\frac{d\omega_{cav}}{dx} = \frac{\omega_{cav}}{L}. \tag{1.8}$$

In the steady state, Eq. (1.5) yields the usual Airy peak for the mean intracavity intensity, with an additional x-dependence due to radiation pressure:

$$\bar{\alpha} = \frac{\sqrt{\kappa}}{\frac{\kappa}{2} - i\Delta - iG\bar{x}} \bar{\alpha}^{in}. \tag{1.9}$$

For a low incident power, this additional dependence can be neglected but in the general case, this nonlinear equation for the intracavity intensity can lead to an *optical bistability* of the cavity.

First observations of such a bistability were performed by H. Walther in 1983 using a cavity with a suspended 60-mg quartz plate as moving mirror, with a few-Hz resonance frequency and a cavity finesse of $\mathcal{F} = \pi/\gamma = 15$ (Fig. 1.4). Observation of the bistable behaviour was made with 1 to 2 W of incident power (Dorsel, McCullen, Meystre, Vignes and Walther, 1983). Such effects were also observed two years later in the microwave regime by A. Gozzini, using a microwave cavity with a movable wall (Gozzini, Maccarrone, Mango, Longo and Barbarino, 1985).

1.2.3   Ultra-sensitive Displacement Sensors

Development of low-loss mirrors and stable laser sources have opened the way to the use of optical interferometry for high-sensitivity measurement of length or displacement. These were first motivated by gravitational-wave detection but now have many applications such as lunar laser ranging (using laser telemetry rather than interferometry),

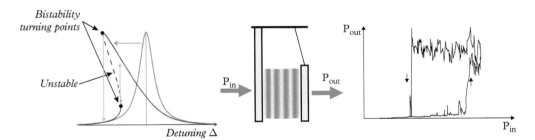

Fig. 1.4 *Principle of the bistable behaviour of an optomechanical cavity (left) and first observation with a cavity composed of a suspended mirror. The bistable hysteresis is clearly visible when the incident intensity is changed up and down. Figure adapted from Dorsel, McCullen, Meystre, Vignes and Walther (1983). Copyright (1983) by the American Physical Society.*

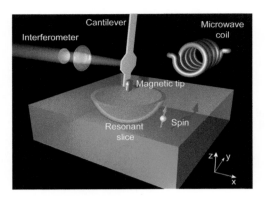

Fig. 1.5 *One example of position measurements using light: the spin flip of a single electron is detected through the oscillations of a cantilever with an attached micro-magnet, with a resolution of about 1 Å. Figure adapted with permission (Rugar, Budakian, Mamin and Chui, 2004), with permission from the authors.*

or the detection of the spin flip of a single electron seen through the interferometric measurement of the oscillations of a cantilever with an attached micro-magnet (Fig. 1.5). As the interference pattern has a periodicity of $\lambda/2$ over the length difference between the two arms, the interferometric measurement is sensitive to a small fraction of the optical wavelength λ. In their experiment, D. Rugar *et al.* reached a sensitivity of about 1 Å, leading to a detection of the magnetic force induced by the spin flip at the attonewton sensitivity level (Rugar, Budakian, Mamin and Chui, 2004).

Sensitivity can be widely improved by the use of a high-finesse cavity, as in gravitational-wave interferometers, which have a cavity in both arms (The LIGO Scientific Collaboration, 2015; The Virgo Collaboration, 2015). Once again, the simplest displacement sensor consists of a Fabry–Perot cavity with 2 mirrors. If one mirror is moving, accurate monitoring of its motion $\delta x(t)$ around its mean position can be obtained by measuring the phase shift of the optical field induced by the motion. The optimal sensitivity is reached at resonance (see Fig. 1.6), as long as the motion frequency lies within the cavity bandwidth. As the phase experiences a 2π-shift over the resonance (corresponding to a displacement δx of the order of $\lambda/2\mathcal{F}$), the phase shift of the reflected light is proportional to the displacement: $\delta\varphi_x^{out} \simeq 8\mathcal{F}\delta x/\lambda$, to be compared to the phase shift in an interferometer $\delta\varphi_x \simeq 4\pi\delta x/\lambda$. The gain is proportional to the cavity finesse $\mathcal{F}$ and the achievable displacement sensitivity can then reach down to the attometer level and below (Hadjar, Cohadon, Aminoff, Pinard and Heidmann, 1999).

One of the first high-sensitivity experiments was performed in 1999, with the scheme shown in Fig. 1.6 based on a compact cavity (0.2-mm long), made of one coupling mirror and one plano-convex mirror of width 1.5 to 2.5 mm, and a diameter of 12 to 30 mm (Hadjar, Cohadon, Aminoff, Pinard and Heidmann, 1999). The geometry of this mirror allowed confined internal acoustic modes with MHz resonance frequencies, masses at the mg level, and high mechanical Qs (10^6). The optical coatings were made of $\lambda/4$ alternating layers of SiO_2/Ta_2O_5, yielding a very high reflectivity. The optomechanical

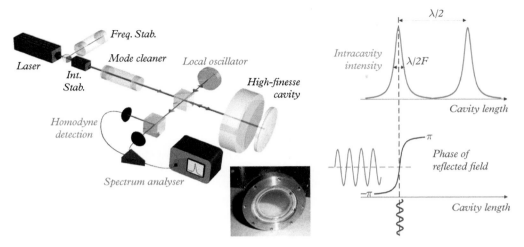

Fig. 1.6 *Optomechanical sensor based on a high-finesse cavity (left) with a moving mirror. The motion here corresponds to internal acoustic modes of a plano-convex mirror. Principle of the measurement (right): the displacement induces a phase shift of the reflected field.*

coupling is slightly different from the standard *mirror with a spring*, but the intracavity light field experiences a phase shift proportional to the longitudinal deformation of the mirror, averaged over the beam waist, of the order of 90 μm, much smaller than the size of the acoustic modes (mostly with a Gaussian shape, and an acoustic waist of 2 to 3 mm). Cavities built have finesses from 40,000 up to 300,000 due to improvement in the layer coatings, driven by research in gravitational-wave interferometry. The incident light is provided by a Titane-Sapphire laser working at 810 nm, frequency-locked to a stable reference cavity, itself locked so that the laser is resonant with the high-finesse cavity. The laser beam is spatially filtered by a mode cleaner cavity (so that only the fundamental TEM$_{00}$ mode is transmitted), leading to a 98% mode-matching with the high-finesse cavity. The phase of the reflected beam is measured by homodyne detection, with a 10-mW local oscillator derived from the incident beam (to be compared to the typical 100μW of the beam entering the cavity). The overall quantum efficiency of the homodyne detection is better than 90%.

The sensitivity is mainly limited by the shot-noise which corresponds to the phase noise of the incident laser beam $\delta\varphi_n^{out} \simeq 1/2\sqrt{\bar{I}^{in}}$, where $\bar{I}^{in}$ is the mean intracavity intensity (we assume here negligible technical noise and lossless mirrors, so that the light is completely reflected by the cavity). Compared to the displacement phase shift, this leads to a minimum measurable displacement (for a unity signal-to-noise ratio):

$$\delta x_{min} \simeq \frac{\lambda}{16\mathcal{F}\sqrt{\bar{I}^{in}}}, \tag{1.10}$$

on the order of 2×10^{-19} m/$\sqrt{\text{Hz}}$ for this experiment.

1.2.4 Optomechanical Sensors: Thermal Noise of Micromirrors

Optomechanical sensors were extensively used to characterize mechanical resonators of different geometries and sizes, either through the observation of their thermal Brownian motion or their response to an external force. Different mirrors were used, from cm to micrometre scales, but it proved difficult to go to smaller dimension and still have a high optical finesse. One of the first examples of optomechanical measurement using a microresonator is shown in Fig. 1.7. It is based on a doubly-clamped, 1-mm long silicon resonator, with a thickness of 60 μm and a width from 400 μm to 1 mm (Arcizet et al., 2006b). The resonance frequencies still are in the MHz range but compared to the previous experiment, the mass goes down to the μg level. The resonators are coated with a similar high-reflectivity and low-loss dielectric coating, now optimized at 1064 nm. The resulting noise spectrum is shown in Fig. 1.7 as well, with thermal peaks for each mechanical resonance.

Thermal noise can be accounted for using the fluctuation-dissipation theorem, which relates the spectral density of the random Langevin force F_T to the imaginary part of the mechanical susceptibility (which corresponds to the mechanical damping):

$$S_{F_T}[\Omega] = -\frac{2k_B T}{\Omega}\mathrm{Im}\left(\frac{1}{\chi[\Omega]}\right),\qquad(1.11)$$

where the spectrum $S_u[\Omega]$ at frequency Ω of a random variable $u[\Omega]$ is defined as the Fourier transform of the correlation function, or for a stationary process to the variance of the random variable in Fourier space:

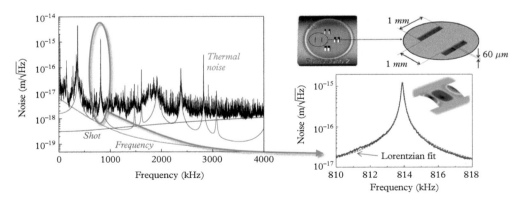

Fig. 1.7 *Thermal noise spectrum of a micromirror measured with an optomechanical sensor (left). The sensitivity is limited by the quantum shot-noise of the laser beam, at a level of 10^{-19} m/$\sqrt{\mathrm{Hz}}$. A close-up on a specific thermal peak (bottom right) clearly shows the Lorentzian behaviour of the thermal noise spectrum of a given mechanical mode, here the third transverse mode of a doubly-clamped 1-mm long beam (see picture and sizes top right).*

$$\langle u[\Omega]u[\Omega'] \rangle = 2\pi\delta\left(\Omega+\Omega'\right)S_u[\Omega], \tag{1.12}$$

where the brackets stand for the average over the random variable statistics.

For a high-Q mechanical resonance at Ω_m, the Langevin force mostly corresponds to a white noise $S_{F_T}[\Omega] = 2m\Gamma k_B T$ and the resulting mirror displacement has a Lorentzian shape:

$$S_x^T[\Omega] = 2m\Gamma k_B T \left| \chi[\Omega] \right|^2, \tag{1.13}$$

which corresponds to the experimental signal observed with the micromirror (see Fig. 1.7).

Other parameters of mechanical modes can be deduced from the measurements made with optomechanical sensors. For instance, actuation of the resonator by the radiation pressure of a modulated auxiliary laser beam provides an accurate and local probe of motion with a spatial sensitivity of a few tens of microns (see Fig. 1.8). This allows one to scan the surface and provides a spatial profile of the mechanical mode. For a plano-convex mirror, among other modes corresponding to the nearly cylindrical mirror case, one gets Gaussian modes confined at the centre with a high Q, as expected for this geometry. We have obtained an excellent agreement for the resonance frequencies as well as for the spatial profiles (see Fig. 1.8) (Briant, Cohadon, Heidmann and Pinard, 2003a).

Also, one can monitor the details of the time evolution by filtering the output signal around a given resonance frequency. Thermal noise then appears as a fast oscillation at the resonance frequency with slowly-varying random amplitude and phase (for a high-Q mode):

$$x(t) = \tilde{x}(t)\cos(\Omega_m t + \tilde{\varphi}(t)), \tag{1.14}$$

$$= X_1(t)\cos(\Omega_m t) + X_2(t)\sin(\Omega_m t), \tag{1.15}$$

where $X_1(t) = \tilde{x}(t)\cos(\tilde{\varphi}(t))$ and $X_2(t) = \tilde{x}(t)\sin(\tilde{\varphi}(t))$.

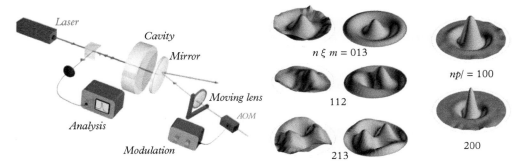

Fig. 1.8 *Actuation of a mechanical resonance by the radiation pressure of an auxiliary laser beam (left) allows one to reconstruct the spatial profile of the mechanical mode for a standard cylindrical mirror (centre left: experimental results, centre right: theoretical model) or for the Gaussian modes of a plano-convex mirror (right).*

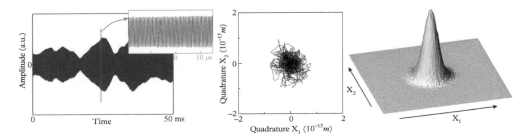

Fig. 1.9 *Temporal evolution of the Brownian motion of a mechanical mode, seen as an oscillation with slowly-varying amplitude and phase (left). The resulting temporal trajectory in phase-space is a bound random walk around the centre (middle) with a corresponding Gaussian distribution (right). Figure adapted from (Briant, Cohadon, Pinard and Heidmann, 2003b), with kind permission from The European Physical Journal (EPJ).*

Phase-space monitoring of the motion clearly demonstrates that Brownian motion is a bound random walk (around the centre for a harmonic oscillator), here with an amplitude of about 10^{-15} m. Reconstruction of the phase-space distribution over long times (typically 10 minutes, to be compared to the damping time $(2\pi/\Gamma \simeq 1/40$ s, so that phase-space is mapped about 25,000 times) leads to a Gaussian distribution with variances $\Delta X_1 = \Delta X_2 = 36 \times 10^{-17}$ m (Briant, Cohadon, Pinard and Heidmann, 2003b), in excellent agreement with the values deduced from the mass inferred from an independent mechanical characterization:

$$\Delta X_1^2 = \Delta X_2^2 = \frac{k_B T}{m\Omega_m^2}. \tag{1.16}$$

A similar analysis can be made in the cold-damped regime (see section 1.5), and shows a more confined random walk and a shrinking of the distribution, in agreement with a thermal equilibrium at lower temperature.

1.3 Optomechanics at the Quantum Level

We will now discuss quantum aspects of optomechanics. From a historical point of view, quantum optomechanics represent a new and emerging domain, although some basic and fundamental concepts have started to be discussed during the early beginning of Quantum Mechanics. Indeed, the quantum behaviour of a mechanical oscillator is a textbook example to introduce the first principles of quantum mechanics, as well as the concepts of quantum measurements and back-action effects of measurements, here due to radiation pressure. In the '80s, thanks to the development of gravitational-wave detection using interferometers, it was discovered that quantum radiation-pressure fluctuations is responsible for fundamental limits in the detection. During the last period, two aspects of quantum optomechanics were mostly considered: quantum noise due to radiation pressure and its impact in high-sensitivity displacement measurements, and the possibility to use optomechanical coupling to control the mechanical state, in particular to put a resonator in its quantum ground state corresponding to a temperature close to absolute zero.

In the following, we will first delineate the simplest theoretical point of view at the quantum level, which is based on two harmonic oscillators coupled by radiation pressure, namely a single mechanical mode of the resonator and an optical mode of the cavity. We will then derive a semiclassical approach which has been successfully used to describe such systems, and we will finally see some consequences of the quantum approach on both light and the mechanical resonator.

1.3.1 The Optomechanical Hamiltonian

We will again consider the simple model depicted in Fig. 1.2 of a single-ended cavity with a moving mirror. The intracavity field is still described by a single mode of frequency ω_{cav} with annihilation and creation operators $\hat{a}$ and $\hat{a}^\dagger$, which are similar to the classical field amplitude $\tilde{\alpha}$ and its complex conjugate $\tilde{\alpha}^\star$ defined in section 1.2.2, except for the commutation relation:

$$[\hat{a}, \hat{a}^\dagger] = 1. \tag{1.17}$$

The operator $N_{cav} = \hat{a}^\dagger \hat{a}$ yields the number of photons in the cavity. For the mechanical motion, we consider a single mechanical mode, described as a harmonic oscillator of resonance frequency Ω_m.

The resonator as a quantum harmonic oscillator

We have already described in section 1.2.2 the classical mechanical behaviour of a moving mirror as a harmonic oscillator of mass m and resonance frequency Ω_m. Within the framework of quantum mechanics, an undamped harmonic oscillator is described by the Hamiltonian:

$$H_x = \frac{\hat{p}^2}{2m} + \frac{1}{2} m \Omega_m^2 \hat{x}^2, \tag{1.18}$$

where $\hat{x}$ and $\hat{p}$ are respectively the position and momentum operators of the mechanical resonator, with $[\hat{x}, \hat{p}] = i\hbar$. Eigenstates $|n\rangle$ correspond to a given integer n of energy quanta (the phonons) with an energy $E_n = \hbar \Omega_m \left(n + \frac{1}{2}\right)$.

The ground state $|n = 0\rangle$ is characterized by a non-zero energy $E_0 = \frac{1}{2}\hbar\Omega_m$. In this state, the mean kinetic energy $\langle \hat{p}^2/2m \rangle$ and the mean potential energy $\langle \frac{1}{2} m \Omega_m^2 \hat{x}^2 \rangle$ are equal. The state is centred at the origin ($\langle \hat{x} \rangle = \langle \hat{p} \rangle = 0$) and the position and momentum dispersions correspond to the minimum allowed by the Heisenberg uncertainty relation:

$$\Delta \hat{x} = \sqrt{\langle \hat{x}^2 \rangle} = x_{zpf}, \ \Delta \hat{p} = \frac{\hbar}{2x_{zpf}}, \quad (1.19)$$

$$\text{with } x_{zpf} = \sqrt{\frac{\hbar}{2m\Omega_m}}, \quad (1.20)$$

where x_{zpf} corresponds to the amplitude of the zero-point position fluctuations.

The Hamiltonian can be written as a function of canonical operators $\hat{b}$ and $\hat{b}^\dagger$ corresponding to the annihilation and creation of a phonon, respectively, using their definitions:

$$\hat{x} = x_{zpf}\left(\hat{b} + \hat{b}^\dagger\right), \quad (1.21)$$

$$\hat{p} = \frac{-i\hbar}{2x_{zpf}}\left(\hat{b} - \hat{b}^\dagger\right). \quad (1.22)$$

The Hamiltonian then takes the usual expression for a harmonic oscillator:

$$H_x = \hbar\Omega_m\left(\hat{b}^\dagger \hat{b} + \frac{1}{2}\right). \quad (1.23)$$

If the resonator is damped via a coupling to a thermal bath, the mechanical state is a linear combination of phonon states $|n\rangle$ with a thermal distribution of phonons. At high temperature, the mean energy is $\langle H_x \rangle = m\Omega_m^2 \Delta \hat{x}^2 \simeq k_B T$ and one recovers the usual classical behaviour of a harmonic oscillator as described in section 1.2.4. According to Eq. (1.23), the mean number of phonons is:

$$n_T \simeq \frac{k_B T}{\hbar\Omega_m}. \quad (1.24)$$

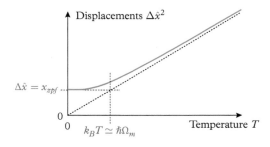

Fig. 1.10 *Variance of the position fluctuations for a mechanical resonator at thermal equilibrium at temperature T. At high temperature ($k_B T \gg \hbar\Omega_m$), the variance is proportional to the temperature (see Eq. 1.16) whereas it goes to the non-zero value x_{zpf} (Eq. 1.20) at low temperature ($k_B T \ll \hbar\Omega_m$).*

The displacement variance $\Delta \hat{x}^2 = \langle H_x \rangle / m \Omega_m^2$ then exhibits a linear dependence with temperature. As shown in Fig. 1.10, this is no longer true at low temperature ($k_B T \ll \hbar \Omega_m$) and the variance reaches a minimum value x_{zpf}^2. This behaviour can be accounted for by a generalized Langevin force as shown in section 1.3.2.

The optomechanical Hamiltonian

Independent cavity and mechanical resonators, both described as harmonic oscillators (with resonance frequencies ω_{cav} for the cavity mode and Ω_m for the mechanical resonator) have the Hamiltonian:

$$H = \hbar \omega_{cav} \hat{a}^\dagger \hat{a} + \hbar \Omega_m \hat{b}^\dagger \hat{b} \tag{1.25}$$

where we have removed the $\frac{1}{2}$-terms which play no role in the following.

The optomechanical coupling can be accounted for by considering that the frequency ω_{cav} of the cavity depends on the cavity length and then on the resonator displacement $\hat{x}$ given by Eq. (1.21). As a general rule we can write:

$$\omega_{cav}\left(\hat{x}\right) \simeq \omega_{cav} + \hat{x} \frac{\partial \omega_{cav}}{\partial x} = \omega_{cav} - G\hat{x}, \tag{1.26}$$

where G depends on the exact geometry of the system. As already discussed in section 1.2.2, G takes a simple expression in the case considered here of a monodimensional system (Eq. 1.8).

From Eqs. (1.25) and (1.26), the coupling Hamiltonian is given by:

$$H_{int} = \hbar g_0 \, \hat{a}^\dagger \hat{a} \left(\hat{b} + \hat{b}^\dagger\right) \tag{1.27}$$

$$\text{with } g_0 = G x_{zpf}. \tag{1.28}$$

g_0 is the *vacuum optomechanical coupling strength*, which corresponds to the cavity frequency shift induced by a resonator displacement equal to its zero-point fluctuations dispersion (Aspelmeyer, Kippenberg and Marquardt, 2014).

It is easy to derive the main consequences of the coupling by looking at the effect of displacement on the field and the reciprocal action. Heisenberg equations for the operators are indeed:

$$\dot{\hat{a}} = \frac{i}{\hbar} \left[H_{int}, \hat{a}\right] = iG\hat{x}\hat{a} \tag{1.29}$$

$$\hat{F}_{rad} = \dot{\hat{p}} = \frac{i}{\hbar} \left[H_{int}, \hat{p}\right] = \hbar G\hat{a}^\dagger \hat{a} \tag{1.30}$$

where $\hat{p}$ is the momentum operator of the mechanical resonator (Eq. 1.22). These equations are the quantum analogue to the classical ones derived in section 1.2.2 (Eqs. 1.5 to 1.7). They describe the phase shift $G\hat{x}$ of the intracavity field induced by the displacement and the radiation pressure exerted upon the mechanical resonator, proportional to the photon number $\hat{a}^{\dagger}\hat{a}$.

Complete quantum description of the optomechanical coupling

The full description of the system also has to include dissipation mechanisms for the field (in particular losses due to the mirrors' reflectivity) and for the mechanical resonator (damping and coupling to a thermal bath).

The losses can be taken into account by adding terms describing the dissipation and including fluctuations and external driving. Most parameters have already been introduced in section 1.2.2: photon decay is given by the cavity damping rate κ, mechanical friction is given by the damping Γ, and mechanical fluctuations are described by phonon statistics. The driving field corresponds to the incident laser beam (and vacuum fluctuations if extra cavity losses are present).

Going to the rotating frame at the laser frequency ω_0 by defining the slowly varying field operator $\tilde{a}(t) = \hat{a}(t)\exp(i\omega_0 t)$, the input–output formalism based on quantum Langevin equations leads to the following equations:

$$\dot{\tilde{a}} = \left(-\frac{\kappa}{2} + i\Delta + iG\hat{x}\right)\tilde{a} + \sqrt{\kappa}\tilde{a}^{in}, \tag{1.31}$$

$$\dot{\hat{b}} = \left(-\frac{\Gamma}{2} - i\Omega_m\right)\hat{b} + ig_0\tilde{a}^{\dagger}\tilde{a} + \sqrt{\Gamma}\hat{b}^{in} \tag{1.32}$$

where G and g_0 are the strengths of optomechanical coupling (see Eqs. 1.8 and 1.28), and Δ is the cavity detuning. Operators $\tilde{a}^{in}$ and $\hat{b}^{in}$ are input terms that represent respectively the driving field incident onto the cavity and the coupling of the mechanical oscillator to a thermal bath. They include quantum and thermal noises, characterized by the following noise correlation functions:

$$\left\langle \tilde{a}^{in}(t)\tilde{a}^{in^{\dagger}}(t')\right\rangle = \delta(t - t'), \tag{1.33}$$

$$\left\langle \hat{b}^{in}(t)\hat{b}^{in^{\dagger}}(t')\right\rangle = (n_T + 1)\delta(t - t') \tag{1.34}$$

where n_T is the mean number of phonons (Eq. 1.24). These equations are similar to the classical ones (Eqs. 1.5 to 1.7) but now deal with operators and quantum input noises instead of classical variables and thermal fluctuations.

1.3.2   Semiclassical Approach

There is a profound similarity between the classical and the quantum descriptions as long as the Hamiltonian is quadratic and fields have large mean values compared to quantum fluctuations, so that a linearization procedure can be applied.

Basic principles

The semiclassical approach is based on the description of the field by its Wigner quasi-probability distribution function. It consists in using semiclassical variables α and α^* to describe the field, randomly distributed according to the Wigner distribution $\mathcal{W}(\alpha, \alpha^*)$. Unlike the Glauber distribution which corresponds to a normal ordering of the quantum operators $\tilde{a}$ and $\tilde{a}^\dagger$, the Wigner distribution is associated with a symmetrical order. In other words, for any symmetrical combination $f_{sym}(\tilde{a}, \tilde{a}^\dagger)$ of the annihilation and creation operators, the quantum mean value is identical to the semiclassical average:

$$\left\langle f_{sym}(\tilde{a}, \tilde{a}^\dagger) \right\rangle = \int d^2\alpha \, \mathcal{W}(\alpha, \alpha^*) f_{sym}(\alpha, \alpha^*). \tag{1.35}$$

As an example, the mean number of photons in a cavity is given by

$$N_{cav} = \left\langle \tilde{a}^\dagger \tilde{a} \right\rangle = \frac{1}{2} \left\langle \tilde{a}^\dagger \tilde{a} + \tilde{a} \tilde{a}^\dagger \right\rangle - \frac{1}{2} = \overline{\alpha \alpha^*} - \frac{1}{2}. \tag{1.36}$$

where the bar symbol describes the mean value over the Wigner distribution. This equation leads to the usual relation between the field energy E_{cav} in the cavity expressed in term of the semiclassical variables, and the number of photons N_{cav}:

$$E_{cav} = \hbar\omega_{cav} \overline{\alpha \alpha^*} = \hbar\omega_{cav} \left(N_{cav} + \frac{1}{2} \right). \tag{1.37}$$

The semiclassical random variables thus include the quantum noise: for a vacuum field ($N_{cav} = 0$), the semiclassical energy $\hbar\omega_{cav}\overline{|\alpha|^2}$ includes the vacuum energy $\hbar\omega_{cav}/2$ associated with the quantum fluctuations of the field.

Semiclassical description of quantum states

It can be shown that for many states of light (vacuum field, coherent states, squeezed states), the quasi-probability distribution $\mathcal{W}(\tilde{a}, \tilde{a}^\dagger)$ is a Gaussian which only takes positive values. In such cases, the Wigner distribution can be considered as a true probability distribution and consequently any value α appears as a possible realization of the quantum field $\tilde{a}$ with a probability $\mathcal{W}(\tilde{a}, \tilde{a}^\dagger)$ (see Fig. 1.11). The semiclassical approach consists of replacing the quantum operator $\tilde{a}$ by its semiclassical counterpart $\alpha = \overline{\alpha} + \delta\alpha$ where $\overline{\alpha} = \langle \tilde{a} \rangle$ is the mean field and $\delta\alpha$ is a random classical variable characterized by the Wigner distribution.

Figure 1.11 shows the phase-space representation of the fields. In the middle and on the right, 3D plots of the Wigner distributions for a coherent and squeezed states are shown: they correspond to Gaussians with equal variances in any direction for the coherent state whereas the squeezing of the distribution leads to different variances along two orthogonal directions for a squeezed state. The left plot is a 2D representation of phase-space, where the horizontal and vertical axes correspond to the real and imaginary parts of the semiclassical variable α, and the dashed areas are isoprobability curves of

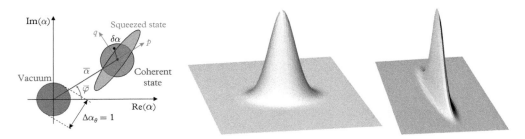

Fig. 1.11 *Phase-space semiclassical representation of the field (left). Each point $\overline{\alpha} + \delta\alpha$ can be considered as a possible realization of the quantum field with a probability given by the Wigner function. For usual states such as vacuum, coherent (middle) and squeezed (right) states, the Wigner distribution is a Gaussian with symmetric or dissymmetric variances along two orthogonal quadratures.*

the Wigner distribution for various states: vacuum, coherent and squeezed states. In the semiclassical approach, points in dashed areas represent the most probable semiclassical realizations of the quantum field, taking into account the quantum noise dispersion.

Semiclassical observables

Almost all measurable quantities can be expressed in terms of the field quadratures defined in the quantum approach by the operator

$$\tilde{a}_\theta = e^{-i\theta}\tilde{a} + e^{i\theta}\tilde{a}^\dagger, \tag{1.38}$$

and by a similar semiclassical variable α_θ, where θ is the angle of the field quadrature. Two quadratures are of particular interest, the amplitude quadrature $p = \alpha_{\overline{\varphi}}$ and the phase quadrature $q = \alpha_{\overline{\varphi}+\frac{\pi}{2}}$, respectively aligned and perpendicular with the mean field $\overline{\alpha}$ of phase $\overline{\varphi}$, where $\alpha = |\overline{\alpha}|e^{i\overline{\varphi}}$ (see Fig. 1.11). In the case of intense light fields, a linear expansion with respect to the relative fluctuations $\delta\alpha / |\overline{\alpha}|$ allows one to relate observables to quadrature fluctuations. As an example the intensity noise is given by $\delta I = |\overline{\alpha}|\,\delta p$, with a noise spectrum $S_I[\Omega] = \overline{I}S_p[\Omega]$. Coherent states, such as the ones depicted in Fig. 1.11, are characterized by equal dispersions $\Delta\alpha_\theta = 1$ in any quadrature direction θ, or equivalently to noise spectra equal to 1, $S_{\alpha_\theta}[\Omega] = 1$. This leads to the usual expression for quantum intensity noise, or shot-noise, $S_I[\Omega] = \overline{I}$.

Semiclassical dynamics

For a system described by a quadratic Hamiltonian such as the one depicted in Fig. 1.12 with various input fields α_i^{in} impinging in the system from different sources (laser, vacuum or squeezing), the semiclassical formalism consists in describing these fields by semiclassical random variables. For laser coherent fields or vacuum, the input noise thus only corresponds to quantum noise with spectra of any quadrature equal to 1:

$$S_p^{in}[\Omega] = S_q^{in}[\Omega] = S_{\alpha_\theta}^{in}[\Omega] = 1 \quad \rightarrow \quad S_I^{in}[\Omega] = \overline{I} \tag{1.39}$$

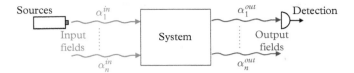

Fig. 1.12 *In the input–output semiclassical formalism, input fields are described by semiclassical variables α_i^{in} with dispersion including the quantum noise and output fields α_j^{out} are obtained from incident ones using the linearized classical equations of the system.*

Input–output relations for the fields are obtained from the time evolution of the Wigner distribution. It can be shown that for quadratic Hamiltonian, it corresponds to the classical equation of evolution. In other words, output fields α_j^{out} are obtained from the incident ones using the linearized classical equations of the system. Examples follow with the semiclassical description of optomechanical systems.

Semiclassical description of a mechanical resonator

The quantum behaviour of a mechanical resonator can be described as well by a semiclassical approach. For a resonator of mass m, resonance frequency Ω_m and damping Γ, the equation of motion (Eq. 1.1) gives the position $x[\Omega]$ as a function of the applied force $F[\Omega]$ and the mechanical susceptibility $\chi[\Omega]$ defined by Eq. (1.2). Quantum fluctuations and coupling to a thermal bath at temperature T are both described by a generalized Langevin force $F_T[\Omega]$ with a spectrum given by:

$$S_{F_T}[\Omega] = m\Gamma\hbar\Omega\coth\left(\frac{\hbar\Omega}{2k_BT}\right). \qquad (1.40)$$

The noise spectrum reduces to the usual thermal expression $S_{F_T}[\Omega] = 2m\Gamma k_BT$ at high temperature ($k_BT \gg \hbar\Omega$) whereas it is limited to $\hbar\Omega/2$ at low temperature ($k_BT \ll \hbar\Omega$, see Fig. 1.13 left). For a high-Q resonator ($\Gamma \ll \Omega_m$), integration over frequency of the mechanical response to the Langevin force (Eqs. 1.1 and 1.40) gives the position fluctuations:

$$m\Omega_m^2\Delta x^2 = \frac{\hbar\Omega_m}{2}\coth\left(\frac{\hbar\Omega_m}{2k_BT}\right) = \hbar\Omega_m\left(n_T + \frac{1}{2}\right), \qquad (1.41)$$

where $n_T = 1/\left(e^{\hbar\Omega_m/k_BT} - 1\right)$ is the mean number of phonons in the thermal state at temperature T: the resonator energy (1.41) reduces to the thermal noise $\hbar\Omega_m n_T$ at high temperature and to the quantum noise $\hbar\Omega_m/2$ at low temperature. The behaviour for different temperatures is shown in Fig. 1.13 (right). Reaching the quantum ground state typically requires a very low bath temperature for a low-frequency resonator, e.g. less than 500 μK for a resonance frequency of 10 MHz.

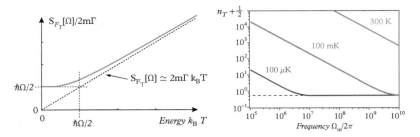

Fig. 1.13 *Noise spectrum $S_{F_T}[\Omega]$ of the generalized Langevin force as a function of the thermal energy $k_B T$ (left) and resulting phonon number $n_T + \frac{1}{2}$ for a high-Q resonator as a function of its resonance frequency Ω_m and for different bath temperatures T (right). Both curves show a thermal behaviour at high temperature ($k_B T \gg \hbar\Omega$ or $\hbar\Omega_m$), which reduces to the quantum noise at low temperature.*

1.3.3   Semiclassical Description of Optomechanics

Reflection on a single movable mirror

We first consider a very simple situation where a light beam is reflected on a moving mirror as depicted in Fig. 1.1. The incident field α^{in} undergoes a phase shift proportional to the displacement x of the mirror. The reflected field α^{out} is then:

$$\alpha^{out}(t) = \alpha^{in}(t) \exp\left(2ikx(t)\right). \tag{1.42}$$

Linearizing with respect to the field fluctuations, one gets the input–output relations for the mean fields, and the amplitude and phase quadratures:

$$\overline{\alpha}^{out} = \overline{\alpha}^{in}, \tag{1.43}$$

$$\delta p^{out} = \delta p^{in}, \tag{1.44}$$

$$\delta q^{out} = \delta q^{in} + 4k\overline{\alpha}^{in}\delta x. \tag{1.45}$$

Eqs. (1.43) and (1.44) indicate that neither the mean intensity nor the intensity fluctuations are altered by the cavity reflection, as expected since we have neglected any retardation effect in Eq. (1.42). However, Eq. (1.45) shows that the phase encodes the mirror motion, with an amplification factor $4k\overline{\alpha}^{in}$. A phase measurement can therefore be used to probe the mechanical motion, despite the additional noise δq^{in}, which is the incident phase noise. The resulting noise spectrum is:

$$S_q^{out}[\Omega] = S_q^{in}[\Omega] + \left(\frac{8\pi}{\lambda}\right)^2 \overline{I}^{in} S_x[\Omega]. \tag{1.46}$$

For an incident coherent field ($S_q^{in}[\Omega] = 1$) and a unity signal-to-noise ratio, the sensitivity is limited to:

$$\delta x_{min} = \frac{\lambda}{8\pi\sqrt{\overline{I}^{in}}}. \tag{1.47}$$

The interpretation of δx_{min} is straightforward in phase space (Fig. 1.14 left): any displacement δx produces a phase shift $\delta\varphi = \delta q^{out}/2\overline{\alpha}^{in} = 4\pi\delta x/\lambda$ which can be resolved only if it is larger than the corresponding quantum noise, i.e. the angle $\delta\varphi_{min} = 1/2\overline{\alpha}^{in}$ in which the Wigner distribution is seen from the origin of phase-space. The sensitivity increases with intensity, as the disc of the coherent state fluctuations is farther from the origin. For a 1-W incident laser at $\lambda = 1\,\mu$m, the intensity is $\overline{I}^{in} = P/\hbar kc \simeq 5 \times 10^{18}$ photons/s, which gives a spectral noise amplitude $\delta x_{min} \simeq 2 \times 10^{-17}$ m/$\sqrt{\text{Hz}}$.

But measurement noise is not the only noise at work in an optical measurement of displacement. Back-action on the mirror is due to radiation pressure given by $F_{rad}(t) = 2\hbar k I^{in}(t)$, with semiclassical fluctuations $\delta I^{in} = \overline{\alpha}^{in}\delta p^{in}$ proportional to the fluctuations of the amplitude quadrature. This noise is conjugate to the measurement noise δq^{in}, with a resulting displacement noise of the mirror scaling as:

$$\delta x_{rad}[\Omega] = 2\hbar k\sqrt{\overline{I}^{in}}\,|\chi[\Omega]|. \tag{1.48}$$

Radiation-pressure noise therefore increases with the incident intensity, contrary to the measurement noise.

In the general case, the measurement is of course simultaneously limited by both noises. For an incident coherent state, i.e. for equal and uncorrelated quadrature noises δp^{in} and δq^{in}, Fig. 1.14 shows the dependence of both noises and the resulting quadratic sum one has to take into account for uncorrelated noises. This overall noise is always larger than the standard quantum limit (SQL), which is the value reached when both

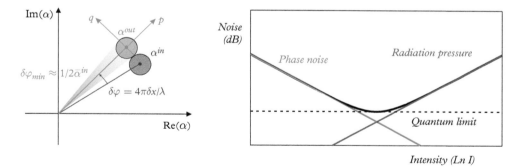

Fig. 1.14 *Interpretation in phase-space of the measurement sensitivity (left): the phase noise $\delta\varphi_{min} \simeq 1/2\overline{\alpha}^{in}$ of the incident coherent field limits the measurement of the phase shift $\delta\varphi = 4\pi\delta x/\lambda$ induced by the mirror displacement δx. The standard quantum limit (right) is a compromise between the phase and radiation-pressure noises.*

noises are equal (Caves, 1980; Jaekel and Reynaud, 1990). Eqs. (1.47) and (1.48) yield:

$$\delta x_{SQL}[\Omega] = \sqrt{\hbar |\chi[\Omega]|}. \tag{1.49}$$

Cavity optomechanics

We will consider now the device of Fig. 1.2: a single-port cavity with a moving mirror. The semiclassical approach still consists in writing the classical equations of the system, linearized around the mean values with injected quantum noises as input fields and generalized Langevin forces. Relevant equations are already written in (1.1) and (1.5) to (1.7). The general linearized equations of the system are then:

$$\left(\frac{\kappa}{2} - i\Omega - i\Delta\right) \delta\alpha[\Omega] = iG\bar{\alpha}\delta x[\Omega] + \sqrt{\kappa}\delta\alpha^{in}[\Omega], \tag{1.50}$$

$$\left(\frac{\kappa}{2} - i\Omega - i\Delta\right) \delta\alpha^{out}[\Omega] = i\sqrt{\kappa}\,G\bar{\alpha}\delta x[\Omega] + \left(\frac{\kappa}{2} + i\Omega + i\Delta\right) \delta\alpha^{in}[\Omega], \tag{1.51}$$

$$F_{rad}[\Omega] = \hbar G|\bar{\alpha}|\delta p[\Omega], \tag{1.52}$$

$$\delta x[\Omega] = \chi[\Omega] \left(F_{rad}[\Omega] + F_T[\Omega] + ...\right). \tag{1.53}$$

Eqs. (1.50) and (1.51) display the dynamics of the intracavity and reflected fields as a function of the mirror displacement δx and of the incident field noise $\delta\alpha^{in}$. The other parameters are G, still the strength of optomechanical coupling (Eq. 1.8), κ the cavity damping rate, and Δ the laser detuning with respect to the cavity resonance. Eqs. (1.52) and (1.53) define the mechanical motion as the response to both the Langevin force and radiation pressure, which depends on the amplitude quadrature δp of the intracavity field.

   These equations are the fundamental equations for such an optomechanical system, including all thermal and quantum noises of the system. As an example, we consider the case of a laser resonant with the cavity ($\Delta = 0$) and we study the ability to use such a system to measure a small mirror displacement. Eqs. (1.50) and (1.51) then yield the following input–output relations for the amplitude and phase quadratures:

$$\left(\frac{\kappa}{2} - i\Omega\right) \delta p^{out}[\Omega] = \left(\frac{\kappa}{2} + i\Omega\right) \delta p^{in}[\Omega], \tag{1.54}$$

$$\left(\frac{\kappa}{2} - i\Omega\right) \delta q^{out}[\Omega] = \left(\frac{\kappa}{2} + i\Omega\right) \delta q^{in}[\Omega] + 2\sqrt{\kappa}\,G\bar{\alpha}\delta x[\Omega]. \tag{1.55}$$

Equation (1.54) shows that the intensity noise is left unchanged ($S_p^{out}[\Omega] = S_p^{in}[\Omega]$), whereas Eq. (1.55) shows that the phase is sensitive to the mirror displacement. As in the case of the reflection on a single mirror presented in the previous subsection, one gets the phase noise spectrum (see Eq. 1.46):

$$S_q^{out}[\Omega] = S_q^{in}[\Omega] + \frac{4\kappa\, G^2 \bar{I}}{\frac{\kappa^2}{4} + \Omega^2} S_x[\Omega], \tag{1.56}$$

with an amplification factor which is now dependent on the parameters of the cavity, and a measurement noise which still corresponds to the incident phase noise. One gets the resulting sensitivity:

$$\delta x_{min} = \sqrt{1 + (2\Omega/\kappa)^2} \, \frac{\lambda}{16\mathcal{F}\sqrt{\bar{I}^{in}}}. \tag{1.57}$$

As compared to the single mirror case (Eq. 1.47), we gain the cavity finesse $\mathcal{F}$, but the optimal measurement sensitivity is now limited to frequencies inside the cavity bandwidth ($\Omega \leq \kappa/2$).

From Eqs. (1.52) and (1.53), we can also compute the displacement δx_{rad} due to radiation pressure, which is proportional to the incident amplitude fluctuations δp^{in}. As in the single-mirror case, measurement noise and back-action noise are uncorrelated for an incident coherent field and taking them both into account leads to an optimal sensitivity corresponding to the SQL (Eq. 1.49).

The existence of such a standard quantum limit to the sensitivity of displacement measurement is a consequence of back-action effects due to radiation pressure. Injection of a squeezed state may reduce radiation pressure (using an intensity-squeezed source) but at the expense of incident phase noise. As a consequence, the SQL still holds. It is however possible to beat it by using a squeezed source with an adapted squeezing angle: this will create correlations between the amplitude and phase quadratures in such a way that the combination of both noises observed in the reflected phase is reduced (Jaekel and Reynaud, 1990). Other back-action evading techniques are explored as well, such as the measurement of a single quadrature of the resonator displacement (instead of measuring both, while back-action noise affects the conjugate quadrature), or the measurement of the mechanical energy instead of the displacement. The former case can be achieved by sending in the cavity an amplitude-modulated laser at the mechanical frequency Ω_m and would allow a noise-free measurement of a single quadrature. The latter case is based on a mechanical membrane inserted in a high-finesse cavity in such a way that the cavity frequency shift becomes sensitive to x^2 rather than x. This would for instance allow for the observation of quantum jumps between phonon Fock states.

1.3.4   Experimental Demonstration of Back-action Noise

Back-action noise has been observed for the first time in C. Regal's group in 2013 (Purdy, Peterson and Regal, 2013*a*), and later in T. Kippenberg's group (Wilson, Sudhir, Piro, Schilling, Ghadimi and Kippenberg, 2015) in 2015. The experiment takes advantage of a thin and high-Q membrane in a high-finesse cavity (membrane in the middle setup, see Fig. 1.15). An intense signal beam at resonance with the cavity generates radiation-pressure effects. Measurement is performed with a weaker meter beam on the tails of the

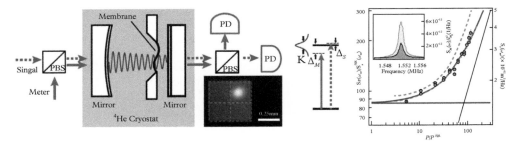

Fig. 1.15 *Experimental demonstration of back-action noise. An intense resonant signal beam and a meter beam are sent in an optomechanical device made of a membrane in the middle of a high-finesse cavity (left). Displacement noise spectra observed on the transmitted meter exhibit extra noise when the intensity of the signal beam is increased (right). Adapted from (Purdy, Peterson and Regal, 2013a). Reprinted with permission from AAAS.*

Airy peak. The meter beam is also used for laser cooling from the cryogenic temperature of 5 K down to 1.7 mK, and to avoid instabilities for large signal intensities by optically damping the membrane.

Displacement noise is observed on the intensity of the transmitted meter: as shown in the insert of Fig. 1.15 (right), the calibrated displacement noise obtained at high signal power (light grey spectrum) is larger than the one obtained at low power (dark grey spectrum). Performing an analysis of noises in the system, the authors can attribute at least 40% of the total displacement spectrum to the quantum back-action noise at the maximum signal beam power.

1.4   Quantum Optics with Optomechanics

Optomechanical coupling induces nonlinear effects in the cavity, as already discussed in section 1.2.2. Due to radiation pressure, the mirror motion depends on the intensity of light. For a cavity with a moving mirror, one gets a nonlinear equation for the intracavity intensity which leads to a bistable behaviour (see Fig. 1.4). Radiation pressure is actually very similar to a Kerr effect: it changes the *physical length* of the cavity $L + x$ whereas a Kerr medium inserted in the cavity changes the *optical length* through the intensity-dependence of its reflective index $n(I)$. Optomechanical systems can thus be used to produce quantum effects already well known in quantum optics, such as squeezed states (Slusher, Hollberg, Yurke, Mertz and Valley, 1985) or quantum non demolition measurement of light (Roch, Vigneron, Grelu, Sinatra, Poizat and Grangier, 1997), which was first studied in the context of optomechanics in the '90s. The main differences with standard quantum optics are the dynamics of the mechanical resonator which induces more complex behaviours, the overall weaker nonlinearities compared to nonlinear media, and the extra noise due to its thermal motion (Fabre, Pinard, Bourzeix, Heidmann, Giacobino and Reynaud, 1994).

Squeezed state generation can be easily understood using the semiclassical approach. Figure 1.16 (left) shows the evolution of the mean fields $\overline{\alpha}$ and the fluctuations $\delta\alpha$ between the incident and reflected fields, for a lossless cavity. The mean field experiences a phase shift, as can be deduced from the classical equations (1.5) and (1.6):

$$\left(\frac{\kappa}{2} - i\Delta\right)\overline{\alpha}^{out} = \left(\frac{\kappa}{2} + i\Delta\right)\overline{\alpha}^{in}, \tag{1.58}$$

where Δ is the cavity detuning and κ the cavity damping rate. For a non-zero cavity detuning, this corresponds to a rotation around the origin in phase-space. Each point in the incident Wigner distribution (shown as a circle) is a possible realization $\overline{\alpha}^{in} + \delta\alpha^{in}$ of the incident field. Due to radiation pressure, this point corresponds to a different incident intensity so that the cavity length is changed by the resulting mirror displacement δx and the detuning becomes $\Delta + G\delta x$. One thus gets a different rotation in phase-space. As the transformation is unitary, the area of the distribution is preserved and at first order (linearization procedure of the fluctuations) it becomes elliptical. This corresponds to a squeezed state, as the variance $\Delta\alpha_{\theta}^{out}$ of the quadrature aligned with the small axis of the ellipse is smaller than 1.

Full calculation can be made in the semiclassical approach using Eqs. (1.50) to (1.53). Figure 1.16 (right) shows output noise spectra of the amplitude quadrature (intensity noise) for different temperatures. At zero temperature, the intensity noise exhibits at low frequency the usual behaviour of a cavity with a Kerr medium: there is no intensity noise reduction at zero frequency due to the conservation of the photon number over a long time, and the squeezing is filtered by the cavity bandwidth. In contrast to the Kerr medium, the spectrum also exhibits strong squeezing around the mechanical resonance

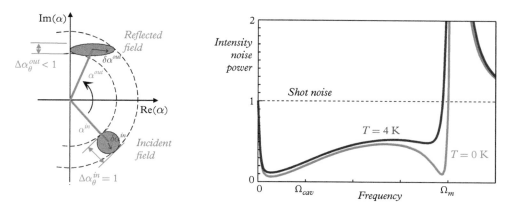

Fig. 1.16 *Principle of the generation of a squeezed state of light by a cavity with a moving mirror (left): due to radiation pressure, the field experiences a phase shift which depends on the intracavity intensity. The incident noise distribution is distorted into an elliptical distribution for the reflected field. Spectra on the right show the intensity noise squeezing that can be expected at different temperatures.*

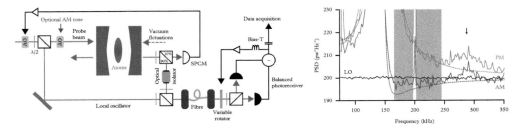

Fig. 1.17 *Experimental demonstration of squeezed state generation. The mechanical element is a cloud of ultracold atoms, inserted in a cavity (left). Noise of the transmitted field is probed with a homodyne detection (local oscillator and balanced photoreceivers) and exhibits a squeezed amplitude quadrature (AM, right) whereas the phase quadrature has extra noise (PM). Adapted by permission from (Brooks, Botter, Schreppler, Purdy, Brahms and Stamper-Kurn, 2012). Copyright (2012) Springer Nature.*

where the dynamics of the resonator plays an important role. At cryogenic temperature (4 K), squeezing is reduced by the thermal motion of the resonator, mainly around the resonance frequency where the thermal noise is peaked.

The first experimental demonstration was made by D. Stamper-Kurn and co-workers in 2012 (see Fig. 1.17), using ultracold atoms trapped in a cavity and acting as a position-dependent refractive medium for the probe beam (Brooks, Botter, Schreppler, Purdy, Brahms and Stamper-Kurn, 2012). The setup included a double-sided cavity (and therefore additional losses) locked at a constant detuning and a homodyne detection of the transmitted probe beam. The quantum noise spectra of the transmitted beam were monitored for amplitude and phase quadratures (Fig. 1.17 right), with a detected squeezing of 1.4%.

Other experimental demonstrations of squeezing were made in 2013, with a silicon micromechanical resonator (zipper cavity)(Safavi-Naeini, Gröblacher, Hill, Chan, Aspelmeyer and Painter, 2013) and a membrane in the middle of a high-finesse cavity (same device as the one used for back-action noise described in section 1.3.4)(Purdy, Yu, Peterson, Kampel and Regal, 2013*b*), leading to an observed squeezing up to 1.7 dB (more than 30% of reduction of the power noise spectrum).

1.5   Control and Cooling of a Mechanical Resonator

Up to now, we have presented the consequences of optomechanical coupling for the field, namely the *quantum limits* in measurement where the field is monitored to detect small displacements, and the *quantum optics effects* produced when radiation pressure is used as a nonlinear mechanism. We now present the consequences of optomechanical coupling on the mechanical resonator itself, in particular the possibility to use the coupling to control and cool the resonator.

1.5.1   Dynamical Effects in a Detuned Cavity

As explained in section 1.2.2, radiation pressure induces both optical spring and damping effects. This can be understood from Fig. 1.18: for a detuned cavity, the working point is on the tails of the Airy peak. Any displacement of the mirror changes the cavity detuning and then the intracavity intensity, resulting in a variation of the radiation pressure exerted onto the mirror. This corresponds to an additional force proportional to the displacement: $F_{rad} \propto \frac{dP}{dL}\delta x$ where $\frac{dP}{dL}$ is the slope of the Airy peak at the working point. This force appears as an additional spring in the equation of motion (1.1) which changes the resonance frequency of the moving mirror.

Cavity dynamics may change this behaviour as the radiation-pressure response to mirror displacement is mediated by the field: the response is delayed by the cavity storage time, or equivalently, low-pass filtered by the cavity bandwidth. For motion frequencies on the order of or larger than the cavity bandwidth, the radiation pressure then acquires a delayed component which appears as a viscous force and changes the damping of the resonator.

The effect can be computed using the semiclassical approach, from Eqs. (1.50) to (1.53). In the general case of a detuned cavity ($\Delta \neq 0$), the first two equations allow one to derive the expression of the intracavity amplitude quadrature δp (the mean intracavity field $\overline{\alpha}$ is here taken real for simplicity):

$$\delta p[\Omega] = -\frac{2G\overline{\alpha}\Delta}{\left(\frac{\kappa}{2} - i\Omega\right)^2 + \Delta^2}\delta x[\Omega] + \sqrt{\kappa}\left(\frac{\delta\alpha^{in}[\Omega]}{\frac{\kappa}{2} - i\Omega - i\Delta} + \frac{\delta\alpha^{in*}[\Omega]}{\frac{\kappa}{2} - i\Omega + i\Delta}\right). \tag{1.59}$$

The first term represents the term of interest, proportional to the mirror displacements. It leads to an additional spring (either positive or negative depending on the sign of Δ) at low frequency ($\Omega \ll \kappa/2$), corresponding to motions slower than the cavity storage time. At high frequency ($\Omega \gg \kappa/2, \Delta$), the force becomes similar to an additional damping,

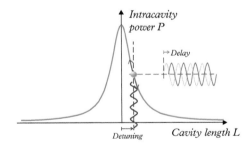

Fig. 1.18 *Principle of dynamical effects in a detuned cavity: when the working point is on the tails of the Airy peak, any displacement of the mirror changes the intracavity power. Radiation pressure then becomes sensitive to mirror displacements, acting as a binding or repulsive force. Due to the cavity dynamics, the response of the force is not instantaneous but is delayed by the cavity storage time so that the force acquires a viscous component.*

either positive or negative according to the sign of Δ. The last terms in Eq. (1.59) represent noises associated with the input noise of the field: this is the back-action effect of the field onto the resonator.

Equations (1.52) and (1.53) then allow one to determine the resulting motion:

$$\delta x[\Omega] = \chi_{eff}[\Omega] \left(\hbar G \overline{\alpha} \sqrt{\kappa} \left(\frac{\delta \alpha^{in}[\Omega]}{\frac{\kappa}{2} - i\Omega - i\Delta} + \frac{\delta \alpha^{in*}[\Omega]}{\frac{\kappa}{2} - i\Omega + i\Delta} \right) + F_T[\Omega] + ... \right), \quad (1.60)$$

where χ_{eff} is the susceptibility of the resonator, modified by the additional optical spring and damping. This effective susceptibility is given by:

$$\chi_{eff}[\Omega]^{-1} = \chi[\Omega]^{-1} + \frac{2\hbar G^2 \overline{\alpha}^2 \Delta}{\left(\frac{\kappa}{2} - i\Omega\right)^2 + \Delta^2}. \quad (1.61)$$

If we have a look at the additional term in this equation, we see that its real part gives a frequency shift, while its imaginary part changes the damping. For a high-Q resonator ($\Gamma \ll \kappa$), the additional term is mainly constant over the frequencies around $\pm\Omega_m$ where the mechanical susceptibility χ is significant. The additional term is maximum when $\Delta = \pm\Omega_m$. In the unresolved sideband regime ($\Omega_m \lesssim \kappa/2$), we have a mix of optical spring and damping (with only a spring effect if $\Omega_m \ll \kappa/2$) whereas in the resolved sideband regime ($\Omega_m \gg \kappa/2$), we mainly have damping.

Such effects were first observed in the '70s, as already presented in section 1.2.2, but also in many recent experiments, in the context of optical cooling (see section 1.5.3). As an example, Fig. 1.21 (centre) shows both optical spring and damping effects: the frequency shift is observed in both directions from the resonant cavity case (curve labeled '300 K'), as well as the shrinking or widening of the resonance associated with negative or positive optical dampings, respectively.

1.5.2   Cold Damping

The mechanism described in the previous section is a *cold-damping* mechanism, as it can increase the damping without adding the associated thermal fluctuations. Indeed, the fluctuation-dissipation theorem (Eq. 1.13 in the classical regime, Eq. 1.40 in the quantum case) dictates that any usual damping mechanism corresponds to the coupling to a thermal bath with an additional Langevin force characterized by a noise spectrum proportional to the damping and to the bath temperature (plus $\frac{1}{2}$ phonon in the quantum case). As a consequence, any damping added to the resonator does not change its temperature, as long as the thermal environment is at the same temperature. This is not the case here, as can be seen from Eq. (1.60): damping is changed but the additional noise terms are associated with the quantum noise of the incident field, which is actually similar to a thermal bath at zero temperature. As a consequence, the resonator is coupled to two baths at different temperatures, through its intrinsic damping and through radiation pressure. Roughly speaking, its temperature is changed to:

$$T_{eff} \simeq \frac{\Gamma T + \Gamma_{add} \times 0}{\Gamma + \Gamma_{add}} \simeq \frac{\Gamma}{\Gamma + \Gamma_{add}} T, \tag{1.62}$$

and becomes small compared to the initial temperature T when the optical damping is larger than the mechanical damping.

Cold damping by feedback cooling

The possibility of cooling a system by cold damping has been studied for a long time, with the first publications on the subject in the 1950s (Milatz, Zolingen and van Iperen, 1953). The first experiment in a cavity optomechanics context was performed by our group in 1999, using a feedback cooling mechanism (see Fig. 1.19). A high-finesse cavity is used to monitor in real-time the position fluctuations of the moving mirror, while a second laser beam is used to cool the mirror: it is indeed intensity-modulated by the feedback loop using the mirror motion as error signal (Cohadon, Heidmann and Pinard, 1999). Parameters are such that quantum noises in the measurement (phase noise and back-action noise) are negligible, so that the measurement faithfully reproduces the thermal fluctuations of the mirror. We choose the feedback gain in such a way that the applied force is viscous:

$$F_{cd}[\Omega] = i\Omega m \Gamma g_{cd} \delta x[\Omega], \tag{1.63}$$

where g_{cd} is the gain of the feedback loop. According to the equation of motion for the resonator (Eqs. 1.1 and 1.2) and to the fluctuation-dissipation theorem (1.13), the resulting motion still corresponds to a thermal equilibrium but with an additional damping Γ_{cd} and a reduced number of phonons n_T^{cd} given by:

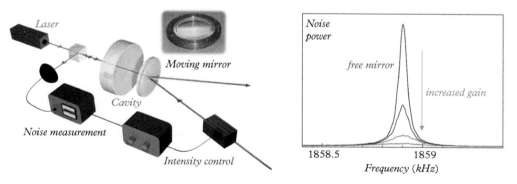

Fig. 1.19 *Former experimental demonstration of cold damping: a high-finesse cavity is used to detect the thermal displacement noise of a mirror coated on a cylindrical plano-convex substrate of gram-scale mass (diameter of 14 mm). The measurement is fed back onto the mirror through the radiation pressure of an intensity-controlled auxiliary laser beam, in such a way that the applied force is viscous. The thermal noise spectra (right) exhibit a strong reduction for increasing gains, corresponding to a cooling of the resonator down to 10 K.*

$$\Gamma_{cd} = g_{cd}\Gamma, \tag{1.64}$$

$$n_T^{cd} = \frac{\Gamma}{\Gamma + \Gamma_{cd}} n_T = \frac{n_T}{1 + g_{cd}}. \tag{1.65}$$

These equations show that the noise spectrum is widened and its height is lowered, in such a way that the temperature, proportional to the area of the Lorentzian resonance, is reduced (see experimental results in Fig. 1.19 right).

Towards the quantum ground state

The observation of a quantum behaviour for macroscopic and massive objects appears as a cornerstone in quantum physics, in the same manner but for different reasons than the understanding of the fundamental limits in quantum measurement which was the main focus of section 1.3. Preparing a macroscopic mechanical resonator in its ground state ($n_T = 0$) is a prerequisite for manipulating and taking advantage of quantum states of mechanical motion and it has been considered for years as the holy grail in the domain. Cavity optomechanics is now a well-established research domain in particular for its successes in this direction, which open the way to a better understanding of decoherence mechanisms and to the frontier between classical and quantum worlds.

Most results have been obtained by optical cooling (see next section), although cold damping by feedback cooling appears as an interesting approach that combines a very sensitive displacement sensor (as the light field is resonant with the cavity), and an efficient cooling mechanism. Equation (1.65) seems to demonstrate that the quantum ground state $n_T^{cd} = 0$ can be reached by increasing the feedback gain g_{cd}. This has to be considered more carefully as we have so far neglected all quantum noises. Including all such noises (measurement noise, quantum back-action, quantum noise of the resonator), one gets the more general expression for the phonon number (Courty, Heidmann and Pinard, 2001):

$$n_T^{cd} + \frac{1}{2} = \frac{\Gamma}{\Gamma + \Gamma_{cd}}\left(n_T + \frac{1}{2} + \zeta + \frac{g_{cd}^2}{16\zeta}\right), \tag{1.66}$$

where $\zeta = 4g_0^2\bar{I}/\kappa\Gamma$ is the *optomechanical cooperativity*, and g_0 is still the vacuum optomechanical coupling strength already defined by Eq. (1.28). The optomechanical cooperativity depends both on mechanical parameters (mass m, resonance frequency Ω_m, damping Γ) and on optical parameters (intensity $\bar{I}$, wavelength λ, cavity damping κ). It corresponds to the ratio between the strength $g_0^2\bar{I}$ of the optomechanical coupling in presence of an intense laser and the dampings of the system, both optical (κ) and mechanical (Γ). It characterizes the coherent quantum dynamics of the system, reached for $\zeta > 1$: in particular, it can be shown that the SQL is reached for $\zeta = 1$.

Equation (1.66) is a generalization of Eq. (1.65), including all quantum noises: quantum fluctuations of the resonator yield the terms $\frac{1}{2}$ in both left- and right-hand sides of the equation, the term ζ corresponds to the back-action noise of the measurement used to drive the feedback loop, and the last term proportional to ζ^{-1} is the measurement noise reinjected by the feedback loop into the mirror motion. For a given cold damping gain g_{cd}, the minimal contamination by quantum noises is reached for $\zeta = g_{cd}/4$. As for the SQL, this condition is the best compromise between back-action effects and measurement contamination, when both noises are equal. The resulting phonon number then reduces to:

$$n_T^{cd} + \frac{1}{2} = \frac{n_T}{1 + g_{cd}} + \frac{1}{2}. \tag{1.67}$$

It is clear from this equation that the phonon number is reduced by the feedback loop, down to the quantum ground state of the mechanical resonator for a large gain. Still, as expected, the quantum noise of the resonator (term $\frac{1}{2}$) is not altered by the loop.

Such a technique was later applied to a number of different systems, including an experiment in 2015 by T. Kippenberg's group in which a nanomechanical resonator was feedback cooled down to an effective temperature of 1 mK corresponding to a mean number of 5 phonons (Wilson, Sudhir, Piro, Schilling, Ghadimi and Kippenberg, 2015).

1.5.3 Optical Cooling

As already discussed in previous sections, optical damping in a detuned cavity plays a role similar to cold damping. This can be easily understood from Eq. (1.60) describing the resonator motion in a detuned cavity: neglecting all quantum noises, this equation reduces to:

$$\delta x[\Omega] = \chi_{eff}[\Omega] F_T[\Omega], \tag{1.68}$$

where χ_{eff} is given by Eq. (1.61). At least at the classical level, the Langevin force F_T applied on the resonator is unchanged whereas an additional damping related to the imaginary part of last term in Eq. (1.61) appears. This is nothing but a cold-damping effect, whose results have been already described by Eqs. (1.64) and (1.65): one gets a widening or shrinking of the thermal spectrum, together with a reduction or increase of the phonon number, depending on the sign of the additional optical damping, i.e. the sign of Δ: from Eq. (1.61), a red-detuned incident laser leads to positive damping which cools the resonator, the reverse being true for a blue-detuned laser.

Stokes and anti-Stokes processes

A more explicit expression of the additional optical damping Γ_{opt} can be deduced from Eq. (1.61), in the case of a high-Q resonator, for which mechanical susceptibilities χ and χ_{eff} only take significant values for frequencies close to $\pm\Omega_m$:

$$\Gamma_{opt}/\Gamma = \frac{i}{m\Gamma\Omega_m} \mathrm{Im}\left(\frac{2\hbar G^2 \overline{\alpha}^2 \Delta}{\left(\frac{\kappa}{2} - i\Omega_m\right)^2 + \Delta^2} \right) \tag{1.69}$$

$$= \zeta \left(\mathcal{A}(\Delta + \Omega_m) - \mathcal{A}(\Delta - \Omega_m) \right), \tag{1.70}$$

where ζ is the optomechanical cooperativity already defined for the feedback cooling (Eq. 1.66) and $\mathcal{A}(\omega)$ is the Airy peak function with respect to frequency ω:

$$\mathcal{A}(\omega) = \frac{\kappa^2/4}{\omega^2 + \kappa^2/4}. \tag{1.71}$$

The optical damping Γ_{opt} then depends on the asymmetry between the heights of the Airy peak evaluated at frequencies $\Delta \pm \Omega_m$. Figure 1.20 presents an interpretation of the damping and cooling mechanisms as a scattering of incident photons (at laser angular frequency ω_0) by the mechanical phonons of the mirror (at frequency Ω_m), giving rise to Stokes (at $\omega_0 - \Omega_m$) and anti-Stokes (at $\omega_0 + \Omega_m$) motional sidebands. Probabilities of these scattering processes are proportional to the resonant condition of the produced photons, that is $\mathcal{A}(\Delta \mp \Omega_m)$, respectively.

For a resonant cavity ($\Delta = 0$), both sidebands have equal amplitudes: they can be used to infer the mirror motion (as a phase modulation of the reflected field) but have no net mechanical effect. Detuning the cavity promotes the scattering of photons to one sideband with respect to the other. For negative detunings, for instance, more photons are scattered at the $\omega_0 + \Omega_m$ frequency than at $\omega_0 - \Omega_m$ and the net effect of the optomechanical coupling is a non-zero energy transfer from the mechanical motion to the optical field, i.e. the cooling of the resonator. Anti-damping and amplification is obtained as well for a positive detuning. The optimal efficiency is very simple to state in the resolved sideband regime ($\Omega_m \gg \kappa$) as the damping and anti-damping processes are resonant for a detuning matching the mechanical resonance frequency: $\Delta = \pm \Omega_m$. Figure 1.20 presents the case where the anti-Stokes process is resonant, both in the non-resolved (centre) and resolved (right) sideband regimes.

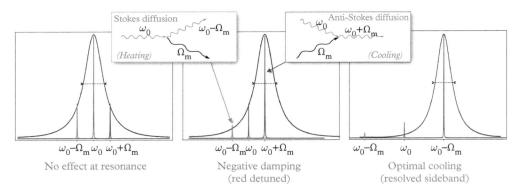

No effect at resonance   Negative damping   Optimal cooling
                              (red detuned)   (resolved sideband)

Fig. 1.20 *Interpretation of optical damping and cooling in terms of anti-Stokes (absorption of a phonon by a photon) and Stokes (creation of a phonon) processes. When the laser is resonant on the cavity (left), both processes are equiprobable and no damping occurs. For a red detuned laser (centre), the anti-Stokes process is dominant ($\mathcal{A}(\Delta + \Omega_m) > \mathcal{A}(\Delta - \Omega_m)$), resulting in a positive damping Γ_{opt} (Eq. 1.70) and a net extraction of phonons by light from the resonator. The process is resonant for $\Delta = -\Omega_m$ and residual phonon heating by Stokes process becomes negligible in the resolved sideband regime $\Omega_m \gg \kappa$ (right).*

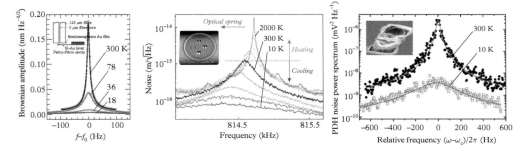

Fig. 1.21 *Various experimental noise spectra demonstrating optical cooling of a moving mirror, obtained by the group of K. Karrai (left) using the photothermal force exerted on a cantilever to reduce its temperature from 300 K down to 18 K, by our group in Paris (centre) using a silicon doubly-clamped 1-mm long beam which is either heated or cooled from room temperature down to 10 K, and by M. Aspelmeyer's group (right) with a low-mass doubly-clamped beam, from ambient down to 10 K. Adapted from (Höhberger Metzger and Karrai, 2004; Arcizet, Cohadon, Briant, Pinard and Heidmann, 2006a; Gigan, Böhm, Paternostro, Blaser, Langer, Hertzberg, Schwab, Bäuerle, Aspelmeyer and Zeilinger, 2006). Copyright (2004 and (2006) Springer Nature.*

Such effects were first demonstrated by Metzger and Karrai using a photothermal force exerted on a cantilever in a low-finesse cavity (see Fig. 1.21 left, (Höhberger Metzger and Karrai, 2004)) . Other spectra in this figure show the results obtained simultaneously in Paris (centre, (Arcizet, Cohadon, Briant, Pinard and Heidmann, 2006*a*)) and in Vienna (right, (Gigan, Böhm, Paternostro, Blaser, Langer, Hertzberg, Schwab, Bäuerle, Aspelmeyer and Zeilinger, 2006)) using radiation pressure (with also a non-negligible contribution of photothermal force in the case of the Vienna experiment). Both experiments use micro-mirrors, i.e. good quality mirrors here coated on doubly-clamped beams, with masses of 100 μg in Paris and of 20 ng in Vienna since most of the substrate was removed after the coating. These micro-mirrors were inserted as end mirror in a high-finesse cavity (respectively 30,000 and 500). Both groups obtained resonator cooling from room temperature down to temperatures of the order of 10 K. Experiment in Paris also demonstrated the optical heating of the resonator by an anti-damping optical force, up to effective temperatures of 2000 K. An unstable regime was even obtained, when $\Gamma + \Gamma_{opt} \to 0$: self-oscillation of the resonator then occurred, with an oscillation amplitude as large as 10 pm, of the order of the cavity bandwidth $\lambda/\mathcal{F}$.

Towards the quantum ground state

Many experiments demonstrated optical cooling with a wide variety of optomechanical systems, going into the resolved sideband regime and closer to the ground state. Indeed, as for feedback cooling, the resonator can be cooled down to its ground state by optical cooling. This can be seen from Eq. (1.60) which contains all the information on the resonator motion. The phonon number is given by:

$$n_T^{opt} + \frac{1}{2} = \frac{\Gamma}{\Gamma + \Gamma_{opt}} n_T + \frac{1}{2} \left(\frac{1 + \zeta \left(\mathcal{A}(\Delta + \Omega_m) + \mathcal{A}(\Delta - \Omega_m) \right)}{1 + \zeta \left(\mathcal{A}(\Delta + \Omega_m) - \mathcal{A}(\Delta - \Omega_m) \right)} \right). \qquad (1.72)$$

Thermal noise can be reduced at will by increasing the optical damping with respect to the intrinsic one. As the Airy function $\mathcal{A}$ is normalized to 1, Eq. (1.70) shows that this requires to work in the red-detuned side ($\Delta \simeq -\Omega_m$), with $\zeta > n_T$. This last condition reflects the fact that the efficiency of the anti-Stokes process must be large enough to absorb the phonons continuously injected in the system via the coupling to the thermal bath, that is n_T phonons during a mechanical coherence time Γ^{-1}.

Finally, reaching the ground state requires that the last term in Eq. (1.72) reduces to $\frac{1}{2}$, which corresponds to the quantum noise of the resonator. In other words, the term in brackets must tend to 1, leading to the condition $\mathcal{A}(\Delta - \Omega_m) \ll \mathcal{A}(\Delta + \Omega_m)$: Stokes processes must be negligible as compared to anti-Stokes ones. This corresponds to the *resolved sideband regime* $\Omega_m \ll \kappa$ (Wilson-Rae, Nooshi, Zwerger and Kippenberg, 2007; Marquardt, Chen, Clerk and Girvin, 2007).

Since the first demonstrations of optical cooling in 2006, followed in 2008 by the first cooling obtained in the resolved sideband regime (Schliesser, Rivière, Anetsberger, Arcizet and Kippenberg, 2008), intense research efforts were made at the international level to reach the quantum ground state of a mechanical resonator. The first experimental demonstration were obtained in 2010 but in a somewhat different context, using only conventional cryogenic refrigeration. It was based on a microwave-frequency mechanical resonator oscillating at 6 GHz (see Fig. 1.22 left), a frequency large enough to reach the fundamental state in a dilution cryostat working at 25 mK: A. N. Cleland and coworkers (O'Connell, Hofheinz, Ansmann, Radoslaw, Bialczak, Lenander, Neeley, Sank, Wang, Weides, Wenner, Martinis and Cleland, 2010) reached a residual phonon number of 0.07 only and made interesting studies by coupling the oscillator to a quantum bit, used

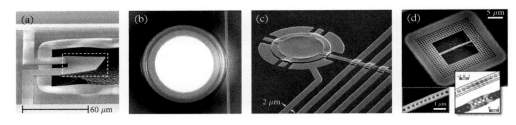

Fig. 1.22 *Four experiments have reached the quantum ground state of a mechanical resonator during the years 2010 and 2011: (a) a very high frequency piezo-electric oscillator inserted in a low-temperature cryostat ($\Omega_m/2\pi = 6$ GHz, $T_{cryo} = 25$ mK, 0.07 phonon); (b) a micro-toroid ($\Omega_m/2\pi = 75$ MHz, $M = 3$ ng, $T_{cool} = 7$ mK, 1.7 phonons); (c) a micro-resonator coupled to a microwave ($\Omega_m/2\pi = 10$ MHz, $M = 50$ pg, $T_{cool} = 150$ μK, 0.3 phonon); and (d) a silicon nanobeam ($\Omega_m/2\pi = 3.7$ GHz, $M = 300$ fg, $T_{cool} = 0.2$ K, 0.8 phonon). Devices (b) to (d) were cooled by optomechanical cooling.*

both to measure the quantum state of the resonator and to swap single excitations from the quantum bit to the mechanical oscillator.

Some time later, during the years 2010 and 2011, three experiments reached or went down very close to the ground state. Remarkably, these results were obtained with very different optomechanical devices, coupled either to microwave or optical fields (see Fig. 1.22): a silica toroidal resonator coupled to the optical field propagating in an optical fibre (T. Kippenberg's group, 1.7 phonons (Verhagen, Deléglise, Weis, Schliesser and Kippenberg, 2012)), a thin aluminium membrane coupled to a superconducting microwave resonant circuit (K. W. Lehnert's group, 0.3 phonon (Teufel, Donner, Dale, Harlow, Allman, Cicak, Sirois, Whittaker, Lehnert and Simmonds, 2011)), and a patterned silicon nanobeam designed to have spatially co-localized optical and acoustic modes (O. Painter's group, 0.8 phonon (Chan, Mayer Alegre, Safavi-Naeini, Hill, Krause, Gröblacher, Aspelmeyer and Painter, 2011)). All these experiments were based on cryogenic and optical (or microwave) coolings, and a detection of the mechanical state by a thermometry technique where the asymmetry between the Stokes and anti-Stokes sidebands is monitored.

1.6   Conclusion

Optomechanics have developed from a small research domain involving only a few groups into one of the most active interdisciplinary fields. During the past few years, researches on optomechanical systems have generated a new community which aims at achieving control over mechanical quantum states, with implications both for foundations and applications of quantum physics.

In this chapter, we presented the basic concepts of optomechanics and the early history of this research field. We tried to give an understanding of radiation-pressure effects using classical and semiclassical descriptions of optomechanical coupling, the latter including quantum noise of light and quantum behaviour of the mechanical resonator. We described different consequences of optomechanical coupling such as back-action noise and cooling of a mechanical resonator, emphasizing their role in quantum measurements and their ability to allow for quantum control of mechanical systems.

We haven't discussed in this introduction the quest for high mechanical quality and low mass resonators which is a major concern in aiming to reach the strong optomechanical coupling regime and to observe quantum signatures. Many kinds and sizes of resonators have been developed and are still continually improved, such as micromirrors, microtoroids, nano-objects inserted in a high-finesse cavity, photonic-crystal nanobeams, cold atoms, moving micro-capacitors coupled to microwave circuits... Figure 1.23 shows the evolution over years of the resonators developed by our group, from a cm-scale plano-convex mirror on the left, to micropillars and nanomembranes on the right. Sizes and masses were reduced by a large amount from 100 mg down to 33 μg (micropillar) and 100 pg (nanomembrane) without degrading the mechanical quality factor (7×10^7 for the micropillar at low temperature) or the optical reflectivity (cavity finesse close to 10^5).

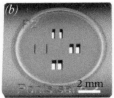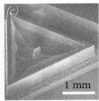

Fig. 1.23 *Examples of optomechanical resonators developed by our group: (a) a massive (100 mg) plano-convex mirror made of fused silica; (b) 4 doubly-clamped 1-mm long beams with a mirror coated over almost all the surface of the silicon wafer; (c) a 1-mm long micropillar made of quartz with a high-reflectivity mirror coated only at the top of the 200-µm large pillar (triangular shape at the centre); and (d) a suspended nanomembrane with a 2D photonic crystal which gives a high optical reflectivity.*

Reaching the strong coupling regime opens the way to very promising developments. Let us mention for example:

- the quantum control of a well-isolated mechanical resonator to study the decoherence over long times of massive mechanical objects and perform fundamental tests of quantum mechanics;
- the coherent transfer of quantum states between light and mechanical resonators, enabling the use of optomechanical systems as hybrid platforms for quantum memory and quantum information processing. Another application is the creation of quantum optomechanical entanglement between light and mechanical objects;
- reaching a regime of single photon optomechanical coupling is at hand, in which one may expect the possibility to prepare single-phonon states, to observe nonlinear quantum effects or to generate non-Gaussian states.

It is clear that mechanical oscillators can now reach a regime where their behaviour is described by quantum mechanics, as was the case in the past for photons, ions, or superconducting circuits. This allows testing of quantum mechanics on a macroscopic scale but it also opens the way to promising applications. Many of these developments have been recently demonstrated or are close to experimental demonstration, as is discussed in the following chapters.

Acknowledgements

We would like to thank all the researchers of the *Optomechanics and Quantum Measurement* team at the Kastler Brossel Laboratory (LKB) who have worked with us since the beginning of our research on optomechanics in the middle of the '90s: the permanent researchers (T. Briant, S. Deléglise, T. Jacqmin, M. Pinard), PhD students (O. Arcizet, T. Caniard, T. Capelle, Y. Hadjar, T. Karassouloff, A. Kuhn, K. Makles, R. Metzdorff, C. Molinelli, L. Neuhaus, A. Tavernarakis, P. Verlot, S. Zerkani), and postdoctoral students

(T. Antoni, M. Bahriz, X. Chen, S. Chua, D. Garcia-Sanchez, J. Teissier). We thank other groups for very fruitful collaborations, in particular the LPN (*Laboratoire de Photonique et de Nanostructures*), the LMA (*Laboratoire des Matériaux Avancés*), the ONERA, and all the Virgo collaboration.

We are also grateful to the constant support of CNRS (*Centre National de la Recherche Scientifique*), UPMC (*Université Pierre et Marie Curie*), ENS (*Ecole Normale Supérieure Paris*), and the financial supports from ANR (*Agence Nationale de la Recherche*, contracts ARQOMM, MiNOToRe, QuRaG and ExSqueez) and the different European Union networks (STREP MINOS and QNEMS, ITN cQOM) we have been involved in.

Bibliography

Arcizet, O., Cohadon, P.-F., Briant, T., Pinard, M., and Heidmann, A. (2006a). Radiation-pressure cooling and optomechanical instability of a micromirror. *Nature*, **444**, 71–4.

Arcizet, O., Cohadon, P.-F., Briant, T., Pinard, M., Heidmann, A., Mackowski, J.-M., Michel, C., Pinard, L., Français, O., and Rousseau, L. (2006b). High-sensitivity optical monitoring of a micromechanical resonator with a quantum-limited optomechanical sensor. *Physical Review Letters*, **97**, 133601.

Aspelmeyer, M., Kippenberg, T. J., and Marquardt, F. (2014). Cavity optomechanics. *Review of Modern Physics*, **86**, 1391–452.

Braginsky, V. B., Manukin, A. B., and Tikhonov, M. Yu. (1970). Investigation of dissipative ponderomotive effects of electromagnetic radiation. *Soviet Physics JETP*, **31**, 829–30.

Briant, T., Cohadon, P.-F., Heidmann, A., and Pinard, M. (2003a). Optomechanical characterization of acoustic modes in a mirror. *Physical Review A*, **63**, 033823.

Briant, T., Cohadon, P.-F., Pinard, M., and Heidmann, A. (2003b). Optical phase-space reconstruction of mirror motion at the attometer level. *European Physical Journal D*, **22**, 131–40.

Brooks, D. W. C., Botter, T., Schreppler, S., Purdy, T.P., Brahms, N., and Stamper-Kurn, D. M. (2012). Non-classical light generated by quantum-noise-driven cavity optomechanics. *Nature*, **488**, 476–80.

Caves, C. M. (1980). Quantum-mechanical radiation-pressure fluctuations in an interferometer. *Physical Review Letters*, **45**, 75–9.

Chan, J., Mayer Alegre, T. P., Safavi-Naeini, A. H., Hill, J. T., Krause, A., Gröblacher, S., Aspelmeyer, M., and Painter, O. (2011). Laser cooling of a nanomechanical oscillator into its quantum ground state. *Nature*, **475**, 89–94.

Cohadon, P.-F., Heidmann, A., and Pinard, M. (1999). Cooling of a mirror by radiation pressure. *Physical Review Letters*, **83**, 3174–7.

The LIGO Scientific Collaboration (2015). Advanced LIGO. *Classical and Quantum Gravity*, **32**, 074001.

The Virgo Collaboration (2015). Advanced Virgo: a second generation interferometric gravitational wave detector. *Classical and Quantum Gravity*, **32**, 024001.

Courty, J.-M., Heidmann, A., and Pinard, M. (2001). Quantum limits of cold damping. *European Physical Journal D*, **17**, 399.

Crawford, F. S. (1985). Running Crooke's radiometer backwards. *American Journal of Physics*, **53**, 1105.

Crookes, W. (1874). On attraction and repulsion resulting from radiation. *Philosophical Transactions of the Royal Society of London*, **164**, 501–27.

Dorsel, A., McCullen, J. D., Meystre, P., Vignes, E., and Walther, H. (1983). Optical bistability and mirror confinement induced by radiation pressure. *Physical Review Letters*, **51**, 1550–3.

Fabre, C., Pinard, M., Bourzeix, S., Heidmann, A., Giacobino, E., and Reynaud, S. (1994). Quantum-noise reduction using a cavity with a movable mirror. *Physical Review Letters*, **49**, 1337–44.

Gigan, S., Böhm, H. R., Paternostro, M., Blaser, F., Langer, G., Hertzberg, J. B., Schwab, K. C., Bäuerle, D., Aspelmeyer, M., and Zeilinger, A. (2006). Self-cooling of a micromirror by radiation pressure. *Nature*, **444**, 67–70.

Gozzini, A., Maccarrone, F., Mango, F., Longo, I., and Barbarino, S. (1985). Light-pressure bistability at microwave frequencies. *Journal of the Optical Society of America B*, **2**, 1841–5.

Hadjar, Y., Cohadon, P.-F., Aminoff, C. G., Pinard, M., and Heidmann, A. (1999). High-sensitivity optical measurement of mechanical brownian motion. *Europhysics Letters*, **47**, 545–51.

Höhberger Metzger, C. and Karrai, K. (2004). Cavity cooling of a microlever. *Nature*, **432**, 1002–5.

Jaekel, M. T. and Reynaud, S. (1990). Quantum limits in interferometric measurements. *Europhysics Letters*, **13**, 301–6.

Lebedev, P. (1901). Untersuchungen über die druckkräfte des lichtes. *Annalen der Physik*, **311**, 433–58.

Locke, C. R., Tobar, M. E., Ivanov, E. N., and Blair, D. G. (1998). Parametric interaction of the electric and acoustic fields in a sapphire monocrystal transducer with a microwave readout. *Journal of Applied Physics*, **84**, 6523–7.

Marquardt, F., Chen, J. P., Clerk, A. A., and Girvin, S. M. (2007). Quantum theory of cavity-assisted sideband cooling of mechanical motion. *Physical Review Letters*, **99**, 093902.

Milatz, J. M. W., Zolingen, J.J., and van Iperen, B.B. (1953). The reduction in the brownian motion of electrometers. *Physica*, **19**, 195–202.

Nichols, E. F. and Hull, G. F. (1903). A preliminary communication on the pressure of heat and light radiation. *Physical Review (Series I)*, **13**, 307–20.

O'Connell, A. D., Hofheinz, M., Ansmann, M., Radoslaw, C., Bialczak, M., Lenander, E. L., Neeley, M., Sank, D., Wang, H., Weides, M., Wenner, J., Martinis, J. M., and Cleland, A. N. (2010). Quantum ground state and single-phonon control of a mechanical resonator. *Nature*, **464**, 697–703.

Purdy, T. P., Peterson, R. W., and Regal, C. A. (2013*a*). Observation of radiation pressure shot noise on a macroscopic object. *Science*, **339**, 801–4.

Purdy, T. P., Yu, P.-L., Peterson, R. W., Kampel, N. S., and Regal, C. A. (2013*b*). Strong optomechanical squeezing of light. *Physical Review X*, **3**, 031012.

Roch, J.-F., Vigneron, K., Grelu, P., Sinatra, A., Poizat, J.-P., and Grangier, P. (1997). Quantum nondemolition measurements using cold trapped atoms. *Physical Review Letters*, **78**, 634–7.

Rugar, D., Budakian, R., Mamin, H. J., and Chui, B. W. (2004). Single spin detection by magnetic resonance force microscopy. *Nature*, **430**, 329–32.

Safavi-Naeini, A. H., Gröblacher, S., Hill, J. T., Chan, J., Aspelmeyer, M., and Painter, O. (2013). Squeezed light from a silicon micromechanical resonator. *Nature*, **500**, 185–9.

Schliesser, A., Rivière, R., Anetsberger, G., Arcizet, O., and Kippenberg, T. J. (2008). Resolved-sideband cooling of a micromechanical oscillator. *Nature Physics*, **4**, 415–19.

Slusher, R. E., Hollberg, L. W., Yurke, B., Mertz, J. C., and Valley, J. F. (1985). Observation of squeezed states generated by four-wave mixing in an optical cavity. *Physical Review Letters*, **55**, 2409–12.

Teufel, J. D., Donner, T., Dale, L., Harlow, J. W., Allman, M. S., Cicak, K., Sirois, A. J., Whittaker, J. D., Lehnert, K. W., and Simmonds, R. W. (2011). Sideband cooling of micromechanical motion to the quantum ground state. *Nature*, **475**, 359–63.

Verhagen, E., Deléglise, S., Weis, S., Schliesser, A., and Kippenberg, T. J. (2012). Quantum-coherent coupling of a mechanical oscillator to an optical cavity mode. *Nature*, **482**, 63–7.

Wilson, D. J., Sudhir, V., Piro, N., Schilling, R., Ghadimi, A., and Kippenberg, T. J. (2015). Measurement-based control of a mechanical oscillator at its thermal decoherence rate. *Nature*, **524**, 325–9.

Wilson-Rae, I., Nooshi, N., Zwerger, W., and Kippenberg, T. J. (2007). Theory of ground state cooling of a mechanical oscillator using dynamical backaction. *Physical Review Letters*, **99**, 093901.

2

Optomechanics for Gravitational Wave Detection: From Resonant Bars to Next Generation Laser Interferometers

David Blair, Li Ju and Yiqiu Ma

School of Physics, The University of Western Australia, Perth, Australia

David Blair

David Blair, Li Ju, and Yiqiu Ma, *Optomechanics for Gravitational Wave Detection: From Resonant Bars to Next Generation Laser Interferometers* In: *Quantum Optomechanics and Nanomechanics*. Edited by: Pierre-Francois Cohadon, Jack Harris, Florian Marquardt, Leticia F. Cugliandolo, Oxford University Press (2020). © Oxford University Press. DOI: 10.1093/oso/9780198828143.003.0002

Chapter Contents

This paper reviews 40 years of research that includes enormous contributions by many PhD students and postdocs at UWA as well as international colleagues. Robin Giffard, Vladimir Braginsky and Bill Hamilton made enormous contributions to the fundamental ideas. Underpinning the research at UWA was the outstanding UWA Physics Workshop where highly skilled technicians made enormous contributions. Ron Bowers, John Devlin, Arthur Woods, Steve Popel, John Moore and Peter Hay made enormous contributions to the construction of the bar detector NIOBE and/or to the construction of the 80m interferometer at the Gingin research centre. John Ferreirinho, Tony Mann, Peter Veitch, Peter Turner, Steve Jones, Eugene Ivanov, Mike Tobar, Ik Siong Heng and Nick Linthorne made enormous contributions to the success of NIOBE, and Peng Hong, Mitsuru Taniwaki and Brett Cuthbertson made huge contributions to transducer development. John Winterflood, Jean-Charles Dumas, Ben Lee and Eujeen Chin built the vibration isolators that underpinned all the research at Gingin. Mark Nottcut helped us develop interferometry at UWA. Chunnong Zhao built the first interferometer and designed almost all the important experiments at Gingin. Slawek Gras, Pablo Barriga and Jerome Degallaix did all the heavy lifting for understanding parametric instability, thermal tuning and cavity mode structures. Yaohui Fan, Sunil Sunsmithan, Qi Fang, and Carl Blair undertook enormously difficult experiments at Gingin while Xu (Sundae) Chen and Jiayi Qin undertook the very difficult small-scale experiments. Haixing Miao and Stefan Danilishin made enormous contributions to our understanding of quantum optomechanics. We owe our thanks too to many, many others whose names are not listed here.

2.1 Introduction

This paper is written at the time of the first direct detection of gravitational waves, one century after the gravitational wave spectrum was first predicted and fifty years after physicists first began to design and construct instruments for this purpose. The discovery was made by the Advanced LIGO detectors which themselves are masterpieces of optomechanical physics and engineering. The detectors are a culmination of half a century of innovation, during which the principles and the technologies of ultrasensitive mechanical measurements were developed, in particular through the development of optomechanics, pioneered by physicists developing earlier generations of gravitational wave detectors.

In 2015 the Advanced LIGO detectors had achieved a factor of 3–4 improvement in amplitude sensitivity. This small step took us across a threshold, from inability to detect astronomical signals, to a regime in which strong signals have become detectable. The next steps in sensitivity will offer enormous scientific rewards.

This paper reviews the 40-year history that led to the first detection of gravitational waves, and goes on to outline techniques which will allow the detectors to be substantially improved. Following a review of the gravitational wave spectrum and the early attempts at detection, it emphasizes the theme of optomechanics, and the underlying physics of parametric transducers, which creates a connection between early resonant bar detectors and modern interferometers and techniques for enhancing their sensitivity.

Developments are presented in a historical context, while themes and connections between earlier and later work are emphasized. We begin by reviewing the gravitational wave spectrum.

2.2   The Gravitational Wave Spectrum

Nature provides us with two fundamental spectra of zero rest-mass wave-like excitations which travel through empty space at the speed of light. The spectrum of electromagnetic waves was predicted by James Clerk Maxwell in 1865 (Maxwell, 1865), 150 years before this Les Houches summer school. It was more than two decades before Heinrich Hertz (Hertz, 1887) succeeded in generating and detecting Maxwell's waves. At the turn of the twentieth century, Marconi and others created radios, but it took another century of innovation to fully harness Maxwell's spectrum, from the lowest frequencies, such as the Schumann resonances of the Earth's ionosphere, for which the photons have energy $\sim 10^{-32}$ J, to the highest energy gamma rays, for which the photon energy approaches 1 Joule.

Fifty years after Maxwell published his electromagnetic field equations, Einstein published the field equations of General Relativity. As shown by Einstein in 1916 and 1918 (Einstein, 1916; Einstein, 1918) the equations predicted gravitational waves which are ripples in spacetime described by the Riemann tensor. Einstein considered the waves to be of academic interest only, because their effects appeared to be too small to measure. Others even suggested that the waves were mathematical artefacts. It was not until 1957 that the theoretical case was made for gravitational waves having firm physical reality, with ability to transport energy and do work. This was demonstrated in a thought experiment by Richard Feynman at a conference in Chapel Hill (Rickles and DeWitt, 2011), that marked the beginning of the modern resurgence of General Relativity.

Even though gravitational waves have firm physical reality, their detection is a daunting challenge because the interaction of gravitational waves with matter is very weak. In this section we will use simple arguments to estimate the amplitude and frequency of gravitational waves. Wave amplitude is measured in dimensionless units that characterize the spatial strain amplitude, that represent the fluctuating spacing ΔL between inertial test masses spaced distance L apart: $h = \Delta L / L$.

In their most compact form, Einstein's equations can be written as $\mathbf{G} = (8\pi G/c^4)\mathbf{T}$, where $\mathbf{G}$ is the Einstein curvature tensor which describes the curvature of spacetime and $\mathbf{T}$ is the stress energy tensor that describes the mass-energy distribution. The coupling constant $8\pi G/c^4$ has a magnitude $\sim 10^{-43}$ in SI units. Time varying stress-energy creates waves of curvature which can be measured as a strain h. Without deriving the wave equation (which can be found in numerous sources) it is obvious that the curvature fluctuations in general must have small amplitude because of the smallness of the coupling constant.

In 1916 Schwarzschild published a solution for Einstein's field equations in the limit of spherical symmetry. His solution describes the spacetime of black holes for which there is a central singularity and an event horizon, of radius now known as the Schwartzchild radius, given by $r_s = 2GM/c^2$. At this time there was no evidence for the physical reality of black holes.

The spacetime curvature for a gravitating source of mass M can be estimated from the ratio of r_s/r, where r is the radius of the source. For the Earth, $r_s/r < 10^{-8}$. At the surface of the Sun, $r_s/r \sim 10^{-6}$. These small factors show that general relativistic effects including the generation of gravitational waves are extremely weak in the solar system.

Einstein showed that the gravitational wave luminosity of a source depends approximately on the square of the third time derivative of the quadrupole moment of the source. The simplest source is a pair of masses orbiting each other. For equal masses M orbiting each other at distance L apart and with orbital frequency ω, the gravitational wave luminosity L_G is given (neglecting constants ~ 1) by

$$L_G \sim \frac{G}{c^5} M^2 L^4 \omega^6. \tag{2.1}$$

The same formula can be expressed in terms appropriate for a binary pair of black holes. In this case it resolves to

$$L_G \sim \frac{c^5}{G} \left(\frac{v}{c}\right)^6 \left(\frac{r_s}{r}\right)^2. \tag{2.2}$$

Here the gravitational wave luminosity depends only on the scale r of the system relative to the Schwarzschild radius, and the velocity of the two masses compared to the speed of light.

Equation (2.2) is remarkably different from Eq. (2.1). The coupling factor has been inverted, so that the gravitational wave power emitted is now scaled by the enormous factor c^5/G, which has magnitude $\sim 10^{53}$. Since two black holes will merge with velocity $v \sim c$ at a spacing $2r \sim 2r_s$, it follows that black hole coalescence can create enormous gravitational wave luminosity $\sim 10^{53}$ Watts, 10^{26} times the solar luminosity. This power output is independent of the system mass. Because the event duration is directly proportional to mass, the total energy output increases with mass. Numerical calculations (Pretorius, 2005) show that the above estimates are roughly correct.

A binary black hole coalescence creates the most powerful energy outbursts since the Big Bang. Typically (depending on the black hole spins and mass ratio) it emits about 5% of the system rest mass in gravitational waves. For this reason such systems have always ranked high amongst the targets of gravitational wave astronomy, but lack of knowledge about formation processes for such systems meant that there was always large uncertainty about the event rate for such coalescences. The first detection represented a gravitational explosion of 3 solar masses of energy, the most powerful transient astronomical event ever observed.

It is easy to use intuitive arguments to estimate the amplitude and frequency of gravitational waves from black hole binary coalescence. To estimate amplitude we use the fact that the spatial perturbations will be maximal at distance $\sim r_s$ from the source at the moment of final merger. At this point, where the curvature is so strong that light paths can be deflected by almost 2π radians, and simultaneously are modulated by the dynamical motion of the black holes, the strain amplitude h reaches a maximal value ~ 1. The escaping waves reduce in amplitude inversely with distance such that the wave intensity (proportional to amplitude squared) follows the usual inverse square law. Thus for coalescence of equal mass black holes the peak amplitude of the waves at the Earth is roughly given by r_s/R where r_s again represents the black hole Schwarzschild radius and R is the distance between the source and the Earth.

For the coalescence of supermassive black holes of mass $10^9 M_{sun}$, at cosmological distances (say 10^{26} metres or 10 billion light-years), the amplitude at Earth would be about 10^{-15}. For 30 solar mass black holes, at 10^{25} metres distance, the amplitude would be 10^{-20}. These numbers are upper limits, because the naive estimate ignores the relativistic corrections to the source dynamics, and ignores the gravitational redshift of the escaping gravitational waves.

Since the Schwarzschild radius depends linearly on mass, it follows that the gravitational wave frequency for binary black hole coalescence depends inversely on the system mass. The peak frequency of the gravitational waves is determined roughly by the orbital period of the last stable orbit ($3\,r_s$). For $10^9 M_{sun}$ black holes this radius $\sim 10^{13}$ metres corresponds to a frequency $\sim 10^{-5}$ Hz or one cycle per day. For stellar mass black holes, say $30\,M_{sun}$, $r_s \sim 10^5$ metres and the peak frequency is a few hundred Hz.

Today we have broad understanding of the expected gravitational wave spectrum. It is summarized in Fig. 2.1. Detectable sources are predicted from 10^{-18} Hz to $\sim 10^4$ Hz, as well as speculative sources at even higher frequency. There are four frequency bands in which there are significant detection efforts. In the cosmological frequency band below 10^{-16} Hz, frozen relic gravitational waves from the Big Bang should create a polarization signature in the cosmic microwave background. Claims of detection in this band in 2014 have been shown as likely to be due to foreground dust (Ade *et al.*, 2015), but future multi-frequency observations may be able to separate dust from the gravitational waves, thereby enabling the testing of theories of inflation. In the nanohertz frequency band, gravitational waves created by supermassive black hole binaries (prior to merger) are potential sources. These could be detected as correlated perturbations in the arrival times of pulses from millisecond pulsars. Detection could allow the growth history of supermassive black holes to be measured.

At millihertz frequencies, gravitational waves from objects falling into intermediate mass black holes, and binary stars systems in the Milky Way are likely to be detectable by future multimillion-km space-based laser interferometers.

This paper focuses on the optomechanics for gravitational wave detection in the audio frequency band. This is the best developed frequency band, offering a rich variety of sources. See (Blair *et al.*, 2012) for a detailed discussion of the range of potential sources. The Advanced LIGO and Advanced Virgo gravitational wave detectors were specifically

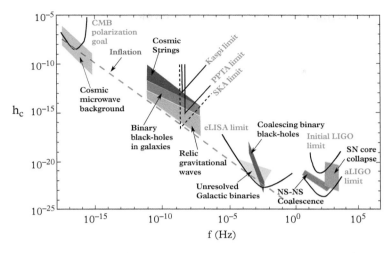

Fig. 2.1 *An outline of the expected spectrum of gravitational waves and detection techniques. The spectrum spans 20 decades, from frozen relic waves at 10^{-16} Hz to the audio frequency band.*

designed to target the coalescence of binary neutron stars because these are a known population with a moderately well predicted event rate.

Since binary black holes had never been observed until the recent detection, the event rate for binary black hole coalescence could only be estimated through astrophysical modelling. They are likely to have originated from very massive Population III stars in the early universe. There are two ways they could have been created. They could have been born from a high-mass Population III binary pair, that evolved rapidly to become black holes. The initial orbital separation of the binary black hole system determines the lifetime to coalescence. If the separation is too large, the time to coalescence can greatly exceed the age of the universe. Observations of coalescence events select for systems born with suitable initial separations.

A more likely source of binary black holes that can be observed merging today, consists of binaries created inside globular clusters. Modelling shows that isolated black holes born from Population III stars in the early universe are likely to sink towards the cores of globular clusters and also into the high-density cusps in the centre of galaxies. Here black hole binaries are likely to be formed by capture events; interactions with other stars are likely to extract sufficient energy from the binary that a significant number will be detectable as coalescence observable today.

It was predicted (Abadie *et al.*, 2010; Press and Teukolsdy, 1977) that such binaries are likely to be detectable at an even greater rate than neutron star coalescences. This prediction has been confirmed by the detection of a 30 solar mass binary black hole by Advanced LIGO. Because waveforms are very sensitive to the black hole spins and masses (which depend on the formation mechanism), as more events are observed it will be possible to determine how the systems were formed, and in this way we will be able to derive information on the ancestors of the stars we see today in the modern universe.

2.3   Gravitational Wave Detection

Because gravitational waves are waves of gravitational tidal force, they apply a differential acceleration to a mass quadrupole (such as a long bar or a pair of separated test masses). Assuming a wave amplitude $h = \Delta L/L$, we can treat L as the effective spacing of the test masses or the length of the resonant bar. Thus the detection of gravitational waves becomes the task of measuring very small differential motions. In a solid bar, a transient differential force can be detected as an acoustic excitation of the fundamental resonant mode. If freely suspended test masses are used, detection requires measurement of differential accelerations of the free masses.

Joseph Weber first considered the design of gravitational wave detectors in a book (Weber, 1961) and a paper published in 1960 (Weber, 1960). He considered the gravitational wave energy absorbed by a resonant mass quadrupole such as a resonant bar. Recognizing the significance of acoustic losses in increasing the effects of thermal noise, he proposed using large resonant masses in the form of discs or cylinders with very high acoustic quality factor. Transient bursts of gravitational waves with millisecond duration (as might be produced by the non-spherical birth of a black hole in a supernova or the final transient from a binary black hole coalescence) could cause excitation of modes with large quadrupole moment such as the fundamental longitudinal mode of a bar. He went on to construct two resonant bar detectors, one of which is shown in Fig. 2.2.

Despite the smallness of the gravitational wave strain expected from likely sources, Weber undertook extended observations with a pair of resonant bars, using coincidence detection to distinguish between local noise sources (expected to be uncorrelated) and signals of cosmic origin, which would be correlated in both detectors.

In 1969, again in 1970, and in subsequent papers (Weber, 1970; Weber *et al.*, 1973), Weber published results claiming coincident excitation of his widely spaced pair of detectors (one was located at the University of Maryland, the other was at Argonne National Laboratory). These observations were claimed to be the first detection of gravitational waves. The wave amplitude of individual bursts was estimated to be $h \sim 10^{-15}$. If valid, they would have represented a huge flux of gravitational waves corresponding to thousands of solar masses being turned into gravitational waves every year in the Milky Way, with each burst representing ~ 1 solar mass of gravitational energy being radiated isotropically.

Weber's claim had enormous repercussions which are still being felt today. Astronomers considered that the enormous flux and event rate implied by Weber's results to be impossible to reconcile with current astronomical knowledge. Many physicists were intrigued, and within a few years approximately 10 laboratories had built or were building Weber-type detectors. By 1972 results began to come in, which failed to confirm Weber's observations. There was anger and disillusionment amongst the physicists attracted to the

Fig. 2.2 *Joseph Weber working on one of his aluminium bars. AIP Emilio Segrè Visual Archives.*

field, but a small subset of the new groups were intent on building substantially improved detectors.

Weber's detectors used massive piezo-electric ceramics (commonly known as PZT) to read out the vibrations of his resonant masses. Sensitivity was limited by electronics noise, and mechanical thermal noise partly due to PZT acoustic losses. By the fluctuation-dissipation theorem, both acoustic and electrical dissipation directly translate to readout noise. Weber's bars represented one of the first examples in macroscopic experimental physics, where *back-action* noise was manifested.

Back-action noise was prominent because piezoelectrics are reciprocal devices. A displacement Δx creates a voltage ΔV and a voltage ΔV creates the same displacement Δx. Linear amplifiers such as operational amplifiers are always characterized by a pair of noise sources, normally described as current noise and voltage noise. Amplifier current noise leads to voltage fluctuations acting on the PZT, which create a noise force which acts directly back onto the mechanical system. This back-action noise acting on the PZT applies forces to the mechanical resonator, which modifies its dynamical state. Because these fluctuations accumulate over time, their effects increase as the integration time is increased (corresponding to reduced measurement bandwidth). This behaviour is opposite to that of voltage noise, which is simply additive, and *reduces* as the measurement bandwidth is reduced. The opposite bandwidth dependence of series noise and back-action noise leads to a classical measurement limit similar to the now well known *standard quantum limit*. There is a minimum detectable energy that occurs at an optimum integration time. Increasing or decreasing the integration time (or its inverse, the measurement bandwidth) degrades performance.

In the early 1970s the standard quantum limit was unknown, but the noise analysis of resonant bars demonstrated a similar measurement limit in the classical regime. The standard quantum limit sets an energy detection limit set by the photon energy hf, for signal frequency f. The classical limit is $k_B T_n$, where T_n is the effective measurement *noise temperature*. In Weber's case the noise temperature was ~ 30K, or about $10^8 hf$. Optimum sensitivity occurred for a bandwidth of about 1Hz. Today back-action noise is most familiar in optomechanics as radiation pressure noise.

Weber type bars at room temperature could only achieve a strain amplitude sensitivity $\sim 10^{-15}$. Cryogenic resonant bars, which we will discuss below, exceeded the sensitivity of Weber's bars by many orders of magnitude, and one of them, NIOBE, implemented the first high-performance optomechanical type readout with microwaves. Laser interferometers went on to make enormous strides in optimizing optomechanical readouts. With these techniques detectors today have exceeded Weber's strain sensitivity by 6–7 orders of magnitude, corresponding to an astonishing 14 orders of magnitude improvement in gravitational wave flux sensitivity.

2.4   Cryogenic Bars and the First Parametric Transducers

Soon after Weber's first claims, William Fairbank and Bill Hamilton proposed the use of cryogenic techniques to create resonant mass detectors of far greater sensitivity. Cryogenics would reduce thermal noise and allow the use of superconducting vibration transducers that in principle should allow extremely high sensitivity. Fairbank and Hamilton led groups at Stanford and Louisiana State University, with the initial goal of creating 5-tonne magnetically levitated resonant bars cooled to 3 mK temperature. Hamilton proposed development of a radio frequency superconducting parametric transducer, while Fairbank proposed using a superconducting quantum interference device (SQUID) based transducer. The program was outlined in a paper published in 1974 (Boughn *et al.*, 1974).

In the above paper two superconducting transducer concepts were presented. One is the parametric transducer in which an acoustic signal modulates a resonant circuit, while the second involves direct modulation of supercurrents by a superconducting ground plane. The two concepts are illustrated in Fig. 2.3.

At Stanford, Ho Jung Paik began development of a SQUID-based transducer system based on superconducting flux conservation (Paik, 1976). Because magnetic flux LI (where L is inductance and I is current) is conserved in a superconducting circuit, and because the inductance of a flat pancake coil depends linearly on spacing to a superconducting ground plane, supercurrent is linearly modulated by proximity of a superconducting test mass in the form of a resonant diaphragm. Paik was the first to

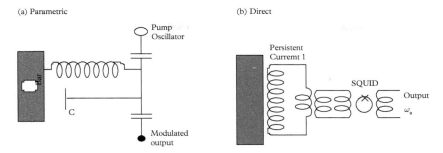

Fig. 2.3 *Two concepts for high sensitivity transducers for gravitational wave detectors. The left side shows a parametric transducer in which motion modulates an LC circuit. The right side shows a superconducting circuit read out by a Superconducting Quantum Interference Device (SQUID) amplifier.*

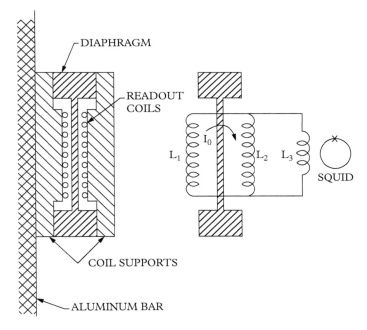

Fig. 2.4 *The Paik transducer design. Motion of the diaphragm causes a difference current to flow through inductance L3 which is coupled to the SQUID. If the diaphragm mode is tuned to the bar frequency, this configuration embodies two-mode impedance matching discussed later.*

observe low acoustic losses $\sim 10^{-7}$ in annealed niobium at low temperatures, which was used both for the test masses and the inductive circuit. Like PZT, this inductive transducer was reciprocal, its forward transductance equal to its reverse transductance. However it obtained high sensitivity by using the extremely low noise of SQUID amplifiers and the low mechanical acoustic loss of the transducer. Figure 2.4 shows a schematic diagram of Paik's transducer.

At Louisiana State University, Blair and Hamilton developed the first superconducting parametric transducer (Blair *et al.*, 1975). Like the Stanford transducer, the
device made use of a superconducting pancake coil, but in this case it was used as
part of a radio frequency LC-circuit, for which the frequency was modulated by the
proximity of the surface of a superconducting test mass. The pancake coil was a thin film
niobium-on-sapphire structure designed to achieve high radio frequency quality factor
at 15–30 MHz. This was the first gravitational wave transducer to make use of the non-
reciprocal properties achievable through parametric upconversion. The acoustic signal
frequency at ~ 1 kHz was up-converted to the radio frequency. The signal appears as
a pair of signal sidebands, which must be demodulated to recover the acoustic signal.
Due to the upconversion process the device is intrinsically non-reciprocal. As discussed
in section 1.7, the forward to reverse transductance ratio is generally proportional to the
frequency upconversion ratio.

Figure 2.5 shows the design of this first parametric transducer. The device used a
magnetically levitated niobium and sapphire spindle, with a sensing surface very close
to the niobium-on-sapphire LC resonator. Magnetic levitation via a high supercurrent
in a single loop of wire provided low acoustic loss. While this device achieved a
sensitivity $\sim 10^{-17}$ m/$\sqrt{\text{Hz}}$ (Blair, 1979; Blair and Hamilton, 1979) detailed theoretical
understanding of such devices was essential to their optimization. This would only occur
with development of microwave parametric devices described in the next section.

These first high-performance superconducting transducers also exposed technical
problems, which for the parametric transducer included the radio frequency power
dependence of the superconducting LC circuit quality factor, and the need for an
ultralow phase noise oscillator with which to excite the transducer.

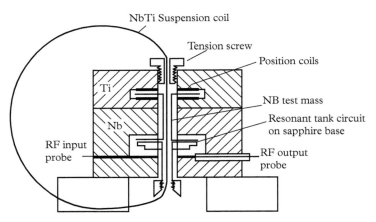

Fig. 2.5 *Implementation of a magnetically levitated parametric transducer. Normal orientation was with
the spindle axis horizontal. Coils at one end allowed the sensing surface to be located very close to a 12-mm
diameter niobium pancake coil etched on a sapphire substrate.*

Fairbank and Hamilton's proposal to build cryogenic gravitational wave detectors had ambitious goals. They aimed a) to reduce the temperature by a factor of 10^5 (300 K to 3 mK) so as to suppress the thermal noise energy, b) to increase the quality factor of the resonant bar by orders of magnitude to suppress the rate of thermal fluctuations, c) to use superconducting magnetic levitation to achieve exceptional vibration isolation, and d) to use superconducting transducers to achieve practically noise-free vibration measurements. They estimated that the signal-to-noise ratio could be increased by a factor of 10^6.

The above ambitious goals were realized to a large extent. While both technical and fundamental issues prevented the full benefits from being realized, the transducers provided a wealth of new physics and techniques that have had enormous impact on experimental physics ever since. Below we summarize the key issues and their outcomes.

Cryogenics and acoustic noise: The feasibility of cryogenic cooling of huge bars to 4 K was quickly demonstrated, and groups in Italy led by Pizella and later Cerdonio successfully cooled 2300 kg Al bars below 100 mK.

Figure 2.6 shows a 2.3 tonne Al antenna, AURIGA, inside a cryostat which successfully cooled the bar to below 100 mK using a dilution refrigerator. NIOBE achieved a quality factor greater than 10^8 at 5 K.

However the benefits of cooling below ~ 2 K were never fully realized due to difficulties in cryogenic vibration isolation. Boiling liquid helium and gas flow in pipes is a significant acoustic noise source. Some detectors used cooling below the liquid helium superfluid transition at 2 K to eliminate the noise of boiling helium. Others created much more elaborate vibration isolation systems that operated within the cryogenic environment. The best systems eliminated cryogenic noise by using very low pressure helium exchange gas as a thermal conductor with negligible acoustic transmission. However below 100 mK the helium vapour pressure is too low to provide thermal conduction— gas effectively ceases to exist at this temperature. Thus at 100 mK mechanical thermal links are required to extract heat and accomplish cooling. Such links also transmit vibrations.

Low acoustic loss systems: It was difficult to construct SQUID based transducer systems without acoustic losses associated with composite structures, so that very high quality factors were not observed except in the detector NIOBE (discussed in more detail below) which used niobium as the detector material, used low acoustic loss bonding, and contained minimal material with high acoustic losses.

Magnetic levitation: Magnetic levitation was only used successfully on small-scale niobium prototypes (up to 67 kg) because no methods were found for attaching high critical field superconducting material to a bar to enable magnetic levitation without unacceptable acoustic loss. Extensive studies in Western Australia demonstrated that the largest possible diameter for magnetic levitation of a niobium bar is about 200mm, limited by the critical magnetic field at which flux penetrates the superconductor.

Amplifier limits: SQUID amplifiers were expected to achieve quantum limited performance (see discussion below) but in practice noise performance was degraded when the SQUID was coupled to an external circuit.

(a)

(b)

Fig. 2.6 *a) The AURIGA aluminium bar that was cooled below 100 mK and used a SQUID passive transducer, and b) the NIOBE detector, made from niobium, with a microwave parametric transducer.*

Standard quantum limit: A major fundamental limitation that was completely over-looked in 1972 was that the proposed strain sensitivity brought detectors into the quantum regime. It came as a shock to the community that measurement of vibrations in highly macroscopic objects such as tonne-scale bars should be governed by quantum mechanics. In 1978 Braginsky *et al.* (Braginsky *et al.*, 1978) demonstrated the existence of the standard quantum limit, which followed much earlier work by Heffner (Heffner, 1962) on the quantum limit to linear amplifiers, later applied by Giffard (Giffard, 1976) to mechanical measurements with transducers. This recognition of the quantum nature of macroscopic measurements marked the beginning of the development of macroscopic quantum mechanics. In the next section we will discuss the first use of non-contacting measurements with microwave re-entrant cavity parametric transducers.

2.5 Non-contacting Superconducting Microwave Parametric Transducers

At Louisiana in 1974, Blair and Hamilton had demonstrated a Q-factor of 6×10^7 in a 6 kg niobium bar. This led to the idea of building a large-scale niobium detector which would combine low acoustic loss with the advantages of superconductivity. In 1975 a project began to develop a very high Q-factor niobium gravitational wave detector in Australia. Following a suggestion by Braginsky, and with much better understanding of the issues associated with parametric transducers, experiments began at UWA to explore the possibilities of using superconducting microwave parametric transducers based on re-entrant cavities, operating around 10 GHz.

A re-entrant cavity is geometrically simple, rigid and robust, and if suspended about 50μm from a superconducting surface could achieve a high Q-factor $\sim 10^5$–10^6 with a very strong tuning coefficient ~ 300 MHz per micron due to capacitive modulation of the gap between the post (which is the main source of inductance) and the sensing surface. Figure 2.7 shows the cross-section of a typical cylindrically symmetrical transducer configuration.

The re-entrant cavity made it possible to create a non-contacting transducer requiring no mechanical connection to the bar, thus helping to minimize the acoustic loss. Two sub-scale prototypes using a fully magnetically levitated bar combined with a similarly levitated transducer demonstrated the potential of this technology. However for a 1.5 tonne detector the transducer required a second mechanical impedance matching stage, and the diameter was too large for magnetic levitation. A non-contacting transducer system was still able to be implemented in a slightly different form as discussed below.

Figure 2.8 shows two examples of a magnetically levitated re-entrant cavity transducer. Both cavity structures include choke sections in the form of a large ring groove which creates a large impedance mismatch for radially guided microwaves, so as to reduce radiation losses. A superconducting coil outside the choke section (in one case) creates a weak repulsive force between the transducer and the levitated bar. When levitated, with appropriate longitudinal magnetic springs, the transducer freely oscillates across resonance. A time trace of the demodulated signal output is shown in Fig. 2.9.

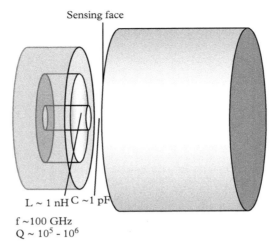

Sensing face

L ~ 1 nH   C ~1 pF

f ~100 GHz
Q ~ 10^5 - 10^6

Fig. 2.7 *A typical re-entrant cavity transducer located near to the face of a cylindrical sensing surface, with a central post about 5 mm long and 1 mm diameter.*

Fig. 2.8 *Two different magnetically levitated re-entrant cavity parametric transducers. Left: a re-entrant cavity with narrow choke section. Right: Transducer with wider choke, superconducting repulsion coil also used for SQUID secondary sensing, and a superconducting frequency stabilization cavity used to suppress phase noise.*

It shows sharp transitions as the re-entrant cavity crosses resonance while oscillating at a frequency ~ 10 Hz with an amplitude of a few nm. A superconducting voice coil provided control forces to lock the transducer at the centre of resonance.

Figure 2.10 shows the levitation cradle for such a transducer, with a 50mm diameter niobium bar in its own levitation cradle. Levitation was achieved by pumping several hundred amps of persistent current into copper clad NbTi coils soldered to a stainless

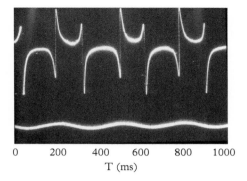

0   200   400   600   800   1000

T (ms)

Fig. 2.9 *Superconducting microwave parametric transducer response as it sweeps through the cavity resonance. The signal is derived from a double balanced mixer with the local oscillator phase adjusted to recover the phase shift across resonance. The sharp transitions represent the phase shift across resonance. This error signal was used to lock the cavity near the centre of resonance via forces applied through the voice coil. The bottom trace shows the sinusoidal motion of the transducer measured by the SQUID readout shown in Fig. 2.11.*

Fig. 2.10 *Magnetic levitation cradles for a small Nb bar and a transducer. The bar diameter is 50 mm. The transducer, supported by a NbTi alloy superconducting shield, is illustrated in Fig. 2.8. Bar and transducer were independently levitated. Radio frequency sense coils monitored the levitation height.*

steel housing. The copper and the stainless steel provide resistive insulation for a zig-zag current loop that supports the levitated masses.

The system illustrated above was excited by a low noise klystron microwave source. It achieved a noise temperature $\sim 10^{-2}$ K, three orders of magnitude better than a room temperature bar, but it also exposed limitations that would need to be solved for a large-scale resonant bar.

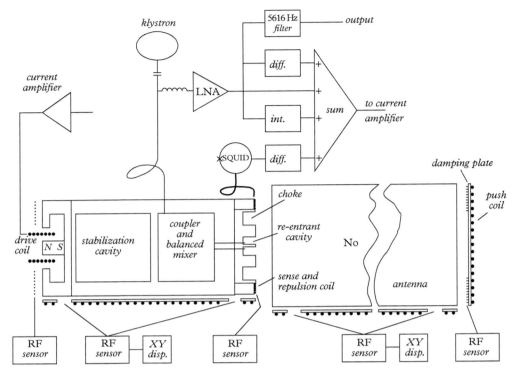

Fig. 2.11 *Schematic diagram of the magnetically levitated gravitational wave detector described in the text above (Blair and Mann, 1981). The transducer required precision active control of the gap spacing since the position bandwidth of the re-entrant cavity was very small (~ 10^{-11} m) and low frequency motion at a few Hz was ~ 10^{-4} m. A radio frequency sensor and a SQUID sensor provided hierarchical sensitivity for the gap spacing, for the purpose of control and calibration.*

2.6 Coupling Coefficients, Thermal Noise and Effective Temperature

In the analysis of resonant bar antennas, the concept of the electromechanical energy coupling coefficient β was introduced to characterize the transducer. While in both types of transducer an acoustic signal in the resonant bar is converted into an electrical signal, the process is somewhat different for a direct transducer, where the electromagnetic energy has the same frequency as the acoustic signal, and the parametric transducer, where the energy is transformed to modulation sidebands.

In a direct transducer with a capacitive sensor with capacitance C, it is straightforward to show that β is given by the ratio of the electrostatic energy divided by the mechanical energy of the mechanical resonator if its vibration amplitude x was equal to the capacitance gap, that is:

$$\beta_1 = \frac{\frac{1}{2}CV^2}{M\omega_a^2 x^2},$$ (2.3)

where ω_a is the resonance frequency of the bar antenna.

For the parametric transducer (Fig. 2.3) we obtain a similar expression, except that the coupling coefficient is increased by the frequency ratio between the mechanical frequency of the bar and the pumping frequency ω_p:

$$\beta = \frac{\omega_p}{\omega_a} \frac{\frac{1}{2}CV^2}{M\omega_a^2 x^2}.$$ (2.4)

This result implies that a large frequency ratio is advantageous. However, because a high resonant frequency generally requires a much smaller capacitor, the advantage occurs as long as the magnitude of the frequency ratio ($\sim 10^7$ for the transducers discussed here) exceeds the capacitance ratio (typically 10^3). Another factor in favour of the parametric transducer is that AC electric field breakdown threshold (which sets the maximum usable voltage across the capacitor) is usually substantially higher than DC breakdown. There are other important advantages of using parametric transducers as discussed further below.

It is important to understand why acoustic losses, or their inverse, the acoustic quality factor, are so important in making low noise measurements of mechanical resonant devices. This is best explained in the context of a phase-space description of the state of the oscillator. We are interested in the measurement of the fundamental mode of a resonant bar because, as discussed above, it has the highest quadrupole moment and couples most strongly to gravitational waves. However, the discussion is relevant for measurements of any mechanical mode. We start with a simple but incomplete classical description of a resonant bar readout, shown in Fig. 2.12. The state of the resonant bar can be expressed in terms of amplitude and phase coordinates, but a much more elegant description uses phase-space quadrature coordinates.

Figure 2.12 shows a bar coupled to a transducer with coupling coefficient β. After amplification, the transducer output is demodulated to zero frequency, using a reference

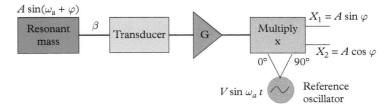

Fig. 2.12 *A classical resonant mass coupled to a transducer, with an amplifier of gain G and demodulated to zero frequency to record the two quadratures A sin φ and A cos φ. It is assumed that the transducer output signal is at the bar antenna frequency ω_a.*

oscillator at the bar frequency ω_a. This process gives rise to two quadrature values X_1 and X_2. A vector from the origin has magnitude equal to the resonator amplitude, and phase angle corresponding to the phase difference between the reference oscillator and the acoustic mode.

This picture, shown in Fig. 2.13, allows us to understand the role of acoustic losses in measurements of mechanical resonators. First suppose the Q-factor is infinite. Then the bar would be a perfect harmonic oscillator with constant amplitude and phase. In this case the phase-space coordinate of the bar is a stationary point with fixed phase and amplitude.

In reality, every mode is coupled to the thermal reservoir: the enormous set of acoustic modes that define the heat capacity of a material, and which contribute fluctuating forces that lead to mechanical thermal noise or Brownian motion. This coupling, which depends inversely on the Q-factor, causes the amplitude and phase of mechanical modes to fluctuate while maintaining a mean energy of $k_B T$. In phase-space, the state of the mode is no longer a fixed point. It undertakes a random walk. If the Q-factor is very high, the velocity of the state vector random walk in phase-space is very low. There is a clear and simple correlation: high Q = weak coupling to the thermal reservoir = slow motion in phase-space.

Now suppose you want to measure an external force which acts on the resonator to change its state vector. If you make a measurement in a short period of time τ_m, compared with the mechanical relaxation time τ_a, the changes in the state vector will be small, so the thermal noise contribution will be minimized. However, if the measurement time τ_m is long compared with τ_a, the noise energy will be comparable to $k_B T$. In general if the measurement takes place in a short time $\tau_m \ll \tau_a$, the noise energy change will be reduced by the factor τ_m/τ_a. The phase-space coordinates are illustrated in Fig. 2.13. The vector (P_1, P_2) represents the state change between two successive measurements. The magnitude of this vector will be reduced if the resonator Q-factor is higher.

Referring to Fig. 2.12, let us ask about the role of the coupling factor β. The coupling factor scales the amount of mechanical resonator energy that appears in the transducer.

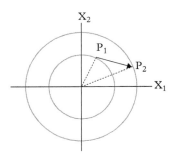

Fig. 2.13 *Phase-space coordinates and state vectors for successive measurements. The state vector for a high-Q resonator undertakes a random walk due to thermal noise, but the length of the difference vector between pairs of measurements (P_1, P_2) is reduced if the Q-factor is high.*

It reduces the total phase-space change. Because there is always additive noise (as we will discuss later), reduction in β causes the measurement time to be increased and the thermal noise contribution to increase. Thus small β = longer measurement time = reduced sensitivity.

Thus the sensitivity of many mechanical measurements depends on the factor β. Today many experiments aim to eliminate thermal noise in mechanical resonator measurements (Underwood *et al.*, 2015; Aasi *et al.*, 2015). These typically need factors of $\beta Q \sim 10^{12}$.

The reduction of thermal noise in high-Q resonators can be expressed by the concept of noise temperature: the magnitude of the energy fluctuations expressed in degrees Kelvin.

Weber's detectors achieved a noise temperature of about one tenth of the ambient temperature, ~ 30 K through use of high Q-factor. The NIOBE detector, discussed below, achieved a noise temperature $\sim 2 \times 10^{-4}$ of the 5 K operating temperature (Blair *et al.*, 1995).

2.7   Impedance Formalism for Transducers

Using a single parameter β to characterize mechanical resonators coupled to transducers, as we considered above, is a gross over-simplification. We did not consider the additive noise from the amplifier, nor did we consider the fact that there is no such thing as a perfect one-way valve. The last concept is behind the uncertainty principle, which appears in any careful analysis, whether classical or quantum.

From a classical standpoint, all of the above is encapsulated in the impedance formalism. The concept of impedance is used extensively in radio frequency electronics and transmission lines (such as the 50 Ω input impedance of many radio frequency devices). If impedances are matched at any junction, then there is maximal power flow without reflection. If there is an impedance mismatch, then energy is reflected at the interface.

However, impedance is a much more general concept that can be applied to all forms of wave energy. A dielectric mirror is a device designed to offer a high impedance mismatch for light, while an anti-reflection coating matches the impedance of free space to the impedance (for light) of a lens or window. A human ear contains complex structures to match the acoustic impedance of the sensing system to the acoustic impedance of the air. We will see that impedance matching was extremely important to the design of cryogenic gravitational wave bar detectors.

Transducers can be described in the following matrix formulation which links the force F and velocity u at the transducer input to the voltage V and current I at the output. Linking these quantities is the 2×2 impedance matrix $\mathbf{Z}$:

$$\begin{bmatrix} F \\ V \end{bmatrix} = \begin{bmatrix} Z_{11} & Z_{12} \\ Z_{21} & Z_{22} \end{bmatrix} \begin{bmatrix} u \\ I \end{bmatrix} \tag{2.5}$$

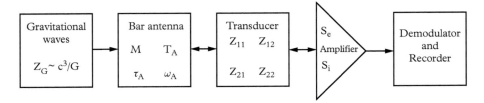

Fig. 2.14 *A schematic diagram showing the measurement chain for a resonant mass gravitational wave detector. Gravitational waves with very high impedance, as discussed in the text, interact with an antenna of mass M, temperature T_A, relaxation time τ_a and resonance frequency ω_a. The transducer is described by four impedance matrix components, with force and velocity at the imput and voltage and current at the output. The transducer is amplified by a linear amplifier with voltage noise S_e and current noise S_i. Both ends of the chain are classical, where back-action can be ignored, but in between back-action plays an important role.*

The four terms of **Z** can be determined for every transducer. A full two-port description of the measurement chain then only needs to have included the effective noise properties of the amplifier to obtain a full theoretical description of system performance. Figure 2.14 shows this schematically. The amplifier has gain G and two noise terms, the voltage noise spectral density S_e and the current noise spectral density S_i. The transducer has a mechanical input impedance Z_{11}. This term has units of force per unit velocity, or kg.s^{-1}, and characterizes the compliance of the transducer input. We will see that this term can be strongly tuned in parametric transducers, whether they be on resonant bars, optomechanical microresonators or laser interferometer gravitational wave detectors. The force and velocity are associated with the interaction between the bar and the transducer.

The second term in the two-port matrix is the forward transductance Z_{21}, which has units of volts per metre per second, Vm^{-1}s^{-1}. This term describes the magnitude of the output voltage due to a mechanical input measured as a velocity.

The terms associated with the output are the output impedance Z_{22} and the reverse transductance Z_{12}. The latter term has units of kg.amp^{-1}. In general, Z_{12} is always finite. It determines how the output circuit, in particular the current noise, is able to act back on the input such that a voltage fluctuation at the output creates a force fluctuation at the input. This force acts back on the mechanical resonator and behaves similarly to the fluctuating forces from thermal noise. It creates a noise term that increases as the measurement integration time increases, and is responsible for a classical uncertainty principle that limits the performance of any classical transduction system. The same physics is described in different language in optomechanical systems (see section 2.11.1). The reverse transduction in optomechanics often arises from quantum radiation pressure noise, or in the classical regime, the intensity noise of the laser light.

The output impedance Z_{22} of the transducer is a simple electrical quantity measured in ohms. The only requirement for this term is that it satisfies an electrical impedance matching condition with the amplifier.

In parametric transducers, which are intrinsically resonant devices, there is large flexibility in their pump frequency, from radio frequency to optical, their tuning, their Q-factors and their linewidths. The transduction matrix contains Lorentzian terms associated with their resonant circuits which can provide strong resonant amplification or suppression of signal sidebands. We will see later that this flexibility allows surprising properties such as electromagnetic or optical springs (the name depends on the nature of the pump radiation), self-cooling, and negative dispersion.

To explore the properties of parametric transducers it is often useful to use mixed pictures, sometimes thinking about them from a quantum picture, and sometimes from a classical circuit standpoint. In the next section, we will review the quantum picture.

2.8   The Quantum Picture for Parametric Transducers

In section 2.3 we looked at the general structure of a parametric transducer, while in section 2.7 we examined a general two-port matrix formulation for transducers. Now we will restrict our analysis to parametric transducers, noting their most important characteristic: they scatter radiation from a pump oscillator into modulation sidebands. Conventionally, parametric transducers create a pair of modulation sidebands. The pump oscillator at frequency ω_p is modulated by the mechanical resonator at frequency ω_a. The modulation sidebands occur at frequency $\omega_p \pm \omega_a$. However, single sideband devices can also exist. One example is the three-mode opto-acoustic parametric amplifier (Zhao *et al.*, 2009) which we will discuss in section 2.12.

If we consider parametric transducers from a quantum mechanical viewpoint, we must consider the mechanical resonator to be a system of phonons and the pump oscillator to provide a stream of photons. In this picture, the transducer scatters photons with phonons as illustrated in Fig. 2.15. In the quantum picture there are two separate processes taking place simultaneously. They can be described by simple vertex diagrams as shown in Fig. 2.16.

Both Figs 2.15 and 2.16 emphasize one key point. If energy is conserved in quantum scattering, and if the pump oscillator is a source of photons entering the transducer, then the signal phonon power flow direction is uniquely determined by the sideband

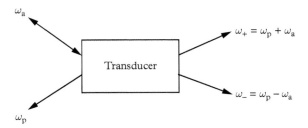

Fig. 2.15 *Diagram of a transducer as a device that scatters mechanical resonator phonons ω_a with pump photons ω_p, to create photons at the upper and lower sideband frequencies $\omega_\pm$.*

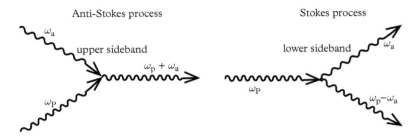

Fig. 2.16 *The processes illustrated in Fig. 2.15 have been broken down into two separate scattering processes, one which creates upper sidebands and the other which creates lower sidebands. Notice that the power flow direction for the phonons has opposite directions.*

frequency. The upper sideband can only be created if a phonon from the mechanical resonator adds to the energy of the pump photons. Therefore, acoustic energy must flow out of the resonator to create an upper sideband. Similarly, a lower sideband photon can only be created if phonon energy ω_a is extracted from ω_p. This phonon energy must flow back into the resonator.

The above ideas were proved classically by Manley and Rowe for lossless nonlinear electrical circuits in the 1950s (Manley and Rowe, 1956). Both the classical and quantum approaches are direct results of energy conservation, but in the quantum picture the conclusions are quite obvious as we saw above.

We have to add just one additional observation to come to a very important conclusion. We started by considering the parametric transducer as a high-Q electrical circuit pumped at resonance. However, the pump frequency does not need to match the transducer electrical resonance. If the transducer cavity frequency (such as the re-entrant cavities discussed earlier) is tuned so that its resonance matches the lower sideband frequency, then naturally the lower sideband will be preferred over the upper sideband. This condition is called blue detuning because the pump oscillator frequency is above the cavity frequency. From a classical standpoint, the lower sideband will have an enhanced amplitude, but from a quantum standpoint excess phonons will flow into the mechanical resonator, increasing its occupation number.

Similarly if the cavity is tuned so that its resonance matches the upper sideband frequency (a red detuned pump oscillator), then more phonons will flow out of the mechanical resonator: its phonon occupation number will decrease.

These changes in occupation number are described as heating or cooling of the mode. Mode cooling, also called self-damping or cold damping, reduces the quality factor of mechanical modes, but in a noiseless way, without addition of classical noise, hence the term cold damping. Mode heating starts by increasing the mechanical resonator Q-factor (but not in a way that increases thermal noise). However, at higher pump power levels it can lead to instability when the phonon injection rate exceeds the loss rate from the intrinsic quality factor of the mechanical resonator: this is called parametric instability.

Cold damping can be very useful. High-Q systems are very difficult to work with because accidental excitation may go on ringing endlessly and can often exceed the dynamic range of the measurement system. Apply cold damping and there is no change in noise performance (because the thermal reservoir is still providing the same fluctuating force) but the modes damp rapidly and the equilibrium signal amplitude is low.

In the above discussion we have ignored zero-point motion. In 1962, Heffner (Heffner, 1962) showed that the uncertainty principle limits the noise temperature of a linear amplifier to

$$T_{n_{\min}} = \frac{\hbar \omega}{k_B \ln 2}. \tag{2.6}$$

This means that a transducer will always experience an additive noise contribution due to the amplifier which follows it, with a minimum possible added noise energy which according to Heffner is ~ 1.4 times the photon energy. Given the quantum limit to the noise energy, the amplifier must always have finite current noise. This acts back on the transducer via the reverse transductance Z_{12}. Thus, a perfect linear transducer must always include back-action noise which acts to disturb the mechanical resonator, and series noise which creates an additive contribution. Even for a lossless transducer with zero classical noise, quantum noise and the thermal noise of the resonator itself set significant limits to measurement.

In section 2.6, we saw that the effective temperature due to thermal fluctuations reduces as the ratio of the measurement time to antenna relaxation time τ_m/τ_a. The measurement noise (expressed as noise temperature T_e) is given by (Blair, 1980)

$$T_e = \frac{2\tau_m}{\tau_a} T_a + \frac{2\hbar}{k_B \ln 2}, \tag{2.7}$$

where T_a is the temperature of the resonant bar antenna.

Remembering that $\tau_a = Q/\omega_a$, we can now specify the Q-value Q_Q required to reach the quantum limit given in Eq. (2.6). It follows that

$$Q_Q = \frac{k_B T_a \tau_m}{2\hbar}. \tag{2.8}$$

Similarly, we can ask what temperature T_Q is required to achieve the quantum limit for a given Q-factor:

$$T_Q = \frac{2\hbar Q}{k_B \tau_a}. \tag{2.9}$$

Any mechanical measurement using a transducer and linear amplifier must satisfy the above conditions to achieve quantum limited performance. For resonant bars, detector noise contributions include acoustic noise generated within the cryostats, non-ideal amplifiers and other noise sources. The best cryogenic resonant bar detectors came within 3 orders of magnitude of the quantum limit. In laser interferometer detectors where the physics of the measurement system is similar, detectors are within an order of magnitude of quantum-limited performance and small-scale optomechanical devices have achieved quantum-limited performance at higher mechanical frequencies.

The above limits set by the Q-factor, temperature and quantum noise are valid for resonant mass gravitational wave detectors, and any resonator-transducer system where it is desired to detect external transient forces in which changes in the state of the resonator are measured in a time τ_m short compared to the mechanical relaxation time τ_a.

2.9   The Impedance Matrix for Parametric Transducers

By treating a parametric transducer as a quantum scattering device, we have seen that there are two separate scattering processes, one which extracts energy from a mechanical resonator, and the other one that returns energy to the resonator. While the two-port model discussed above is valid, it is useful to extend the formalism by treating the sidebands separately. This leads to a new impedance matrix in which we separate all the impedances into upper and lower sideband components, corresponding to the two diagrams in Fig. 2.16. The new impedance matrix, which identifies upper and lower sideband components by $+/-$ subscripts is given as follows

$$
\begin{bmatrix} V_-(\omega) \\ F_1(\omega) \\ V_+(\omega) \end{bmatrix} = \begin{bmatrix} Z_{--} & Z_{-1} & 0 \\ Z_{1-} & Z_{11} & Z_{1+} \\ 0 & Z_{+1} & Z_{++} \end{bmatrix} \begin{bmatrix} I_- \\ u_1 \\ I_+ \end{bmatrix} \tag{2.10}
$$

Here F_1 and u_1 are the force and velocity at the transducer input while $V_\pm$ and $I_\pm$ are the upper and lower sidebands voltage and current at the output.

In the impedance matrix, we see forward transductance components $Z_{\pm 1}$, which describe the signal transduction to each sideband. Similarly, each sideband can independently act back on the input via $Z_{1\pm}$. Each sideband has a separate output impedance $Z_{--/++}$. However, the input impedance does not separately access the sidebands, so this is given by Z_{11}. Here we summarize an analysis of parametric transducers presented by Blair (Blair, 1980), based on this approach. The analysis was based on the equivalent circuit shown in Fig. 2.17.

The LC circuit in Fig. 2.17 assumes a parametric transducer with capacitive sensing, driven by a pump oscillator with frequency ω_p close to the LC resonance frequency ω_0. Normally, we are interested in low loss transducers for which R is small. Tuning of the pump frequency relative to the cavity frequency changes the magnitude and phase of the impedance components, and has particularly strong effects in the high electrical Q-factor limit, which is often described as the resolved sideband limit. In particular, the input impedance and the forward transductance are very strongly tuned by the tuning of the pump frequency. Of particular interest is the extreme tunability of the mechanical input impedance.

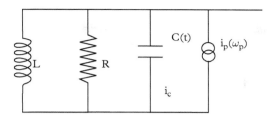

Fig. 2.17 *Equivalent circuit for a parametric transducer.* $i_p(\omega_p)$ *represents the pump oscillator.*

The mechanical input impedance is given by the sum of two complex terms, one for each sideband, as given below.

$$Z_{11} = -\frac{\frac{1}{2}C_0 V_p^2}{\omega_a \omega_0 x^2}\left[\frac{\omega_+ Q(1 - 2j\Delta_+ Q)}{1 + 4Q^2\Delta_+^2} - \frac{\omega_- Q(1 + 2j\Delta_- Q)}{1 + 4Q^2\Delta_-^2}\right], \qquad (2.11)$$

where C_0 is the rms value of the capacitance of the transducer, V_p is the pumping voltage amplitude, Q is the quality factor of the transducer, ω_a is the resonance frequency of the bar antenna and ω_0 is the resonance frequency of the cavity. In this equation, the delta terms that describe the frequency detuning are given by

$$\Delta_\pm = \frac{1}{2}\left[\frac{\omega_\pm}{\omega_0} - \frac{\omega_0}{\omega_\pm}\right]. \qquad (2.12)$$

Equation (2.11) can be re-written in terms of absolute detuning of pumping microwave field with respect to the cavity resonance δ, coupling coefficient β, and the cavity damping rate γ defined as $\gamma = \omega_0/Q$. As we will see later, this formula can be directly mapped to the input mechanical impedance of laser interferometer detectors Eq. (2.22), given by:

$$\begin{aligned}Z_{11} &= -j\frac{C_0 V_p^2}{\omega_0 x^2}\frac{\omega_p}{\omega_a}\frac{\delta}{(j\gamma/2 + \delta - \omega_a)(-j\gamma/2 + \delta + \omega_a)}, \\ &= -2j\frac{\beta\delta}{(j\gamma/2 + \delta - \omega_a)(-j\gamma/2 + \delta + \omega_a)}, \quad j = \sqrt{-1}, \qquad (2.13)\end{aligned}$$

where the second equality comes from the definition of coupling coefficient β in Eq. (2.4).

If the Q-factor rises high enough that the linewidth is small compared to the mechanical modulation frequency, the input impedance experiences a strong peak when the transducer is pumped at the sideband frequency.

While this resolved sideband behaviour was discovered in the context of resonant bar gravitational wave detectors described in the next section, it was not applied in this context because the electrical Q-factors were not high enough. However, the

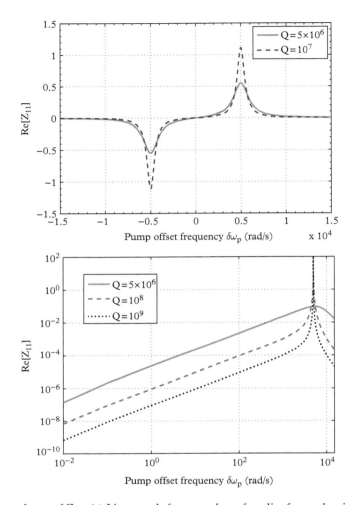

Fig. 2.18 *The real part of Z_{11}. (a) Linear scale for two values of quality factor, showing the impedance changing between negative and positive values. For zero detuning the impedance is near zero. (b) With log scales. We see that high Q-factor suppresses the detuned input impedance except for when the detuning matches the sideband frequency. The sideband frequency used here was chosen to match the resonance frequency of the NIOBE detector.*

same physics is relevant to laser interferometer gravitational wave detectors and to optomechanical devices.

The real part of the input impedance is shown in Fig. 2.18 as a function of the detuning frequency $\delta\omega_p = \omega_p - \omega_0$. The real component is analogous to a resistive load in a transmission line. However in the parametric transducer it can be varied between large negative and positive values, and can be tuned to zero for small negative detuning.

Remembering that an impedance ratio at an interface determines the transmitted and reflected power, negative input impedance describes a system for which the reflected mechanical signal power is greater than the incident mechanical driving power. Microwave parametric amplifiers created in the 1960s exploited similar negative resistance in varactor diodes to enable reflective amplification of microwaves. In the context of transducers, negative impedance can lead to instability. However, in the context of laser interferometers, mirror acoustic mode amplification by three mode interactions (a form of resolved sideband parametric transducer) have been shown to be useful as a means of predicting parametric instability at higher power (Ju *et al.*, 2014).

Large positive mechanical input impedance describes a situation where the mechanical input impedance can be tuned to match the mechanical output impedance of a mechanical resonator, under which conditions the transmission of acoustic power is optimized. In the case of the NIOBE gravitational wave detector, a parametric transducer was matched to a secondary coupled mechanical resonator attached to the resonant bar, as discussed in the next section. The third interesting case, zero input impedance, describes the case where the transducer has minimal interaction with the mechanical signal source. It draws no net energy from the mechanical resonator.

In 2015 Ma and Blair *et al.* applied an analysis similar to the impedance analysis above, to a laser interferometer gravitational wave detector (Ma *et al.*, 2015*a*), to answer the question: do laser interferometers absorb energy from gravitational waves? While the analysis was presented in a modern quantum framework, its derivation was equivalent to the 1979 results discussed here. Laser interferometers today operate with near zero detuning and hence near zero mechanical input impedance. As the above analysis shows, the impedance can be raised by using detuning, such that the energy absorbed from gravitational waves is maximized.

This important result replaces the resonant bar with spacetime itself. The impedance mismatch of interest is that between spacetime and the suspended mirrors of a laser interferometer. The laser interferometer is a parametric transducer, which has tunable input impedance. If the input impedance is raised, more of the gravitational wave signal will be absorbed by the interferometer. Intuitively one would expect that this could increase the signal-to-noise ratio but this conjecture has not yet been proved.

The imaginary part of the input impedance is also very important. This is like a reactance in a transmission line and for the transducer it represents electromagnetic stiffness, which in optomechanics is described as an optical spring.

In general, the reactive component of the input impedance can strongly tune the resonance frequency of the mechanical resonator, while the resistive components tune the damping. The slope of the reactance curve determines whether the spring constant is positive or negative, while the magnitude of the real part determines the damping.

If the negative spring of the transducer exceeds the positive spring of the mechanical resonator, it can create dynamical instability similar to that of an inverted pendulum.

Figure 2.19 shows the reactive component of the input impedance for two values of the quality factor. The slope of these curves determine the strength and the sign of the optical spring. For zero detuning, the spring term is zero but for a detuning equal to the sideband frequencies, strong positive and negative springs appear.

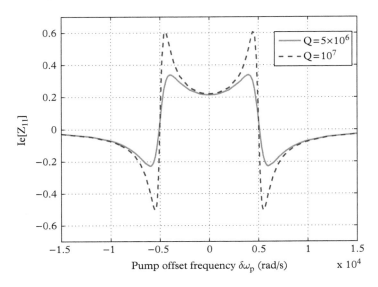

Fig. 2.19 *The imaginary components of the input impedance represent the 'optical' spring properties of parametric transducers which create positive or negative damping and cause the transducer to change the mechanical frequency of the resonator as shown experimentally in Figs 2.25 and 2.26.*

A very important property of the springs is apparent from comparison of the above curves: positive springs are associated with negative damping, while positive damping is associated with negative springs.

In the next section, we will see results from the resonant bar NIOBE that confirmed the above behaviour. Interestingly, the strength of the electromagnetic spring was sufficient that a few milliwatts of microwave power could significantly tune the resonant frequency of a 1.5-tonne mechanical resonator.

We now go on to discuss the forward transductance of the parametric transducer. The two sideband components (indicated by the $+$ and $-$ signs) of the forward transductance are given by:

$$Z_{\pm 1} = \pm \frac{V_{\mathrm{p}} Q}{2x\omega_0}\left(\frac{\omega_{\mathrm{p}}}{\omega_{\mathrm{a}}} \pm 1\right)\left(\frac{1 - 2j\Delta_{\pm}Q}{1 + 4Q^2\Delta_{\pm}^2}\right). \tag{2.14}$$

Normally, both sidebands of the transducer will be amplified and demodulated together. Then it is appropriate to combine both sideband transductances into the combined forward transductance which is equivalent to the term Z_{21} of Eq. (2.5). Then the total signal voltage for double sideband detection V_{DSB} is given by

$$V_{\mathrm{DSB}} = (|Z_{+1}| + |Z_{-1}|)u_1. \tag{2.15}$$

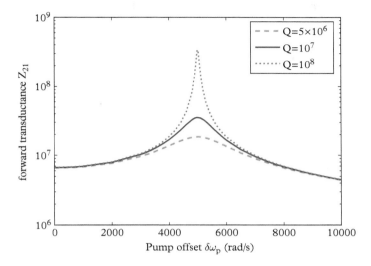

Fig. 2.20 *The forward transductance is not strongly dependent on detuning except in the high Q-limit, where it can be substantially enhanced.*

As a function of detuning the forward transductance shows a resonant peak at the sideband frequencies, especially for high Q-factor. However, Fig. 2.20 shows also that the transductance is only weakly dependent on detuning for small detuning. This means that a transducer in which there is, say, weak seismic noise modulation causing the gap spacing to be modulated, will have negligible changes in transductance if the modulations are small compared with the sideband frequency. However, such motion would be accompanied by significant input impedance variations as shown in Figs. 2.18 and 2.19.

This phenomenon of impedance variation due to the offset frequency dependence of optical springs is very important in ground-based gravitational wave detectors. Low frequency seismic-induced motions are difficult to avoid. They always give rise to seismic frequency modulation. The phenomenon was clearly observed in the NIOBE detector and is commonly observed in laser interferometer detectors. In NIOBE low frequency tracking was used to minimize this effect.

2.10 The Design of NIOBE, a 1.5-tonne Resonant Bar with Superconducting Parametric Transducer

In the previous sections, we have seen the implementation of non-contacting re-entrant cavity transducers controlled in three dimensions and vibration isolated by magnetic levitation. The impedance formalism for parametric transducers was presented, and results were used to illustrate the complex behaviour of such systems.

The need for impedance matching has been emphasized in the preceding sections. Matching to a lower impedance increases the mechanical amplitude and decreases the applied force. This acts to increase the signal-to-noise ratio if the transducer performance is dominated by series noise, such as the noise generated by frequency fluctuations of the pump oscillator.

The output impedance of a mechanical resonator such as a large bar is given by $M\omega_a$. Its large magnitude is difficult to match to a small transducer. The mechanical input impedance is determined by the transducer parameters as discussed above, and is tunable by changing the magnitude of the pump signal. To obtain reasonably good impedance matching for NIOBE a secondary resonator mass of ~ 1 kg was desirable.

The simplest means of impedance matching is to couple a low mass harmonic oscillator to the high mass oscillator. This provides narrow band impedance matching, because the coupled harmonic oscillator transfers energy between the resonator components on a relatively long timescale. Broadband impedance matching was exquisitely developed in the days of the acoustic gramophone. Devices such as levers and acoustic horns were used to couple the high impedance of a stylus needle moving in a record groove to the low impedance of the air. Devices based on tapered whip-like cantilevers also allow broadband resonant gain. If several tuned mass harmonic oscillators are connected in a chain with successively lower masses it is possible to achieve N-stages of impedance matching. Figure 2.21 shows a block diagram of a two-stage system, which can easily be elaborated to include additional stages (Blair *et al.*, 1987).

Figure 2.22 shows the predicted bandwidth of various single, two and three stage impedance matching networks. In the presence of significant noise sources, both sensitivity and bandwidth are increased by using extra stages. Yet these curves also emphasize the intrinsic narrow band nature of resonant bar detectors, compared with laser interferometers that achieve ~ 1 kHz bandwidth.

The concept for the readout of NIOBE was chosen to facilitate a) impedance matching; b) a means for a simple low acoustic loss mechanical design for the re-entrant cavity transducer; and c) a non-contacting coupling as illustrated in Fig. 2.23. The need

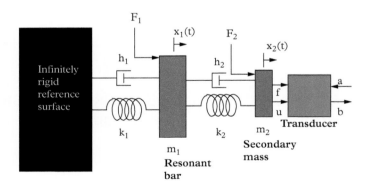

Fig. 2.21 *Mechanical model of a two-mode antenna with a transducer. Such models can easily be extended to multi-element impedance matching stages. In general impedance matching allows increase in bandwidth.*

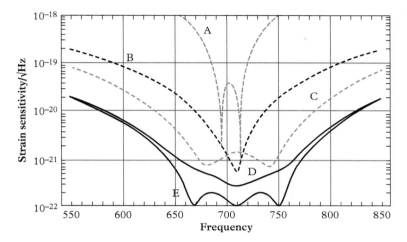

Fig. 2.22 *Strain sensitivity curves showing the possibility of improving both bandwidth and sensitivity of resonant mass gravitational wave detectors by modified impedance matching. Curve A shows the sensitivity of NIOBE. Curve E shows the ultimate achievable sensitivity that could have been achieved using three mechanical modes for improved impedance matching combined with a quantum-limited amplifier and reduced oscillator phase noise. Possible intermediate cases are also shown.*

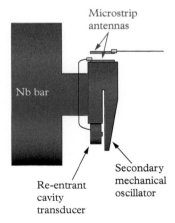

Fig. 2.23 *Schematic diagram of the NIOBE transducer system showing the re-entrant cavity parametric transducer and the non-contacting readout system. The base side of the transducer was thicker than shown here to prevent epoxy-induced distortions from deforming the gap spacing during cooldown. Figure 2.21 shows a mechanical model of the secondary resonator.*

for simple mechanical integrity was emphasized because complex mechanical elements such as nuts and bolts have always been well known as sources of acoustic loss.

A simple single impedance matching element consisting of a pure niobium tapered bending flap on a massive base was used (Fig. 2.23). It was attached to the bar using very thin layers of epoxy resin bonding pads to create a very low loss acoustic bond. It was this innovation that allowed the antenna to be operated without input and output cables. Instead, a pair of microstrip antennas enabled radiative coupling of the microwave signal into and out of the transducer. This eliminated a critical source of noise: acoustic vibration from the noisy cryogenic environment transmitted as transverse acoustic waves through flexible cables.

The transducer shown in Fig. 2.23 had a single short semi-rigid cable between the transducer and the microstrip antenna. It was fixed in location to prevent induced vibration and losses. A simplified diagram is shown in Fig. 2.24.

To obtain optimum noise performance a cryogenic microwave high electron mobility transistor (HEMT) amplifier with noise temperature just a few times larger than the quantum limit ($T_n \sim 6$ K) was chosen to amplify the signal from the re-entrant cavity. Such amplifiers, designed for radio astronomy, have excellent noise performance but only if the signal level is very small.

While the expected signal sidebands from the vibration amplitude of the niobium bar would be small, it would normally be almost impossible to set the microwave couplings at

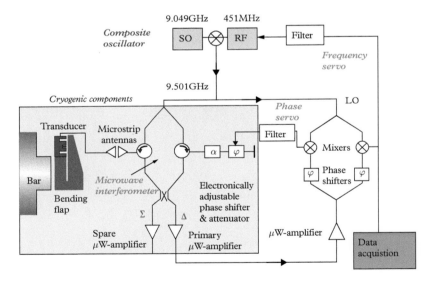

Fig. 2.24 *The readout system for NIOBE which included a cryogenic interferometer and a sapphire-loaded cavity oscillator (SO), a non-contacting readout system to prevent cryostat acoustic noise from reaching the bar, and a servo system for compensating for phase changes.*

each interface to be so close to unity that a significant amplitude of carrier signal would not be reflected back into the amplifier.

To overcome the consequent amplifier saturation problem, a cryogenic interferometer was used to suppress the microwave carrier signal. The signal was obtained at the differential 'dark port' of the interferometer. This allowed the HEMT microwave amplifier signal level to be maintained at less than about -60 dBm. Electronically adjustable phase shifters and attenuators were required to provide the correct amplitude and phase of pump signal such that the the the pump power at the amplifier was sufficiently suppressed.

The phase shifters and attenuators themselves had to be specially developed for the NIOBE transducer. They consisted of high permeability low loss ferrite-loaded stripline circuits for which the permeability could be tuned using a superconducting solenoid. Tuned beyond their transmission line cut-off, the devices acted at attenuators, while in the linear regime they acted as phase shifters. Three such devices in series were required to access 2π phase shift and up to 20 dB attenuation. They operated with fixed trapped supercurrents. A fourth device was used to provide a small servo controlled phase correction signal to compensate for low frequency changes in the spacing across the non-contacting antennas due to residual pendulum motion of the niobium bar (Ivanov *et al.*, 1993). This was required because the bar, suspended by a long chain of vibration isolation components, acted like a simple pendulum with a rocking frequency below 1 Hz.

An unavoidable problem was caused by having the re-entrant cavity fixed in location relative to the sensing surface of the bending flap, rather than being magnetically levitated in the configurations discussed earlier. Because the microwave resonance frequency of the transducer could not be adjusted, it was impossible for this frequency to be matched to the frequency of a fixed frequency low phase noise oscillator. Precision bonding of the cavity to the transducer assembly to about 0.3 microns resolution caused 100 MHz errors in the transducer frequency.

As part of the NIOBE project, ultralow phase noise sapphire-loaded cavity oscillators had been created, with quality factors $\sim 10^{10}$. These oscillators had exceptionally low noise, but could not be tuned to the transducer frequency (Tobar and Blair, 1992). A variable frequency low phase noise source was required as the pump oscillator.

When oscillator signals are mixed to create a new frequency, the noise performance is generally dominated by the worst oscillator. The best option at that time was to choose a frequency in the frequency band where the best possible frequency synthesizers could be used in conjunction with an ultralow phase noise oscillator of fixed frequency. The composite oscillator so created consisted of a sapphire-loaded cavity oscillator (Tobar and Blair, 1992), combined with the best Hewlett-Packard frequency synthesizer operating at around 450 MHz.

The entire microwave assembly shown in Fig. 2.24 combined very high-performance cryogenic microwave electronics consisting of the supercurrent tuned phase shifters, attenuators, circulators and microwave amplifiers, operating in a vacuum at 5 K. The room temperature demodulation section included two mixers for deriving the main signal output and the servo control signal used to compensate for the low frequency rocking of the suspended niobium bar. The entire system worked remarkably well over many years. Thanks to masterful microwave and electronic engineering by Eugene Ivanov, the phase servo worked so well that only in very large earthquakes, when rocking amplitudes rose to ∼ 5 mm, did the servo control systems lose lock.

One of the first experimental problems encountered when working with very high Q-factor mechanical resonators is that they can be excited accidentally and maintain very large amplitudes for many hours as shown in Fig. 2.25(a). This makes experiments difficult and can lead to saturation of amplifiers and consequent excess noise. A simple solution to this problem is the use of self-damping or mode cooling through red-detuning of the pump oscillator to the 3 dB point of the transducer cavity response. Self damping of the 1.5-tonne Nb bar was very effective. The Q-factor of mechanical modes at 694 Hz and 713 Hz were reduced by 3 orders of magnitude using a few hundred microwatts of incident microwave power as shown in Fig. 2.25(b). In Fig. 2.26, full data for the electromagnetic spring effect in NIOBE is shown. Because it takes some time for parametric instability to build up, it was possible to explore the electromagnetic springs in the blue-detuning regime where the Q-factor becomes infinite.

NIOBE operated for long periods of time between 1993 and 2000 in the five-detector International Gravitational Events Collaboration. Four of the detectors used SQUID

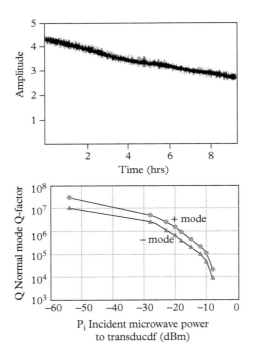

Fig. 2.25 *(a) A ringdown curve for NIOBE, showing amplitude decay of the 700 Hz fundamental mode by about 25% in six hours. (b) Self cooling as a function of microwave pump power. The Q-factor of the fundamental acoustic modes of the 1500 kg Nb bar-transducer system was reduced by a factor of 10^3 using a few hundred microwatts of input power.*

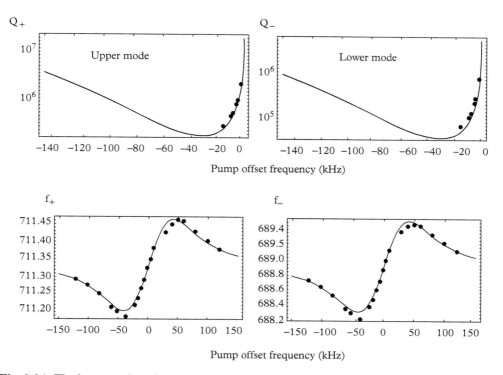

Fig. 2.26 *The frequency dependence of the real and imaginary parts of the input impedance act on the two longitudinal normal modes of NIOBE to change both the quality factor (top curves) and the mode frequency (bottom curves). Red detuning lowers both the quality factor and the mode temperature (i.e. the thermally excited amplitude distribution) because it enhances the coupling to the upper sideband which absorbs energy from the resonator. Detuning also creates an electromagnetic spring through the imaginary components of the input impedance. A few hundred microwatts of microwave power is sufficient to cause one mode of the 1500 kg bar to be detuned by more than 1 Hz.*

transducers. NIOBE alone used a parametric transducer. The detector array was able to set new upper bounds on the strength and the rate of gravitational wave bursts (Allen *et al.*, 2000). First experimental search for gravitational wave bursts was conducted by a network of resonator bar detectors, at a sensitivity $\sim 4 \times 10^{-21}/\sqrt{\text{Hz}}$.

During the development of NIOBE, the UWA group was also exploring other transducer technologies. Of particular interest here is the sapphire dielectric transducer (Peng, Blair and Ivanov, 1994; Cuthbertson, Tobar, Ivanov and Blair, 1998). The transducer element consisted of a pair of whispering gallery mode sapphire dielectric resonators structured as a face-to-face pair of mushrooms. They operated at microwave frequencies. The transducer was tuned by the interaction of the evanescent waves between the resonators. This transducer had a high Q-factor $\sim 10^6$ at room temperature, and could be expected to have an extremely high Q-factor $Q \sim 10^9$ at cryogenic temperatures. While

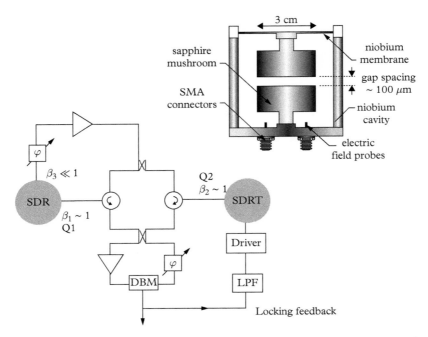

Fig. 2.27 *The sapphire dielectric cavity transducer. High Q-factor whispering gallery modes interact through their evanescent fields. A solid spindle sapphire-loaded cavity configured as a loop oscillator provides a microwave field and acts as pump oscillator for the interferometric pair of resonators. The solid resonator has high Q and low phase noise while the split resonator is sensitive to acceleration due to its membrane suspension. The transducer resonator is locked to the microwave source via a very low frequency feedback loop which applies a weak force to the membrane.*

its tuning coefficient was lower than that of a re-entrant cavity, this was outweighed by its high Q-factor. The entire device is shown in Fig. 2.27. It used an interferometric readout in which one arm contained a reference sapphire-loaded cavity resonator. In the cryogenic regime the device operated in the resolved sideband limit for audio frequency signals. A room temperature version was tested on a Virgo superattenuator, and yielded very impressive sensitivity, able to see low frequency acoustic modes that were hidden by the noise of conventional accelerometers, as shown in Fig. 2.28 (Peng and Blair, 1994).

While gravitational waves were not detected by resonant bar detectors, they set important upper limits. NIOBE uncovered the beautiful physics of optomechanics with parametric transducers, and was able to make displacement measurements $\sim 10^{-19}$ m. The best noise temperature of about 100 μK was achieved in the final run in 2000. Of particular significance, NIOBE contributed understanding and knowhow which could be applied to laser interferometer gravitational wave detectors, from high-performance vibration isolation to the fundamental optomechanics of the devices themselves.

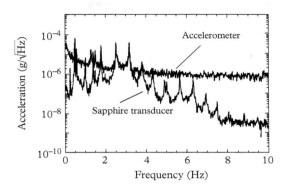

Fig. 2.28 *The calibrated acceleration responses of a sapphire parametric transducer compared with a high-quality accelerometer. Both devices were suspended on the bottom of a high-performance multistage 'superattenuator' vibration isolator developed for the Virgo laser interferometer gravitational wave detector. Numerous acoustic modes in the 0 − 10 Hz band were only discernable with the sapphire transducer accelerometer.*

2.11 Advanced Laser Interferometer Gravitational Wave Detectors

In previous sections, we have seen the conceptual similarity between the designs of resonant bar detectors with parametric transducer and the laser interferometers. Many important design ideas of modern laser interferometric gravitational wave detectors, such as interferometric readout for balancing the technical noises, resonant cavities, frequency stabilization, etc., can be found in the design of the readout system of the NIOBE resonant bar detector. Moreover, many of the important physical principles in modern laser interferometric detectors, such as the quantum limit and modification of the system dynamics by optomechanical interactions, were manifest in the physics of bars with parametric transducers. This similarity is not surprising. Both the principles of interferometric detectors and the parametric transducer readout of a resonant bar are based on the coupling between mechanical degrees of freedom and electromagnetic fields. Both are optomechanical parametric transducers.

The differences between the bar detectors and the interferometric detectors are also important to highlight, because they deepen our conceptual understanding of both systems. First, as we saw earlier, the bar detector used the extremely high-Q acoustic vibration of the bar to suppress thermal noise. This led to a relatively narrow detection bandwidth. Laser interferometers also use high-Q resonance to suppress thermal noise. However, by measuring the spacing between nearly free masses, the laser interferometer achieves relatively large detection bandwidth especially if it works in the so-called resonant sideband extraction mode as used by Advanced LIGO.

In the resonant bar, the gravitational wave tidal forces do work on the bar, thereby exciting the fundamental internal longitudinal acoustic mode of the resonant bar. In interferometers, the centres of mass of the mirrors respond like test particles to the incoming gravitational tidal forces. At the technical level, the electromagnetic fields used for sensing the motion of test masses is at the optical frequency in laser interferometers, where as we saw above, microwave fields were used for NIOBE. This 'simple' change of electromagnetic wave frequency makes a big difference for detector sensitivity, since advanced optical coating technology allows the creation of ultra-high optical power (several hundred kWs). In the late 1990s, the interferometer designs finally won the race for sensitivity.

On 14 September 2015, the Advanced LIGO interferometers (AdvLIGO) detected the gravitational waves from a black hole binary coalescence event (Abbott *et al.*, 2016). This detection was a great triumph of optomechanics. Advanced laser interferometers are very sophisticated parametric transducers that directly receive gravitational wave signals. The improvement of LIGO by a factor ~ 3 in strain sensitivity (compared with initial LIGO in 2006) brought signals over an audibility threshold. Below threshold, we could not resolve signals. Above threshold, we could recognize signals and obtain a remarkable amount of information regarding black holes that were probably created by the first massive stars in the universe. The discovery confirmed the theory of detection in which laser beams measure spatial strains between freely suspended test masses. It also gave us a calibration for astrophysical signals which allows firm predictions of the benefits of sensitivity improvements. It opened the era of gravitational wave astronomy.

Now that signals have been detected, the payoff in increasing LIGO sensitivity is enormous. Each factor of 3 improvement increases the accessible volume of the universe by approximately $3^3 = 27$ times. This has multiple benefits: a) It increases the event rate for the type of sources already detected ($\sim 30 - 30$ solar mass black holes) by 27 times allowing statistical studies to begin; b) it greatly increases the signal-to-noise ratio of nearer events (such as those already detected), allowing much greater resolution of system parameters and deeper testing of general relativity; and c) it increases the probability of detecting new sources such as stochastic backgrounds and continuous wave sources.

The first detection of gravitational waves combined the optomechanics of kilometre-scale high optical power dual recycling interferometers with the minimization of thermal noise by use of very high Q-factor mirrors and suspensions, and prevention of unwanted optomechanical three-mode interactions that lead to instability. In this section we will focus on specific aspects of interferometric detectors: a) their fundamental nature as parametric transducers; b) the parametric interactions that can cause instability; and c) the use of optomechanics to create devices able to enhance the sensitivity of gravitational wave detectors.

2.11.1   Laser Interferometer as a Parametric Transducer

The schematic diagram of Advanced LIGO is shown in Fig. 2.29. In essence, the device measures changes in distance between two widely spaced mirrors, through the phase shift of a very high quality factor optical mode, due to cavity length changes. The cavity quality factor is approximately the cavity finesse (set by the mirror reflectivity and transmissivity) times the number of free-space wavelengths in the cavity. Typically the quality factor is 10^{12}.

In practice, single cavity detection is not currently possible, due to technical noise sources. A Michelson interferometer configuration allows the detector to become a differential device like a Wheatstone bridge, in which technical noise sources such as intensity and frequency noise are balanced out. The microwave interferometer in the NIOBE transducer readout (Fig. 2.24) played a similar role. The 4 km arm cavities in the interferometer are designed to enhance the signal by multi-reflections. Theses two arms are carefully matched so that the laser field will destructively interfere at the dark port (see Fig. 2.29). Under the influence of gravitational waves, one cavity is lengthened when the other cavity is shortened and the beamsplitter extracts this signal at the dark port.

A real detector contains a succession of serial and nested optical cavities with optical linewidths varying from Hz to MHz. While complex, we want to emphasize that fundamentally the detectors are parametric transducers similar to the transducers we discussed earlier. Similar to the bar detector, the basic structure of laser interferometers

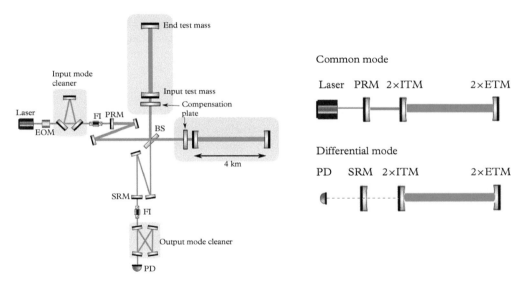

Fig. 2.29 *A simplified schematic diagram of Advanced LIGO showing the main optical cavities. Left panel: the dual recycling interferometer configuration. This configuration can be described by two effective single cavities as shown in the right panel. The common mode cavity corresponds to equal length changes in both arms. The differential mode cavity is sensitive only to differential length changes between the two arms, such as those created by gravitational waves. The common mode cavity is normally chosen to be resonant with the pump laser frequency for maximal build-up of intracavity power. The resonance frequency of the differential mode cavity is detuned by tuning the signal recycling mirror. Since the common and differential modes both involve motion in the two interferometer arms, these equivalent representations require the mass of test mass mirrors and the intracavity power to be doubled compared to a real detector.*

can be described by the general block diagram we saw earlier (Fig. 2.14) for a resonant bar with a parametric transducer. Instead of using a resonant bar as an intermediary device between the gravitational waves and the transducer, the antenna and transducer are integrated into a single device. The test mass mirrors and the Fabry–Perot Michelson interferometer form the parametric transducer. The pump frequency is the stabilized laser light which is detected by homodyne detection at the beamsplitter.

In the local Lorentz frame, the optomechanical interaction between the cavity field and the test mass differential motion transduces the mechanical signal to the optical signal by phase modulation of the optical cavity laser field. As with the other parametric transducers we considered, strong optomechanical coupling is required to obtain a large transduced signal.

The method for increasing the optomechanical interaction is to enhance the intra-cavity laser field intensity by adding a power recycling mirror (PRM). Because the interferometer operates with a dark fringe at the output, the laser input power is normally reflected. The power recycling mirror resonates the pump laser light within the interferometer, allowing substantial power build-up as long as the technical power losses from scattering and absorption are low. The invention of power recycling was a breakthrough that allows the creation of very high optical power interferometers without the need for extremely powerful laser sources. For example, in current running AdvLIGO, a 20 W laser source can create ~ 150 kW of intracavity power, enhanced by a factor of 7500. This power recycling concept was invented by Ron Drever at a Les Houches Summer School in 1981 (Drever *et al.*, 1981). While increasing the optomechanical coupling, the high power also decreased quantum shot noise.

The power recycling interferometer can be treated as a simple Fabry–Perot interferometer with a much higher power laser. In the presence of gravitational waves of frequency Ω, the intracavity laser field is perturbed by relative test mass motion at this frequency. This perturbation generates two signal sidebands with frequencies $\omega_0 \pm \Omega$. The signal is extracted at the dark port of the interferometer.

The gravitational wave signal frequency Ω appears only at the dark port of the interferometer, because of the asymmetric action of the gravitational waves. Common path effects in the two arms are cancelled at the interferometer dark port. It is therefore possible to resonantly enhance the signal sidebands at the dark port by introducing a signal recycling mirror (SRM) (Meers, 1988). The signal recycling mirror sees a composite cavity created by the two input test mass mirrors.

The design of advanced laser interferometer detectors combines power recycling and signal recycling, and is normally described as a dual recycling interferometer. This dual recycling interferometer can be described using two effective single cavities as shown in Fig. 2.29. For details, see (Buonanno and Chen, 2003).

Earlier, we saw how the behaviour of parametric transducers could be modified by detuning the pump frequency relative to the cavity resonance frequency. In a very similar manner, many new possibilities are introduced to laser interferometer detectors by tuning the signal recycling cavity. For example, the tuning of the position of the signal recycling mirror changes the phase of the equivalent mirror created by the signal recycling mirror and input test mass mirror. For this reason, adjusting the position of the signal recycling

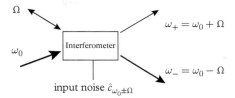

Fig. 2.30 *Transducer diagram of interferometer systems. The thick arrow here is the pumping laser field with frequency ω_0. It is modulated by the test mass motion driven by gravitational waves at frequency Ω. The input quantum vacuum also enters the system and contributes quantum noise.*

mirror is equivalent to the detuning of the parametric transducer in section 1.9, which leads to optomechanical changes to the dynamics of the differential motion of the test masses, through the action of optical springs.

For parametric transducers we used a 3×3 impedance matrix to describe the relationship between the input and output quantities as given in Eq. (2.10). Here we use a similar formalism to describe laser interferometers. Specifically we need to be able to describe how the intracavity laser field is disturbed by noise sources, such as the quantum fluctuations that enter from the dark port, stochastic motion of the test mass mirrors driven by thermal fluctuations and seismic motion, and classical optical noise due to light scattering and other processes. These disturbances contribute noise that contaminates the signal field. Here, for simplicity, we only consider the quantum noise entering the dark port of the interferometer and develop an impedance matrix similar to Eq. (2.10).

Impedance measures how a physical system responds to external excitation: mechanical impedance describes the velocity response to the external forces; electronic impedance describes the voltage generated by the injected current. In optical resonator systems, the impedance tells us how the cavity responds to inputs, which may be mechanical motion or optical fields. The impedance matrix formalism presented here is not normally used. However, the individual input–intracavity field relations are widely used in computing the sensitivity of gravitational wave detectors, for example see (Kimble *et al.*, 2001).

The impedance matrix we present below in Eq. (2.16) actually relates the input fields and the relative velocity of the test masses to the intracavity fields of the interferometer and the force acting on the test masses. For comparison, we used the same impedance subscripts as in Eq. (2.10). This new impedance matrix, together with a simple equation (Eq. (2.17)) that relates the intracavity fields to the outgoing fields, provides a useful tool for analysing the quantum behaviour of the laser interferometer.

In Eq. (2.16) we denote the sideband frequency component of the input vacuum fields that enters the dark port of the interferometer by $\hat{a}^{(\dagger)}_{\omega_0\pm\Omega}$.[1] These fields can also be modified by squeezing.

---

[1] $\hat{a}$ and $\hat{a}^\dagger$ are the annihilation and creation operator of light quanta.

The intracavity sideband fields, denoted by $\hat{A}^{\dagger}_{\omega_0 \pm \Omega}$, are connected to the input fields by the output impedance Z_{++} and Z_{--}. The intracavity fields are also connected to the test mass velocity v_{Ω} by the forward transductances Z_{+1} and Z_{-1}.

Unlike the previous impedance matrix Eq. 2.10, which is based on a simplified classical model given in Fig. 2.17, here we have non-zero impedance matrix elements Z_{+-} and Z_{-+} which connect the upper and lower sideband fields. These terms arise because radiation pressure creates correlations between the sidebands, described as ponderomotive squeezing.

The force acting on the test mass F_{Ω} is connected to the input vacuum fluctuation fields by the reverse transductance terms $Z_{1\pm}$. This term, as before, describes back-action or radiation pressure noise. The connection between the force acting on the test masses, and their relative velocity is given by the detector input impedance Z_{11}.[2]

All the above relations are summarized in the 3×3 impedance matrix given below:

$$
\begin{pmatrix} \hat{A}_{\omega_0+\Omega} \\ F_{\Omega} \\ \hat{A}^{\dagger}_{\omega_0-\Omega} \end{pmatrix} = \begin{pmatrix} Z_{++} & Z_{+1} & Z_{+-} \\ Z_{1+} & Z_{11} & Z_{1-} \\ Z_{-+} & Z_{-1} & Z_{--} \end{pmatrix} \begin{pmatrix} \hat{a}_{\omega_0+\Omega} \\ v_{\Omega} \\ \hat{a}^{\dagger}_{\omega_0-\Omega} \end{pmatrix}.
\tag{2.16}
$$

The connection between the intracavity field and the output field is given by:

$$
\hat{b}_{\omega_0 \pm \Omega} = -\hat{a}_{\omega_0 \pm \Omega} + \sqrt{2\gamma}\,\hat{A}_{\omega_0 \pm \Omega},
\tag{2.17}
$$

where γ is the bandwidth of the optical resonance peak of the interferometer differential mode, which is determined by the parameters of input test masses, end test masses and the signal recycling mirror.

Now we will present the formula of the individual impedance matrix elements in detail. First, the matrix elements Z_{++} and Z_{--} with dimension $[\text{Hz}]^{-1/2}$ can be written as:

$$
Z_{++/--}(\Omega) = \pm j\sqrt{\frac{\gamma}{2}}\,\frac{\alpha + 2\Omega^2(\mp j\gamma + \delta \mp \Omega)}{\Omega^2[\delta^2 + (\gamma - j\Omega)^2] - \alpha\delta}.
\tag{2.18}
$$

Combined with Eq. (2.17) which connects the intracavity fields to the output fields, Z_{++} and Z_{--} can be effectively considered as output impedance. Here $\alpha = 8\omega_0 I_c / mLc$ represents the optomechanical interaction strength, where $\omega_0, I_c, m, L, c, j$ are the pumping laser frequency, intracavity power strength, mirror mass, cavity length, speed of light, and $\sqrt{-1}$ respectively. The optical detuning δ is the frequency difference between optical resonance peak of the interferometer differential mode and the pumping laser frequency ω_0) determined by the parameters of the signal recycling cavity.

---

[2] Note that here we use Ω to represent the mechanical frequency, different from the notation ω_a used in the discussion of bar detectors.

Second, the forward transductances Z_{+1} and Z_{-1} with dimension $[\text{m}^{-1}\text{Hz}^{-1}]$ are given by

$$Z_{\pm 1}(\Omega) = \frac{j(\sqrt{\alpha}/L)[\pm(\Omega + j\gamma) - \delta]}{\Omega^2[\delta^2 + (\gamma - j\Omega)^2] - \alpha\delta}. \tag{2.19}$$

They describe the signal transduction to each intracavity sideband field. Because photon number is dimensionless, the transductance quantities are slightly different from the voltage and current transductances of the classical formalism.

In the quantum regime, the input fields $a_{\omega_0 \pm \Omega}$ are the quantum fluctuations entering into the cavity. This means that the force generated through reverse transductances Z_{1+} and Z_{1-} acts back on the test mass motion. This is the back-action noise that comes from the radiation pressure force fluctuations, similar to the back-action noise due to fluctuation currents in the previous transducer formalism, Eq. (2.10). The reverse transductance Z_{1+} and Z_{1-} has dimension $[\text{N} \cdot \text{Hz}^{-1/2}]$ and is given by:

$$Z_{1\pm}(\Omega) = \frac{\sqrt{\hbar m \gamma \alpha}}{2} \frac{1}{\gamma - j(\Omega \pm \delta)}. \tag{2.20}$$

The most interesting optomechanical physics manifest themselves in the matrix elements Z_{-+} and Z_{+-}, and in the mechanical input impedance Z_{11}. As mentioned above, the former describes ponderomotive squeezing while the input impedance describes optical spring effects. The optomechanical interaction mixes the upper and lower sideband fields through impedance matrix elements Z_{-+} and Z_{+-}:

$$Z_{-+/+-}(\Omega) = \mp \frac{j\alpha\sqrt{\gamma/2}}{\Omega^2[\delta^2 + (\gamma - j\Omega)^2] - \alpha\delta}. \tag{2.21}$$

Note that these terms vanish when we turn off the optomechanical interaction ($\alpha = 0$). This ponderomotive mixing establishes a correlation between the phase and amplitude fluctuations of the vacuum light, resulting in the generation of squeezed light. The squeezing angle and degree of squeezing are determined by the optomechanical coupling strength. The physical process of this correlation is that the radiation pressure force noise (characterized by the optical amplitude fluctuations) drives the motion of test mass, then enters into the phase of the optical field (Kimble *et al.*, 2001) through phase modulation.

Ponderomotive squeezing is one of the most important characteristics of optomechanical systems. This squeezing effect exists even for basic optomechanical systems, such as a light beam interacting with a reflective mirror without any cavity structure. Ponderomotive squeezing was first observed by (Purdy *et al.*, 2013). It could lead to new sources of squeezed light as long as the mechanical resonator is not corrupted by thermal noise (Corbitt *et al.*, 2006; Purdy *et al.*, 2013).

The second important characteristic of optomechanical systems, as we saw on NIOBE, is their optical springs. This arises, as we saw earlier, from optomechanical modification of the input mechanical impedance Z_{11} with dimension $[\text{N} \cdot \text{s} \cdot \text{m}^{-1}]$, represented by:

$$Z_{11}(\Omega) = jmL\Omega - \frac{jm\alpha}{4\Omega L^2} \frac{\delta}{(\Omega - \delta + j\gamma)(\Omega + \delta + j\gamma)}. \qquad (2.22)$$

Here the first term represents the mechanical impedance of the test mass system. This term does not appear in Eq. (2.11) because in the case of the resonant bar this mechanical impedance is the impedance to which we are trying to match. For the interferometer, the gravitational wave is the signal source, that sees both the mechanical impedance of the test masses and the additional impedance due to the transducer. It is clear that the second term of Eq. (2.22) has the same form as Eq. (2.13), which shows the unity of the physics of the microwave–bar system and the laser–mirror system. In detuned laser interferometers, the optical spring can actually enhance the sensitivity in a narrowband way since the response of the test masses is increased when the gravitational wave frequency is close to the optical spring frequency.

Current Advanced LIGO is built based on a non-detuned configuration with $\delta = 0$. Therefore, the test mass dynamics is not modified and the previous impedance matrix elements can be greatly simplified, making the physics behind these formulae more transparent. For example, in the non-detuned case the transductance Z_{++} and Z_{--} can be written as:

$$Z_{++/--}^{\delta=0}(\Omega) = \frac{\sqrt{2\gamma}}{\gamma - j\Omega} \pm \frac{j\alpha\sqrt{\gamma/2}}{\Omega^2(\gamma - j\Omega)^2}. \qquad (2.23)$$

It is clear that the first term in Eq. (2.23) comes from the direct response of a static cavity to the incoming field $\hat{a}_{\omega_0+\Omega}, \hat{a}^\dagger_{\omega_0-\Omega}$, which reveals the fact that the cavity field can be effectively mapped to a mechanical motion under the effect of a friction force.[3] The second term comes from the contribution of the stochastic motion of the test mass driven by the radiation pressure noise, which is equal to the transductance $Z_{+-}^{\delta=0}$ and $Z_{-+}^{\delta=0}$ when there is no detuning. Some interferometric gravitational wave detectors, such as the Japanese cryogenic detector KAGRA, plan to work in the detuned region where the test mass dynamics are modified by optical spring (Aso *et al.*, 2013).

The modification of input mechanical impedance Z_{11} provides new insights about energy interactions in laser interferometers. They can be understood in a simple picture that only considers the pump beam and the sideband signals. Upper and lower sidebands are created by gravitational waves. The sidebands beat with the high power carrier light. The beating signal creates radiation pressure forces that act on the interferometer test masses. Figure 2.31 shows the sideband energy flow. The key point is that the radiation pressure force from one sideband has the opposite phase to the force from the other sideband. Treating the gravitational wave itself as a part of the system, as shown in

---

[3] Written down explicitly, we have $\hat{A} = \sqrt{2\gamma}/(\gamma - j\Omega)\hat{a}$. This frequency domain formula corresponds to the equation of motion of intracavity field $\hat{A}$ in the time domain as $\dot{\hat{A}} = -\gamma\hat{A} + \sqrt{2\gamma}a$, which is analogous to Newton's second law: $\dot{v} = -\gamma v + f$, where v is the velocity, f and γv are respectively the driving force and friction force per unit mass.

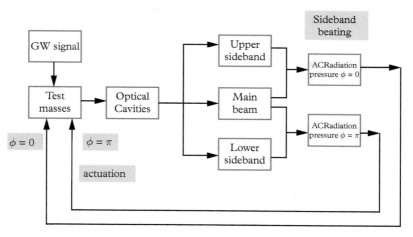

Fig. 2.31 *Radiation pressure feedback in a laser interferometer. Gravitational waves create signal sidebands. Each sideband beats with the carrier creating a radiation pressure force which acts on the test masses. This radiation pressure force is either in-phase or in anti-phase. Acting together, the two forces cancel so that the test mass remains as a free mass. If unbalanced the gravitational wave does work against the radiation pressure force. The radiation pressure forces determine the real and imaginary parts of the detector input impedance as we saw in sections 2.9 and 2.10. They act to change the dynamical response of the detector to gravitational waves but can also lead to instability.*

Fig. 2.14, it follows that one sideband extracts energy from the gravitational wave, while the other sideband returns energy to the wave. Without detuning, the gravitational wave detector has zero input impedance. No gravitational wave energy is absorbed. However, the input impedance can be tuned to a high energy absorbing value by detuning the detector.

With appropriate parameters, detuning increases the transductance and the input impedance. Given that huge amounts of energy are available in gravitational waves, it appears self-evident that extracting more of this energy should allow substantial increases in detector sensitivity. However, to this date, it has not been proved that such configurations can increase the signal-to-noise ratio.

It is useful to ask what is the mechanism by which detuning increases the energy extraction. The answer is that detuning, as observed experimentally in NIOBE, creates an optical spring. A detuned interferometer ceases to be one in which test masses float freely. The optical spring creates a rigidity against which the gravitational wave has to do work. The fraction of gravitational wave energy coupled into the detector is set by the impedance mismatch ratio between the detector and the waves. While this ratio will always be tiny, due to the enormouse impedance c^3/G of free space to gravitational waves, it is possible to increase the energy absorption by ~ 6 orders of magnitude, as shown by (Ma *et al.*, 2015*a*) which derived the energy absorbed by the double optical spring interferometer (Rehbein *et al.*, 2008).

Modification of optomechanical dynamics not only happens for centre of mass motion of test masses. It can also happen for mirror internal acoustic modes. In the next section, we will discuss this effect, which is of great importance in interferometer design. We will see that the modification of the input mechanical impedance for the mirror internal acoustic modes can make it difficult to achieve the required high optical power. Solution of this problem is an important frontier of research on advanced interferometers.

2.12   Three-Mode Interactions and Parametric Instability

All the parametric transducer systems discussed so far involve devices that create a pair of signal sidebands. In most devices there is an intrinsic symmetry that allows both sidebands to simultaneously exist, although their relative amplitudes can be varied by detuning.

In 2001, Braginsky *et al.* predicted three-mode interactions in which a single sideband frequency is resonant in an optical cavity transverse mode (Braginsky *et al.*, 2001). Such interactions are likely in long optical cavities because of their intrinsically asymmetric mode structure.

In the three-mode interaction, photons from the pump laser are scattered into transverse optical modes by acoustic modes of the mirror test masses. If the scattering transition is to a lower frequency optical mode, it creates a phonon in the test mass. Braginsky's group showed that this interaction could generate parametric instability.

Essentially it was realized that intense laser light can scatter inelastically from macroscopic acoustic modes of a mirror such that the photon energy is divided between a lower frequency transverse optical photon and an acoustic phonon in the mirror, as illustrated in Fig. 2.32(b). If the acoustic power injected by this mechanism exceeds the acoustic losses of the mirror, the mirror acoustic amplitude will grow exponentially, steadily increasing over seconds or minutes, until a very large amplitude (of say 1 nanometre) causes saturation of amplifiers and failure of the instrument.

In 2005, Zhao *et al.* undertook a detailed 3D simulation of parametric instability in Advanced LIGO type detectors. This led to a prediction (Zhao *et al.*, 2005) that detectors like Advanced LIGO would indeed experience three-mode optomechanical instability, involving tens of acoustic modes across the four main interferometer test masses. Thereafter many sophisticated simulations and experiments in specially designed optical cavities at Gingin, Western Australia were used to study the phenomenon.

It takes the extreme technology of long baseline laser interferometers to enter the regime where three-mode instability can occur: 40 kg scale mirrors with ultralow acoustic losses, 4 km long optical cavities, and very high optical power $\sim$ hundreds of kilowatts. On small scales, the much larger free spectral range of shorter cavities causes optical modes to be more widely spaced. Simultaneously, smaller test masses have a lower acoustic mode density so that the probability of coincidences between the mode gap (the frequency difference between a cavity TEM_{00} mode and a higher order TEM_{mn} mode) and an acoustic mode becomes small. On the 80 m scale of the Gingin high optical power facility (Ju *et al.*, 2004), only specific radii of curvature give conditions for instability.

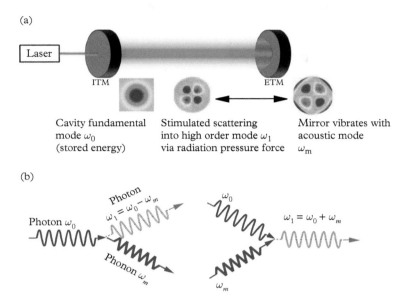

Fig. 2.32 *(a) Classical picture of a cavity three-mode interaction: the main high power cavity optical mode beats with a transverse mode generated by light scattered from a mirror acoustic mode. The beat frequency causes a time-varying radiation pressure force in phase with the acoustic mode. This drives the mirror and excites the acoustic mode as long as the transverse optical and acoustic mode shapes are similar. (b) The quantum picture of the three-mode parametric interaction treats it as a photon–phonon scattering process in which a carrier photon scatters from the acoustic phonon on the mirror surface to create a transverse optical mode photon. The acoustic mode frequency is exactly equal to the difference between the two optical frequencies.*

Using the 80 m long cavities with carefully designed kg-scale test masses and high optical power, it was possible to study most aspects of three-mode interactions: their tuning, their suppression, and their use as high sensitivity transducers for monitoring test masses.

In 2009, Zhao *et al.* showed that three-mode parametric interactions could be created on a table-top scale using low mass acoustic resonators in the MHz frequency range (Zhao *et al.*, 2009). A parallel program of table-top studies of three-mode interactions was begun, using specially designed resonators and silicon nitride membranes. These experiments led to the first observation of three-mode parametric instability in free space cavities in 2014 (Chen *et al.*, 2015).

Despite the successful demonstration of three-mode parametric instability in a small-scale device, the 2009 paper promised the creation of a new class of non linear optomechanical parametric devices based on three-mode interactions. This has not yet been realized, mainly due to the difficulty in tuning optical modes. Opto-acoustic parametric amplifiers (Zhao *et al.*, 2009), with one acoustic channel and two optical channels, offer the possibility of being general versatile devices like optical parametric amplifiers. They can allow creation of new sensors that could operate with quantum-limited sensitivity, for the detection of weak forces and fields. Their successful implementation may need the development of other optomechanical techniques such as optical dilution discussed below.

In 2014, Zhao *et al.* observed parametric instability in a cavity designed to be comparable to those of advanced interferometers (Zhao *et al.*, 2015). Later in 2014, the Advanced LIGO interferometer in Louisiana observed parametric instability (Evans *et al.*, 2015), at an optical power level consistent with the 2005 prediction. Preliminary estimates (Gras *et al.*, 2010) indicate that about 40 acoustic modes may be unstable in Advanced LIGO at full optical power unless control methods are implemented.

Because the frequency of transverse optical modes depends strongly on the mirror radius of curvature, any method that modifies the mirror radius of curvature can allow parametric instability to be detuned. Thermal tuning uses surface heating to change the mirror radius of curvature (by say 10 m in 2000 m). The method is now fully confirmed and has been used frequently at the Gingin facility to tune three-mode interactions. Because parametric gain is linearly proportional to power, there is always a power threshold at which the gain passes unity. Thermal tuning allowed the laser power threshold for instability to be raised from 5% to about 12% in the Advanced LIGO detectors. Unfortunately, in a long baseline interferometer the mode density is so high that thermal tuning generally tends to transfer the instability from one mode to another. Thus additional control techniques are required.

There are various other methods for instability control. The methods can be summarized under four headings (Ju *et al.*, 2009): passive damping (Gras *et al.*, 2009; Gras *et al.*, 2015) acoustic feedback (Miller *et al.*, 2011), optical feedback (Fan *et al.*, 2010;

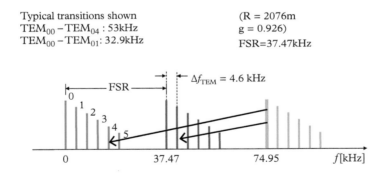

HOM not symmetric: Upconversion or down conversion

Fig. 2.33 *An example of the cavity mode structure in a 4 km LIGO-type arm cavity, showing the transverse mode families relative to the cavity free spectral range and typical transitions from the pump mode to certain transverse modes that could cause parametric instability if a suitable acoustic mode existed in the test mass. Because the transverse mode frequencies are strongly tuned by radius of curvature changes and typically have linewidths of kHz, the parametric gain for such transitions varies with the thermal conditions.*

Zhang *et al.*, 2010), and detuning (Degallaix *et al.*, 2007). Keeping in mind the fact that laser interferometer gravitational wave detectors are the most sensitive instruments ever created, with displacement noise sensitivity $\sim 10^{-20}\text{m}/\sqrt{\text{Hz}}$, the risk in any method of instability control is that noise forces will degrade the exquisite sensitivity of the instrument. For example, noise can come from photon radiation pressure, scattered light, thermal Brownian motion, electronics noise, and inhomogeneities in mirror coatings.

Many suppression schemes such as ring dampers (Gras *et al.*, 2009) and mechanical mode dampers (Gras *et al.*, 2015) were shown by modelling to cause significant sensitivity degradation. Others such as optical feedback suppression (Fan *et al.*, 2010) were demonstrated at Gingin, but involved optical configurations that were complex and risked introduction of noise. C. Blair *et al.* have shown that 15.5 kHz instabilities can be controlled in LIGO by direct electrostatic feedback to the test masses using electrostatic drive plates installed in Advanced LIGO (Blair *et al.*, 2016).

In 2014, the UWA team discovered a new method (Zhao *et al.*, 2015) of suppressing instability consisting of low frequency radius of curvature modulation of test masses using modulated radiant infrared heating. Suppression is achieved by diluting the parametric gain across many modes in a time-dependent fashion, using a modulation frequency below the sensitive signal band. In principle this method can suppress the parametric gain by about an order of magnitude and in theory it should not introduce significant noise.

Three-mode interactions have been shown to have very high sensitivity as a readout for thermally excited acoustic modes (Ju *et al.*, 2014). Because the mode amplitude depends strongly on three-mode interaction gain conditions, simple monitoring of ultrasonic acoustic modes through their three-mode interactions can be used to predict which modes will become unstable when the power is increased. The technique was demonstrated at Gingin.

As the optical power in advanced interferometers is stepped up, acoustic mode monitoring offers a useful tool for defining the instabilities that will have to be controlled at the next step up in power. This can give enough time for suppression methods to be ready at the appropriate time. While proven at Gingin, for advanced detectors where the ultrasonic acoustic mode density is 1000 times higher, this approach requires extensive measurements on thousands of acoustic modes, combined with detailed modelling. (Ju *et al.*, 2014) shows that it can be developed into a powerful tool for monitoring interferometer test masses with unprecedented precision.

2.13 White Light Optomechanical Cavities for Broadband Enhancement of Gravitational Wave Detectors

We have seen in previous sections how optomechanics has been used to create gravitational wave detectors, and how three-mode optomechanical interactions lead to parametric instability. In this final section, we show how optomechanics can be used to create a new route to improved sensitivity, in the form of an optomechanical white light signal recycling cavity. This type of scheme has been proven theoretically

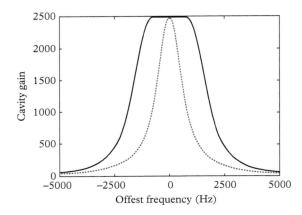

Fig. 2.34 *The white light cavity frequency response (solid line) compared with that of a conventional cavity (dotted).*

(Miao *et al.*, 2015) and the associated negative dispersion have been proven in the classical regime(Qin *et al.*, 2015). The white light cavity scheme offers possibly the only efficient means of significantly increasing interferometer sensitivity in a broad frequency band around 1–2 kHz. However, it faces formidable challenges involving beautiful new implementations of optomechanics.

Signal recycling is now a standard technique used in gravitational wave detectors to resonantly enhance the signal sidebands. Throughout physics, resonance allows a narrow band of frequencies to be enhanced, at cost of reduced bandwidth. The white light cavity breaks the inverse relationship between resonant gain and bandwidth. It allows resonant build-up of a broad band of frequencies. This is achieved by creating a low loss optomechanical cavity with negative dispersion that compensates the normal frequency dependence of phase accumulated in the main interferometer. The white light cavity is a cavity in which the effective light velocity depends on wavelength such that all frequencies are simultaneously resonant across the frequency band of interest.

In current advanced gravitational wave interferometers, the signal recycling gain is low to prevent loss of bandwidth. White light cavities are designed to create broadband signal recycling with high gain, allowing a substantial improvement in sensitivity for signal frequencies between 200 Hz and 2 kHz as shown in Fig. 2.35.

The white light cavity concept was first demonstrated using atomic media (Wicht *et al.*, 2002). Research by Salit *et al.* (Salit and Shahriar, 2010), and by Pati *et al.* (Pati *et al.*, 2007) confirmed the concepts. However Ma *et al.* (Ma *et al.*, 2015b) demonstrated that those configurations based on atomic media have fundamental noise limitations.

The breakthrough came with the recognition that the same physics can be realized with optomechanics. A mm-scale mirror resonator coupled to a light field and pumped with blue-detuned laser light creates a cavity with negative dispersion for a signal beam, as shown in Fig. 2.36. The configuration is intrinsically unstable. The key to making it practical is to use feedback through intensity modulation of the pump beam as proposed by Miao *et al.* (Miao *et al.*, 2015).

White light cavity technology with noise at the quantum level can increase the sensitivity of advanced interferometers such as LIGO by threefold at 200 Hz and sevenfold at 1–2 kHz, compared with realistic estimates of the improvements already anticipated. Operated in conjunction with squeezed light techniques, white light technology could give sensitivity as shown in Fig. 2.35.

Gravitational wave event rates for binary black holes similar to the source already detected increase as the cube of the strain sensitivity in the relevant frequency band. For curve e) in Fig. 2.35, the benefits of squeezing are combined with the white light cavity. Event rates could increase to ∼ 600 events per day! Other sources such as 5–10 solar mass black hole binaries and the quasi-normal modes of the newborn black hole could become observable within a large fraction of the observable universe.

The greatest challenge in creating a white light cavity system is in creating a suitable low optical loss resonant mirror with ultralow acoustic losses. Acoustic losses create unacceptable thermal noise and need to be reduced by a factor ∼ 10^6. While this sounds extremely challenging, an extensive analysis and design study by Page *et al.* (Page *et al.*, 2016) indicates that it is possible.

The attainment of minimal thermal noise at room temperature requires the combination of specially constructed resonators called cat-flap resonators, and quantum noise cancelling optical dilution systems in which the mechanical resonance is created by optical springs in a novel new cavity design called a Double End Mirror Sloshing

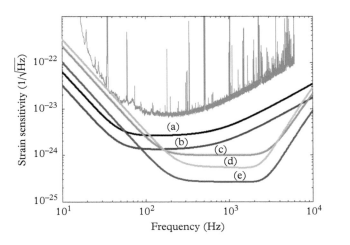

Fig. 2.35 *Current Advanced LIGO sensitivity (top curve) and predicted sensitivity after a) increase of laser power to the design level, b) implementation of frequency-dependent optical squeezing, c) and d) white light signal recycling at two levels of the noise parameter TQ^{-1} (see text), and e) white light signal recycling combined with frequency-dependent squeezing.*

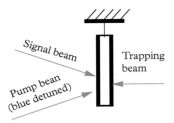

Fig. 2.36 *The white light cavity (WLC) combines three laser frequencies interacting with a single high Q-factor resonator. One beam (arbitrary frequency, ~ 4mW) creates the optical trap which creates a mechanical resonator (~ 200 kHz) with ultralow loss. The second beam (a few mW) tuned on the blue side of the optical cavity resonance creates a negative dispersion filter. The signal beam is the interferometer output signal. All beams are inside optical cavities.*

cavity or DEMS cavity. In the devices we propose, optical dilution minimizes the thermal fluctuations and optical cooling (self-cooling) suppresses the thermal amplitude.

These two techniques can allow thermal fluctuations to be suppressed below the zero-point fluctuations, thereby enabling devices that are free of thermal noise, in which the quantum behaviour of macroscopic objects can be explored. In the next paragraph we will re-visit optomechanical self-cooling which was already introduced in the context of NIOBE, but this time in the context of small-scale resonators. Optical dilution techniques will be discussed in detail in the next section.

One of the first applications of self-cooling was the cold damping of NIOBE discussed above. In 2006, Cohadon *et al.* (Arcizet *et al.*, 2006) used optomechanics to cool a micro-resonator from 300 K to 10 K and suggested that optomechanical cooling to the quantum ground state might be possible. In 2011, Cohadon's prediction was realized when Chan *et al.* (Chan *et al.*, 2011) used optomechanics to successfully cool a 3.68 GHz mechanical resonator from 20 K to the quantum ground state, and Teufel *et al.* (Teufel *et al.*, 2011) used microwave optomechanics to cool a 10 MHz resonator from 15 mK to the ground state. This work proved that thermal-noise-free mechanical resonators were indeed attainable.

Regarding white light cavity technology, in 2014 Qin *et al.* experimentally demonstrated (Qin *et al.*, 2015) tuneable linear negative dispersion, created in an optomechanical cavity with a blue-detuned doublet of control beams. The system could have very low optical losses dominated by a few ppm optical coating loss (Rempe *et al.*, 1992). A related scheme, but using a single control beam, and feedback stabilization, was proven theoretically by Miao et al. (Miao *et al.*, 2015)

White light cavity technology opens up a broad band at the high end of the spectrum as shown in Fig. 2.35. Once brought to the quantum noise level, it could increase the sensitivity of LIGO substantially as discussed above. In the very high event rate regime, detectors could be switched between optimum high frequency sensitivity using the white light cavity (Fig 2.35, curves c–e), and optimum low frequency sensitivity using frequency-dependent squeezing (Fig 2.35, curve b).

Table 2.1 *Technical requirements for improvement of gravitational wave detectors using white light cavities.*

CONCEPT	FUNCTION	BENEFIT
Optical dilution, Optical trapping	Use elastic stiffness created by radiation pressure forces to replace mechanical springs to enable thermal fluctuations to be suppressed by the ratio of optical to mechanical spring constants.	Allows creation of high frequency resonators $\sim$ 200 kHz with very weak mechanical spring stiffness and very low thermal fluctuations.
Quantum noise suppression	Destructive interference of vacuum fluctuations to cancel quantum radiation pressure noise.	Creates optomechanical resonators free of quantum radiation pressure noise.
Optical cooling	Extract mechanical energy from resonator.	Reduces thermal noise amplitude without cryogenic cooling.
White light cavity	Optomechanical interaction creates a negative dispersive filter. In an optical cavity, a band of frequencies is simultaneously resonant.	Allows resonant enhancement of a broad band of signal frequencies in gravitational wave detectors.
Double end-mirror sloshing cavity DEMS	Cavity scheme for creating strong stable optical dilution with negative dispersion response.	Allows stable operation of the system without deteriorating the signal-to-noise ratio.
Cat-flap resonator	A mini-pendulum consisting of a sub-millimetre scale low loss mirror supported by a nano-scale membrane or nanowire suspension.	Reduces mechanical coupling to the thermal reservoir to allow high optical dilution, in a device suitable for low loss coupling to large-scale optics.

The lower five curves in Fig. 2.35 assume that test mass thermal noise will be suppressed by use of cryogenics, (like the Japanese detector KAGRA now under construction), silicon test masses, and low loss (low thermal noise) optical coatings such as those developed by Crystalline Mirror Systems.

The key concepts that must be implemented to attain practical white light cavity devices are summarized in Table 2.1.

2.13.1 Optical Trapping, Optical Dilution and Quantum Noise

Optical traps were first used in manipulating microscopic objects such as molecules and biological cells, usually called 'optical tweezers'. An optical trap is a deep potential well created by radiation pressure that confines a mechanical resonator. This enables the losses to be *optically diluted*.

Using an optical trap, mechanical springs can be largely replaced by optical springs. The dilution factor is the ratio of the elastic energy stored in the optical field to the elastic energy stored in the mechanical spring. Strong optical dilution was demonstrated by Corbitt *et al.* (Corbitt *et al.*, 2007) who achieved a dilution factor $\sim 10^4$.

The mechanics behind optical dilution is simple. If a lossless spring k_{opt} is placed in parallel with a lossy spring of spring constant k_{int}, then the Q-factor of the final resonator is given by $Q_{dil} = Q_{int}(k_{opt}/k_{int})$ assuming structural damping. Since $k \sim \omega^2$, the dilution can be estimated from the square of the ratio of optical spring frequency to free resonator frequency. For maximum optical dilution we need to create resonators with the lowest possible k_{int} (i.e. the lowest possible zero-gravity frequency). In practice, the quality factor of diluted resonators is less interesting than the noise. Very high Q-factors are difficult to measure and normally optical cooling will simultaneously reduce the Q-factor and the mode amplitude towards the quantum ground state.

The problems with simple optical springs are a) negative damping and b) quantum radiation pressure noise. Strong dilution requires a strong optical field acting on the mechanical resonator. By beating with vacuum fluctuations, this creates strong radiation pressure noise, which drives the resonator, thereby injecting extra noise and setting limits on the maximum dilution.

In 2012, a Caltech team led by Kimble achieved optical spring trapping in a configuration that avoided quantum radiation pressure noise (Ni *et al.*, 2012). They demonstrated optical trapping at 145 kHz, and a 50-fold increase in quality factor, consistent with a prediction by Chang *et al.* (Chang *et al.*, 2012). The mirror-to-flexure mass ratio limited their Q-factor, while torsional compliance and internal acoustic modes led to optically induced angular instabilities.

In 2014, Ma *et al.* (Ma *et al.*, 2014) and Miao *et al.* (Miao *et al.*, 2015) analysed alternative dilution schemes; one in which the resonator mirror sits in the middle of an optical cavity, the other an optomechanical cavity with double optical spring. It was shown that a high reflectivity end mirror enables total destructive quantum interference to cancel the radiation pressure noise (Marquardt *et al.*, 2009), while instabilities from negative damping are cancelled.

The Double End Mirror Sloshing (DEMS) cavity (Fig. 2.37) is topologically equivalent to the membrane in the middle cavity. It has the same Hamiltonian, the same noise cancellation and allows the use of a double-sided high reflectivity cat-flap resonator. The DEMS cavity creates optical springs in such a way that the negative damping terms cancel, to prevent spring instabilities, and also allows quantum radiation pressure noise to be suppressed (Ma *et al.*, 2014; Page *et al.*, 2016). Note that the DEMS configuration and the mirror-in-the-middle configuration are equivalent. The former allows separate control of transmission and reflectivity and allows for use of an opaque substrate.

The key to creating ultra-high Q-factors is to create resonators with minimal surface density, low acoustic loss suspensions. Page *et al.* ((Page *et al.*, 2016)) showed that dilution factors of $10^8 - 10^{12}$ can be achieved in cat-flap resonators, with suspension elements and adequate low loss materials as discussed below.

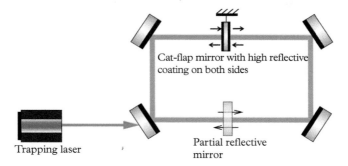

Fig. 2.37 *The Double End Mirror Sloshing (DEMS) cavity, that applies optical spring forces to trap the cat-flap resonator while suppressing quantum noise through interference. The two surfaces of the cat-flap mirror are the end mirrors of a two-mode Fabry–Perot cavity. The two modes are created by the partially reflective mirror.*

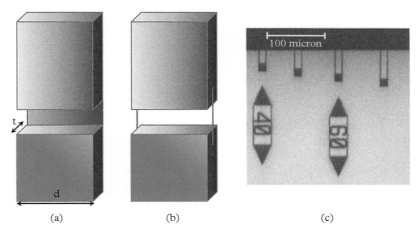

Fig. 2.38 *The cat-flap resonator: a 0.5 mm × 0.5 mm low loss mirror suspended by a) a 10–50 nm silicon nitride membrane; b) two nanowires. c) Image of micro-scale devices created by TNO (Netherlands Organisation for Applied Scientific Research).*

The cat-flap resonator: The cat-flap is a miniature flexure pendulum with a flat square optical test mass with linear dimensions between 0.1 mm and 1 mm. Current devices use a silicon nitride membrane or nanowires as the flexure (Fig. 2.38). Future devices could use graphene or nanotube suspensions. Optical dilution can allow extraordinarily high quality factors, limited by two main loss mechanisms as discussed below.

a) *Acceleration loss.* The optical force from the optical spring acts mainly on the mirror surface, and deforms the mirror as it accelerates, thereby coupling the resonator to internal loss mechanisms. This loss reduces as $(\omega_{\text{int}}/\omega_{\text{opt}})^2$, thus requiring the resonator to be small and ω_{opt} not to be too high.

b) *Suspension losses*. These are mainly due to violin string modes in the suspension fibres and require very high tension in low mass density suspensions. This limits the maximum length of the suspension fibre to keep violin string frequencies > 1 MHz. Centre of percussion tuning of the trapped resonator, by adjusting the laser spot position, can be used to reduce suspension losses threefold by ensuring that translational reaction forces do not act at the suspension point, as demonstrated by Braginsky *et al.* (Braginsky *et al.*, 1999).

Silicon nitride flexures can be constructed in the 10–50 nm thickness range. Since the flexure spring constant depends on the cube of the flexure thickness/length ratio, there is large benefit in minimizing the thickness. The suspension materials have high tensile strength. We have designed for safety factors ~ 5. SiN tensile membranes show $Q \sim 5 \times 10^7$ (Chakram *et al.*, 2014) at room temperature. For our design, the internal membrane modes are about ten times higher than the resonator frequency to reduce the acceleration loss due to the coupling between the optical spring and the mirror internal modes. Smaller resonators can also be suspended with silicon nitride nanowires.

The intrinsic (gravity-free) resonance frequency of the silicon nitride cat-flap is ~ 20 Hz. If this is diluted to 200 kHz we have typical dilution factors of $\sim 10^8$. The Q-enhancement will be less than the dilution factor due to the loss mechanisms discussed above. We estimate that $Q \sim 10^{11}$ could be observable at 200 kHz for a 0.5 mm square and 0.1 mm thick resonator.

2.13.2 Optomechanical Devices for Gravitational Wave Detectors

While experiments on optomechanical devices have proved that the quantum ground state is attainable, this has only been attained under special conditions of very low mass, high mechanical frequency and cryogenic cooling—see the 2014 review (Aspelmeyer *et al.*, 2014). Devices suitable for enhancing gravitational wave detectors have not yet been developed. They must have low optical losses and ideally should operate at room temperature, but should have mechanical noise close to the quantum ground state, therefore requiring optical cooling.

Figure 2.40 shows a gravitational wave interferometer with an optomechanical white light cavity at the back end. The DEMS cavity that creates the optically diluted resonator is integrated with a triangular negative dispersion cavity.

Aspelmeyer *et al.* pointed out that the minimum requirement for quantum optomechanics at room temperature is the product of Q-factor $\times$ frequency $> 6 \times 10^{12}$. Optomechanical devices for enhancing gravitational wave detectors have additional requirements which have not been met in any devices to date. The requirement is determined by the detector bandwidth and leads to even higher Q-factor requirements. Figure 2.39 shows the gain improvement ratio at 2 kHz as a function of resonator Q-factor. This shows that Q-factors must be in the range 10^{10}–10^{12} (Miao *et al.*, 2015) to achieve up to sevenfold improvement in detector sensitivity.

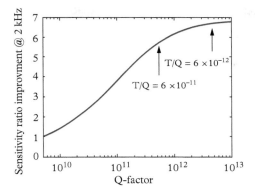

Fig. 2.39 *The gain improvement ratio at 2 kHz detection frequency as a function of resonator Q-factor. The values of TQ^{-1} in Fig. 2.35 curves c) and d) are indicated.*

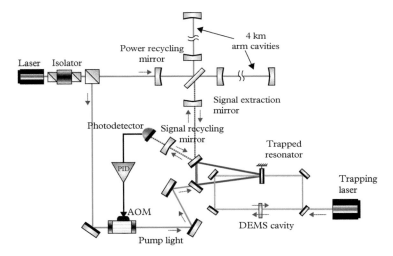

Fig. 2.40 *Laser interferometer gravitational wave detector with a white light signal recycling cavity. Negative dispersion is created in the triangular dispersion cavity. This acts as a dispersive mirror, changing the length of the dotted optical path. The optically trapped resonator in Fig. 2.37 is the key component in the negative dispersion cavity, which is activated through injection of blue-detuned pump light. Feedback forces are applied by intensity modulation of the pump light using an acousto-optic modulator (AOM).*

2.14 Conclusion

We have surveyed concepts of optomechanics from the first implementation of superconducting parametric transducers to current research challenges aiming to create optomechanical devices at room temperature that are largely free of thermal noise. In between, we reviewed the analysis of parametric transducers that first predicted interesting resolved sideband phenomena and the nulling of the mechanical input impedance. We went on to examine the use of microwave re-entrant cavity transducers and their application on the detector NIOBE which was an extremely large-scale application of optomechanics using microwaves. This experiment demonstrated strong self-damping and strong optical springs. Then we went on to examine the same physics applied to laser interferometer gravitational wave detectors and the optomechanics of three-mode interactions which are a significant instability mechanism in the current advanced gravitational wave detectors. The final section described a broad research program now underway aiming to create noise-free optomechanical devices suitable for low loss light processing, able to create long white light cavities. While our focus has been in improving gravitational wave detectors, optical dilution has the possibility of broad applications from macroscopic quantum mechanics to novel ultra-high sensitive devices for measuring forces and fields.

Bibliography

Aasi, J. *et al.* (2015). Advanced LIGO. *Classical and Quantum Gravity*, **32**(7), 074001.

Abadie, J., Abbott, B. P., Abbott, R., Abernathy, M., Accadia, T .and Acernese, F., Adams, C., Adhikari, R., Ajith, P., Allen B. *et al.* (2010). Predictions for the rates of compact binary coalescences observable by ground-based gravitational-wave detectors. *Classical and Quantum Gravity*, **27**(17), 173001.

Abbott, B. P. *et. al.* (2016). Observation of gravitational waves from a binary black hole merger. *Physical Review Letters*, **116**, 061102.

Ade, P. A. R., Aghanim, N., Ahmed, Z. *et.al.* (2015). Joint analysis of BICEP2/ *Keck Array* and *Planck* data. *Physical Review Letters*, **114**, 101301.

Allen, Z. A., Astone, P., Baggio, L. *et al.* (2000). First search for gravitational wave bursts with a network of detectors. *Physical Review Letters*, **85**, 5046–50.

Arcizet, O., Cohadon, P.-F., Briant, T., Pinard, M., and Heidmann, A. (2006). Radiation-pressure cooling and optomechanical instability of a micromirror. *Nature*, **444**, 71–4.

Aso, Y., Michimura, Y., Somiya, K., Ando, M., Miyakawa, O., Sekiguchi, T., Tatsumi, D., and Yamamoto, H. (2013). Interferometer design of the KAGRA gravitational wave detector. *Physical Review D*, **88**(043007).

Aspelmeyer, M., Kippenberg, T. J., and Marquardt, F. (2014, Dec). Cavity optomechanics. *Review of Modern Physics*, **86**, 1391–452.

Blair, C., Zhao, C., Liu, J., Blair, D., Betzwieser, J., Frolov, V., DeRossa, R., Adams, C., O'Reilly, B., Lormand, M., Grote, H., Miller, J., Fritschel, P., Gras, S., and Evans, M. (2016). First demonstration of electrostatic damping of parametric instability at Advanced LIGO. *LIGO Doc P1600090-v4*.

Blair, D., Howell, E., Ju, L., and Zhao, C. (2012). *Advanced Gravitational Wave Detectors*. Cambridge University Press.

Blair, D. G. (1979). Superconducting accelerometer using niobium-on-sapphire rf resonator. *Review of Scientific Instruments*, **50**(3).

Blair, D. G. (1980). Gravity wave antenna-transducer systems in gravitational radiation. In *Collapsed Objects and Exact Solutions*, Ed C Edwards, *Lecture Notes in Physics*, Volume 124, p. 299. Springer-Verlag, Berlin.

Blair, D. G., Bermat, T. P., and Hamilton, W O (1975). Superconducting accelerometer for use in gravity wave experiment. In *Proceedings of the 14th International Conference on Low Temperature Physics*, Volume 50, pp. 254–7. North-Holland.

Blair, D. G., Giles, A., and Zeng, M. (1987). Impedance matching element for a gravitational radiation detector. *Journal of Physics D: Applied Physics*, **20**(2), 162.

Blair, D. G. and Hamilton, W. O. (1979). Fabrication and properties of rf niobium–on–sapphire superconducting resonators. *Review of Scientific Instruments*, **50**(3), 279–85.

Blair, D. G., Ivanov, E. N., Tobar, M. E., Turner, P. J., van Kann, F., and Heng, I. S. (1995). High sensitivity gravitational wave antenna with parametric transducer readout. *Physical Review Letters*, **74**, 1908–11.

Blair, D. G. and Mann, A. G. (1981). Low-noise temperature gravitational-radiation antenna-transducer system. *Nuovo Cimento B*, **61B**(1), 73–81.

Boughn, S., Fairbank, W. M., McAshan, M., Paik, H. J., Taber, R. C., Bernat, T. P., Blair, D. G., and Hamilton, W. O. (1974). The use of cryogenic techniques to achieve high sensitivity in gravitational wave detectors. In *Proc. of the IAU. Conference on Gravitational Radiation and Gravitational Collapse*, Volume 64, pp. 40–51. Reidel. 1973 Warsaw, Poland.

Braginsky, V. B., Levin, Yu., and Vyatchanin, S. P. (1999). How to reduce suspension thermal noise in LIGO without improving the Q of the pendulum and violin modes. *Measurement Science and Technology*, **10**(7), 598.

Braginsky, V. B., Strigin, S.E., and Vyatchanin, S.P. (2001). Parametric oscillatory instability in Fabry–Perot interferometer. *Physics Letters A*, **287**, 331.

Braginsky, V. B., Vorontsov, Yu. I., and Khalili, F. Ya. (1978). Optimal quantum measurements in gravitational-wave detectors. *Soviet Physics-JETP*, **27**, 276.

Buonanno, A. and Chen, Y. (2003). Scaling law in signal recycled laser-interferometer gravitational-wave detectors. *Physical Review D*, **67**, 062002.

Chakram, S., Patil, Y. S., Chang, L., and Vengalattore, M. (2014). Dissipation in ultrahigh quality factor SiN membrane resonators. *Physical Review Letters*, **112**, 127201.

Chan, J., Alegre, T. P. M., Safavi-Naeini, A. H., Hill, J. T., Krause, A., Gröblacher, S., Aspelmeyer, M., and Painter, O. (2011). Laser cooling of a nanomechanical oscillator into its quantum ground state. *Nature*, **478**, 89–92.

Chang, D. E., Ni, K. K., Painter, O., and Kimble, H. J. (2012). Ultrahigh-Q mechanical oscillators through optical trapping. *New Journal of Physics*, **14**(4), 045002.

Chen, X., Zhao, C., Danilishin, S., Ju, L., Blair, D., Wang, H., Vyatchanin, S. P., Molinelli, C., Kuhn, A., Gras, S., Briant, T., Cohadon, P. F., Heidmann, A., Roch-Jeune, I., Flaminio, R., Michel, C., and Pinard, L. (2015). Observation of three-mode parametric instability. *Physical Review A*, **91**, 033832.

Corbitt, T., Chen, Y., Khalili, F., Ottaway, D., Vyatchanin, S., Whitcomb, S., and Mavalvala, N. (2006). Squeezed-state source using radiation-pressure-induced rigidity. *Physical Review A*, **73**, 023801.

Corbitt, T., Wipf, C., Bodiya, T., Ottaway, D., Sigg, D., Smith, N., Whitcomb, S., and Mavalvala, N. (2007). Optical dilution and feedback cooling of a gram-scale oscillator to 6.9 mK. *Physical Review Letters*, **99**, 160801.

Cuthbertson, B. D., Tobar, M. E., Ivanov, E. N., and Blair, D. G. (1998). Sensitivity and optimization of a high-Q sapphire dielectric motion-sensing transducer. *IEEE Transactions on Ultrasonics, Ferroelectrics, and Frequency Control*, **45**(5), 1303.

Degallaix, J., Zhao, C., Ju, L., and Blair, D. G. (2007). Thermal tuning of optical cavities for parametric instability control. *Journal of the Optical Society of America B*, **24**(6), 1336–43.

Drever, R. W. P., Hough, J., Munley, A. J., Lee, S. A., Spero, R. E., Whitcomb, S. E., Ward, H., Ford, G. M., Hereld, M., Robertson, N. A., Kerr, I., Pugh, J. R., Newton, G. P., Meers, B. J., Brooks III, E. D., and Gürsel, Y. (1981). Gravitational wave detectors using laser interferometers and optical cavities: Ideas, principles and prospects. In *Quantum Optics, Experimental Gravity, and Measurement Theory, Proceedings of the NATO Advanced Study Institute, NATO ASI Series B, (Plenum Press, New York, 1983)*, Volume 94, p. 503–14.

Einstein, A. (1916). Näherungsweise Integration der Feldgleichungen der Gravitation. *Sitzungsberichte der Königlich Preussischen Akademie der Wissenschaften Berlin*, **part 1:**, 688–96.

Einstein, A. (1918). Näherungsweise Integration der Feldgleichungen der Gravitation. *Sitzungsberichte der Königlich Preussischen Akademie der Wissenschaften Berlin*, **part 2:**, 154–67.

Evans, M., Gras, S., and Fritschel, P. *et al.* (2015). Observation of parametric instability in advanced LIGO. *Physical Review Letters*, **114**, 161102.

Fan, Y., Merrill, L., Zhao, C., Ju, L., Blair, D., Slagmolen, B., Hosken, D., Brooks, A., Veitch, P., and Munch, J. (2010). Testing the suppression of opto-acoustic parametric interactions using optical feedback control. *Classical and Quantum Gravity*, **27**(8), 084028.

Giffard, R. P. (1976). Ultimate sensitivity limit of a resonant gravitational wave antenna using a linear motion detector. *Physical Review D*, **14**, 2478–86.

Gras, S., Blair, D. G., and Zhao, C. (2009). Suppression of parametric instabilities in future gravitational wave detectors using damping rings. *Classical and Quantum Gravity*, **26**(13), 135012.

Gras, S., Fritschel, P., Barsotti, L., and Evans, M. (2015). Resonant dampers for parametric instabilities in gravitational wave detectors. *Physical Review D*, **92**, 082001.

Gras, S., Zhao, C., Blair, D. G., and Ju, L. (2010). Parametric instabilities in advanced gravitational wave detectors. *Classical and Quantum Gravity*, **27**(20), 205019.

Heffner, H. (1962). The fundamental noise limit of linear amplifiers. In *Proceedings of the IRE*, Volume 50, pp. 1604–8.

Hertz, H. R. (1887). Über sehr schnelle elektrische Schwingungen. *Annalen der Physik*, **267**(7), 421–44.

Ivanov, E. N., Turner, P. J., and Blair, D. G. (1993). Microwave signal processing for a cryogenic gravitational radiation antenna with a noncontacting readout. *Review of Scientific Instruments*, **64**(11), 3191–7.

Ju, L., Aoun, M., Barriga, P. *et.al.*(2004). ACIGA's high optical power test facility. *Classical and Quantum Gravity*, **21**(5), S887.

Ju, L., Blair, D. G., Zhao, C., Gras, S., Zhang, Z., Barriga, P., Miao, H., Fan, Y., and Merrill, L. (2009). Strategies for the control of parametric instability in advanced gravitational wave detectors. *Classical and Quantum Gravity*, **26**(1), 015002.

Ju, L., Zhao, C, Blair, D. G., Gras, S., Susmithan, S., Fang, Q., and Blair, C. D. (2014). Three mode interactions as a precision monitoring tool for advanced laser interferometers. *Classical and Quantum Gravity*, **31**(18), 185003.

Kimble, H. J., Levin, Yu., Matsko, Andrey B., Thorne, K. S., and Vyatchanin, S. P. (2001). Conversion of conventional gravitational-wave interferometers into quantum nondemolition interferometers by modifying their input and/or output optics. *Physical Review D*, **65**(022002).

Ma, Y., Blair, D. G., Zhao., C, and Kells, W. (2015a). Extraction of energy from gravitational waves by laser interferometer detectors. *Classical and Quantum Gravity*, **32**(1), 015003.

Ma, Y., Danilishin, S. L., Zhao, C., Miao, H., Korth, W. Z., Chen, Y., Ward, R. L., and Blair, D. G. (2014). Narrowing the filter-cavity bandwidth in gravitational-wave detectors via optomechanical interaction. *Physical Review Letters*, **113**, 151102.

Ma, Y., Miao, H., Zhao, C., and Chen, Y. (2015*b*). Quantum noise of a white-light cavity using a double-pumped gain medium. *Physical Review A*, **92**, 023807.

Manley, J. M. and Rowe, H. E. (1956). Some general properties of nonlinear elements-part I. general energy relations. In *Proceedings of the IRE*, Volume 44, pp. 904–13.

Marquardt, F., Girvin, S. M., and Clerk, A. A. (2009). Quantum noise interference and backaction cooling in cavity nanomechanics. *Physical Review Letters*, **102**, 207209.

Maxwell, J. C. (1865, June). A dynamical theory of the electromagnetic field. *Philosophical Transactions of the Royal Society*, **155**, 459–512.

Meers, B. J. (1988). Recycling in laser-interferometric gravitational-wave detectors. *Physical Review D*, **38**, 2317–26.

Miao, H., Ma, Y., Zhao, C., and Chen, Y. (2015). Enhancing the bandwidth of gravitational-wave detectors with unstable optomechanical filters. *Physical Review Letters*, **115**, 211104.

Miller, J., Evans, M., Barsotti, L., Fritschel, P., MacInnis, M., Mittleman, R., Shapiro, B., Soto, J., and Torrie, C. (2011). Damping parametric instabilities in future gravitational wave detectors by means of electrostatic actuators. *Physics Letters A*, **375**(3), 788–94.

Ni, K.-K., Norte, R., Wilson, D. J., Hood, J. D., Chang, D. E., Painter, O., and Kimble, H. J. (2012). Enhancement of mechanical Q factors by optical trapping. *Physical Review Letters*, **108**, 214302.

Page, M., Ma, Y., Zhao, C., Blair, D., Ju, L., Pan, H.-W., Chao, S., Mitrofanov, V., and Sadeghian, H. (2016). Towards thermal noise free optomechanics. Journal of Physics D: Applied Physics, **49**, 45.

Paik, H. J. (1976). Superconducting tunable-diaphragm transducer for sensitive acceleration measurements. *Journal of Applied Physics*, **47**(3), 1168.

Pati, G. S., Salit, M., Salit, K., and Shahriar, M. S. (2007). Demonstration of a tunable-bandwidth white-light interferometer using anomalous dispersion in atomic vapor. *Physical Review Letters*, **99**, 133601.

Peng, H. and Blair, D. G. (1994). Test of an interferometric sapphire transducer with the super attenuator of the VIRGO gravitational wave antenna. *Physics Letter A*, **189**(5), 141–4.

Peng, H., Blair, D. G., and Ivanov, E. N. (1994). An ultra high sensitivity transducer for vibration measurement. *Journal of Physics D: Applied Physics*, **27**, 1150–5.

Press, W. H. and Teukolsdy, S. A. (1977). On formation of close binaries by 2-body tidal capture. *The Astrophysical Journal*, **213**(17), 183–92.

Pretorius, F. (2005). Evolution of binary black-hole spacetimes. *Physical Review Letters*, **95**, 121101.

Purdy, T. P., Peterson, R. W., and Regal, C. A. (2013). Observation of radiation pressure shot noise on a macroscopic object. *Science*, **339**(6121), 801–4.

Qin, J., Zhao, C., Ma, Y., Ju, L., and Blair, D. G. (2015). Linear negative dispersion with a gain doublet via optomechanical interactions. *Optics Letters*, **40**(10), 2337–40.

Rehbein, H., Müller-Ebhardt, H., Somiya, K., Danilishin, S. L., Schnabel, R., Danzmann, K., and Chen, Y. (2008). Double optical spring enhancement for gravitational-wave detectors. *Physical Review D*, **78**, 062003.

Rempe, G., Lalezari, R., Thompson, R. J., and Kimble, H. J. (1992). Measurement of ultralow losses in an optical interferometer. *Optics Letters*, **17**(5), 363–5.

Rickles, D. and DeWitt, C. M. (2011). *The Role of Gravitation in Physics, Report from the 1957 Chapel Hill Conference*. Edition Open Access.

Salit, M. and Shahriar, M. S. (2010). Enhancement of sensitivity and bandwidth of gravitational wave detectors using fast-light-based white light cavities. *Journal of Optics*, **12**(10), 104014.

Teufel, J. D., Li, D., Allman, M. S., Cicak, K., Sirois, A. J., Whittaker, J. D., and Simmonds, R. W. (2011). Circuit cavity electromechanics in the strong-coupling regime. *Nature*, **471**(7337), 204–8.

Tobar, M. E. and Blair, D. G. (1992). Phase noise of a tunable and fixed frequency sapphire loaded superconducting cavity oscillator. In *Microwave Symposium Digest, 1992, IEEE MTT-S International*, Volume 1, pp. 477–80.

Underwood, M., Mason, D., Lee, D., Xu, H., Jiang, L., Shkarin, A. B., Børkje, K., Girvin, S. M., and Harris, J. G. E. (2015). Measurement of the motional sidebands of a nanogram-scale oscillator in the quantum regime. *Physical Review A*, **92**, 061801.

Weber, J. (1960). Detection and generation of gravitational waves. *Physical Review*, **117**(1), 306.

Weber, J. (1961). *General Relativity and Gravitational Waves*. Wiley-Interscience, New York.

Weber, J. (1970). Anisotropy and polarization in the gravitational-radiation experiments. *Physical Review Letters*, **25**, 180–4.

Weber, J., Lee, M., Gretz, D. J., Rydbeck, G., Trimble, V. L., and Steppel, S. (1973). New gravitational radiation experiments. *Physical Review Letters*, **31**, 779–83.

Wicht, A., Rinkleff, R.-H., Spani Molella, L., and Danzmann, K. (2002). Comparative study of anomalous dispersive transparent media. *Physical Review A*, **66**, 063815.

Zhang, Z., Zhao, C., Ju, L., and Blair, D. G. (2010). Enhancement and suppression of opto-acoustic parametric interactions using optical feedback. *Physical Review A*, **81**, 013822.

Zhao, C., Ju, L., Degallaix, J., Gras, S., and Blair, D. G. (2005). Parametric instabilities and their control in advanced interferometer gravitational-wave detectors. *Physical Review Letters*, **94**, 121102.

Zhao, C., Ju, L., Fang, Q., Blair, C., Qin, J., Blair, D. G., Degallaix, J., and Yamamoto, H. (2015). Parametric instability in long optical cavities and suppression by dynamic transverse mode frequency modulation. *Physical Review D*, **91**, 092001.

Zhao, C., Ju, L., Miao, H., Gras, S., Fan, Y., and Blair, D. G. (2009). Three-mode optoacoustic parametric amplifier: A tool for macroscopic quantum experiments. *Physical Review Letters*, **102**, 243902.

3

Optomechanical Interactions

Ivan Favero

Matériaux et Phénomènes Quantiques,
Université Paris Diderot, CNRS, France

Ivan Favero

Ivan Favero, *Optomechanical Interactions* In: *Quantum Optomechanics and Nanomechanics*. Edited by: Pierre-Francois Cohadon,
Jack Harris, Florian Marquardt, Leticia F. Cugliandolo, Oxford University Press (2020). © Oxford University Press.
DOI: 10.1093/oso/9780198828143.003.0003

Chapter Contents

Light exerts mechanical action on matter through various mechanisms. Of course, the most famous to physicists is radiation pressure, with the associated picture of a photon bouncing on a perfectly reflective movable mirror and transferring twice its momentum. If this picture is incomplete (if the photon bounces back with the opposite momentum and same energy, where does the kinetic energy gained by the mirror upon the impact come from?) it allows one to easily discuss some consequences of radiation pressure. Unfortunately, this simplification can lead the experimentalist to wrong conclusions: historically, the Crooks radiometer rotated in the direction opposite to expectations (Hull, 1948; Woodruff, 1968), and it is only in the early twentieth century that Lebedev, Nichols and Hull (Lebedev, 1901; Nichols and Hull, 1901) could clearly unravel radiation pressure in the laboratory. Still today, unambiguously observing the effects of radiation pressure without employing expensive experiments remains a challenge. In the quantum domain, the radiation pressure interaction between a moving mirror and light stored in a cavity accepts a simple Hamiltonian formulation (Law, 1995). Some consequences of this quantized interaction have been identified for more than 30 years (Caves, 1980) and were just recently observed, after intense efforts of a whole community of physicists. If this Hamiltonian description is concise, it is again sometimes oversimplified, and still today some publications underestimate or miss other mechanical effects of light accompanying radiation pressure in experiments.

In this chapter, we will not only address radiation pressure but also other relevant optical forces such as the optical gradient force, electrostriction, or the photothermal and optoelectronic forces. Our researches have taught us that these interactions are key in micro- and nanoscale devices. In our view, they must all be controlled on an equal footing to fully harness the technological and scientific potential of miniature optomechanical systems.

3.1 Optically Induced Forces

3.1.1 Radiation Pressure

In free space, the per-photon momentum associated to a plane wave of wavelength λ_0 has an amplitude $\hbar k_0$, with $\hbar =$ the Planck constant and $k_0 = 2\pi/\lambda_0$ the free space wavenumber. When this photon impinges and bounces on a perfectly reflecting rigid mirror at normal incidence, conservation of momentum dictates a momentum exchange of amplitude $2\hbar k_0$ between the photon and the mirror (see Fig. 3.1a). If the mirror is mechanically compliant, there is actually some energy transferred to the mirror's motional energy upon such impact, such that the reflected photon experiences a subtle redshift analogous to a Doppler effect. This redshift is typically a correction of relativistic order and has never been measured to date, even though it may have observable consequences for state-of-the-art devices (Karrai, Favero and Metzger, 2008). We will neglect it below. In this case, for a perfectly reflecting mirror illuminated under orthogonal incidence by a plane wave carrying an optical power P, the force associated to radiation pressure has an amplitude:

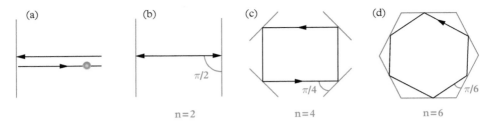

Fig. 3.1 *(a) Momentum transfer as a photon strikes a perfectly reflecting rigid mirror. (b), (c) and (d): Schematic view of a circular optical cavity composed of n=2, 4 or 6 mirrors.*

$$F = 2P/c \qquad (3.1)$$

with c the speed of light. In a circular optical resonator where a photon makes round-trips by total internal reflection, momentum conservation also allows to easily express radiation pressure. A photon confined inside a polygonal cavity with n sidewalls will strike the sidewalls n times per round-trip at an angle of π/n, each time transferring a momentum of amplitude $2\hbar k \sin(\pi/n)$ in the radial direction (see Fig. 3.1 b, c, d). If the vectorial transfer sums-up to zero upon a round-trip, it does not once projected onto the radial direction. The radial transfer per photon and per round-trip in a circular resonator is the limit:

$$2\hbar k \lim_{n\to\infty} n \sin(\pi/n) = 2\pi\hbar k \qquad (3.2)$$

with k the wavenumber of the photon inside the bulk of the resonator medium. The associated force per photon is therefore the momentum transfer per round-trip (Eq. 3.2) divided by the round-trip time τ_{rt}. For a circular resonator of radius R, and a photon in the transparency region of the medium, the force has an amplitude:

$$F = \underbrace{\frac{2\pi\hbar k}{2\pi R n_{opt}/c}}_{\tau_{rt}} = \frac{\hbar k c}{n_{opt}R} = \frac{\hbar k_0 c}{R} \qquad (3.3)$$

where n_{opt} is the index of refraction inside the resonator, be it vacuum or a dielectric material, and where the photon momentum $\hbar k$ in the material is written in the Minkowski formulation $\hbar k = \hbar n_{opt} k_0$. The force per photon given by Eq. (3.3) appears to be independent of the refractive index.

Beyond these considerations in simple geometrical cases, the radiation pressure exerted by photons in arbitrary geometries can be computed by means of the Maxwell Stress Tensor (MST). The MST formalism stems again from momentum conservation arguments, this time applied to interacting charges and electromagnetic fields (Jackson, 1998; Rakich, Davids and Wang, 2010). In a dielectric medium of relative permittivity $\varepsilon_r(x,y,z)$ and permeability μ_r, the ij components of the MST are given by Eq. (3.4):

$$T_{ij} = \varepsilon_0 \varepsilon_r(x,y,z) \left[E_i E_j - \frac{1}{2} \delta_{ij} |E|^2 \right] + \mu_0 \mu_r \left[H_i H_j - \frac{1}{2} \delta_{ij} |H|^2 \right] \qquad (3.4)$$

where $\varepsilon_0 = 8.85 \cdot 10^{-12}$ F·m^{-1} and $\mu_0 = 4\pi \cdot 10^{-7}$ H·m^{-1} are respectively the vacuum permittivity and permeability, δ_{ij} is Kronecker's delta and $E_i(H_i)$ is the ith electric (magnetic) field component. With the choice of notations of Eq. (3.4) (similar to (Rakich, Davids and Wang, 2010)), the radiation-pressure induced stress σ_{ij}^{rp} (applied along the j direction onto the face normal to the i direction) is expressed as a function of the MST element T_{ij} as $\sigma_{ij}^{rp} = -T_{ij}$. The MST approach not only allows computing the normal stresses in a structure (for which the stress is normal to the face, i.e. σ_{ii}, like in the mirror or circular resonator cases discussed above), but also the shear stresses ($\sigma_{ij}, i \neq j$) due to the presence of photons. Since typical optomechanical devices cannot respond to rapidly varying forces at optical frequencies (in the 10^{14} Hz range), the time-averaged value $\langle ... \rangle_{opt}$ of the stress over an optical cycle is generally computed.

3.1.2 Optical Gradient Force

The optical gradient force is generally treated as distinct from radiation pressure, while both have the same microscopic origin. The historical development of optical manipulation of polarizable micro- and nano-particles by Ashkin (Ashkin, 1970; Ashkin, Dziedzic, Bjorkholm and Chu, 1986), followed by the extension of these methods to atoms and ions, has strengthened this conceptual distinction between the two types of force in many minds. The distinction is easily grasped for a two-level atom placed near the focus of a laser beam. The atom absorbs incoming directional photons and re-emits them all around with no directionality, resulting in a net transfer of momentum to the atom in the laser propagation direction $\hat{\mathbf{x}}$: this is radiation pressure. A complementary picture is drawn in the transverse direction by considering the polarizability of the atom, turning it into a point dipole. If the atom, in the direction transverse to $\hat{\mathbf{x}}$, does not sit perfectly centred in the focus but on the intensity flank of the spot, the dipole attempts to minimize its energy by moving to the exact centre of the spot where the time-averaged electric field is maximal. Simply stated, the atom moves to the high-field regions, attracted by a force scaling with the gradient of the time-averaged optical energy density: this is the optical gradient force.

One gains insight by recalling that both radiation pressure and optical gradient force are simply different facets of the Lorentz forces acting on charges (again the interaction between light and moving charges is the elementary microscopic process). In the dipolar approximation where the size of the polarizable dielectric particle is smaller than the incoming radiation wavelength, and where all electromagnetic fields and their derivatives can be taken as constant over the volume of the particle, the total optical force can be derived exactly starting from the standard microscopic expression of Lorentz forces. In this approximation, the general expression for an arbitrary monochromatic light field illuminating the particle is (Moine, 2005):

$$F = \varepsilon_0 \varepsilon_r k_0 \alpha_1 \langle |E|^2 \rangle_{opt} \hat{\mathbf{x}} + \frac{\varepsilon_0 \varepsilon_r}{2} \alpha_0 \nabla \langle |E|^2 \rangle_{opt} + \dots \tag{3.5}$$

where the complex polarizability of the particle is $\alpha = \alpha_0 + i\alpha_1$ (in the atomic-dipole case α_1 accounts for the absorption-emission cycles) and k_0 is the wavenumber of the impinging laser light. The first term proportional to the local optical energy density is radiation pressure (sometimes also called scattering force) while the second, proportional to the gradient of the energy density, is the optical gradient force. There is also a third corrective force term that forbids rigorous decomposition of the total optical force as the sum of radiation pressure and optical gradient force. This corrective force is not always negligible in applications like optical trapping and manipulation, and adopts a more 'scattering' or 'optical gradient' nature under specific illumination configurations. Accurate discussions of this third term can be found in the specialized literature (Moine, 2005) in the cases of a focused Gaussian beam, of a propagating plane wave or of a standing plane wave. The mere existence of this corrective force shows that the widespread distinction between radiation pressure and optical gradient force, if sometimes convenient, must not be considered as a fundamental feature.

In the regime where the particle size exceeds the optical wavelength, a geometric optics approach can be employed to describe optical forces. Light rays are refracted and deviated by the particle, resulting again in a momentum transfer. If very pictorial, there are however very few cases where the geometric approach permits a quantitative description. This includes the illumination of a sphere by a propagating plane wave and the illumination of a very large object by a collimated laser, which renders the single ray limit exact. In the intermediate regime where the particle's size is commensurate to the wavelength, both the dipolar or geometric approaches break down and the Mie theory is best employed, leading to exact expressions in the specific case of a spherical particle.

Finally, we see that there are different regimes for optical forces acting on polarizable particles, which have all been relevant in the advent of optical tweezers, optical atomic traps or light-trapped mirrors for astronomy (Labeyrie, 1996). In some limit cases an analytic force expression does exist, with an associated nickname such as 'optical gradient force' or 'scattering force'. But it is important to appreciate that, all these forces being Lorentz forces, they are naturally encompassed in a general treatment of momentum transfer from light to matter. In consequence, the MST formalism and its tensor expression (3.4) is the most general tool to model optical forces acting on a rigid dielectric body. It should be used whenever analytical models fail and an exact solution is requested.

3.1.3 Electrostriction

Now instead of considering the dielectric body as rigid, let us make it compliant and able to internally deform. Under illumination, the optical forces acting on its elementary constitutive elements can induce its deformation and the appearance of a strain field. This mechanism is dubbed electrostriction (or sometimes photostriction) and is well known in fluids of polarizable molecules or in dielectric materials: under an applied electric field E, the material elastically contracts (or expands) in proportion of E^2, not to be confused with a piezo-electric strain proportional to E. If E is an optical frequency field, electrostriction can be seen as a light-induced stress. Indeed, just like radiation pressure and the optical gradient force, it is associated to the optical energy density. But in contrast to them, it directly produces a strain field within the deformable bulk material, when the usual description of the two others relies on internally rigid refractive bodies. In a fluid of polarizable molecules, the electrostrictive mechanism can be understood from an optical gradient force picture: a beam of light illuminating the fluid attracts the particles to the high-field zones, producing an increase of density in these areas. Hence the reader is free to conceptually distinguish or not electrostriction from radiation pressure and the optical gradient force. They are all Lorentz forces acting on charges.

Electrostriction naturally couples optical fields and (acoustic) density waves in solids, and plays an important role in Brillouin scattering processes. In dielectrics, the local density change produced by electrostriction finds its origin in the mechanism of photo-elastic coupling (in other words the dependence of a substance optical susceptibility on its density), which is captured at a basic level by the nineteenth century Clausius–Mossoti relation. In practice, the sign and amplitude of electrostrictive effects vary for different materials, depending on their microscopic structure and symmetry. The link between electrostriction and photoelasticity may not be immediately intuitive. One simplified way of seeing it is to consider a body suddenly subject to strain. This strain leads to a change in the material's permittivity $\Delta\varepsilon$, via photoelastic effects. Provided some photons were stored in the body at the time, the stored optical energy (proportional to εE^2) is changes due to the change in permittivity $\Delta\varepsilon$. This change in energy ΔE can be seen as the opposite of the work of the electrostrictive force during the displacement leading to the strained state.

Following such energetic arguments, the electrostrictive stress can be expressed in terms of the material photoelastic tensor p_{ijkl} (Feldman, 1975; Rakich, Davids and Wang, 2010) that links the strain field in the material S_{ij} to a change in the material's dielectric tensor ε_{ij} (reminder: for a non-magnetic and transparent material $\varepsilon_{ij} = \varepsilon_{ji}$):

$$\varepsilon_{ij}^{-1}(S_{kl}) = \varepsilon_{ij}^{-1} + \Delta\left(\varepsilon_{ij}^{-1}\right) = \varepsilon_{ij}^{-1} + p_{ijkl}S_{kl} \tag{3.6}$$

The general relation is:

$$\sigma_{ij}^{es} = -\frac{1}{2}\varepsilon_0(\varepsilon_{km}p_{mnij}\varepsilon_{nl})E_kE_l \tag{3.7}$$

where E_k is the real amplitude of the kth component of the rapidly oscillating electric field **E**. The photoelastic tensor in principle has $3^4 = 81$ different elements. However, these reduce to only three independent coefficients in the case of amorphous materials and cubic semiconductor crystals such as GaAs and silicon (Newnham, 2004). These three parameters are p_{11}, p_{12} and p_{44}, written here in contracted notation, where $11 \to 1$; $22 \to 2; 33 \to 3; 23, 32 \to 4; 31, 13 \to 5; 12, 21 \to 6$. In this case, thanks to the diagonal nature of the dielectric tensor, the electrostrictive stresses are expressed as functions of the electric field components in the following way:

$$
\begin{pmatrix}
\sigma_{xx}^{es} \\
\sigma_{yy}^{es} \\
\sigma_{zz}^{es} \\
\sigma_{yz}^{es} = \sigma_{zy}^{es} \\
\sigma_{xz}^{es} = \sigma_{zx}^{es} \\
\sigma_{xy}^{es} = \sigma_{yx}^{es}
\end{pmatrix}
= -\frac{1}{2}\varepsilon_0 n_{opt}^4
\underbrace{
\begin{pmatrix}
p_{11} & p_{12} & p_{12} & 0 & 0 & 0 \\
p_{12} & p_{11} & p_{12} & 0 & 0 & 0 \\
p_{12} & p_{12} & p_{11} & 0 & 0 & 0 \\
0 & 0 & 0 & p_{44} & 0 & 0 \\
0 & 0 & 0 & 0 & p_{44} & 0 \\
0 & 0 & 0 & 0 & 0 & p_{44}
\end{pmatrix}
}_{\text{photoelastic tensor}}
\begin{pmatrix}
E_x^2 \\
E_y^2 \\
E_z^2 \\
E_y E_z \\
E_x E_z \\
E_x E_y
\end{pmatrix}
\tag{3.8}
$$

Since optomechanical systems do not respond fast enough to forces at optical frequencies, it is again common to consider the time-averaged stresses in response to time-averaged electric fields.

3.1.4   Photothermal Forces

Up to now, we have considered forces induced by the non-dissipative coupling between electromagnetic radiation and moving charges attached to mechanical bodies. Of course, these forces of Lorentz type do not account for all processes found in the solid-state. For example, absorbed radiation can be turned into heat in matter. In classical electrodynamics, this process is described by the Joule effect accompanying the dynamics of moving charges. In many mechanical devices, this optical heat production gives rise to a thermoelastic displacement and/or strain under steady-state illumination. Such a situation is equivalent to the presence of a stress field acting on the device, and is usually associated with the existence of (optically generated) photothermal forces.

Photothermal forces rely on the ability of constitutive materials to expand or contract under thermal changes. They are best expressed using a thermal stress tensor form σ_{ij}^{th}:

$$
\sigma_{ij}^{th} = C_{ijkl}\beta_{kl}\Delta T
\tag{3.9}
$$

where C_{ijkl} is the stiffness tensor and β_{kl} the thermal expansion tensor relating a temperature increase ΔT to a thermally generated strain field. For an isotropic elastic medium of Young modulus Y and Poisson ratio ν, the thermal stress is diagonal and adopts the simple form $\sigma_{ij}^{th} = \delta_{ij}\frac{Y}{1-2\nu}\beta\Delta T$, where β is the thermal expansion coefficient.

In terms of generated strain, photothermal effects (also sometimes named bolometric effects) can be orders of magnitude larger than radiation pressure, as can be appreciated in the simple case of a moving mirror of velocity **v** orthogonal to the mirror plane, illuminated by a plane wave with normal incidence. In the elementary process associated with radiation pressure, the photon is reflected by the mirror and its energy E shifted by an amount of order $(v/c)E$ by the Doppler effect. The mechanical energy given by the photon to the mirror during this process is hence extremely small. In contrast, in the elementary process associated with the photothermal force, the photon is absorbed and its whole energy E is transferred to the mirror free energy. Provided that the mirror thermoelastic properties are optimized, this can result in a large displacement and/or strain of the mirror, hence a large effective force acting on the (Restrepo, Gabelli, Ciuti, and Favero, 2011). In many experimental situations, the photothermal force can hence overcome radiation pressure, and can actually even adopt the opposite direction. For these reasons, historically, photothermal effects dominated the behaviour in the Crooks radiometer and made it even rotate in the wrong direction (Hull, 1948; Woodruff, 1968). Since even the best mirrors in the world always experience some level of residual absorption, photothermal effects are always present in optomechanical experiments, be they dominant or not. If correctly harnessed, they can lead to enhanced possibilities of optical manipulation of mechanical systems.

3.1.5   Optoelectronic Forces

In material bodies possessing optical resonances, the incoming light can resonantly transfer electrons to excited states. Because of the existence of electron–phonon couplings, the generated excited electronic population induces a stress in the structure, which we name here the optoelectronic stress σ^{oe}. In semiconductor materials where the band structure is accurately known, one can express this stress as a function of measurable macroscopic physical parameters. For example, if we consider the case of a semiconductor optically excited at its bandgap energy E_g, the relevant electrons and holes form an out-of-equilibrium population in the bottom of the conduction band and top of the valence band. Each generated electron-hole pair sits in this case at an energy about E_g above the Fermi sea. In an isotropic crystal, energetic considerations allow to express the isotropic stress appearing with the excited electron-hole pair as a function of the derivative dE_g/dp of the bandgap energy with an applied hydrostatic pressure p. This stress, summed in a unit-volume, is $\sigma^{oe} = BdE_g/dp$ ($\sim 8\,\mathrm{eV}$ in the GaAs crystal) with $B = \frac{Y}{3(1-2v)}$ the isotropic elastic bulk modulus. When $\frac{dE_g}{dp}$ is negative (as in silicon), the optoelectronic stress has its sign opposite to that of the photothermal stress and will tend to make the semiconductor crystal contract. When $\frac{dE_g}{dp} > 0$ (as in GaAs), the optoelectronic stress is of the same sign as the thermal stress and will tend to expand the crystal. In the presence of a population of such electron-hole pairs, one gets:

$$\sigma_{ij}^{oe} = B\frac{dE_g}{dp}\delta_{ij}n_e \tag{3.10}$$

where n_e is the free electron concentration (half of the free carrier concentration in the case of a band-edge optical excitation of electron-hole pairs in an intrinsic semiconductor).

3.1.6   Relation Between Stress Field and Point-force

In the above discussion of optically induced forces acting on material bodies, we have given general expressions for stress tensors. These tensors depend on the spatial coordinates $\mathbf{r} = (x, y, z)$ and can be combined with the tensor formalism of elasticity theory to describe arbitrary optomechanical devices, as long as they consist of a continuous mechanical medium and are operated in the elastic regime. Now, in optomechanics, one is often confronted with situations where a single optical mode and single mechanical mode are coupled. Since the mechanical eigenmode dimensionless spatial profile $\mathbf{U}(\mathbf{r})$ is generally known, a useful strategy is to parameterize the time-dependent mechanical displacement by the variable $u(t)$ such that the time-varying displacement profile becomes $u(t)\mathbf{U}(\mathbf{r})$. By choosing an arbitrary spatial reduction point $\mathbf{r}_0$ and setting the field normalization such that $|\mathbf{U}(\mathbf{r}_0)| = 1$, we identify $u(t)$ to the displacement amplitude at $\mathbf{r}_0$. With these conventions, the physical dimension of the displacement is expressed in the $u(t)$ variable alone, and the considered 3D mechanical problem projects into an effective 1D harmonic oscillator equation:

$$m_{eff}\ddot{u}(t) + m_{eff}\Gamma_M \dot{u}(t) + K u(t) = \sum_k F^k(t) \tag{3.11}$$

with F^k the optically induced forces previously discussed and with the effective mass defined by the relation:

$$\frac{1}{2}m_{eff}\dot{u}(t)^2 = \iiint_V \frac{1}{2}\rho(\mathbf{r})(\dot{u}(t)\mathbf{U}(\mathbf{r}))^2 d^3\mathbf{r}. \tag{3.12}$$

In a consistent manner, the effective spring for the considered mode and considered reduction point is $K = m_{eff}\Omega_M^2$ with Ω_M the mechanical eigenfrequency. The point-forces expressed for the chosen reduction point $\mathbf{r}_0$ are obtained via:

$$F^k(t) = \iiint_V \frac{\sigma_{ij}^k(\mathbf{r}, t)S_{ij}(\mathbf{r}, t)}{u(t)} d^3\mathbf{r} \tag{3.13}$$

where S_{ij} is the strain-field associated to the spatial profile $u(t)\mathbf{U}(\mathbf{r})$ (again with the convention $|\mathbf{U}(\mathbf{r}_0)| = 1$). Since $u(t)$ naturally cancels in this force expression (3.13), the only left-over time dependence stems from the stress term. Note that even for the same stress, the sign of the effective force F^k depends on the considered mechanical mode though the associated strain field S_{ij}. For the same mechanical mode, the amplitude of the force also varies depending on the choice of the reduction point $\mathbf{r}_0$, the force being maximal when choosing a reduction point with minimal displacement amplitude in the considered mechanical mode, and minimal when the displacement at the chosen reduction point is large. The effective force can also change sign if the optical illumination conditions vary, producing a change in the optical stress field. The dependences of 1D optical forces on the considered mechanical mode and illumination conditions, illustrated by Eq. (3.13), have been studied in different optomechanical devices in the past, see (Ortlieb, 2006; Jourdan, Comin and Chevrier, 2008; Metzger, Ludwig, Neuenhahn, Ortlieb, Favero, Karrai and Marquardt, 2008b).

3.2 Light Influenced by Mechanical Motion

In the previous section, we have listed several optically induced forces acting on mechanical bodies. They correspond to different manners for light to impact mechanical motion. In return, the mechanical motion of bodies affects light interacting with them. In optomechanics, where light is stored in an optical cavity, the cavity is deformed by mechanical motion, impacting both the cavity decay rate κ (dissipative optomechanical coupling) and the cavity eigenfrequency ω_o (dispersive coupling). In this section, we will neglect the impact of motion on the decay rate. Discussions of such dissipative (sometimes named absorptive) coupling can be found in the literature, for example in experimental studies on mechanical nanowires vibrating in the middle of an optical cavity (Favero and Karrai, 2008; Favero, Stapfner, Hunger, Paulitschke, Reichel, Lorenz, Weig and Karrai, 2009; Favero, Sankey and Weig, 2014) and in early quantum theoretical investigations (Elste, Girvin and Clerk, 2009). As regards the mechanical modification of the cavity frequency ω_o (dispersive coupling), a common way to parameterize the coupling is to introduce a dispersive frequency-pull parameter $g_{om} = -\frac{d\omega_o}{du}$ where u is the displacement parameter introduced in the previous section (with $|\mathbf{U}(\mathbf{r}_0)| = 1$, $u(t)$ as the amplitude of displacement at the reduction point $\mathbf{r}_0$). In the simple Fabry–Perot case, u can be identified to the cavity length, and a length increase producing a negative optical frequency shift, g_{om} adopts a positive value (in Hz/m).

If the optical cavity is pumped by an external laser, tuned to a flank of a cavity resonance, the phase and intensity of the cavity output field are modulated by mechanical motion. In the limit of a small amplitude motion, a linear relation exists between variation in u and variations in optical phase and intensity, in proportion of g_{om}. This regime is particularly useful to optically read-out the mechanical motion. Figure 3.2 depicts the optical measurement of mechanical motion through acquisition of intensity fluctuations at the cavity output. In this regime of 'linear optomechanical read-out', it is obvious that a larger value of g_{om} permits a larger optical sensitivity to motion.

In this section we provide the reader with useful formulas to compute g_{om} in commonly encountered cases. The displacement $u(t)$ induces a change in the optical cavity geometry G: change of length in a Fabry–Perot, change of radius in a circular cavity, or any more complex geometric change δG in more exotic cases. This geometric change modifies the electromagnetic boundary conditions of the cavity and shift its eigenfrequencies. But in addition, the displacement $u(t)$ is also accompanied by a strain field S within the body of the cavity, which in the case of a constitutive photoelastic material also produces an optical shift through a change in the refractive index. In terms of frequency-pull parameters this means that the total g_{om} can be split into two independent contributions related to each of these two mechanisms:

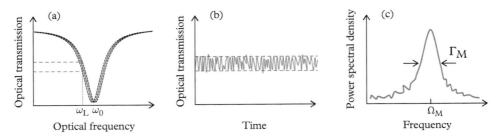

Fig. 3.2 *Side-of-the-fringe optical read-out of mechanical motion. (a) Lorentzian optical transmission resonance (solid blue curve). Due to the mechanical fluctuations changing the cavity length, the optical resonance frequency continuously fluctuates around its equilibrium position at ω_0 (dashed black lines). Keeping the pump laser frequency ω_L fixed, the optical transmission fluctuates in time between two bounding values (dashed red lines). (b) Optical transmission versus time, with the laser tuned to the optical resonance flank. (c) The optical noise power spectral density of the transmission reveals the presence of a mechanical mode at frequency Ω_M.*

$$g_{om} = \frac{d\omega_0\,(G,\varepsilon)}{du} = \underbrace{\frac{\partial\omega_0}{\partial G}\frac{\partial G}{\partial u}}_{geometric\,g_{om}^{geo}} + \underbrace{\frac{\partial\omega_0}{\partial\varepsilon}\frac{\partial\varepsilon}{\partial u}}_{photoelastic\,g_{om}^{pe}} \qquad (3.14)$$

Here ε is the material's permittivity, which is no longer necessarily isotropic nor uniform inside the body of the cavity. The photoelastic contribution g_{om}^{pe} is obviously unique to resonators where light is confined inside a solid body, like a disc, toroid or photonic crystal, and would not appear in an empty Fabry–Perot optomechanical cavity. For this reason, it has long been disregarded in early works in optomechanics that employed macroscopic mirror cavities.

3.2.1   Geometric Contribution g_{om}^{geo}

The geometric frequency-pull parameter adopts very simple approximate expressions in the case of an ideal Fabry–Perot cavity of length L, $g_{om}^{geo} = \omega_0/L$ (here the reduction point is the coordinate of the moving back mirror, the front mirror being rigid), and in the case of a circular dielectric cavity of radius R supporting whispering-gallery modes. In the latter case, under the assumption that the in-plane and out-of-plane dependences of the electric field can be treated separately (a standard assumption in the effective index method), and considering a purely radial mechanical displacement of the disc, on can rigorously demonstrate that $g_{om}^{geo} = \omega_0/R$ (Ding, Baker, Senellart, Lemaitre, Ducci, Leo and Favero, 2010) (the reduction point being in this case the radial coordinate of the disc peripheral boundary).

For more involved geometric configurations, the problem is to compute how an optical resonator of arbitrary geometry, experiencing an arbitrary infinitesimal deformation, has its optical eigenfrequencies shifted. After some incorrect literature on the topic, this problem was finally solved (Johnson, Ibanescu, Skorobogatiy, Weisberg, Joannopoulos and Fink, 2002), by means of a self-consistent perturbation theory for Maxwell's equations in the case of shifting material boundaries. From this theory, the term g_{om}^{geo} can be calculated as a surface integral of the (unperturbed) optical and mechanical modes that are interacting, taken over the unperturbed dielectric resonator interface, see Eq. (3.15):

$$g_{om}^{geo} = \frac{\omega_0}{4} \iint_{\text{resonator}} (\vec{q} \cdot \vec{n}) \left[\Delta\varepsilon_{12} |\vec{e}_{\parallel}|^2 - \Delta\left(\varepsilon_{12}^{-1}\right) |\vec{d}_{\perp}|^2 \right] dA \qquad (3.15)$$

Here $\vec{q}$ and $\vec{n}$ are, respectively, the normalized mechanical displacement vector and surface normal vector. $\vec{e}_{\parallel}$ (resp. $\vec{d}_{\perp}$) is the component of the electric field (electric displacement field) parallel (orthogonal) to the surface of the resonator. $\vec{q}$ and $\vec{e}$ are normalized such that $\max|\vec{q}|=1$ and $\frac{1}{2} \int\int\int \varepsilon|e|^2 dV = 1$.

$\Delta\varepsilon_{12} = \varepsilon_1 - \varepsilon_2$ is the difference in permittivity between the materials on either side of the surface boundary and $\Delta\left(\varepsilon_{12}^{-1}\right) = \varepsilon_1^{-1} - \varepsilon_2^{-1}$. In dielectric cavities filled by a uniform material of refractive index n_{opt}, ε_1 is simply n_{opt}^2 over the entire resonator, while $\varepsilon_2=1$ for air or vacuum outside the cavity. From Eq. (3.15), g_{om}^{geo} is numerically calculated as follows. First the mechanical eigenmode of the resonator is calculated using the Finite Element Method (FEM). The deformation profile of the mechanical mode is imported into a FEM simulation of the targeted *unperturbed* optical mode of the resonator. The optomechanical coupling g_{om}^{geo} is then calculated by computing the surface integral of Eq. (3.15) over the resonator boundaries. Some examples of this approach applied to semiconductor disc cavities can be found in (Ding, Baker, Senellart, Lemaitre, Ducci, Leo and Favero, 2010; Baker, Hease, Nguyen, Andronico, Ducci, Leo and Favero, 2014).

Another case of interest where a similar integral expression can be obtained for g_{om}^{geo} is the case of a perturbing dielectric element of index n_{opt} inserted into the optical mode of an originally empty optical cavity, represented by an unperturbed electric field E. The dielectric element produces a dispersive shift of the cavity mode $\delta\omega_0$ (Waldron, 1960):

$$\delta\omega_0 = -\frac{\omega_0}{2} \frac{\iiint_{\text{element}}(n_{opt}^2 - 1)E^2 d^3 r}{\iiint_{\text{cavity}} E^2 d^3 r}. \qquad (3.16)$$

With this equation in hands, one can compute the variations in $\delta\omega_0$ as the position of the dielectric element changes by an infinitesimal amount δu, in order to obtain g_{om}. This was, for example, employed to describe experiments where a flexible nanowire or nanotube is inserted in a miniature rigid optical Fabry–Perot cavity (Favero, Stapfner, Hunger, Paulitschke, Reichel, Lorenz, Weig and Karrai, 2009; Stapfner, Ost, Hunger, Reichel, Favero and Weig, 2013; Stapfner, Favero, Hunger, Paulitschke, Reichel, Karrai and Weig, 2010).

Let us stress the direct connection between the geometric coupling g_{om}^{geo} on one hand, and radiation pressure and optical gradient forces on the other hand. In the previous sections, we have associated these latter Lorentz forces to internally rigid moving dielectric bodies (moving mirror in a Fabry–Perot, moving polarizable dielectric particle, dielectric cavity with moving boundaries). Imagine that N_{cav} photons are stored in the cavity optical mode, and that the system is lossless. The stored optical energy

$N_{cav}\hbar\omega_0$ changes due to the change in optical frequency $\delta\omega_0 = -g_{om}^{geo}\delta u$ associated with an infinitesimal displacement δu experienced by the moving system. This change in stored optical energy is the opposite of the work of the optical force on the system during the displacement, such that the force can be expressed as:

$$F^{rp} = N_{cav}\hbar g_{om}^{geo} \qquad (3.17)$$

In the case of a circular dielectric cavity with moving boundaries, such as optomechanical semiconductor discs (Favero, 2014), this 'geometric' optical force resembles radial radiation pressure pushing the walls of the disc apart. In the case of a nanowire or nanotube sitting in the middle of an optical cavity, this force resembles more of an optical gradient ponderomotive type of interaction, where the polarizable nanowire or tube is attracted to the high-field regions within the optical cavity.

3.2.2 Photoelastic g_{om}^{pe}

If the dielectric body is now free to internally deform, its photolelastic properties will also lead to a shift of the cavity optical eigenfrequency, produced by the strain-induced changes of the refractive response. This is the photoelastic contribution to g_{om}. To numerically compute such a contribution, the unperturbed optical resonance frequency of the resonator is first computed by means of a FEM simulation. Second, the desired mechanical eigenmode is solved for in another FEM simulation, providing the complete deformation profile and strain distribution inside the resonator. We then use Eq. (3.6) to model how the strain distribution inside the resonator modifies the dielectric tensor. At this stage, note that the dielectric tensor is now both anisotropic and non-uniform inside the resonator, even if the rigid resonator originally possesses a uniform and isotropic dielectric response. The problem of finding the new (perturbed) optical resonance frequency of the deformed resonator under these conditions can be solved through another FEM simulation. This provides the photoelastic frequency shift associated to the known mechanical displacement δu.

We have seen in section 3.1.3 that simple energy arguments directly relate the electrostrictive force to photoelastic effects. For a small displacement δu associated with a strain δS, the work W done by electrostriction is given by Eq. (3.18):

$$W = \iiint_{\text{resonator}} \langle \sigma^{es} \rangle : \delta S\, d^3\mathbf{r} \qquad (3.18)$$

If we equal this work to the work of the electrostrictive force $F^{es} = N_{cav}\hbar g_{om}^{pe}$ like we have done before for radiation pressure and optical gradient interactions, we finally obtain an integral expression for the photoelastic frequency-pull parameter, with the same normalization conventions as in (3.15), expressed for an isotropic or cubic material, and a normalized modal strain field S:

$$g_{om}^{pe} = -\frac{\omega_0}{4} \iiint_{\text{resonator}} \varepsilon_0 n_{opt}^4 \cdot e_i p_{ijkl} S_{kl} e_j \cdot d^3\mathbf{r} \qquad (3.19)$$

Examples of computation of photoelastic couplings g_{om}^{pe} in semiconductor disc resonators are given in (Baker, Hease, Nguyen, Andronico, Ducci, Leo and Favero, 2014).

3.3   Equations of Optomechanics

In the previous section, we have given the conceptual and technical tools to describe how light influences mechanical motion through optically induced forces, and reciprocally how mechanical motion impacts light trapped in a cavity. With these two mutual aspects of optical-mechanical interactions in hands, we are now ready to describe the dynamics of optomechanical devices.

3.3.1   Delayed Force Model

The delayed force model is the simplest form of dynamical optomechanical equation that can be written, but it already grasps the majority of dynamical optomechanical effects. It consists of stating that all dynamical effects can be accounted for by the non-instantaneous nature of optical forces. If the movable mirror is displaced in a laser-driven optomechanical cavity, photons need a finite amount of time to reach a new equilibrium state in the cavity. If we consider that Lorentz forces, whose response time is associated with dielectric phenomena, are instantaneous at the time scale of mechanical devices (with resonance frequencies in the kHz–GHz range), the finite response time of the force is simply the cavity response time $\tau_{cav} = 1/\kappa$ (κ is the energy decay rate of the optical cavity mode). This reasoning applies to radiation pressure, electrostriction and optical gradient forces. For photothermal forces in contrast, it is the thermal response time of the mechanical device τ_{th} that sets the delay of the optically induced force. Such thermal time is typically longer than the optical cavity response time: $\tau_{th} > \tau_{cav}$. In the case of optoelectronic forces associated to an out-of-equilibrium free-carrier population generated by a laser, it is the free-carrier lifetime τ_{fc} that sets the response time of the optically induced stress. Finally, even if the microscopic origin of all these forces may differ, the delayed force model offers a powerful unified description, provided the delay time τ of the force is known. In that case, the model's equation is purely mechanical, taking the form of a harmonic oscillator subject to a delayed optical force that accounts for the dynamical behaviour of all other degrees of freedom (cavity field, temperature, free carrier population) (Metzger, Favero, Ortlieb and Karrai, 2008a):

$$m_{eff}\ddot{u}(t) + m_{eff}\Gamma_M \dot{u}(t) + Ku(t) = F(u(t)) + F_L(t) \qquad (3.20)$$

where

$$F(u(t)) = F(u_0) + \int_{-\infty}^{t} \frac{dF(u(t'))}{dt'} h(t-t')dt' \qquad (3.21)$$

where h is the delay function of the force F, u_0 the equilibrium position of the oscillator under the static force $F(u_0)$, set by the balance between the optical force and the restoring force of the mechanical spring, and F_L the fluctuating Langevin force. The simplest delay function to consider is an exponential function $h(t) = 1 - \exp(-t/\tau)$, which can represent a great variety of dynamical responses. The expression (3.21) of the force can be understood from the geometrical picture of Fig. 3.3. Adopting a simple integration by parts in the integral of (3.21), the reader may also prefer an expression in terms of force values at past time $F(u(t'))$ rather than the time derivative of the force (Restrepo, Gabelli, Ciuti and Favero, 2011).

Under the exponential delay hypothesis, and under a linearization whereby one considers small amplitude motion δu around the equilibrium position $u = u_0 + \delta u$, established models (Metzger, Favero, Ortlieb and Karrai, 2008a) show how the harmonic mechanical oscillator subject to the time-delayed force F becomes an effective harmonic oscillator δu with effective frequency Ω_{eff} and modified damping rate Γ_{eff}:

$$\delta\ddot{u}(t) + \Gamma_{\text{eff}}\delta\dot{u}(t) + \Omega_{\text{eff}}^2\delta u(t) = \frac{F_L(t)}{m_{eff}} \tag{3.22}$$

with

$$\Omega_{\text{eff}}^2 = \Omega_M^2\left(1 - \frac{1}{1 + \Omega_M^2\tau^2}\frac{\nabla F}{K}\right) \tag{3.23}$$

where Ω_M and K are respectively the unperturbed mechanical mode frequency and stiffness, while $\nabla F = \partial F/\partial u|_{u_0}$ represents the change in the steady-state optical force for a small displacement δu of the mechanical resonator around the equilibrium position. Likewise, the modified damping Γ_{eff} is given by Eq. (3.24) (Metzger, Favero, Ortlieb and Karrai, 2008a):

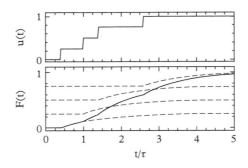

Fig. 3.3 *Displacement u and force F as a function of time. u(t) varies in steps in this example. In the force plot, the dashed lines represent the time-evolution for the force after each displacement step, progressively converging towards a steady-state force value. The black line represents the sum of these different contributions delayed in time. This geometric picture allows us to represent the force at time t as a temporal sum of elementary force increments at previous times t', weighted by the temporal delay, leading to Eq. (3.21).*

$$\Gamma_{\text{eff}} = \Gamma_M \left(1 + Q_M \frac{\Omega_M \tau}{1 + \Omega_M^2 \tau^2} \frac{\nabla F}{K} \right) \qquad (3.24)$$

where Q_M and $\Gamma_M = \Omega_M / Q_M$ are respectively the unperturbed mechanical quality factor and damping.

The origin of the modification of the mechanical frequency can be easily understood from the optical spring picture. The optically induced force being proportional to the optical intensity in the cavity, it also depends on the cavity resonance ω_0, at least when the cavity is driven by a laser of fixed wavelength. Because ω_0 depends on the position u in an optomechanical device, the force also depends on u. At first order this gives rise to a spring-type of optical force proportional to δu. This force can be of restoring type, giving birth to an added rigidity of optical origin, which produces an effective mechanical frequency increase. But it can also adopt the opposite sign, becoming an anti-spring, and producing a frequency decrease. Both cases are encompassed in Eq. (3.23), with the sign being set by the sign of ∇F. The origin of the modified mechanical damping, which can also become an anti-damping, depending again on the sign of ∇F, can be grasped from the delayed force picture. Indeed the delayed force produces positive or negative work on the mechanical motion with an efficiency depending on delay conditions. A pictorial discussion is given in Fig. 3.4.

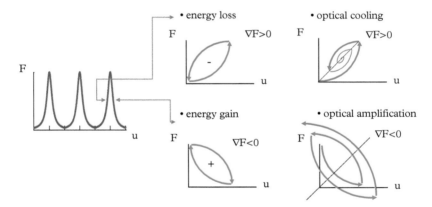

Fig. 3.4 *Work produced by a time-delayed optical force. Left panel: the force F as a function of displacement u. The force is proportional to the number of photons stored in the cavity and hence follows the cavity optical resonances. The cavity is operated on a flank of an optical resonance with a positive or negative ∇F, corresponding respectively to a red- or blue-detuned laser. Middle panels: upon mechanical oscillation between two points, the optical force F varies with a temporal lag with respect to the displacement u. As a consequence of this lag, the F(u) curve encompasses a surface of finite area, corresponding to either negative (-) or positive (+) work of the optical force on mechanical motion. Right panels: when looking at subsequent mechanical oscillations, the net mechanical energy loss or gain induced by optical forces results in a motion of damped amplitude (cooling) or increasing amplitude (amplification).*

Under these optically modified conditions, the mechanical resonator still follows a kind of equipartition theorem; however, it behaves as if it was fluctuating at an effective temperature T_{eff} set by (Metzger, Favero, Ortlieb and Karrai, 2008a):

$$\frac{T_{eff}}{T} = \frac{\Gamma}{\Gamma_{eff}} \tag{3.25}$$

This shows that effective cooling of the mechanical motion can be produced by the damping effect of delayed optomechanical forces. Conversely, the delayed force model and Eq.(3.24) also predict that a situation of zero effective mechanical damping can be reached in optomechanical devices, in which $\Gamma_{eff} = 0$. In this latter case, the mechanical system adopts a quasi-sinusoidal time trajectory, which can be seen as an optically sustained harmonic oscillation. The situation can also be understood as mechanical lasing, or parametric oscillation, where mechanical dissipation is compensated by coherent parametric amplification of motion by optical forces.

The delayed force model hence predicts both cooling and amplification of motion, and correctly grasps many aspects encountered in optomechanics. But because it oversimplifies the temporal dynamics of optical, thermal and other degrees of freedom, it may miss some regimes of interest. To go further, the sections below present complementary models.

3.3.2   Coupled Optical and Mechanical Equations

One shortcoming of the delayed force model introduced above is that it misses the exact dynamical impact of the detuning between the pump laser and the optical cavity. Indeed, the time-evolution of the intracavity optical intensity cannot always be described by a simple exponential delay, and a more complete description is in principle required to fully understand the effects of Lorentz forces that are proportional to this intensity. This shortcoming has limited consequence as long as $\Omega_M \ll \kappa$ (the so-called *bad cavity* or *unresolved sideband* limit) but it becomes important when $\Omega_M \gg \kappa$ (*good cavity* or *resolved sideband* limit). A more general approach to consistently describe all possible situations is to simultaneously consider two coupled differential equations that account for the optical and mechanical degrees of freedom in parallel. The dynamics of the complex cavity field amplitude $a(t)$ and the mechanical motion amplitude $u(t)$ are given by (Aspelmeyer, Kippenberg and Marquardt, 2014):

$$\dot{a} = -\frac{\kappa}{2}a + i(\Delta + g_{om}u)\, a + \sqrt{\kappa_{ex}}a_{in} \tag{3.26}$$

$$m_{eff}\ddot{u} + m_{eff}\Gamma_M\dot{u} + m_{eff}\Omega_M^2 u = F^{opt} = \hbar g_{om}|a|^2 \tag{3.27}$$

with

$$\kappa = \kappa_0 + \kappa_{ex} \tag{3.28}$$

where κ_0 is the intrinsic cavity energy decay rate and κ_{ex} the extrinsic cavity energy decay rate (for example coupling through the cavity ports). $\Delta = \omega_L - \omega_0$ is the detuning between the laser angular frequency and the cavity angular eigenfrequency when $u = 0$. A linear relation is assumed here for simplicity $\omega_0(0) - \omega_0(u) = g_{om}u$. The cavity field is normalized such that $|a|^2 = N_{cav}$ is the number of photons stored in the cavity mode. a_{in}^2 is the rate of photon number impinging on the cavity, equal to $P/(\hbar\omega_L)$, with P the incoming optical power at the input port. In Eq. (3.27), we have made use of the fact that for Lorentz forces such as radiation pressure, optical gradient force and electrostriction, the total optical force can be expressed as $F^{opt} = F^{rp} + F^{es} = N_{cav}\hbar g_{om}$.

Starting from these coupled differential equations, it is again possible to proceed with a linearization of the dynamics around average values both for the cavity complex field $a(t) = \langle a \rangle + \delta a(t)$ and for the mechanical displacement $u(t) = u_0 + \delta u(t)$, this time without any specific assumption about the temporal dynamics of both variables:

$$\delta\dot{a} = (i\Delta - \frac{\kappa}{2})\delta a + ig_{om} < a > \delta u \tag{3.29}$$

and

$$m_{\text{eff}}\delta\ddot{u} + m_{\text{eff}}\Gamma_M\delta\dot{u} + m_{\text{eff}}\Omega_M^2\delta u = \hbar g_{om}(< a >^* \delta a + < a > \delta a^*) \tag{3.30}$$

Such a linearization allows for solving the two coupled equations in the Fourier space, and describing again the mechanical oscillator dressed by light as having an effective mechanical frequency $\Omega_{eff} = \Omega_M + \delta\Omega_M$ and an effective damping $\Gamma_{eff} = \Gamma_M + \Gamma_{om}$ (Aspelmeyer, Kippenberg and Marquardt, 2014):

$$\delta\Omega_M = < a >^2 g_{om}^2 \frac{\hbar}{2m_{eff}\Omega_M} \left[\frac{\Delta - \Omega_M}{\kappa^2/4 + (\Delta + \Omega_M)^2} + \frac{\Delta + \Omega_M}{\kappa^2/4 + (\Delta - \Omega_M)^2} \right] \tag{3.31}$$

and

$$\Gamma_{om} = < a >^2 g_{om}^2 \frac{\hbar}{2m_{eff}\Omega_M} \left[\frac{\kappa}{\kappa^2/4 + (\Delta + \Omega_M)^2} - \frac{\kappa}{\kappa^2/4 + (\Delta - \Omega_M)^2} \right] \tag{3.32}$$

In this form it clearly appears that both optomechanical effects, on frequency and on dissipation, scale with the average number of photons stored in the cavity $\langle a \rangle^2$. In contrast to the delayed force model, a direct dependence with the detuning Δ is now present. However, in the *bad cavity* limit and for detuning smaller than the cavity linewidth, these expressions reduce to the delayed force model results.

3.3.3   Coupled Optomechanical Equations with Photothermal Forces

The coupled equations analysis presented above does not allow to account consistently for photothermal forces, and is restricted to systems driven by Lorentz forces like radiation pressure, optical gradient force and electrostriction. The delayed force model has the strength of being able to account for any delayed interaction, independently of its microscopic origin, but as we have seen it is confronted to some shortcomings when the *good cavity* limit is approached. A way out is to deal with three coupled differential equations on optical, thermal and mechanical degrees of freedom. If reducing the thermal degrees of freedom to a single-parameter temperature increase ΔT, these equations read, with the same approximations as above:

$$\dot{a} = -\frac{\kappa}{2}a + i\left(\Delta + g_{om}u\right)a + \sqrt{\kappa_{ex}}a_{in} \tag{3.33}$$

$$m_{\text{eff}}\ddot{u} + m_{\text{eff}}\Gamma_M \dot{u} + m_{\text{eff}}\Omega_M^2 u = \hbar g_{om}|a|^2 + F^{pth} \tag{3.34}$$

$$\frac{d\Delta T}{dt} = -\frac{\Delta T}{\tau_{th}} + \frac{R_{th}\hbar\omega_L\kappa_{abs}}{\tau_{th}} \tag{3.35}$$

where the photothermal force F^{pth}, proportional to ΔT, can be computed from Eqs. (3.9) and (3.13). The thermal resistance R_{th} (with dimension of K/W) expresses the linear relation between the absorbed optical power in the cavity and the consequent temperature increase ΔT of the resonator in the steady state. Note that the cavity energy loss rate now includes an optical absorption channel κ_{abs}, responsible for the optical heating of the resonator, and singles out the cavity decay rate due to non-absorptive intrinsic photon loss κ_0:

$$\kappa = \kappa_0 + \kappa_{abs} + \kappa_{ex} \tag{3.36}$$

Thermo-optical effects shifting the optical cavity frequency under temperature elevation can also be included in these equations by adding a term $i\frac{\omega_0}{n_{opt}}\frac{dn_{opt}}{dT}\Delta T$ to the right-side of the first equation. Just as for the previous cases, these equations can be linearized, leading to exact solutions in the Fourier space and analytic expressions for the optomechanically modified frequency and damping of the harmonic mechanical oscillator dressed by light. They complement conventional coupled differential equations of radiation-pressure optomechanics.

3.3.4   Coupled Optomechanical Equations with Optoelectronic Forces

In the same spirit as above, one can go one step further by studying semiconductor optomechanical cavities where multi-photon absorption processes do play a role. This extension is particularly relevant in nanoscale resonators operated at large circulating optical powers, where two-photon absorption modifies dissipation mechanisms, and creates a population of free carriers that generates an optoelectronic stress. This population also induces free-carrier absorption (FCA) and dispersion (FCD), in proportion to the free-carrier density N:

$$\dot{a} = -\frac{\kappa}{2}a + i\left(\Delta + g_{om}u + \frac{\omega_0}{n_{opt}}\frac{dn_{opt}}{dT}\Delta T + \frac{\omega_0}{n_{opt}}\frac{dn_{opt}}{dN}N\right)a + \sqrt{\kappa_{ex}}a_{in} \qquad (3.37)$$

with the mechanical equation now embedding an optoelectronic force F^{oe}, also proportional to the carrier density N and computable through Eqs. (3.10) and (3.13):

$$m_{\text{eff}}\ddot{u} + m_{\text{eff}}\Gamma_M\dot{u} + m_{\text{eff}}\Omega_M^2 u = \hbar g_{om}|a|^2 + F^{pth} + F^{oe} \qquad (3.38)$$

and two equations to describe the thermal and carrier dynamics:

$$\frac{d\Delta T}{dt} = -\frac{\Delta T}{\tau_{th}} + \frac{R_{th}|a|^2\hbar\omega_L(\kappa_{abs} + \kappa_{\text{TPA}} + \kappa_{\text{FCA}})}{\tau_{th}} \qquad (3.39)$$

$$\frac{dN}{dt} = -\frac{N}{\tau_{fc}} + \frac{\beta_{TPA}c^2\hbar\omega_L}{2n^2 V_{\text{TPA}}^2}|a|^4 \qquad (3.40)$$

with τ_{fc} the free-carrier relaxation time, β_{TPA} the conventional TPA coefficient of the semiconductor material and V_{TPA} the effective volume of the TPA interaction (Johnson, Borselli and Painter, 2006). The cavity energy loss rate now includes two additional loss channels corresponding to two-photon absorption (TPA) and free-carrier absorption (FCA):

$$\kappa = \kappa_0 + \kappa_{abs} + \kappa_{ex} + \kappa_{\text{TPA}} + \kappa_{\text{FCA}} \qquad (3.41)$$

This set of four equations now contains several nonlinear couplings, both optomechanical and merely optical. These equations can again be linearized to analyse fluctuations around mean values, but a fully numerical approach is more suited to exploring the novel regimes offered by such optomechanical cavities combining optomechanical and optical nonlinearities. This is especially true in the regime of amplification where large amplitude motion disrupts the linearized approach. We will not describe in details these regimes here, but they underline the enrichment of the dynamics of miniature optomechanical devices as additional degrees of freedom are progressively involved in interactions.

Acknowledgements

The author acknowledges contributions to the figures from Christophe Baker and Khaled Karrai, and financial support of the French ANR through the Nomade, QDOM and Olympia projects, as well as from the European Research Council through the and Nombi projects.

Bibliography

Ashkin, A. (1970). Acceleration and trapping of particles by radiation pressure. *Physical Review Letters*, **24**, 156.

Ashkin, A., Dziedzic, J. M., Bjorkholm, J. E., and Chu, S. (1986). Observation of a single-beam gradient force optical trap for dielectric particles. *Optics Letters*, **11**, 288.

Aspelmeyer, M., Kippenberg, T. J., and Marquardt, F. (2014). Cavity optomechanics. *Review of Modern Physics*, **86**, 1391.

Baker, C., Hease, W., Nguyen, D. T., Andronico, A., Ducci, S., Leo, G., and Favero, I. (2014). Photoelastic coupling in gallium arsenide optomechanical disk resonators. *Optics Express*, **22**, 14072.

Caves, C. M. (1980). Quantum-mechanical radiation-pressure fluctuations in an interferometer. *Physical Review Letters*, **45**, 75.

Ding, L., Baker, C., Senellart, P., Lemaitre, A., Ducci, S., Leo, G., and Favero, I. (2010). High frequency GaAs nano-optomechanical disk resonator. *Physical Review Letters*, **105**, 263903.

Elste, F., Girvin, S. M., and Clerk, A. A. (2009). Quantum noise interference and backaction cooling in cavity nanomechanics. *Physical Review Letters*, **103**, 149902.

Favero, I. (2014). Gallium Arsenide Disks as Optomechanical Resonators. In: Aspelmeyer M., Kippenberg T., Marquardt F. (eds) *Cavity Optomechanics. Quantum Science and Technology.* Springer, Berlin, Heidelberg.

Favero, I. and Karrai, K. (2008). Cavity cooling of a nanomechanical resonator by light scattering. *New Journal of Physics*, **10**, 095006.

Favero, I., Sankey, J., and Weig, E. M. (2014). Mechanical Resonators in the Middle of an Optical Cavity. In: Aspelmeyer M., Kippenberg T., Marquardt F. (eds) *Cavity Optomechanics. Quantum Science and Technology.* Springer, Berlin, Heidelberg.

Favero, I., Stapfner, S., Hunger, D., Paulitschke, P., Reichel, J., Lorenz, H., Weig, E. M., and Karrai, K. (2009). Fluctuating nanomechanical system in a high finesse optical microcavity. *Optics Express*, **15**, 12813.

Feldman, A. (1975). Relations between electrostriction and the stress-optical effect. *Physical Review B*, **11**, 5112.

Hull, G. F. (1948). Concerning the action of the Crookes radiometer. *American Journal of Physics*, **16**, 185.

Jackson, J. D. (1998). *Classical Electrodynamics*. John Wiley & Sons, New York.

Johnson, S. G., Ibanescu, M., Skorobogatiy, M.A., Weisberg, O., Joannopoulos, J.D., and Fink, Y. (2002). Perturbation theory for Maxwell equations with shifting material boundaries. *Physical Review E*, **65**(6), 066611.

Johnson, T. J., Borselli, M., and Painter, O. (2006). Self-induced optical modulation of the transmission through a high-Q silicon microdisk resonator. *Optics Express*, **14**, 817.

Jourdan, G., Comin, F., and Chevrier, J. (2008). Mechanical mode dependence of bolometric backaction in an atomic force microscopy microlever. *Physical Review Letters*, **101**, 133904.

Karrai, K., Favero, I., and Metzger, C. (2008). Doppler optomechanics of a photonic crystal. *Physical Review Letters*, **100**, 240801.

Labeyrie, A. (1996). Resolved imaging of extra-solar planets with future 10-100 km optical interferometric arrays. *Astronomy and Astrophysics Supplement Series*, **118**, 517.

Law, C. K. (1995). Interaction between a moving mirror and radiation pressure: A Hamiltonian formulation. *Physical Review A*, **51**, 2537.

Lebedev, P. (1901). Untersuchungen über die druckkräfte des lichtes. *Annalen der Physik*, **311**(11), 433.

Metzger, C., Favero, I., Ortlieb, A., and Karrai, K. (2008*a*). Optical self cooling of a deformable Fabry–Perot cavity in the classical limit. *Physical Review B*, **78**(3), 035309.

Metzger, C., Ludwig, M., Neuenhahn, C., Ortlieb, A., Favero, I., Karrai, K., and Marquardt, F. (2008*b*). Self-induced oscillations in an optomechanical system driven by bolometric backaction. *Physical Review Letters*, **101**(13), 133903.

Moine, O. (2005). *Modélisation de forces optiques*. PhD thesis, Université d'Aix-Marseille.

Newnham, R. E. (2004). *Properties of Materials*; Anisotropy, Symmetry, Structure. Oxford University Press, New York.

Nichols, E. F. and Hull, G. F. (1901). A preliminary communication on the pressure of heat and light radiation. *Physical Review*, **13**, 307.

Ortlieb, A. (2006). *Laserkuhlung nanomechanischer Resonatoren*. DiplomArbeit thesis, LMU, Muenchen.

Rakich, P. T., Davids, P., and Wang, Z. (2010). Tailoring optical forces in waveguides through radiation pressure and electrostrictive forces. *Optics Express*, **18**(14), 14439.

Restrepo, J., Gabelli, J., Ciuti, C., and Favero, I. (2011). Classical and quantum theory of photothermal cavity cooling of a mechanical oscillator. *Comptes Rendus Physique*, **12**, 860.

Stapfner, S., Favero, I., Hunger, D., Paulitschke, P., Reichel, J., Karrai, K., and Weig, E. M. (2010). Cavity nano-optomechanics: a nanomechanical system in a high finesse optical cavity. *Proceedings of SPIE*, **7727**, 772706.

Stapfner, S., Ost, L., Hunger, D., Reichel, J., Favero, I., and Weig, E. M. (2013). Cavity-enhanced optical detection of carbon nanotube brownian motion. *Applied Physics Letters*, **102**, 151910.

Waldron, R.A. (1960). Perturbation theory of resonant cavities. *Proceedings of the IEE - Part C: Monographs*, **107**(12), 272–4.

Woodruff, A. E. (1968). The radiometer and how it does not work. *The Physics Teacher*, **6**, 358.

4

Quantum Optomechanics: from Gravitational Wave Detectors to Macroscopic Quantum Mechanics

Yanbei Chen

Burke Institute for Theoretical Physics, and the Institute for
Quantum Information and Matter (IQIM),
California Institute of Technology, Pasadena, California, USA

Yanbei Chen

Yanbei Chen, *Quantum Optomechanics: from Gravitational Wave Detectors to Macroscopic Quantum Mechanics* In: *Quantum Optomechanics and Nanomechanics*. Edited by: Pierre-Francois Cohadon, Jack Harris, Florian Marquardt, Leticia F. Cugliandolo,
Oxford University Press (2020). © Oxford University Press. DOI: 10.1093/oso/9780198828143.003.0004

Chapter Contents

4.1 Overview and Basic Notions

4.1.1 Overview: Two Aspects of Quantum Measurement

The quantum measurement process connects the quantum world and the classical world. The phrase 'quantum measurement' can have two meanings, as we discuss below.

Measurement of a weak classical force

Let us consider measuring a weak classical force, by first letting it act on a macroscopic test object, and then read out the object's displacement caused by the force. In this context, we need to minimize the noises that arise due to quantum fluctuations of both the mass and the light. We will also have to isolate the system from other classical forces, e.g. thermal fluctuations. As we shall see soon in this chapter, the quantum noise level consists of the shot noise of light, which can be viewed as arising from the inaccuracy of repetitive measurements, and the radiation-pressure noise, which enforces the measurement-induced back action. Together, the trade-off between these two types of noise gives rise to the so-called *Standard Quantum Limit* (SQL), which was first realized by Vladimir Braginsky in the 1960s when he was contemplating the ultimate limit of his precision measurement devices that use macroscopic test masses (Braginsky and Khalili, 1992). The SQL is not a limitation for the measurement precision, after all— but it provides the magnitude in which we must consider both measurement precision and measurement-induced back action. Sections 4.1–4.3 will be devoted to this thread of thought.

In the rest of section 4.1, we will develop a linear quantum measurement formalism that clarifies in which context the SQL would apply, and see how we might in fact circumvent to the SQL. We will also introduce the necessary notations.

In section 4.2, we will present several configurations that allow the interferometer to circumvent the SQL—we finally realize that optical losses, which destroy quantum coherence, are the major obstacles toward sub-SQL interferometers.

Given all these possibilities, what are the better configurations that are most immune to optical losses? Is there a fundamental limit that incorporates losses? These questions are not yet answered, but with these questions in mind, we will move on to the quest for a more systematic understanding of such interferometers, which will be presented in section 4.3.

Quantum mechanics of macroscopic objects

Alternatively, especially in recent years, we can study a second topic, which is to try to *prepare*, *manipulate* and *characterize* the quantum state of a macroscopic quantum object—through quantum measurement.

As we will show in section 4.4, the free-mass SQL actually provides a benchmark for the 'quantum-ness' of the system. If a device has classical sources of noise much below the SQL, then these classical noises will not significantly affect the preparation, manipulation and characterization of the quantum state of the test mass. Furthermore, the quantum noise of the device determines its efficiency in preparing and characterizing the quantum state of the test mass. We will then show that a sub-SQL device can be used to prepare nearly pure quantum states, mechanical entanglement, and non-Gaussian quantum states that have no classical counterparts.

Finally, in section 4.5, we will study 'tests of quantum mechanics'. Quantum mechanics contains two parts: (i) unitary evolution for the joint wavefunction of a quantum system, and (ii) Born's rule that converts the wavefunction into probability distributions that we can observe. We will mainly focus on testing (i), although it is actually interesting that issue (ii) also arises during our discussion.

4.1.2   Origin of the Standard Quantum Limit

For a test mass whose position is monitored by light, the Standard Quantum Limit can be derived in two different ways.

Focusing on optical fluctuations

Let us first motivate the SQL from the optical-field point of view, by considering the sensing noise and the back-action noise. For an optical field trying to detect a displacement x, we have

$$\delta\phi = \frac{2\omega_0 x}{c} \tag{4.1}$$

where ω_0 is the angular frequency of light. Suppose this sensitivity is achieved from N photons, we will have

$$\delta\phi \sim \sqrt{\frac{1}{N}}, \tag{4.2}$$

which leads to

$$\delta x \sim \frac{c}{2\omega_0}\sqrt{\frac{1}{N}}. \tag{4.3}$$

For a duration of τ, a light beam with power P provides a photon number of

$$N = \frac{P}{\hbar\omega_0}\tau, \tag{4.4}$$

where ω_0 is the angular frequency of light. This leads to

$$\delta x \sim \frac{c}{2\omega_0}\sqrt{\frac{\hbar\omega_0}{P}}\sqrt{\frac{1}{\tau}}. \tag{4.5}$$

Note that for a random process A with a white spectrum S_A, if we measure it for a duration of τ, taking the time average, then the standard deviation of that mean is given by

$$\delta A_\tau \approx \sqrt{\frac{S_A}{\tau}}. \tag{4.6}$$

From this, we can extract a steady-state sensing noise spectrum of

$$S_x^{\text{sens}} \sim \frac{c^2 \hbar}{4\omega_0 P}. \tag{4.7}$$

In a similar fashion, we can also derive the back-action noise. For a duration of τ, the uncertainty of the number of photons hitting the mirror is given by

$$\delta N \sim \sqrt{N} = \sqrt{\frac{P}{\hbar \omega_0} \tau}, \tag{4.8}$$

which means the force acting on the mirror has an uncertainty of

$$\delta F = \frac{2 \hbar \omega_0}{c} \delta N \sim \frac{2}{c} \sqrt{\frac{\hbar \omega_0 P}{\tau}}, \tag{4.9}$$

which corresponds to a spectrum of

$$S_F \sim \frac{4 \hbar \omega_0 P}{c^2}. \tag{4.10}$$

This force noise spectrum causes a back-action noise

$$S_x^{\text{BA}} = \frac{1}{M^2 \Omega^4} \frac{4 \hbar \omega_0 P}{c^2} \tag{4.11}$$

in the displacement. If we superimpose these two types of noise, we obtain

$$S_x^{\text{tot}} = S_x^{\text{sens}} + S_x^{\text{BA}} = \frac{c^2 \hbar}{4\omega_0 P} + \frac{1}{M^2 \Omega^4} \frac{4 \hbar \omega_0 P}{c^2} \geq \frac{2\hbar}{M\Omega^2}. \tag{4.12}$$

This turns out to be equal to the free-mass Standard Quantum Limit. Note that the SQL is at the 'same scale' as the zero-point fluctuations of a harmonic oscillator with eigenfrequency Ω.

Focusing on mechanical quantum state

Let us now turn to the test mass. Suppose, that we would like to measure the position of the mass successively at different moments of time. For this, let us consider the Heisenberg operator $\hat{x}_H(t)$ of a free mass

$$\hat{x}_H(t) = \hat{x}_0 + \frac{\hat{p}_0 t}{M}. \tag{4.13}$$

From this, we obtain

$$\left[\hat{x}_H(t), \hat{x}_H(t')\right] = \frac{i\hbar(t' - t)}{M}. \tag{4.14}$$

These Heisenberg operators do not commute. We cannot 'simultaneously measure' these operators! More precisely speaking, for different moments of time, t and t', we cannot prepare the test mass into a special quantum state which has vanishing quantum uncertainties at both t and t'. In fact, for any test-mass quantum state,

$$\Delta x(t) \cdot \Delta x(t') \geq \frac{\hbar}{2M} |t' - t|. \tag{4.15}$$

If we measure with an interval of τ, and keep Δx the same at all times, we will obtain

$$\Delta x = \sqrt{\frac{\hbar \tau}{2M}}, \tag{4.16}$$

which is consistent with the SQL.

Quantum Non-Demolition (QND)

Both ways toward the SQL have some loopholes. In the first approach, we did not consider the intrinsic dynamics of the test mass, but only the dynamics as driven by light; we did not consider possible correlations between the sensing and the back-action noise, either. We will quickly fix these problems, and arrive at a formalism that will allow us to compute the quantum noise correctly—and find ways to surpass the SQL.

The second approach was adopted earlier in history. It does not seem to be the most efficient way to obtain a computational tool for the noise spectrum, but this argument highlights the problem: if we are to 'measure' non-commuting observables successively, we must add noise. There are *two* ways to see why 'measuring non-commuting observables' is not only tricky but in fact the phrase does not have a unique meaning, until we specify how the measurement is done.

First, from an *information point of view*. If we have $\hat{x}(t)$, with $[\hat{x}(t), \hat{x}(t')] \neq 0$, and a quantum state of the system $|\psi\rangle$, we do not have a unique way to obtain the probability distribution for $\tilde{x}(t)$, the measurement record. In fact, we cannot transcribe $\hat{x}(t)$ for different values of t into different memory units. By contrast, if we had $\hat{Z}(t)$ with

$[\hat{Z}(t), \hat{Z}(t')] = 0$, we could simply project the $|\psi\rangle$ into simultaneous eigenstates of $\{\hat{Z}(t)\}$, and that would provide us with the probability distribution for the measuring record. In this way, the *hand-off* from quantum to classical cannot be done at the level of $\hat{x}(t)$.

Second, from the *dynamics point of view*. If we couple x to another quantum system for 'further processing', e.g. via the interaction Hamiltonian of

$$V = -xf \tag{4.17}$$

we will have, for a linear system (or perturbatively for any system)

$$\hat{x}^{(1)}(t) = \hat{x}^{(0)}(t) + \frac{i}{\hbar} \int_0^t dt' \left[\hat{x}(t), \hat{x}(t')\right] f(t') \tag{4.18}$$

This means the evolution of the actual $\hat{x}(t)$, after being coupled to the measuring device, depends on the details of the device—unless we specify the details of the device, we cannot obtain a reliable description of the measurement process. From the Schrödinger picture, this also means even if we prepare an initial state for the probe, this state will be 'demolished' by the measuring device, in a way that depends on the details of the device.

Historically, one way to circumvent the above problem was to construct and measure commuting observables, or Quantum Non-Demolition (QND) observables, of the test mass. For a free test mass, this will be

$$\hat{x} - \frac{\hat{p}t}{m}, \quad \hat{p} \tag{4.19}$$

and for oscillators, the quadrature operators

$$\hat{x}\cos\omega_m t - \frac{\hat{p}}{m}\sin\omega_m t, \quad \hat{x}\sin\omega_m t + \frac{\hat{p}}{m}\cos\omega_m t. \tag{4.20}$$

This has recently been realized by Keith Schwab's research group (Suh, Weinstein, Lei, Wollman, Steinke, Meystre, Clerk and Schwab, 2014). In addition, since

$$\left[\hat{x}(t), \hat{x}(t')\right] = \frac{i\hbar \sin[\omega_m(t' - t)]}{m\omega_m}, \tag{4.21}$$

one can perform so-called stroboscopic experiments, at moments of time separated by half a period of oscillation. This same strategy has been used in 'Pulsed Optomechanics' (Vanner, Hofer, Cole and Aspelmeyer, 2013). These will be excellent ways to quantify and manipulate the quantum state of the test mass. However, as we shall see later, explicitly constructing QND observables at the oscillator level is not the only way toward improvement of sensitivity.

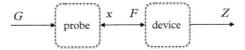

Fig. 4.1 *The probe and the device.*

The full treatment of the problem will always be that the test mass and the light together form a system that is being measured. For this system, we measure the outgoing field at different moments of time, which *automatically commute*, and are trivially QND observables. This does not mean we do not have noise—it just means there does not need to be an additional noise than those already in the optomechanical quantum state.

We have to emphasize that these commuting observables can be found when our aim is to measure classical quantities—or quantum operators which commute. If one insists on 'measuring non-commuting quantum observables that do not commute', then one must deal with the uncertainties that have to arise. One way of quantifying the uncertainties was inequalities derived by Ozawa (Ozawa, 2003).

4.1.3 Linear Quantum Measurement Theory

Let us develop the quantum measurement theory for linear systems more precisely, following Braginsky and Khalili (Braginsky and Khalili, 1992).

Equation of motion

Suppose now we have a system that comprises a probe (test mass) and a device (optical field), each associated with a different Hilbert space, with the total system described by the product space. Suppose we have the Hamiltonian of

$$\hat{H} = \hat{H}_P + \hat{H}_D - \hat{x}\hat{F} - \hat{x}G \tag{4.22}$$

where $\hat{x}$ belongs to the probe (test mass), and $\hat{F}$ belongs to the device (light). Here G is the classical force that we would like to measure. We can write

$$\hat{Z}^{(1)}(t) = \hat{Z}^{(0)}(t) + \frac{i}{\hbar} \int_0^t dt' \, C_{ZF}(t - t') \hat{x}^{(1)}(t'). \tag{4.23}$$

$$\hat{x}^{(1)}(t) = \hat{x}^{(0)}(t) + \frac{i}{\hbar} \int_0^t dt' \, C_{xx}(t - t') \left[G(t') + \hat{F}^{(1)}(t') \right]. \tag{4.24}$$

$$\hat{F}^{(1)}(t) = \hat{F}^{(0)}(t) + \frac{i}{\hbar} \int_0^t dt' \, C_{FF}(t - t') \hat{x}^{(1)}(t'). \tag{4.25}$$

Here we have defined

$$C_{AB}(t - t') = \left[\hat{A}(t), \hat{B}(t') \right]. \tag{4.26}$$

We suppose that $\hat{Z}(t)$ is an outgoing field that can be measured, with

$$\left[\hat{Z}(t), \hat{Z}(t')\right] = 0 \qquad (4.27)$$

Let us consider the special case where

$$C_{ZF} = i\hbar\delta(t - t') \qquad (4.28)$$

so that the outgoing field has a nearly instantaneous response to the motion of the test mass. Let us also assume that

$$C_{FF}(t - t') = \left[\hat{F}(t), \hat{F}(t')\right] = 0. \qquad (4.29)$$

This is as if we have a series of independent harmonic oscillators; each comes in, interacts instantaneously with the test mass for a single moment, and then leaves. In this case, we find

$$\hat{Z}^{(1)}(t) = \hat{Z}^{(0)}(t) + \hat{x}^{(1)}(t), \qquad (4.30)$$

$$\hat{x}^{(1)}(t) = \hat{x}^{(0)}(t) + \frac{i}{\hbar} \int_0^t dt'\, C_{xx}(t - t')[G(t') + \hat{F}^{(0)}(t')]. \qquad (4.31)$$

or

$$\hat{Z}^{(1)}(t) = \hat{x}^{(0)}(t) + \hat{Z}^{(0)}(t) + \frac{i}{\hbar} \int_0^t dt'\, C_{xx}(t - t')\hat{F}^{(0)}(t')$$

$$+ \frac{i}{\hbar} \int_0^t dt'\, C_{xx}(t - t')G(t'). \qquad (4.32)$$

The Heisenberg Uncertainty Relation and the Standard Quantum Limit

Relations (4.27)–(4.29) give rise to the Heisenberg Uncertainty Relation in the spectral domain:

$$S_{ZZ}S_{FF} - |S_{ZF}|^2 \geq \hbar^2. \qquad (4.33)$$

If we calculate the noise spectrum, we will have

$$S_x = S_x^{\mathrm{zp}} + S_{ZZ} + 2\mathrm{Re}\,[R_{xx}S_{ZF}] + |R_{xx}|^2 S_{FF}, \qquad (4.34)$$

where S_x^{zp} is the zero-point fluctuations of the test mass, while the rest is *optical noise* that arises from fluctuations of the incoming optical field. Here we have defined

$$R_{xx}(\Omega) = \frac{i}{\hbar} \int_0^{+\infty} C_{xx}(t) e^{i\Omega t} dt. \tag{4.35}$$

Note that our definition of spectral density is slightly different from the one by Clerk et al. (Clerk, Devoret, Girvin, Marquardt and Schoelkopf, 2010). We have a symmetrized spectrum:

$$\frac{1}{2} S_{AB}(\Omega) 2\pi \delta(\Omega - \Omega') = \left\langle \frac{A(\Omega)B^\dagger(\Omega') + B^\dagger(\Omega')A(\Omega)}{2} \right\rangle \tag{4.36}$$

For a mechanical oscillator with eigenfrequency ω_m, we have

$$R_{xx} = -\frac{1}{m(\omega^2 - \omega_m^2)} \tag{4.37}$$

Here we have taken the limit of an ideal oscillator with an infinite quality factor. The zero-point part is focused near the eigenfrequency of the oscillator. If we measure off-resonance, then we will only be limited by the optical noise, although this noise does bear some imprint from the dynamics of the test mass, since it depends on R_{xx}. In the case when $S_{ZF} = 0$, we will have

$$S_x = S_{ZZ} + |R_{xx}|^2 S_{FF} \geq 2\hbar |R_{xx}|, \tag{4.38}$$

which means

$$S_x^{\mathrm{SQL}} = 2\hbar |R_{xx}| = \frac{2\hbar}{m|\omega^2 - \omega_m^2|}. \tag{4.39}$$

We also have, for a force measurement

$$S_G^{\mathrm{SQL}} = \frac{2\hbar}{|R_{xx}|} = 2\hbar m|\omega^2 - \omega_m^2| \tag{4.40}$$

As we can see here, if we are near the resonance of a harmonic oscillator, the force SQL is particularly low. However, there we will have to be careful about the zero-point fluctuations, as well as thermal noise, in practice. That we will postpone till section 4.4, when we deal with the quantum state of the oscillator. In the rest of the sections, we will mainly focus on the fluctuations of the light—away from the oscillator's eigenfrequency. As we see from the above, if we build the appropriate correlations between Z and F, then the SQL is not enforced. In other words, the test mass is only one degree of freedom, while the light has an infinite number of degrees of freedom. In the frequency domain, all the quantum-ness of the test mass is concentrated near its eigenfrequency—when we are away from that, we will only have to deal with fluctuations of light.

4.2   Various Configurations that Circumvent the SQL

In this section, I will discuss various ways the SQL can be surpassed in GW detectors. With this, we will illustrate concepts such as back-action evasion and the optical spring effect.

4.2.1   Quantization of Light: a Brief Review

In this section, I will discuss the basics of field quantization.

Field operators

For a field propagating along the x direction with speed c, we write, in the Heisenberg picture,

$$\hat{E}(t,x) = \int_0^{+\infty} \frac{d\omega}{2\pi} \sqrt{\frac{2\pi\hbar\omega}{\mathcal{A}c}} \left[a_\omega e^{ikx-i\omega t} + a_\omega^\dagger e^{-ikx+i\omega t} \right] \tag{4.41}$$

For simplicity, here we have taken a transverse mode that is unity within a cylinder with area $\mathcal{A}$, and zero amplitude outside. Here we note that $a_\omega^\dagger$ is the creation operator for a spatial mode whose wavevector is given by $k = \omega/c$, with a spatial wavefunction of e^{ikx}. The creation and annihilation operators satisfy

$$\left[\hat{a}_\omega, \hat{a}_{\omega'}^\dagger \right] = 2\pi\delta(\omega - \omega'), \left[\hat{a}_\omega, \hat{a}_{\omega'} \right] = \left[\hat{a}_\omega^\dagger, \hat{a}_{\omega'}^\dagger \right] = 0. \tag{4.42}$$

The normalization here follows a Gaussian unit system, with energy density given by $E^2/(4\pi)$.

If we study phenomena governed by fields close to a central frequency ω_0, we will be focusing on operators $a_{\omega_0\pm\Omega}$ and $a_{\omega_0\pm\Omega}^\dagger$, where Ω is limited within a small range. It is often convenient to re-organize the operators into quadrature operators:

$$\hat{a}_1(\Omega) = \frac{a_{\omega_0+\Omega} + a_{\omega_0-\Omega}^\dagger}{\sqrt{2}}, \quad \hat{a}_2(\Omega) = \frac{a_{\omega_0+\Omega} - a_{\omega_0-\Omega}^\dagger}{\sqrt{2}i}. \tag{4.43}$$

We can also define, in the time domain,

$$\hat{a}_{1,2}(t) = \int_0^\Lambda \frac{d\Omega}{2\pi} \left[\hat{a}_{1,2}(\Omega)e^{-i\Omega t} + \hat{a}_{1,2}^\dagger(\Omega)e^{i\Omega t} \right], \tag{4.44}$$

where Λ is a cut-off that is much less than ω_0 but much higher than the bandwidth of the process we care about. In this way, field around ω_0 can be written as

$$\hat{E} = \hat{a}_1(t)\cos\omega_0 t + \hat{a}_2(t)\cos\omega_0 t. \tag{4.45}$$

In general, we can define the θ-quadrature as

$$a_\theta = a_1\cos\theta + a_2\sin\theta. \tag{4.46}$$

This can be done either in the time domain or in the frequency domain. This quantization of the quadrature representation of EM fields is also called *two-photon* quantum optics, first formulated by Caves and Schumaker (Caves and Schumaker, 1985; Schumaker and Caves, 1985).

Gaussian states

The simplest quantum states of light are Gaussian states—those whose Wigner functions are Gaussian. The vacuum state is a Gaussian pure state, which satisfies

$$\hat{a}_\omega|0\rangle = 0. \tag{4.47}$$

Using the definition of quadratures and their commutation relations, we can show that in a vacuum state,

$$S_{a_1 a_1} = S_{a_2 a_2} = 1, \quad S_{a_1 a_2} = 0. \tag{4.48}$$

All other Gaussian pure states can be formed through unitary displacement and squeezing operators:

$$D[\alpha(\omega)] = \exp\left\{ \int \frac{d\omega}{2\pi} \left[\alpha(\omega)a_\omega^\dagger - \alpha^*(\omega)a_\omega \right] \right\} \tag{4.49}$$

$$S[\beta(\omega_1,\omega_2)] = \exp\left\{ \iint \frac{d\omega_1}{2\pi}\frac{d\omega_2}{2\pi} \left[\beta(\omega_1,\omega_2)a_{\omega_1}^\dagger a_{\omega_2}^\dagger - \beta^*(\omega_1,\omega_2)a_{\omega_1} a_{\omega_2} \right] \right\} \tag{4.50}$$

It is often convenient to carry out a unitary transformation, which brings the quantum state back to vacuum, but generates a linear transformation on field operators. For coherent states, by 'returning to vacuum state', we simply add coherent amplitude to the field operators,

$$\hat{a}_\omega \to \alpha(\omega) + \hat{a}_\omega, \tag{4.51}$$

while for squeezed states, it is possible to write

$$\hat{a}_\omega \to \int \frac{d\omega'}{2\pi} \left[\mu(\omega,\omega') \hat{a}_{\omega'} + \nu(\omega,\omega') \hat{a}_{\omega'}^\dagger \right] \qquad (4.52)$$

Here μ and ν must be chosen in order to preserve the commutation relations (4.42).

Special coherent and squeezed states in the quadrature representation

As we analyse an optomechanical system, we often restrict it to a special coherent state for the incoming optical field, which only has a coherent amplitude at a single frequency ω_0. After the unitary transformation back to vacuum state, the electric field operator can be written as

$$\hat{E}(t,x) = \sqrt{\frac{4\pi\hbar\omega_0}{Ac}} \left[\left(\sqrt{\frac{2P}{\hbar\omega_0}} + \hat{a}_1(t) \right) \cos\omega_0 t + \hat{a}_2(t) \sin\omega_0 t \right] \qquad (4.53)$$

Physically, the coherent amplitude corresponds to a light beam with power P and angular frequency ω_0; this is often referred to as the carrier field, and ω_0 the carrier frequency. Note that a_1 and a_2 are now amplitude and phase modulations to the carrier. In this way, sometimes they are referred to as the amplitude and phase quadratures.

We shall consider special squeezed states that corresponds to transformations like

$$\hat{a}_\phi \to e^{+\xi} \hat{a}_\phi, \quad \hat{a}_{\phi+\pi/2} \to e^{-\xi} \hat{a}_{\pi/2+\phi}. \qquad (4.54)$$

This can be done either in the time domain or in the frequency domain. In the next section, we will use a particular type of squeezed states, with

$$\hat{S} = \exp\left[\int_0^\infty \frac{d\Omega}{2\pi} \left(\chi_\Omega \hat{a}_{\omega_0+\Omega}^\dagger \hat{a}_{\omega_0-\Omega}^\dagger - \chi_\Omega^* \hat{a}_{\omega_0+\Omega} \hat{a}_{\omega_0-\Omega} \right) \right] \qquad (4.55)$$

By defining $\chi_\Omega \equiv \xi_\Omega e^{-2i\phi_\Omega}$ we have a frequency-dependent squeezing of the quadratures:

$$\begin{pmatrix} \hat{a}_1 \\ \hat{a}_2 \end{pmatrix} = \begin{pmatrix} \cosh\xi + \sinh\xi\cos 2\phi & -\sinh\xi\sin 2\phi \\ -\sinh\xi\sin 2\phi & \cosh\xi - \sinh\xi\cos 2\phi \end{pmatrix} \begin{pmatrix} \hat{a}_1 \\ \hat{a}_2 \end{pmatrix} \qquad (4.56)$$

On the other hand, S generates pairs of photons, and it is not hard to show that each pair has a relative time separation whose distribution is given by the inverse Fourier transform of χ. In this way, if χ is to vary within a narrow bandwidth, this means the two photons are separated by a long time. If we detect one of those photons, then we will be heralding a photon with a long spatial mode.

4.2.2   Light Reflection Off a Moving Mirror

Let us now consider the coupling between light and mirror (Caves, 1980; Kimble, Levin, Matsko, Thorne and Vyatchanin, 2001). Suppose the incident beam is reflected off a mirror which has a displacement of $X(t)$. Suppose X is much less than the wavelength of light, and that $\dot{X}$ is much less than the speed of light; we can then write

$$\hat{E}_{\text{out}}(t) = \hat{E}_{\text{in}}[t - 2X(t)/c] \tag{4.57}$$

We shall consider two special cases where we different linearizations of the dynamics are appropriate. To motivate them, it is instructive to consider the following Hamiltonian for a single optical mode coupled to a movable mirror:

$$\hat{V} \propto -\hat{E}^2 \hat{x} \tag{4.58}$$

In order to convert this cubic term into a quadratic one, we can either write

$$\hat{E} \to \langle \hat{E} \rangle + \delta\hat{E} \tag{4.59}$$

and expand in $\delta\hat{E}$ and $\hat{x}$, which is a linear, optomechanical coupling, or we can write

$$x \to \langle x \rangle \tag{4.60}$$

leaving the $\hat{E}^2$ term, which modifies the optical-field Hamiltonian. It will be for different situations where these two approximations apply separately, as we shall see below.

Strong carrier, small motion: sensing of motion

Let us write, in the quadrature representation

$$\hat{E}_{\text{in}}(t,x) = \sqrt{\frac{4\pi\hbar\omega_0}{Ac}} \left[\left(\sqrt{\frac{2P}{\hbar\omega_0}} + \hat{a}_1(t) \right) \cos\omega_0 t + \hat{a}_2(t)\sin\omega_0 t \right] \tag{4.61}$$

and

$$\hat{E}_{\text{out}}(t,x) = \sqrt{\frac{4\pi\hbar\omega_0}{Ac}} \left[\left(\sqrt{\frac{2P}{\hbar\omega_0}} + \hat{b}_1(t) \right) \cos\omega_0 t + \hat{b}_2(t)\sin\omega_0 t \right] \tag{4.62}$$

In this case, we expand the t inside cos and sin, and obtain

$$\hat{b}_1(t) = \hat{a}_1(t), \quad \hat{b}_2(t) = \hat{a}_2(t) + \frac{2\omega_0\hat{X}(t)}{c}\sqrt{\frac{2P}{\hbar\omega_0}} \tag{4.63}$$

Now, if we detect b_2, we have a shot noise of

$$S_Z = \frac{c^2 \hbar}{8 \omega_0 P},$$
(4.64)

which we obtain by normalizing b_2 such that the coefficient in front of $\hat{X}(t)$ is unity, and taking the spectrum of the error term. We can also compute the spectrum of the back-action force onto the mirror, which is given by

$$F = \frac{2\mathcal{A}}{c} \overline{\frac{\hat{E}^2}{4\pi}} = \frac{2\hbar\omega_0}{c^2} \sqrt{\frac{2P}{\hbar\omega_0}} \hat{a}_1$$
(4.65)

Here the time average extracts only the low frequency components of $\hat{E}^2$. This gives

$$S_F = \frac{8\hbar\omega_0 P}{c^2}$$
(4.66)

Since $\hat{a}_1$ and $\hat{a}_2$ are uncorrelated at the vacuum state, we obtain $S_{FZ} = 0$. This recovers the result given at the beginning of this section. However, we do see that S_{FZ} can be non-zero for squeezed states.

No carrier, big motion: Dynamical Casimir Effect

It is also interesting to consider the case without carrier, but with a displacement that is not very small (but still smaller than the wavelength of light). In fact, let us suppose

$$X(t) = X_0 \cos(2\omega_0 t)$$
(4.67)

We can write

$$
\begin{aligned}
E_{\text{out}}(t) &= \int \frac{d\omega}{2\pi} \sqrt{\omega} a_\omega e^{-i\omega(t - 2X/c)} + \text{h.c.} \\
&= \int \frac{d\omega}{2\pi} \sqrt{\omega} \left[a_\omega e^{-i\omega t} + \frac{i\omega X_0}{2c} a_\omega e^{-i(\omega + 2\omega_0)t} + \frac{i\omega X_0}{2c} a_\omega e^{-i(\omega - 2\omega_0)t} \right] \\
&+ \int \frac{d\omega}{2\pi} \sqrt{\omega} \left[a_\omega^\dagger e^{+i\omega t} - \frac{i\omega X_0}{2c} a_\omega^\dagger e^{+i(\omega + 2\omega_0)t} - \frac{i\omega X_0}{2c} a_\omega^\dagger e^{+i(\omega - 2\omega_0)t} \right]
\end{aligned}
$$
(4.68)

Here we see a conversion between different frequencies. However, in quantum field theory, conversions between positive and negative frequencies have the profound consequence of particle creation! This takes place for all frequencies less than $2\omega_0$.

If we restrict ourselves to the $\omega_0 + \Omega$ component of the outgoing field, with $\Omega \ll \omega_0$, we can write

$$b_{\omega_0 + \Omega} = a_{\omega_0 + \Omega} + \sqrt{\frac{\omega_0 - \Omega}{\omega_0 + \Omega}} \left[\frac{-i(\omega_0 - \Omega)X_0}{2c} \right] a_{\omega_0 - \Omega}^\dagger \approx a_{\omega_0 + \Omega} - \frac{i\omega_0 X_0}{2c} a_{\omega_0 - \Omega}^\dagger$$
(4.69)

This means

$$b_{\pi/4} = \left(1 - \frac{\omega_0 X_0}{2c}\right) a_{\pi/4}, \quad b_{3\pi/4} = \left(1 + \frac{\omega_0 X_0}{2c}\right) a_{3\pi/4} \qquad (4.70)$$

This is squeezing! The squeeze factor is

$$r = \frac{\omega_0 X_0}{2c} = \frac{V_0}{2c} \qquad (4.71)$$

In other words, as we shake the mirror at $2\omega_0$, pairs of photons at $\omega_0 \pm \Omega$ will be created.

This is the Dynamical Casimir Effect: as we shake anything that has a non-zero polarizability, it will radiate photon pairs (Nation, Johansson, Blencowe and Nori, 2012). Another way to understand this is the following: any object with non-zero polarizability couples to the surrounding electromagnetic field, so that the state of the surrounding electromagnetic field is not exactly vacuum, but, instead, polarized by that object. As we shake that object, these polarization photons will escape, causing the Dynamical Casimir Effect.

In reality, it is perhaps very difficult to shake the mirror fast enough in order to detect the photons. However, one way to achieve a similar effect is to use a nonlinear crystal with $\chi^{(2)}$ nonlinearity. Suppose we have a second harmonic of our carrier frequency ω_0, driving a crystal with $\chi^{(2)}$ nonlinearity, its refractive index will be modulated at $2\omega_0$,

$$n = n_0 + n_1 E \cos 2\omega_0 t \qquad (4.72)$$

where n_0 is the refractive index with zero field, n_1 the coefficient of nonlinearity, and E the amplitude of the second harmonic. This creates the same effect of a mirror moving at $2\omega_0$, producing squeezed vacuum. In the terminology of quantum optics, this is squeezing generation via parametric down-conversion (Wu, Kimble, Hall and Wu, 1986).

Extensions of the Dynamical Casimir Effect

The fundamental reason for particle creation that we discussed above is a non-uniform map from emission to arrival time. Any such process will cause particle creation. These types of processes take place in the early universe, when particles are created during inflation; they also take place during gravitational collapse. The collapse of a star into a black hole distorts the space-time geometry in an ultimate way, when a boundary forms between those rays that escape toward infinity and those that do not escape. The Dynamical Casimir Effect in this case gives Hawking Radiation (Hawking, 1975).

Notably, during the early days of studying Hawking Radiation, Bill Unruh showed that an accelerating object in Minkowski vacuum will feel a non-zero temperature, which arises from such particle creations due to the non-uniform map in time between points with fixed spatial locations in the Minkowski coordinate system and the proper time of the accelerating observer (Unruh, 1976). This is called the Unruh Effect. Bill Unruh was also the first one to realize that squeezing can be used to improve detector sensitivity beyond the Standard Quantum Limit (Unruh, 1983).

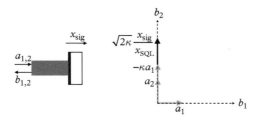

Fig. 4.2 *Left panel: free mirror probed by light beam. Right panel: phasor diagram illustrating the signal and noise content of the output field.*

4.2.3 Free Mirror Probed by Light Beam

In this section, we shall compute the quantum noise spectrum for the simplest optomechanical device.

Single movable mirror

For a free test-mass mirror being probed by a light beam, where a constant external force balances the static force due to the light beam, we simply need to express the mirror's displacement X in terms of the fluctuating part of the back-action force. In the frequency domain, we have

$$\hat{X}(\Omega) = x_{\text{sig}}(\Omega) - \frac{\hat{F}(\Omega)}{M\Omega^2} \tag{4.73}$$

where x_{sig} is the signal displacement. Using this equation and Eq. (4.63), we can write the quadrature input–output relation as

$$\hat{b}_1 = \hat{a}_1 \tag{4.74}$$

$$\hat{b}_2 = \hat{a}_2 - \kappa \hat{a}_1 + \sqrt{2\kappa}\,\frac{x_{\text{sig}}}{x_{\text{SQL}}} \tag{4.75}$$

where

$$x_{\text{SQL}} \equiv \sqrt{S_x^{\text{SQL}}} = \sqrt{\frac{2\hbar}{m\Omega^2}} \tag{4.76}$$

and

$$\kappa = \frac{4\omega_0 P}{mc^2\Omega^2} \tag{4.77}$$

We can use the phasor diagram in Fig. 4.2 to illustrate the situation. Basically, the signal is within the phase quadrature—but in that quadrature we also have the back-

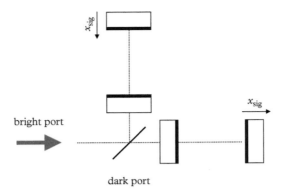

Fig. 4.3 *Fabry–Perot Michelson interferometer for gravitational-wave detection.*

action noise. For a vacuum input state, we can show that the sensing and back-action noises together enforce the free-mass SQL:

$$S_x = \left[\frac{1}{\kappa} + \kappa\right] \frac{S_x^{\text{SQL}}}{2} \geq S_x^{\text{SQL}} \tag{4.78}$$

However, already from here, we see how to circumvent the SQL. If we squeeze the linear combination $a_2 - \kappa a_1$, we will be able to suppress the entire noise by the squeeze factor. If we detect a quadrature other than the phase quadrature, we can also allow the back-action to partially cancel, resulting in noise below the SQL. In both cases, we will have shot noise correlated with radiation-pressure noise. The challenge seems to be that κ is frequency-dependent—hence we will need to have frequency-dependent squeezed vacuum (as mentioned in the previous section) and/or frequency-dependent homodyne detection.

Fabry–Perot Michelson interferometers

Before we go into the details of how to realize such frequency-dependence, let us present a more realistic configuration that involves interferometers.

First of all, we have a Michelson configuration that is at the dark port, in order for classical laser noise to cancel. In this way, the common mode of the interferometer provides the driving, while the differential mode is where the quantum dynamics takes place. When injecting squeezed vacuum, it will be into the dark port. The signal also emerges from the dark port; it can be detected via homodyne or heterodyne detection. We shall assume the former, which basically measures one particular quadrature. If we superimpose the outgoing field with a local oscillator at the θ-quadrature, and then detect the intensity of light, we will obtain b_θ.

Another improvement that we need is resonant cavities in the arms, plus the power recycling cavity (which Ron Drever invented at Les Houches in 1983). Ingredients for obtaining the input–output relation involve the input–output relations of: (i) movable mirror, as already given above, (ii) propagation through free space for a distance of L,

$$\begin{bmatrix} b_1(\Omega) \\ b_2(\Omega) \end{bmatrix} = e^{i\Omega L/c} \begin{pmatrix} \cos\phi & \sin\phi \\ -\sin\phi & \cos\phi \end{pmatrix} \begin{bmatrix} a_1(\Omega) \\ a_2(\Omega) \end{bmatrix} \tag{4.79}$$

where $\phi = \omega_0 L/c$, and (iii) transmission and reflection at a mirror:

$$d_j = \sqrt{R}a_j + \sqrt{T}b_j, \quad c_j = \sqrt{T}a_j - \sqrt{R}b_j. \tag{4.80}$$

In the case of gravitational waves with amplitude h, propagation direction orthogonal to the detector plane and with the optimal polarization, it creates an arm length difference of Lh for free test masses. The input–output relation for a Fabry–Perot Michelson interferometer in presence of this wave is given by

$$b_1 = e^{2i\beta} a_1 \tag{4.81}$$

$$b_2 = e^{2i\beta} [a_2 - \mathcal{K}a_1] + e^{i\beta} \sqrt{2\mathcal{K}} \frac{h_{\mathrm{GW}}}{h_{\mathrm{SQL}}} \tag{4.82}$$

Here

$$\beta = \arctan \frac{\Omega}{\gamma} \tag{4.83}$$

with

$$\gamma = Tc/(4L) \tag{4.84}$$

the cavity bandwidth, and

$$\mathcal{K} = \frac{2\gamma\Theta^3}{\Omega^2(\Omega^2 + \gamma^2)}, \quad \Theta^3 = \frac{8\omega_0 P_c}{mLc}, \tag{4.85}$$

where P_c is the power circulating in the arms. We also have

$$h_{\mathrm{SQL}} = \sqrt{\frac{8\hbar}{m\Omega^2 L^2}} \tag{4.86}$$

This is after accounting for four equal test masses, and response to the GW. The structure of this input–output relation is the same as before—just with a more complex $\mathcal{K}$, which contains not only the poles of a free mass, but also that of the cavity.

4.2.4 Surpassing the SQL: frequency-dependent squeezing and homodyne detection

Let us turn back to the issue of generating the correct frequency-dependence in the squeezed vacuum and performing the frequency-dependent homodyne detection.

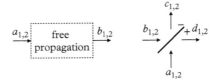

Fig. 4.4 *Propagation through free space and reflection/transmission at a mirror.*

The required frequency-dependence and the optical filters

It is not hard to work out the required frequency-dependence. For input squeezing, suppose we need a device that applies rotation $\phi(\Omega)$ to the input fields, with

$$
\begin{pmatrix} a_1 \\ a_2 \end{pmatrix} = \underbrace{\begin{pmatrix} \cos\phi & \sin\phi \\ -\sin\phi & \cos\phi \end{pmatrix}}_{\text{rotation}} \underbrace{\begin{pmatrix} e^{-r} & \\ & e^{+r} \end{pmatrix}}_{\text{squeezing}} \begin{pmatrix} c_1 \\ c_2 \end{pmatrix}
\tag{4.87}
$$

In order for $a_2 - \mathcal{K}a_1$ to be squeezed, we need to require

$$
\tan\phi_{\text{sqz}} = \frac{1}{\mathcal{K}}
\tag{4.88}
$$

At low frequencies, we squeeze a_1, while at high frequencies, we squeeze a_2.

For frequency-dependent homodyne detection, we first apply frequency-dependent quadrature rotation to the outgoing field, and then make a frequency-independent homodyne detection of the first quadrature:

$$
\begin{pmatrix} d_1 \\ d_2 \end{pmatrix} = \underbrace{\begin{pmatrix} \cos\phi & \sin\phi \\ -\sin\phi & \cos\phi \end{pmatrix}}_{\text{rotation}} \begin{pmatrix} a_1 \\ a_2 \end{pmatrix}
\tag{4.89}
$$

This requires

$$
\tan\phi_{\text{var}} = \frac{1}{\mathcal{K}}
\tag{4.90}
$$

as well. It is also possible to show that such a rotation can be achieved with detuned Fabry–Perot cavities. In fact, we can show that a series of n cavities can always achieve rotation with the following form:

$$
\tan\phi = \frac{\sum\limits_{j=0}^{n} a_j \Omega^{2j}}{\sum\limits_{j=0}^{n} b_j \Omega^{2j}}, \quad a_n b_n \neq 0
\tag{4.91}
$$

Role of optical losses

The above schemes are in principle able to allow the interferometer to evade back-action, and improve sensitivity indefinitely. The limitation will be optical losses. Schemes that are mathematically equivalent without losses could be very different when losses are considered.

First of all, in order to achieve frequency-dependence, the filter cavities must have a bandwidth that is comparable to the detection bandwidth—which for GW detection will be below kHz. Second, in particular for the variational readout scheme, the back-action evading quadrature has very little signal, increasing its susceptibility to losses.

In the variational readout scheme, sensitivity of the interferometer will be highly contaminated by loss. Farid Khalili derived a SQL-beating limit due to optical losses, and it is given by

$$\sqrt{S_h} \geq (e^{-2r}\epsilon)^{1/4}\sqrt{S_h^{\text{SQL}}} \tag{4.92}$$

where ϵ is the loss in power, and e^{-2r} is the power squeeze factor that is injected into the interferometer. For 10 dB squeezing and 1% loss, we can only hope to beat the SQL by 5. However, in the case when $\mathcal{K}$ is too large, the beating factor of 5 is only a maximum—performance at lower frequencies tends to be much worse.

4.2.5 Speed Meters: Modifying the Optical Transfer Function

Velocity versus momentum

The speed meters (Braginsky and Khalili, 1990) are another strategy. The motivation was that momentum is a QND observable—but its advantage really lies in other places.

Momentum is a conserved quantity of a free mass, hence a QND observable. However, let us note that once we couple the velocity of a mass to an external field, there will be the distinction between canonical and kinetic momentum.

$$L = \frac{m\dot{x}^2}{2} \Rightarrow p = m\dot{x}, \tag{4.93}$$

$$L = \frac{m\dot{x}^2}{2} - \alpha\dot{x}a_1 \Rightarrow p = m\dot{x} - \alpha a_1. \tag{4.94}$$

This means, even though canonical momentum is still conserved, it will not be velocity; there is back-action on the velocity! However, there could still be some practical advantages, though.

Optical speed meters

In order for the interferometer to measure the speed of the mirror, we can either add an additional cavity into the dark port (Purdue and Chen, 2002), or build a Sagnac interferometer (Chen, 2003; Danilishin, 2004), as shown in Fig. 4.5. There is in fact a

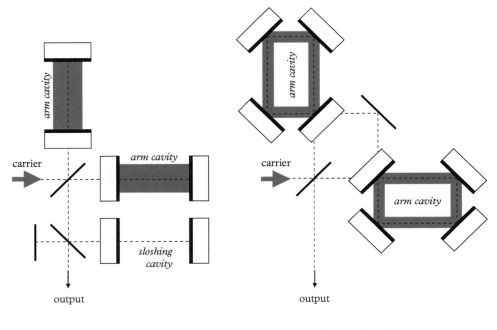

Michelson Speed Meter   Sagnac Speed Meter

Fig. 4.5 *Michelson (left) and Sagnac (right) optical speed meters.*

mathematical mapping between these two types of interferometers, because in both cases we have two optical oscillators coupled with one free mass. In this case, we can achieve

$$\mathcal{K} = \frac{2\gamma\Theta^3}{(\Omega^2 - \omega_s^2)^2 + \gamma^2\Omega^2} \tag{4.95}$$

Here ω_s is the new resonant frequency of the system, and γ the new line width. At low frequencies, $\mathcal{K}$ is a constant, and this represents a response to the velocity of the mass. By choosing ω_s and γ that are comparable, we can have this $\mathcal{K}$ to be approximately constant at low frequencies—this allows the interferometer to already beat the SQL without variational readout or frequency-dependent squeezing, although mathematically the optimal will be to use both schemes above. However, if the same level of loss and squeezing can be achieved, the $\mathcal{K}$ for a speed meter is more advantageous when it comes to the susceptibility to losses.

4.2.6 Optical Spring: Modifying Mechanical Response

Let us now comment on relaxing the assumption (4.29) in Secs 4.1.3. In general, we have $[F(t), F(t')] \neq 0$. This means the optical field can modify the dynamics of the test mass. In fact, if F responds to the motion of the mirror, this is really a spring constant. This effect is familiar to this audience.

Optical transfer function

Historically, in GW detection, the signal recycling technique was introduced to modify the optical transfer function. An additional mirror is placed at the dark port of the interferometer, feeding signal light back into the arm cavities with a phase shift (Meers, 1988; Mizuno, Strain, Nelson, Chen, Schilling, Rüdiger, Winkler and Danzmann, 1993). This phase shift makes displacement signals with a non-zero frequency to be resonant with the interferometer, thereby shifting the location of maximum sensitivity in the frequency domain. One can simply view the differential optical mode as propagating in a single detuned cavity, with bandwidth γ and detuning frequency Δ, or $\omega_0 = \omega_c + \Delta$ (Buonanno and Chen, 2003). Assuming that the mirrors do not move under radiation pressure, this type of configuration will have a sensitivity that is peaked at $\Omega = \Delta$, with a width of γ.

Dynamics

As power is increased in signal recycling interferometers, as it turns out, in addition to having a peak sensitivity at a non-zero frequency, the dynamics of the mirrors are also modified (Buonanno and Chen, 2001; Buonanno and Chen, 2002). This comes from

$$\hat{F}^{(1)}(t) = \hat{F}^{(0)}(t) + \frac{i}{\hbar} \int_0^t dt' C_{FF}(t-t')\hat{x}^{(1)}(t'), \tag{4.96}$$

noting that the dependence of the radiation pressure force on (the history of) the location of the mirror makes an optical spring, which will modify the mirror's dynamics. As we compute $[\hat{F}(t), \hat{F}(t')]$, we obtain the spring constant,

$$\frac{K}{m} = -\frac{\Theta^3 \Delta}{(\Omega + i\gamma)^2 - \Delta^2} \tag{4.97}$$

Near $\Omega \approx 0$, we can always expand

$$K(\Omega) = K_{\text{opt}} - i\gamma_{\text{opt}}\Omega \tag{4.98}$$

where K_{opt} is spring constant at DC, and γ_{opt} is optomechanical damping.

For the 'blue-detuned' case, $\Delta > 0$, we have

$$K_{\text{opt}} > 0, \quad \gamma_{\text{opt}} < 0. \tag{4.99}$$

This is a restoring force with anti-damping. For the 'red-detuned' case, $\Delta < 0$, we have

$$K_{\text{opt}} < 0, \quad \gamma_{\text{opt}} > 0. \tag{4.100}$$

This is a restoring force with anti-damping. This means one cannot make a stable optical trap for a *free mass* with a single optical spring, but can with two (Corbitt, Chen, Innerhofer, Müller-Ebhardt, Ottaway, Rehbein, Sigg, Whitcomb, Wipf and Mavalvala, 2007).

Quantum noise spectrum

It is straightforward to compute the noise spectrum of a signal recycling interferometer in the frequency domain. One simply combines the input–output relation of a single mirror with a detuned cavity. Even if the system is unstable, one can show that if we stabilize its dynamics using feedback, we can still achieve the same sensitivity as calculated in the frequency domain.

For the blue-detuned case, the noise spectrum has a dip near the 'optomechanical resonance', which is the shifted mirror eigenfrequency. In the red-detuned case, the noise spectrum does not have a dip. One can resort to the argument that since we have an oscillator with eigenfrequency in the detection band, sensitivity around that frequency is improved.

4.3   A More Systematic Approach toward Further Sensitivity Improvements

Having seen section 4.2, we know that the SQL can be beaten. However, do we have limits beyond the SQL? We devote this section to a more general bound, which roughly speaking can be seen as arising from the energy–time uncertainty.

4.3.1   The Mizuno Theorem for Interferometers with Free Masses

Let us summarize our understanding here. It seems that without back-action, or optical spring, or in cases with back-action evasion, we have a clear picture. In fact, here I quote the so-called Mizuno Theorem, which states that

$$\int \frac{d\Omega}{2\pi} \frac{1}{S_h(\Omega)} \propto \mathcal{E}, \tag{4.101}$$

that is, this quantity is independent from the optical bandwidth, or detuning; it is simply proportional to the energy stored in the cavity, $\mathcal{E}$. Here we can also see that the figure of merit above is the interferometer's SNR for a pulse with zero width. In fact, if we add squeezing, we might write

$$\int \frac{d\Omega}{2\pi} \frac{1}{S_h(\Omega)} \propto \mathcal{E}e^{2r}, \tag{4.102}$$

where the squeeze factor is added by hand. As one can check, all configurations where test-mass dynamics remain unchanged satisfy this relation. In cases with optical springs,

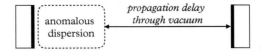

Fig. 4.6 *White-light cavity.*

Haixing Miao and Rana Adhikari have argued that it also applies, if we change e^{2r} into the ponderomotive squeezing due to the optical spring.

The expression (4.101) has also been referred to as the 'Energetic Quantum Limit' (Braginsky, Gorodetsky, Khalili and Thorne, 1999). In fact, in this section we shall show that this is intimately related to the Quantum Cramer–Rao bound (Helstrom, 1967).

4.3.2   White-light Cavities

The Mizuno Theorem was proposed many years ago in the Ph.D. thesis of J. Mizuno at the University of Hannover, when he was studying the general features of signal recycling interferometers. Interest in this theory has recently been revived since it was seemingly broken by *white-light cavities*.

White-light cavities are the ultimate weapon for improving sensitivity of gravitational-wave detectors. As we have seen in previous sections, a GW detector must have a long arm length to have a large displacement signal. However, a long arm gives rise to a delay, which ultimately limits the number of round trips the light can take before the GW signal stops cumulating—there is a trade-off between peak sensitivity and bandwidth. Ultimately, the arm length still wins, but that is why SNR only cumulates like L, instead of L^2.

White-light cavities are cavities that have high finesse but also high bandwidth, thanks to the magic of negative-dispersion devices. They were proposed within the gravitational-wave community in the past (Wicht, Danzmann, Fleischhauer, Scully, Müller and Rinkleff, 1997), with the most recent proposals by Shahriar *et al.* (Zhou, Zhou and Shahriar, 2014). These devices provide a frequency-dependent phase shift of

$$\frac{d\Phi(\omega)}{d\omega} < 0. \qquad (4.103)$$

As we have such a device, the round trip phase of a signal with sideband frequency Ω will be given by

$$\frac{\omega_0 L}{c} + \frac{\Omega L}{c} + \left.\frac{d\Phi(\omega)}{d\omega}\right|_{\omega_0} \Omega \qquad (4.104)$$

If we manage to have

$$\frac{d\Phi}{d\omega} = -\frac{L}{c} \qquad (4.105)$$

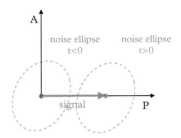

Fig. 4.7 *Instantaneous state displacement due to a δ-function signal.*

then the resonant bandwidth of the cavity will be dramatically enhanced with the same input mirror reflectivity. This will have to violate the Mizuno Theorem—at least the non-squeezed version.

Are they too good to be true?

In fact, if we have a device that has $d\Phi/d\omega = -L/c$ and does not have damping of the signal amplitude, it necessarily has the consequence that a Gaussian wavepacket must come out before it comes in—because, as we might recall, $d\Phi/d\omega$ is also the *group delay* (Wise, Quetschke, Deshpande, Mueller, Reitze, Tanner, Whiting, Chen, Tünnermann, Kley *et al.*, 2005). This is quite interesting!

Classically, for a band-limited signal, it is possible for the peak to come out before it comes in. As we are given the beginning of the signal, we can analyse it, and deduce its further shape, and produce the peak at the output port before the peak even comes in. Nevertheless, for a quantum wave packet, we cannot have the same packet come out before it comes in, because we cannot *clone* a quantum state. This means, what ever negative-dispersion medium we build, if it does provide an output that comes out before the input, it must be highly damped. If we need it not to be damped, we can amplify it, but with additional noise.

At this moment, we might conclude that a white-light cavity is a bad idea—it might create the correct signal transfer function, but the additional noise created might as well cancel the gain due to the opening of the bandwidth. However, we should not jump to conclusions too fast!

4.3.3   Proof of the Mizuno Theorem and the Quantum Cramer–Rao Bound

Thought experiment for the Mizuno Theorem

Our approach is to prove the Mizuno Theorem rigorously. In this discussion, we shall ignore quantum mechanical motion of x, but simply consider the shot-noise-limited sensitivity of our detector.

In order to do so, we imagine that the interferometer is at a steady state without signal, then it has a $\delta(t)$ in displacement. Let us write an interaction Hamiltonian,

$$V = -\hbar G_0(a + a^\dagger)\delta(t) \equiv -\hbar G_0\sqrt{2}\hat{A}\delta(t) \qquad (4.106)$$

where $\hat{A}$ is the amplitude quadrature of the cavity mode, and

$$G_0 = \omega_0\sqrt{\bar{n}} = \sqrt{\omega_0\mathcal{E}/\hbar} \qquad (4.107)$$

where $\bar{n}$ is the mean number of photons in the cavity, and $\mathcal{E}$ the (mean) energy stored in the cavity. Before and after this impulse, the steady quantum state of the cavity mode will be translated by the signal amplitude, yet its shape unchanged, as shown in Fig. 4.7. The unitary evolution is given by

$$U = \exp[i\sqrt{2}G_0\hat{A}] \qquad (4.108)$$

This is a displacement in the phase quadrature

$$\hat{P} \equiv \frac{\hat{a} - \hat{a}^\dagger}{\sqrt{2}i} \qquad (4.109)$$

by $\sqrt{2}G_0$. However, as we detect the outgoing field, we can only extract the impulse from the conditional quantum state of the cavity mode—after accounting for all the outgoing fields.

That accuracy is no better than the conditional fluctuation of the phase quadrature of the cavity mode, yet that is higher than the inverse of the fluctuations in the amplitude quadrature. In this way, we prove the Mizuno Theorem, although we note that the more fluctuations in the amplitude quadrature, the better the SNR.

$$\int \frac{d\Omega}{2\pi} \frac{1}{S_x(\Omega)} = \frac{2G_0^2}{V_{PP}} < 8G_0^2 V_{AA} = \frac{8\omega_0\mathcal{E}V_{AA}}{\hbar} \qquad (4.110)$$

Now, coming back to the white-light cavity case: the additional noise is indeed going to increase fluctuations in the amplitude quadrature—isn't this consistent with the Mizuno Theorem? In fact, the better question to ask is: how do we know that we will have a conditional pure state—and that we can condition it in such a way not to decrease the amplitude-quadrature variance? If we think about it, we can always have a pure state, if we disregard optical losses and thermal noises, and if we account for all the outgoing channels. In the case of the negative-dispersion medium with additional noise, that just means we have an additional outgoing channel, which, after detection and appropriate conditioning, will allow us to go to a pure state for the cavity mode.

To make the above argument rigorous, let us look at the situation with operators at $t = 0^+$. We have the incoming field, the outgoing field, and the cavity mode operators; basically, we have

Fig. 4.8 *Illustrating the information content of fields in the Schrödinger picture.*

$$a_{1,2}(x < 0), \quad a_{1,2}(x > 0), \quad \hat{A}, \quad \hat{P} + \sqrt{2}G_0. \tag{4.111}$$

Here we have used $x < 0$ to label incoming fields, and $x > 0$ outgoing fields. Now, if we detect the entire signal from the output port, we will be measuring some linear combination of the above four entities, e.g.

$$\cos\beta\hat{A} + \sin\beta(\hat{P} + \sqrt{2}G_0) + \int f(x)a_{\theta(x)}\,dx \tag{4.112}$$

This is certainly no better than $\sqrt{2}G_0\sin\beta$ via measuring a conditional fluctuation of $\hat{A}\cos\beta + \hat{P}\sin\beta$. The error is therefore no better than

$$\begin{aligned}
\epsilon &= \frac{V_{AA}^c\cos^2\beta + 2V_{AP}^c\sin\beta\cos\beta + \sin^2\beta\,V_{PP}^c}{2G_-^2\sin^2\beta} \\
&\geq \frac{1}{2G_0^2}\left[V_{PP}^c - \frac{V_{AP}^c V_{AP}^c}{V_{AA}^c}\right] \geq \frac{1}{V_{AA}^c} \geq \frac{1}{8G_0^2 V_{AA}}
\end{aligned} \tag{4.113}$$

Even though this is a lower bound for the error, it does seem reachable in the case where we do not consider back-action.

In this context, the white-light cavity might work, exactly due to the additional noise it imposes. However, we must identify the additional outgoing channels that carry away information that we really need in order to complete the conditioning of the cavity mode. Furthermore, the recent proposal by Peano *et al.* (Peano, Schwefel, Marquardt and Marquardt, 2015), where an internal squeezer in the optical cavity de-amplifies the signal quadrature and amplifies the amplitude quadrature, is right along this approach!

The classical Cramer–Rao Bound

Let us introduce the so-called Quantum Cramer–Rao bound. First of all, there is the classical Cramer–Rao bound, which says that the estimation is related to how the probability distribution changes with respect to the parameter θ.

$$E(\hat{\theta} - \theta)^2 \leq \frac{1}{\mathcal{I}_\theta} \tag{4.114}$$

with

$$\mathcal{I}_\theta = E\left[\left(\frac{\partial \log p(\mathbf{x}|\theta)}{\partial \theta}\right)^2 \middle| \theta\right] = E\left[-\frac{\partial^2 \log p(\mathbf{x}|\theta)}{\partial \theta^2} \middle| \theta\right] \tag{4.115}$$

referred to as the Fisher Information. Now suppose we have

$$x = \theta s(t) + n(t) \tag{4.116}$$

then

$$p(x|\theta) \propto \exp\left[-\frac{1}{2}\langle x - \theta s | x - \theta s\rangle\right] \tag{4.117}$$

with

$$\langle A|B\rangle \equiv \int \frac{d\Omega}{2\pi} \frac{A^*(\Omega)B(\Omega)}{S_n(\Omega)} \tag{4.118}$$

In this way, we have

$$\frac{\partial^2 \log p(\mathbf{x}|\theta)}{\partial \theta^2} = -\langle s|s\rangle \tag{4.119}$$

so the bound is given by

$$E(\hat{\theta} - \theta)^2 \leq \frac{1}{\langle s|s\rangle} \tag{4.120}$$

The Quantum Cramer–Rao bound

Let us write down the quantum version and prove it (Helstrom, 1967). Suppose we have a quantum system with density matrix $\hat{\rho}(\theta)$, which depends on a parameter θ. We would like to provide an estimator $\hat{X}$ of θ, and an unbiased one, with

$$\text{tr}\left[\hat{X}\hat{\rho}(\theta)\right] = \theta \tag{4.121}$$

We will show that there exists a minimum bound for the error we can achieve, which only depends on $\hat{\rho}(\theta)$.

Let us first define the *logarithmic derivative* for $\hat{\rho}$, namely, we write

$$i\frac{\partial \hat{\rho}}{\partial \theta} = \hat{L}\hat{\rho} - \hat{\rho}\hat{L} \tag{4.122}$$

where $\hat{L}$ is a Hermitian operator. (It becomes clear soon why we define in this way.) Then, taking the derivative of (4.121), we obtain

$$\mathrm{tr}\left[-i\hat{X}\hat{L}\hat{\rho} + i\hat{X}\hat{\rho}\hat{L}\right] = 1 \tag{4.123}$$

We also write

$$\mathrm{tr}\left[-i(\hat{X}-\theta)\hat{L}\hat{\rho} + i(\hat{X}-\theta)\hat{\rho}\hat{L}\right] = 1 \tag{4.124}$$

Note that

$$\mathrm{tr}\left[-i(\hat{X}-\theta)\hat{L}\hat{\rho}\right] = \left(\mathrm{tr}\left[i(\hat{X}-\theta)\hat{\rho}\hat{L}\right]\right)^{*} \tag{4.125}$$

This actually means

$$\left|\mathrm{tr}[(\hat{X}-\theta)\hat{L}\hat{\rho}]\right|^{2} \geq \frac{1}{4} \tag{4.126}$$

Using the fact that

$$\mathrm{tr}[A^{\dagger}A]\,\mathrm{tr}[B^{\dagger}B] \geq \left|\mathrm{tr}[A^{\dagger}B]\right|^{2} \tag{4.127}$$

We can write

$$\mathrm{tr}\left[\sqrt{\rho}(\hat{X}-\theta)(\hat{X}-\theta)\sqrt{\rho}\right]\mathrm{tr}\left[\sqrt{\rho}LL\sqrt{\rho}\right] \geq \left|\mathrm{tr}\left[\sqrt{\rho}(\hat{X}-\theta)L\sqrt{\rho}\right]\right|^{2} \geq 1/4. \tag{4.128}$$

or

$$\mathrm{tr}[\rho(\hat{X}-\theta)^{2}] \geq \frac{1}{4\mathrm{tr}[\rho L^{2}]} \tag{4.129}$$

In our case,

$$\hat{\rho}_{0+} = U(\theta)\hat{\rho}_{0-}U^{\dagger}(\theta) \tag{4.130}$$

with

$$U(\theta) = \exp[i\sqrt{2}G_{0}\theta\hat{A}] \tag{4.131}$$

In this way, L is simply given by

$$L = -\sqrt{2}G_{0}\hat{A} \tag{4.132}$$

This gives an error of

$$\epsilon \geq \frac{1}{8 G_0^2 V_{AA}} \tag{4.133}$$

As a more general case than the discussion here, Tsang *et al.* obtained a bound for each frequency Ω—SNR for each frequency is bounded by the anti-squeezing of the cavity mode at that frequency (Tsang, Wiseman and Caves, 2011). In this way, in terms of improving SNR without consideration of optical losses, one must squeeze the quadrature where the signal lives in, and anti-squeeze the amplitude quadrature.

Another interpretation of the Cramer–Rao bound

In the context of linear measurement, we can have another justification why the sensitivity is governed by the fluctuations in the amplitude quadrature—using a reciprocity argument, which was proposed earlier by Yuri Levin, and also discussed by Yiqiu Ma, Bill Kells and David Blair, as well as Peter Saulson. Basically, gravitational waves and optical fields are coupled, via an optomechanical and gravitational interaction Hamiltonian. The mode of the gravitational wave and the optical field interact, mutually inject information. It is possible to argue that the SNR for detecting the gravity signal is related to the power injected to the gravitational field, due to the gravitational radiation of the detector: the more efficient the detector is, in radiating gravitational waves, the more sensitive it potentially can be, given that the correct outgoing field is detected. This argument also extends to the optical-spring case; in fact, Haixing Miao has recently shown that interferometer sensitivity involving the optical spring can also be related to the Cramer–Rao bound, with V_{AA} given by ponderomotive squeezing.

4.3.4 Realizing White-light Cavities using Unstable Filters

Even with white-light cavities in principle possible, we found that the most straightforward implementation involves an unstable filter. For example, Haixing Miao and Yiqiu Ma considered the situation where a cavity is pumped by blue-detuned light $\omega_p = \omega_c + \omega_m$, and sending in the carrier at around $\omega_0 = \omega_c$. One has, approximately (when $\gamma_{\text{opt}} \gg \gamma_m$)

$$\hat{a}_{\text{out}\,1,2}(\Omega) \approx \frac{\Omega + i\gamma_{\text{opt}}}{\Omega - i\gamma_{\text{opt}}} \hat{a}_{\text{in}\,1,2}(\Omega) \tag{4.134}$$

This filter, in the frequency domain, cancels the delay of the arm cavity, and opens up the bandwidth. The fact that the filter is unstable on its own is not a fundamental problem—one can detect an outgoing field from the entire interferometer, and then use feedback to stabilize the system. This control system does not need to introduce noise—but it makes the frequency-domain sensitivity formula (which has a broad bandwidth and high peak sensitivity) still valid (Miao, Ma, Zhao and Chen, 2015).

4.3.5 Summary

In this section, we have shown that the gain in sensitivity to gravitational waves can simply be achieved in two steps:

1. For the optical field to couple strongly to the gravitational wave field, we need to anti-squeeze the intracavity amplitude as much as possible—this can either be injected squeezing or ponderomotive squeezing. Strong coupling increases the Cramer–Rao bound for signal-to-noise ratio.

2. In order to take full advantage of the optomechanical-gravitational coupling, saturating the Cramer–Rao bound, we need to readout all the outgoing optical fields.

4.4 Quantum State Preparation and Verification

In this section, I will discuss how a linear quantum measurement device can also be used to prepare and verify the quantum state of the macroscopic test masses.

4.4.1 Zero-Point Fluctuation of an Oscillator, the Fluctuation-Dissipation Theorem and Optical Cooling

In optomechanics experiments, it is often hard to start with a temperature at which the equilibrium state of the mechanical object is already nearly a pure state (e.g. with thermal occupation number below unity), which requires

$$T_m < \frac{\hbar \omega_m}{k_B T} \tag{4.135}$$

where T_m is the environmental temperature of the mass. Here we have to introduce another property of the oscillator, its quality factor, Q_m, which characterizes the rate at which it relaxes to thermal equilibrium. As we couple an optical field to the mass, its dynamics, as well as its steady state, will change. As we will show in this section, the larger the Q_m, the slower heat transfers to the mass from the bath, and therefore the better chance we can use the optical field to bring the mass to a purer quantum state.

Zero-point fluctuations and the FDT

This will be already covered in Chapter 5 by A. Clerk. I will not go into the details, but simply mention that a way to look at this is that all oscillators can be viewed to have their zero-point fluctuations driven by external fields (see e.g. (Khalili, Miao, Yang, Safavi-Naeini, Painter and Chen, 2012)). The fluctuation-dissipation theorem dictates that, if an oscillator is damped, with

$$m \left(\ddot{x} + 2\gamma_m \dot{x} + \omega_m^2 x \right) = F \tag{4.136}$$

then at thermal equilibrium,

$$S_F = 4m\gamma \hbar\omega \coth\left(\frac{\hbar\omega}{2k_B T}\right) = 8M\gamma\hbar\omega\left(\bar{n} + \frac{1}{2}\right) \qquad (4.137)$$

Here we have assumed velocity damping, where γ is a constant. In the high-temperature limit, we have

$$S_F = 8m\gamma k_B T, \quad k_B T \gg \hbar\omega_m \qquad (4.138)$$

One way to gain some insight into this is to look at the Heisenberg operator of $x(t)$, which reads

$$x_H(t) = \left[x_0(t)\cos\omega_m t + \frac{p_0(t)}{m\omega_m}\cos\omega_m t\right]e^{-\gamma t} + \frac{1}{m\omega_m}\int_0^t dt' \sin\omega_m(t-t')e^{-\gamma(t-t')}F(t')$$

$$(4.139)$$

The momentum, in the Heisenberg picture, will have a similar structure. In this way, unless F has a non-zero two-time commutator, the canonical commutator $[x_H(t), p_H(t)]$ will vanish over time, which is unacceptable.

As Clerk mentions in Chapter 5, Eq. (4.139) has the profound consequence that the zero-point fluctuations of any oscillator are always driven by fluctuations from other degrees of freedom. What about the case when $\gamma = 0$? This does not happen because as soon as the oscillator is coupled to the external field, it will have non-zero damping and non-zero intrinsic linewidths.

Coupling to multiple baths and optical cooling

If we have two baths simultaneously coupled to the test mass, the temperature of the mass will be determined jointly by the two baths:

$$n = \frac{\sum\limits_j \gamma_j n_j}{\sum\limits_j \gamma_j} \qquad (4.140)$$

This was applied by Hirakawa to lower the noise of the readout system of resonant-bar gravitational-wave detectors (Hirakawa, Hiramatsu and Ogawa, 1977). As we couple the optical cavity to the mass, we now have two baths, the optical one and the mechanical one. The optical field is nearly vacuum, since the frequency of the photon is much higher than $k_B T$. The more we replace the bath by the optical one, the less the temperature is the oscillator; this is *damping* and *cooling*. If the oscillator mainly damps to the optical bath, with $\gamma_{\text{opt}} \gg \gamma_m$, we will have

$$n \approx n_m \frac{\gamma_{\text{m}}}{\gamma_{\text{opt}}} + n_{\text{opt}} \qquad (4.141)$$

Here n_m is the starting occupation number of the mass, while n_{opt} is the occupation due to the optical bath, which should be computed from quantum fluctuations of light. In order to achieve damping, we need to have the pumping field red detuned from the cavity resonance, see Eq. (4.97). The case with $\omega_p = \omega_c - \omega_m$ is the most intuitive, and in this case, for low optical powers (Marquardt, Chen, Clerk and Girvin, 2007),

$$n_{\text{opt}} = \frac{1}{4}\left(\frac{\gamma}{\Delta}\right)^2 \tag{4.142}$$

Low optical power actually applies approximately till $\gamma_{\text{opt}} \approx \omega_m$. In this case, the final oscillator has a very low quality factor, but this is when the thermal occupation number is a minimum, which is n_m/Q_m. In this way, it is often said that resolved-sideband cooling can reach occupation number below unity if

$$\frac{\hbar\omega_m Q_m}{k_B T_m} > 1. \tag{4.143}$$

For this reason, the 'Q-f product' is often quoted as a figure of merit for the oscillator's potential to be prepared into nearly pure quantum states.

However, when optomechanical coupling is strong, the Q-f product can *significantly underestimate* the potential. In addition to producing damping, the optical field can also increase the eigenfrequency of the mechanical oscillator, to a new ω_{opt}, which is much greater than ω_m. As we do so, if the mechanical damping rate is kept the same, the contribution of thermal occupation will decrease by $\omega_{\text{opt}}^2/\omega_{\text{m}}^2$. Note that one factor of $\omega_{\text{opt}}/\omega_m$ arises from the increase of the resonant frequency, while the other arises from an increase of the Q value. In other words, the requirement for the Q-f product is lowered by a factor of $\omega_{\text{opt}}^2/\omega_{\text{m}}^2$. This can be referred to as optical *trapping and cooling* (Bhattacharya and Meystre, 2007).

It is interesting to draw the analogy between optical trapping and the so-called 'dilution' effect in suspended pendulums (Saulson, 1994). Note that a pendulum suspended from an elastic wire gets its restoring force mainly from gravity (in most textbooks the eigenfrequency of a simple pendulum does not depend on the property of the suspension wire)—only a small fraction of the pendulum's potential energy is stored in elastic energy, which is lossy. In this way, a pendulum's eigenfrequency is much larger than the frequency determined by the elasticity of the wire and the mass of the test mass, and its Q value is much larger than the intrinsic Q factor of the elastic material.

4.4.2 Quantum Measurement of the Mass: Stochastic Schrödinger/Master Equations

The cooling and cooling-trapping mentioned in the previous section can bring the oscillator into nearly pure quantum states; the optical field modifies the dynamics of the oscillator and provides an additional, very-low-temperature bath.

Another way to bring the test mass into a nearly pure state is through quantum measurement. Such a strategy also exists in classical statistical physics. Suppose we have a classical oscillator at thermal equilibrium with temperature T_m; if we measure its position and momentum at time $t = 0$, then momentarily we know its state precisely. At $t > 0$, position and momentum uncertainties of the mass will start off being very low, growing only gradually, reaching equilibrium value only at the relaxation timescale, Q_m/ω_m. A continuous measurement will allow us to maintain low values of uncertainty; the mass will appear to have a low temperature. Mathematically, the distribution of position and momentum uncertainty, taking into account the measurement result, is called *conditional distribution*. In this way, we are performing a *conditional* state preparation on the mass.

Starting from this section, we will study the quantum mechanical version of conditional state preparation, where we have a trade-off between sensitivity in position and back-action in momentum. In this section, we shall review the general stochastic formalism for treating a continuous quantum measurement processes.

The form of the stochastic Schrödinger/master equations

For a system undergoing a Markovian measurement process on the operator $\hat{x}$ with measurement strength α, we can write the stochastic Schrödinger equation (SSE) (Wiseman and Milburn, 2010):

$$d|\psi\rangle = -\frac{i}{\hbar}\hat{H}|\psi\rangle dt + \frac{\alpha(\hat{x} - \langle\hat{x}\rangle)}{\sqrt{2}}|\psi\rangle dW - \frac{\alpha^2}{4}(\hat{x} - \langle\hat{x}\rangle)^2|\psi\rangle dt \qquad (4.144)$$

$$dy = \alpha\langle\hat{x}\rangle + dW/\sqrt{2} \qquad (4.145)$$

Here dW is the 'Wiener increment'; basically dW/dt is a white noise. For a density matrix, we can also replace Eq. (4.144) above by the stochastic master equation (SME):

$$d\hat{\rho} = -\frac{i}{\hbar}\left[\hat{H}, \hat{\rho}\right]dt + \frac{\alpha}{\sqrt{2}}\left\{\hat{x} - \langle\hat{x}\rangle, \rho\right\}dW - \frac{\alpha^2}{4}\left[\hat{x}, [\hat{x}, \hat{\rho}]\right]dt. \qquad (4.146)$$

The third term on the right-hand side of the SME (4.146) is the Lindblad term. If we throw away the measurement data (i.e. taking expectation value over dW), we will be left with a master equation with a Lindblad term, which is usually used to describe decoherence. Note that this term describes the effect of a force noise with a white spectrum.

Derivation of the SSE/SME

Let us recall how the SME/SSE are derived, as illustrated by the left panel of Fig. 4.9. For an interval from t to $t + dt$, we will consider the initial state of the system and the optical field at t. Note that the optical field is those degrees of freedom that will interact with the system during this small interval. We shall assume that the initial state of the joint system is a product state,

$$\rho(t) = \rho_{\text{sys}}(t) \otimes \rho_{\text{opt}}(t) \qquad (4.147)$$

During the interval, the joint system evolves into a new system that contains the mass and the outgoing field. The incoming field disappears—the state of these degrees of freedom are replenished by the newly incoming field. If we measure the outgoing field projectively, we will project the mass into a new state.

It is important to remember that the SSE/SME are only good at dealing with Markovian systems. For a non-Markovian scenario, the system can be quantum mechanically entangled with the environment, and we cannot simply treat its evolution independently. One example is when we use single photon states to drive a cavity, yet the photons have coherence time longer than the storage time of the cavity—in this situation, the photon can be 'half inside the cavity and half outside' (see the right panel of Fig. 4.9). One can always enlarge the system in order to make it Markovian. In this example situation, one needs to incorporate not only the cavity, the mechanical oscillator, but also the continuum of field modes outside of the cavity.

Two physical contexts

There are two contexts where we get to use the SME/SSE. In the first one, we use these equations to simulate the statistical distribution of possible outcomes of the evolution of the quantum state under continuous measurement. Here the dW needs to be randomly generated.

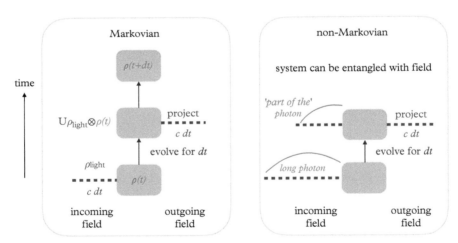

Fig. 4.9 *Markovian versus non-Markovian measurement processes illustrated in a space-time diagram (time goes up, incoming field on the left side, while outgoing field on the right side). In a Markovian process (left panel), the system has a stand-alone density operator at each given time t. Incoming field that enters the system from t to t + dt is independent from other degrees of freedom; these 'new' degrees of freedom will be 'removed' at the end, as we trace out the outgoing fields that exit the system during the same interval. In a non-Markovian process (right panel), the system can become entangled with the incoming optical field— the system's density matrix does not have a self-contained evolution. For example, if a 'long photon' with an extended wavefunction is to enter the system, then the system must be treated together with the incoming field.*

In the other one, we use these equations to help us achieve an ongoing estimation of the quantum state of the system during an experiment. Here we keep an estimate of the density matrix or quantum state of the system, and use it to compute the expectation of the measurement outcome; the difference between the true outcome and the expected outcome will yield dW, which we use to update our estimate of the system's quantum state.

4.4.3 Wiener Filter and Conditional Quantum State Preparation

If we have a linear system, we have analytical solutions to the equations of motion. In this case, we can solve the conditioning problem using a different approach. At each moment in time, to obtain the conditional quantum state of the mass, we need to estimate the distribution of $(\hat{x}(t), \hat{p}(t))$ based on the history of measurements of some output field $\{\hat{y}(t') : t' < t\}$. Classically, for normal distributions, we can achieve this through linear regression. This turns out also to work for Gaussian states in the quantum situation, since the operators $\{\hat{y}(t') : t' < t\}$ all commute with each other, and they all commute with both $\hat{x}(t)$ and $\hat{p}(t)$.

Wigner function and linear regression

For this, we need to study the Wigner function, which is a quasi-probability distribution for the position and momentum of the mass. Formally, we define

$$W(x,p) = \iint \frac{d\mu}{2\pi} \frac{d\nu}{2\pi} e^{-i(\mu x + \nu p)} \text{tr}\left[e^{i\mu\hat{x} + i\nu\hat{p}} \hat{\rho} \right] \tag{4.148}$$

which takes the same form as a joint distribution for x and p. We can show that any *marginal distribution* of $\alpha x + \beta p$, obtained from W, is the same as the quantum distribution of $\alpha\hat{x} + \beta\hat{p}$. Conversely, if we obtain all the distributions of $\alpha\hat{x} + \beta\hat{p}$, we can construct $W(x,p)$, using the Radon transform, in the same way a computed tomography (CT) scan can be constructed. We note that for some quantum states, the Wigner function will not be positive for all values of (x,p). Those states can be viewed as having no classical counterparts.

Let us now digress and discuss linear regression in classical statistics. Suppose we have a random variable x, and a list of random variables $z_1, \ldots, z_N$. If these variables follow a Gaussian distribution, we can then use a linear combination of $z_1, \ldots, z_N$ to best estimate x:

$$\sum_j k_j z_j \tag{4.149}$$

Here this linear combination turns out to be the conditional expectation of x, given $z_1, \ldots, z_N$, or

$$E[x|z_1,\ldots,z_N] = \sum_j k_j z_j \qquad (4.150)$$

and this turns out to be the linear estimator with the minimum mean-square error

$$E\left[\left(x - \sum_j k_j z_j\right)^2\right] = \min_{\alpha_j} E\left[\left(x - \sum_j \alpha_j z_j\right)^2\right] \qquad (4.151)$$

and this also satisfies

$$E\left[\left(x - \sum_j k_j z_j\right) z_m\right] = 0, \quad \forall m \qquad (4.152)$$

One way to obtain the coefficients k_j is to use a two-step procedure. First, obtain linear combinations of z_j, or re-define z_j so that they satisfy

$$\langle z_j | z_k \rangle = \delta_{jk} \qquad (4.153)$$

and then define

$$k_j = \langle x | z_j \rangle \qquad (4.154)$$

Gaussian quantum state preparation

Armed with tools of linear regression and the Wigner function, Gaussian quantum state preparation can be studied in a similar manner to classical Gaussian statistics (Müller-Ebhardt, Rehbein, Schnabel, Danzmann and Chen, 2008; Müller-Ebhardt, Rehbein, Li, Mino, Somiya, Schnabel, Danzmann and Chen, 2009). For outgoing field $\hat{y}(t)$, obtained from $-\infty < t' < t$, we can use it to construct the best estimator for x. In this way, given the measurement result of $y(t)$, we obtain

$$x = \int_{-\infty}^t K_x(t - t')y(t')dt' \qquad (4.155)$$

Similarly, we obtain

$$p = \int_{-\infty}^t K_p(t - t')y(t')dt' \qquad (4.156)$$

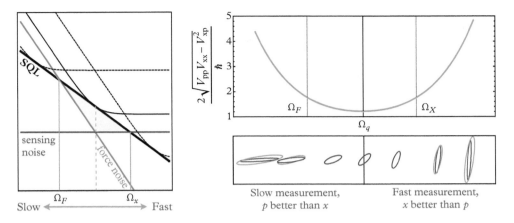

Fig. 4.10 *Illustration of conditional state preparation for a free test mass, in presence of white force noise (causing a $1/\Omega^2$ noise in displacement) and white sensing noise. In the left panel, we illustrate the situation where both classical noise sources together leave a window below the SQL, with force-induced noise crossing the SQL at Ω_F, and sensing noise crossing SQL at Ω_x, $\Omega_F < \Omega_x$. The quantum noise of the light touches the SQL at a frequency Ω_q. In the right panel, we show that the conditional state is nearly pure when Ω_q lies within the window of $[\Omega_F, \Omega_x]$.*

The conditional covariance of $\mu x + \nu p$ is obtained as the variance of

$$\mu \hat{x} + \nu \hat{p} - \int_{-\infty}^{t} [\mu K_x(t-t') + \nu K_p(t-t')]\hat{y}(t') \tag{4.157}$$

From this, we can reconstruct the Wigner function of the conditional state.

Example

As an example, for a free mass under measurement from a light beam in a coherent state, we have

$$-M\Omega^2 \hat{x} = \hbar \alpha \hat{a}_1, \quad b_2 = a_2 + \alpha x, \quad \alpha^2 = M\Omega_q^2. \tag{4.158}$$

Suppose we measure b_2, the output field quadrature that contains maximum signal, we can have

$$K_x(t) = \sqrt{2}\Omega_q e^{-\Omega_q t/\sqrt{2}} \cos \frac{\Omega_q t}{\sqrt{2}} \tag{4.159}$$

$$K_p(t) = \sqrt{2}M\Omega_q e^{-\Omega_q t/\sqrt{2}} \cos \left(\frac{\Omega_q t}{\sqrt{2}} + \pi/2 \right) \tag{4.160}$$

and the conditional covariance matrix is given by

$$
\mathbf{V} = \begin{pmatrix} \dfrac{\hbar}{\sqrt{2}M\Omega_q} & \hbar/2 \\ \hbar/2 & \hbar M\Omega_q/\sqrt{2} \end{pmatrix}
\tag{4.161}
$$

Note that the scale of the quantum state is given by Ω_q, the frequency at which measurement noise touches the free-mass SQL.

In the case with both classical force and sensing thermal noise—assuming both are white—we can use two further frequencies Ω_F and Ω_x, to represent the frequencies at which these noise spectra are equal to the SQL. If $\Omega_F < \Omega_x$, we will have a window in which classical noise is below the SQL. In this region, the conditional state we obtain is nearly pure, with

$$
\frac{V_{xx}V_{pp} - V_{xp}^2}{\hbar^2/2} = \left(1 + \frac{2\Omega_F^2}{\Omega_q^2}\right)\left(1 + \frac{2\Omega_q^2}{\Omega_x^2}\right)
\tag{4.162}
$$

Extensions of conditional Gaussian state preparation

Feedback control

One extension of conditional Gaussian state preparation is to consider feedback control—we can take the measurement result and feed it back onto the mass as a classical force, achieving a much lower uncertainty on the actual motion of the mass (Miao, Danilishin, Müller-Ebhardt and Chen, 2010*a*). Since the feedback control generates a predictable outcome that is superimposed onto the motion of the system, the conditional uncertainty of the test mass motion remains the same—however, the actual motion of the mass becomes much quieter. However, the state achieved by control is often not as pure as the conditional state. Note that since $p = m\dot{x}$, at a steady state, we must have $\langle xp + px \rangle = 0$. If the conditional state has a non-zero covariance between x and p, as is often the case (see the above example), this constraint necessarily means that the steady state produced by control must have strictly larger uncertainty than the conditional state. In fact, one can show that all possible states achievable by control are those whose covariance matrix has no x-p correlation and is strictly larger than the conditional covariance matrix (i.e. when viewed as quadratic forms).

The limitation of control can also be seen as arising from the fact that for a force feedback, with a term $-f_c\hat{x}$ in the Hamiltonian, it can directly modify the momentum of the mass, but only indirectly modify the position, via changing momentum. This makes it generally impossible to instantaneously actuate onto both the position and momentum of the test mass and make its state agree with the conditional state. If we were to add another control term $-g_c\hat{p}$, we might be able to achieve a controlled state that has the same covariance matrix as the conditional state. Note here that the above steady-state argument will not apply since with a Hamiltonian that includes a linear term in p,

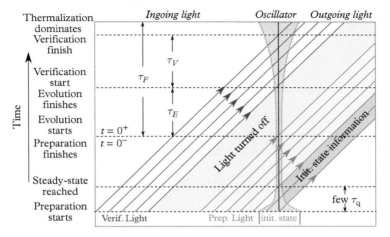

Fig. 4.11 *Space-time diagram of the preparation-evolution-verification process. Ideally, one can prepare the oscillator into a pure quantum state, let it evolve unitarily for a duration, and then independently measure its state at the end of the evolution via a verification process.*

p will no longer be proportional to $\dot{x}$, hence x and p can have non-zero correlations at a steady state.

Experimentally, feedback control has been applied to optomechanical systems with various scales, from gravitational wave detectors (Abbott, Abbott, Adhikari, Ajith, Allen, Allen, Amin, Anderson, Anderson, Arain *et al.*, 2009) to nanomechanical oscillators (Gavartin, Verlot and Kippenberg, 2012; Krause, Blasius and Painter, 2015).

Entangling macroscopic test masses.

As we have multiple macroscopic degrees of freedom, they can be prepared into entangled states. One example is to measure common and differential modes of a Michelson interferometer with different strengths—this will be an EPR-type entanglement (Müller-Ebhardt, Rehbein, Schnabel, Danzmann and Chen, 2008; Müller-Ebhardt, Rehbein, Li, Mino, Somiya, Schnabel, Danzmann and Chen, 2009; Schnabel, 2015).

Preparation of non-Gaussian quantum states

Macroscopic test masses can also be prepared into non-Gaussian states, in particular ones with Wigner functions that are not positive definite. Gaussian states are natural for linear systems driven by—and projected onto—Gaussian quantum states, which include coherent states, squeezed states and thermal states. In this way, in order to generate non-Gaussian states, we can either modify the input optics (e.g. injecting of non-Gaussian states) (Khalili, Danilishin, Miao, Müller-Ebhardt, Yang and Chen, 2010), the output optics (e.g. measuring operators not linear in field amplitudes), and/or the dynamics of the system.

4.4.4   Quantum State Tomography

Through conditional quantum state preparation, we obtain our best knowledge of the quantum state of the object we have prepared. More specifically, at time t, based on our model for the device and the various noise spectra, we can compute the conditional state of the system from the measurement outcome during $t' < t$. If we measure the outgoing field during $t' > t$, we will have additional knowledge about the quantum state of the system at t. In fact, if we repeatedly prepare the same state (according to our model of the system), and measure the distributions of a set of variables of the system via the future outgoing field, the future data will provide an independent test of the state prepared (Miao, Danilishin, Müller-Ebhardt, Rehbein, Somiya and Chen, 2010*b*). This process is shown in the space-time diagram in Fig. 4.11.

Let us consider a simple model, namely a harmonic oscillator under measurement with a beam, and ignore all classical noise. In the time domain, we write

$$b_1 = a_1, \quad b_2 = a_2 + \alpha x(t) \tag{4.163}$$

while

$$x(t) = x(0)\cos\omega_m t + \frac{p(0)\sin\omega_m t}{m\omega_m} + \frac{1}{m\omega_m}\int_0^t \sin\omega_m(t - t')\alpha a_1(t'). \tag{4.164}$$

We suppose that the distribution of $(x(0), p(0))$ is obtained through conditioning on data during $t < 0$. For $t > 0$, we can measure one outgoing quadrature at each moment in time, and linearly combine these results with a weight, resulting, in general, with

$$\xi \equiv \int_0^{+\infty} [g_1(t)b_1(t) + g_2(t)b_2(t)]\, dt \tag{4.165}$$

Here,

$$\tan\theta(t) \equiv g_2(t)/g_1(t) \tag{4.166}$$

determines the quadrature we measure at each moment, while the overall normalization can be tuned by post-processing. We note that ξ contains the test-mass quadrature of

$$\left[\int_0^{+\infty} \alpha g_2(t)\cos\omega_m t\right] x(0) + \left[\int_0^{+\infty} \alpha g_2(t)\frac{\sin\omega_m t}{m\omega_m}\right] p(0) \tag{4.167}$$

and the noise content of

$$\int_0^{+\infty} dt \left[g_1(t)a_1(t) + g_2(t)a_2(t) + g_2(t)\int_0^t dt'\frac{\alpha\sin\omega_m(t - t')}{m\omega_m}a_1(t')\right] \tag{4.168}$$

Note that the first two terms correspond to sensing (shot) noise, while the third corresponds to back-action (radiation-pressure) noise.

From Eqs. (4.166) and (4.167), we see that, given the same $\theta(t)$, we can extract all possible test-mass quadratures via post-processing; i.e. re-scaling $g_{1,2}(t)$ together with the same time-dependent factor. However, the error for the joint distribution of x and p contained in ξ will necessarily be larger than Heisenberg uncertainty, because we are inferring x and p using the same set of measurements. In order to achieve sub-Heisenberg error, we will need to measure different sets of outgoing quadratures. A particularly interesting approach can be obtained by re-arranging the third term in Eq. (4.168), which indicates that there exists a $g_1(t)$ which makes the first and third terms cancel each other. This will give us a back-action-evading measurement of the test-mass quadrature. Note that such a time-dependent homodyne detection strategy had been considered earlier as a strategy for the back-action-evading detection of a signal with known shape and arrival time (Vyatchanin and Zubova, 1995).

In this way, we simply need to focus on optimizing g_2 for each particular quadrature— keeping in mind that g_1 will always be chosen to cancel the back-action noise. Our particular model here will lead to an infinitesimal error for all test-mass quadratures if we allow an infinite measurement time: signal increases linearly in time, while noise only with the square root. In realistic situations, integration time will be limited by thermal noise, which makes long-time integration unfavorable (Miao, Danilishin, Müller-Ebhardt, Rehbein, Somiya and Chen, 2010*b*).

4.5   Testing Quantum Mechanics

In this final section, let us discuss how to 'test' quantum mechanics, exploring how to look for small deviations from quantum mechanics.

4.5.1   Collapse Models

Motivations

Collapse models are constructed to provide a common framework for the reduction of quantum states—so that states of macroscopic systems will spontaneously reduce, in an objective way (Ghirardi, Rimini and Weber, 1986; Bassi, Lochan, Satin, Singh and Ulbricht, 2013). From the relative-state, or 'many-world' interpretation of quantum mechanics, this means one of the worlds gets chosen, among all possible ones that are contained in the global wavefunction of the universe. Because we do not know *which* one of the worlds is chosen, this approach does not get rid of the randomness. The way to detect such a choice is to use the consequence of the state reduction. More specifically, if positions of particles are the observables which are localized in the reduction process, then the momenta of these particles will be disturbed, causing additional randomness—that is, additional randomness because we are not given the way the choice is made. This process is therefore modelled by a measurement process on certain variables—with measurement results unknown to us. This will therefore appear as an anomalous decoherence process.

In general, we can model decoherence by adding Lindblad terms into the master equation. For example, for a single observable, we can write

$$\dot{\rho} = \frac{i}{\hbar}[H,\rho] - \frac{\lambda}{2\hbar^2}[L,[L,\rho]] \tag{4.169}$$

This easily generalizes to a list of observables,

$$\dot{\rho} = \frac{i}{\hbar}[H,\rho] - \frac{1}{2\hbar^2}\sum_{jk}\lambda_{jk}[L_j,[L_k,\rho]] \tag{4.170}$$

Dynamically, this corresponds to adding the following term

$$\sum_j f_j(t)L_j \tag{4.171}$$

into the Hamiltonian, with f_j stochastic forces that have the following form of correlation functions:

$$\langle f_j(t)f_k(t')\rangle = \lambda_{jk}\delta(t-t'). \tag{4.172}$$

In the frequency domain, these forces have white cross-spectra. This is consistent with the fact that master equations model Markovian processes. In the single-sided convention for spectral density,

$$S_{f_jf_k} = 2\lambda_{jk} \tag{4.173}$$

In other words, λ_{jk} is 1/2 the single-sided cross-spectrum, and is equal to the double-sided cross-spectrum.

The Continuous Stochastic Localization (CSL) model

The so-called Continuous Stochastic Localization (CSL) model was first proposed by Ghirardi, Rimini and Weber (Ghirardi, Rimini and Weber, 1986), and was later discussed in the literature with many different modifications (Bassi, Lochan, Satin, Singh and Ulbricht, 2013). The most recent attempt to explore the consequence of CSL in optomechanics was carried out by Nimmrichter, Hornberger and Hammerer (Nimmrichter, Hornberger and Hammerer, 2014) and Diosi (Diósi, 2015). In that study, they have the following Lindblad term:

$$\mathcal{L}_{\text{CSL}} = \frac{\lambda_{\text{CSL}}}{\pi^{3/2} r_{\text{CSL}}^3 \text{amu}^2} \int d^3 \mathbf{s} \left[\hat{m}(\mathbf{s}) \rho \hat{m}(\mathbf{s}) - \frac{1}{2} \left\{ \rho, \hat{m}^2(\mathbf{s}) \right\} \right],$$

$$\hat{m}(\mathbf{s}) = \sum_n m_n \exp \left[-\frac{(\mathbf{s} - \mathbf{r}_n)^2}{2 r_{\text{CSL}}^2} \right] \tag{4.174}$$

One can slightly modify m as dependent on matter distribution, with

$$\hat{m}(\mathbf{s}) = \int \hat{\mu}(\mathbf{z}) e^{-(\mathbf{s}-\mathbf{z})^2/(2 r_{\text{CSL}}^2)} d^3 \mathbf{z} \tag{4.175}$$

where $\hat{\mu}(\mathbf{z})$ is the mass density operator at position $\mathbf{z}$. Note that the mass density operator depends on the position operators of all the ingredients of the mass. Suppose the density matrix for the entire system is a product between the centre-of-mass (CM) density matrix and internal-motion density matrix,

$$\rho = \rho_{\text{int}} \otimes \rho_{\text{CM}} \tag{4.176}$$

we can expand the field operator $\hat{m}(\mathbf{s})$ around small CM motion around 0,

$$\hat{m}(\mathbf{s}) = \hat{m}_0(\mathbf{s}) - \sum_{j=1}^{3} \frac{\partial \hat{m}_0(\mathbf{s})}{\partial s_j} \hat{x}_j \tag{4.177}$$

Here $\hat{m}_0(\mathbf{s})$ is $\hat{m}(\mathbf{s})$ when the CM is fixed at 0, and $\hat{m}_0$ only depends on the internal state of the object, which oscillates at a much higher frequency, and is independent from CM motion. We will first average over the internal motion, and use the expectation value of $\hat{m}_0(\mathbf{s})$

$$\bar{m}_0(\mathbf{s}) \equiv \langle \hat{m}_0(\mathbf{s}) \rangle_{\text{int}} \tag{4.178}$$

This will allow us to write

$$\mathcal{L}_{\text{CSL}} \rho_{\text{CM}} = -\frac{1}{2\hbar^2} \sum_{jk} \Lambda_{jk}^{\text{CSL}} [\hat{x}_j, [\hat{x}_k, \hat{\rho}_{\text{CM}}]], \tag{4.179}$$

with

$$\Lambda_{jk}^{\text{CSL}} = \frac{\lambda_{\text{CSL}} \hbar^2}{\pi^{3/2} r_{\text{CSL}}^3 \text{amu}^2} \int d^3 \mathbf{s} \frac{\partial \bar{m}_0(\mathbf{s})}{\partial s_j} \frac{\partial \bar{m}_0(\mathbf{s})}{\partial s_k} \tag{4.180}$$

Here amu stands for the atomic mass unit. From this, one can see that the contribution to the decoherence rate only comes from boundaries across different materials. More specifically, for any uniform object (i.e. with atomic separation much less than r_{CSL})

whose geometrical features are at lengths larger than r_{CSL}, the field m can only be non-vanishing near the surface, pointing opposite to the outgoing normal direction of the surface. In fact, we can show that

$$\nabla m_0(\mathbf{s}) = -2\pi r_{\text{CSL}}^2 \mu_0 e^{-l^2/r_{\text{CSL}}^2} \mathbf{n} \tag{4.181}$$

where l is the distance between $\mathbf{s}$ and the surface, $\mathbf{n}$ is the outgoing normal of the object, and μ_0 is the mean density of the object. In this way, for an object with a shape describable by an area A translating with a distance D along its normal direction, the Λ^{CSL} for the object moving along the axis is given by

$$\Lambda^{\text{CSL}} = \frac{8\pi A r_{\text{CSL}}^2 \lambda_{\text{CSL}} \mu_0^2 \hbar^2}{\text{amu}^2} \tag{4.182}$$

This leads to a force spectrum of

$$S_F^{\text{CSL}} = \frac{16\pi A r_{\text{CSL}}^2 \lambda_{\text{CSL}} \mu_0^2 \hbar^2}{\text{amu}^2} \tag{4.183}$$

The Diosi–Penrose model

The Diosi–Penrose model differs in the term (Penrose and Jorgensen, 2006; Diósi, 2007),

$$\mathcal{L}\rho = -\frac{1}{32\pi^2 \hbar G} \sum_{jk} \delta_{jk} \int d^3\mathbf{s} \left[g_j(\mathbf{s}), [g_k(\mathbf{s}), \rho] \right] \tag{4.184}$$

but it can also be converted into a CM equation

$$\mathcal{L}_{\text{DP}}\rho_{\text{CM}} = -\frac{1}{2\hbar^2} \sum_{jk} \Lambda_{jk}^{\text{DP}} [x_j, [x_k, \rho_{\text{CM}}]] \tag{4.185}$$

where

$$\Lambda_{jk}^{\text{DP}} = \frac{\hbar}{4\pi} \int d^3\mathbf{s} \frac{\partial^2 \bar{\Phi}}{\partial r_j \partial r_k} \bar{\mu} \tag{4.186}$$

where the bar indicates the gravitational potential and the mass density averaged over internal motion. In fact, the trace of Λ_{jk}^{DP} is given by

$$\Lambda^{\text{DP}} = \delta^{jk} \Lambda_{jk}^{\text{DP}} = \hbar G \int d^3\mathbf{s} \bar{\mu}^2 \tag{4.187}$$

Assuming isotropy, this leads to

$$\Lambda_{xx}^{\mathrm{DP}} = \frac{1}{3} \Lambda^{\mathrm{DP}} = \frac{\hbar G M m}{24 \pi^{3/2} x_{\mathrm{ZPF}}^3} \tag{4.188}$$

or

$$S_F^{\mathrm{DP}} = 2 \Lambda_{xx}^{\mathrm{DP}} = \frac{\hbar G M m}{12 \pi^{3/2} x_{\mathrm{ZPF}}^3} \tag{4.189}$$

where x_{ZPF} is the zero-point fluctuation in the position of the atoms near their lattice equilibrium.

Proposed values for decoherence rates

The collapse models are proposed with the aim of explaining the phenomenology of the macroscopic world. For this reason, their decoherence rates must have a lower bound. This was discussed for the CSL model, by Adler, and then by Bassi *et al.* In these discussions, one must impose a cut-off scale for a particular type of process, in which the CSL must have enough collapse rate in order to prevent the formation of superpositions of macroscopically distinguishable states. For example, Adler (Adler, 2007) analysed the formation of a chemical photograph, while Bassi *et al.* analysed the human eye (Bassi, Deckert and Ferialdi, 2010). They reached a value of $\lambda_{\mathrm{CSL}} \sim 10^{-10 \pm 2}\,\mathrm{s}^{-1}$.

As one can see from the postulate of the CSL, it is the motion of nucleons which are relatively massive, not electrons which are very light, that will cause the decoherence. In this way, if we consider other approaches to photography, e.g. digital photography, which only involves moving electrons and holes in semiconductors, but not ions, and other senses of the human being, which may not involve the motions of ions, we may have to *increase the CSL rate* in order to make those senses also classical.

Low-frequency mechanical oscillators

As Nimmrichter *et al.* have argued, it is crucial to use low-frequency oscillators to test the DP and the CSL models. However, in actual experiments, it is often the case that high-mass oscillators can achieve lower resonant frequencies. Here we explore torsional pendulums, which, as it turns out, can be used to put interesting bounds on CSL, while achieving better bounds on the Diosi–Penrose model than many nanomechanical devices, but still not reaching the rate given above.

The key is to compare the force spectrum from the collapse models to the suspension thermal noise of the torsional pendulums (Saulson, 1994). Here we assume the thermal noise to arise from 'structural damping', which makes the frequency-domain spring constant of the torsional pendulum a complex number,

$$k \rightarrow k(1 + i\phi) \tag{4.190}$$

where ϕ is the loss angle, which remains roughly constant over frequencies. The angle corresponds to the fraction of the energy dissipated per radian of oscillation. Note that the loss angle here is not just the loss angle of the elastic moduli of the suspension fiber, but *diluted* by a significant factor, since potential energy of a torsional pendulum is mostly stored in gravitational potential energy (which is lossless).

Structural damping leads to the following form of thermal noise:

$$S_F(\omega) = 8 \frac{m\omega_m}{Q_m} k_B T \frac{\omega_m}{\omega} \tag{4.191}$$

where ω_m is the eigenfrequency of the pendulum. Note that the force spectrum decreases as ω becomes larger than ω_m, making the pendulum more suitable for measuring collapse models. However, other types of noise eventually become dominant over the suspension thermal noise.

Just to illustrate the situation, we quote numbers from a torsional pendulum experiment at the Australian National University. It seems one can comfortably reach the CSL region if we produce cryogenic pendulums, and if we have alternating material layers. As for the DP model, it is more challenging. One can reach the 10^{-14} m cut-off scale comfortably, but requires the combination of low temperature, high Q and dense material in order to reach our DP scale based on matter distribution.

Connection to test of General Relativity

Both the CSL and DP models have additional fields coupled to matter distribution over space-time. One can regard these fields as gravitational. If we generalize the coupling here from Markovian to non-Markovian, we will start modifying the dynamics of the system as if gravitational interaction is modified.

4.5.2 Semiclassical Gravity

In this and the next section, we discuss using optomechanical systems to test the nature of gravity. As we shall see, what we will be testing here will be quite subtle, since we do not yet have a fully consistent treatment of a theory in which gravity is classical. However, by going through this exercise, we identify possible parameter regimes where physical effects could take place, and re-examine various possibilities for formulating nonlinear quantum mechanics.

Motivation

It is interesting to study gravitational interaction with macroscopic objects. We note that, for a system of particles, even if we write a joint Schrödinger equation, with a Newtonian gravitational potential,

$$i\hbar\partial_t \psi(t; \mathbf{x}_1, \ldots, \mathbf{x}_n) = \hat{H}_0 \psi(t; \mathbf{x}_1, \ldots, \mathbf{x}_n) - \frac{1}{2}\sum_{j \neq k} \frac{GM_j M_k}{|\mathbf{x}_j - \mathbf{x}_k|} \psi(t; \mathbf{x}_1, \ldots, \mathbf{x}_n). \tag{4.192}$$

this still amounts to quantizing gravity, since this can be written as a two-step process: (i) the operators of the near-zone gravitational field depend on the position operators of the particles,

$$\hat{\phi}(\mathbf{x}) = -\sum_j G\frac{M_j}{|\mathbf{x} - \hat{\mathbf{x}}_j|} \tag{4.193}$$

and (ii) the particles interact with the quantized gravitational field

$$\hat{V} = -\sum_j \frac{1}{2}M_j\hat{\phi}(\hat{\mathbf{x}}_j) \tag{4.194}$$

In step (i), we can further regularize the problem by using matter density to drive the gravitational field. In this way, if one can transfer quantum information via Newtonian gravitational interaction, it would demonstrate the quantumness of gravity. However, directly demonstrating such a transfer would be quite difficult, as has been pointed out by Kafri, Taylor and Milburn (Kafri, Taylor and Milburn, 2014).

One way to see why it is hard to transfer quantum information from gravity is to estimate the timescales. Suppose we have two oscillators, each with frequency ω_0, interacting gravitationally. One can show that the split in the frequency is given by a scale

$$\Delta \approx \frac{\omega_g^2}{2\omega_0} \tag{4.195}$$

where

$$\omega_g \approx \sqrt{G\rho} \tag{4.196}$$

Even for the most dense material, osmium,

$$\omega_g^{\mathrm{OS}} = 2\pi \times 0.2\,\mathrm{mHz}, \tag{4.197}$$

and this is a very low frequency. The rate for quantum state transfer will be very low.

In this section, we will provide a toy model in which we assume gravity to be classical (see (Yang, Miao, Lee, Helou and Chen, 2013) and references therein), and then explore how self-gravitating systems can behave. It has even been shown that this is a mean-field approximation for the Correlated World-Line (CWL) model of quantum gravity (Stamp, 2015). We will use the so-called Schrödinger–Newton equation, which has the following form

$$i\hbar\partial_t\psi(t;\mathbf{x}_1,\ldots,\mathbf{x}_n)=\hat{H}_0\psi(t;\mathbf{x}_1,\ldots,\mathbf{x}_n)-\frac{1}{2}\sum_j M_j\phi(t,\mathbf{x}_j)\psi(t;\mathbf{x}_1,\ldots,\mathbf{x}_n). \qquad (4.198)$$

Here, ϕ is a potential evaluated from the expectation values of the mass density operator,

$$\phi(\mathbf{x})=-\int d^3\mathbf{y}\frac{G\langle\hat{\rho}(\mathbf{y})\rangle}{|\mathbf{x}-\mathbf{y}|} \qquad (4.199)$$

The difference between quantum gravity and classical gravity, in this limit, can be appreciated as we look at the evolution of the wavepacket of a single particle. In quantum gravity, the gravitational field is entangled with the position of the particle. However, in the classical case, the wavepacket of the wavefunction creates a potential in which the particle would evolve. This will lead to a 'self potential' for the spread of the packet, which does not act on the central location of the packet. This feature is also present in the CWL theory.

Schrödinger–Newton equation for the centre of mass

Yang *et al.* obtained the CM version of the SN equation, in a way similar to those used as we discussed collapse models. If the displacement of the CM is small compared to the zero-point mass distribution of matter near lattice sites, we have

$$i\hbar\frac{\partial\Psi_{\mathrm{CM}}}{\partial t}=\left[-\frac{\hbar^2\nabla^2}{2M}+\frac{1}{2}\omega_{\mathrm{CM}}^2 x^2+\frac{1}{2}M\omega_{\mathrm{SN}}^2(x-\langle x\rangle)^2\right]\Psi_{\mathrm{CM}} \qquad (4.200)$$

Here ω_{CM} is the frequency we observe in classical experiments, which probe the centre of the wavepackets, while ω_{SN} is a correction frequency that only acts on the spread of the wavepacket. Here $\langle x\rangle$ is the expectation value of x on the CM wavefunction, making this equation nonlinear. The value of ω_{SN} depends on the matter distribution around lattice sites, which are almost zero-point fluctuations at temperatures much below the Debye temperature. In this case, we have

$$\omega_{\mathrm{SN}}=\frac{Gm}{12\sqrt{\pi}x_{\mathrm{ZPF}}^3} \qquad (4.201)$$

This frequency is certainly much larger than ω_g, since the density here is not the mean density of the material, but the density of the material near the lattice site. However, this is *not* nuclear density. For osmium, the SN frequency is

$$\omega_{\mathrm{SN}}^{\mathrm{Os}}\approx 2\pi\times 64\,\mathrm{mHz}. \qquad (4.202)$$

which is above the eigenfrequency of many torsional pendulum experiments.

Quantum measurement for nonlinear quantum mechanics

As we confront the above theory with experiments, we will necessarily have to formulate how the wavefunction converts to the statistics of measurement data. It is known that nonlinear quantum mechanics, when combined with quantum state reduction, can lead to superluminal signal propagation (Polchinski, 1991). This has historically suppressed interest in nonlinear quantum mechanics.

In fact, we can show here that a more serious issue exists for nonlinear quantum mechanics. In standard, linear quantum mechanics, there exists a symmetry between the initially prepared state $|\psi_{\text{ini}}\rangle$, determined by the property and the initial state of the state preparation device/process, and the finally detected state $|\psi_{\text{fin}}\rangle$, determined by the detector. Although one usually computes the probability using

$$\langle \psi_{\text{fin}} | \hat{U} | \psi_{\text{ini}} \rangle \tag{4.203}$$

with the intention of applying the evolution operator $\hat{U}$ onto the initial state, and then projecting it onto the final state, one can also interpret the above process as taking ψ_{fin}, evolving it backwards, and then projecting it onto the initial state. Such an idea was explicitly illustrated by the two-time formulation of quantum mechanics, proposed by Aharonov and collaborators (Reznik and Aharonov, 1995); it was also followed in the so-called 'transactional interpretation' of quantum mechanics (Cramer, 2009).

If we keep standard quantum mechanics, it would then seem that both formulations are possible, yet equivalent. This freedom is broken if we consider nonlinear quantum mechanics, since here $\hat{U}$ depends on the state of the system. It would be interesting to see whether choosing the appropriate prescription disallows the superluminal signal transfer—or at least limits it to a low-enough therefore yet untested level.

Acknowledgements

The author's research in areas this chapter covers was supported by the US National Science Foundation, via the CAREER Grant PHY-0956189, and Grant PHY-1506453, also by the David and Barbara Groce Startup Fund at the California Institute of Technology. These lecture notes contain recent (some yet to be published) results in collaboration with Rana Adhikari, Bassam Helou, Farid Khalili, Mikhail Korobko, Yiqiu Ma, Haixing Miao, Belinda Pang, Nicolas Smith-Lefebvre and Christopher Wipf. These notes have also been updated according to questions and comments during the lectures.

Bibliography

Abbott, B., Abbott, R., Adhikari, R., Ajith, P., Allen, B., Allen, G., Amin, R., Anderson, S. B., Anderson, W. G., Arain, M. A. et al. (2009). Observation of a kilogram-scale oscillator near its quantum ground state. *New Journal of Physics*, **11**(7), 073032.

Adler, S. L. (2007). Lower and upper bounds on CSL parameters from latent image formation and GM heating. *Journal of Physics A: Mathematical and Theoretical*, **40**(12), 2935.

Bassi, A., Deckert, D.-A., and Ferialdi, L. (2010). Breaking quantum linearity: Constraints from human perception and cosmological implications. *EPL (Europhysics Letters)*, **92**(5), 50006.

Bassi, A., Lochan, K., Satin, S., Singh, T. P., and Ulbricht, H. (2013). Models of wave-function collapse, underlying theories, and experimental tests. *Reviews of Modern Physics*, **85**(2), 471.

Bhattacharya, M. and Meystre, P. (2007). Trapping and cooling a mirror to its quantum mechanical ground state. *Physical Review Letters*, **99**(7), 073601.

Braginsky, V. B. and Khalili, F. Y. (1990). Gravitational wave antenna with qnd speed meter. *Physics Letters A*, **147**(5–6), 251–6.

Braginsky, V. B., Gorodetsky, M. L., Khalili, F. Y., and Thorne, K. S. (1999). Energetic quantum limit in large-scale interferometers. *arXiv preprint gr-qc/9907057*.

Braginsky, V. B. and Khalili, F. Y. (1992, September). *Quantum Measurement*. Cambridge University Press.

Buonanno, A. and Chen, Y. (2001, Jul). Quantum noise in second generation, signal-recycled laser interferometric gravitational-wave detectors. *Physical Review D*, **64**, 042006.

Buonanno, A. and Chen, Y. (2002, Jan). Signal recycled laser-interferometer gravitational-wave detectors as optical springs. *Physical Review D*, **65**, 042001.

Buonanno, A. and Chen, Y. (2003). Scaling law in signal recycled laser-interferometer gravitational-wave detectors. *Physical Review D*, **67**(6), 062002.

Caves, C. M. (1980). Quantum-mechanical radiation-pressure fluctuations in an interferometer. *Physical Review Letters*, **45**(2), 75.

Caves, C. M. and Schumaker, B. L. (1985). New formalism for two-photon quantum optics. i. quadrature phases and squeezed states. *Physical Review A*, **31**(5), 3068.

Chen, Y. (2003, Jun). Sagnac interferometer as a speed-meter-type, quantum-nondemolition gravitational-wave detector. *Physical Review D*, **67**, 122004.

Clerk, A. A., Devoret, M. H., Girvin, S. M., Marquardt, F., and Schoelkopf, R. J. (2010, Apr). Introduction to quantum noise, measurement, and amplification. *Rev. Mod. Phys.*, **82**, 1155–208.

Corbitt, T., Chen, Y., Innerhofer, E., Müller-Ebhardt, H., Ottaway, D., Rehbein, H., Sigg, D., Whitcomb, S., Wipf, C., and Mavalvala, N. (2007). An all-optical trap for a gram-scale mirror. *Physical Review Letters*, **98**(15), 150802.

Cramer, J. G. (2009). Transactional interpretation of quantum mechanics. In *Compendium of Quantum Physics*, pp. 795–798. Springer.

Danilishin, S. L. (2004). Sensitivity limitations in optical speed meter topology of gravitational-wave antennas. *Physical Review D*, **69**(10), 102003.

Diósi, L. (2007). Notes on certain Newton gravity mechanisms of wavefunction localization and decoherence. *Journal of Physics A: Mathematical and Theoretical*, **40**(12), 2989.

Diósi, L. (2015). Testing spontaneous wave-function collapse models on classical mechanical oscillators. *Physical Review Letters*, **114**(5), 050403.

Gavartin, E., Verlot, P., and Kippenberg, T. J. (2012). A hybrid on-chip optomechanical transducer for ultrasensitive force measurements. *Nature Nanotechnology*, **7**(8), 509–14.

Ghirardi, G., Rimini, A., and Weber, T. (1986). Unified dynamics for microscopic and macroscopic systems. *Physical Review D*, **34**(2), 470.

Hawking, S. W. (1975). Particle creation by black holes. *Communications in Mathematical Physics*, **43**(3), 199–220.

Helstrom, C. W. (1967). Detection theory and quantum mechanics. *Information and Control*, **10**(3), 254–91.

Hirakawa, H., Hiramatsu, S., and Ogawa, Y. (1977). Damping of Brownian motion by cold load. *Physics Letters A*, **63**(3), 199–202.

Kafri, D., Taylor, J. M., and Milburn, G. J. (2014). A classical channel model for gravitational decoherence. *New Journal of Physics*, **16**(6), 065020.

Khalili, F., Danilishin, S., Miao, H., Müller-Ebhardt, H., Yang, H., and Chen, Y. (2010). Preparing a mechanical oscillator in non-Gaussian quantum states. *Physical Review Letters*, **105**(7), 070403.

Khalili, F. Y., Miao, H., Yang, H., Safavi-Naeini, A. H., Painter, O., and Chen, Y. (2012, Sep). Quantum back-action in measurements of zero-point mechanical oscillations. *Physical Review A*, **86**, 033840.

Kimble, H. J., Levin, Y., Matsko, A. B., Thorne, K. S., and Vyatchanin, S. P. (2001, Dec). Conversion of conventional gravitational-wave interferometers into quantum nondemolition interferometers by modifying their input and/or output optics. *Physical Review D*, **65**, 022002.

Krause, A. G., Blasius, T. D., and Painter, O. (2015). Optical read out and feedback cooling of a nanostring optomechanical cavity. *arXiv preprint arXiv:1506.01249*.

Marquardt, F., Chen, J. P., Clerk, A. A., and Girvin, S. M. (2007, Aug). Quantum theory of cavity-assisted sideband cooling of mechanical motion. *Physical Review Letters*, **99**, 093902.

Meers, B. J. (1988). Recycling in laser-interferometric gravitational-wave detectors. *Physical Review D*, **38**(8), 2317.

Miao, H., Danilishin, S., Müller-Ebhardt, H., and Chen, Y. (2010a). Achieving ground state and enhancing optomechanical entanglement by recovering information. *New Journal of Physics*, **12**(8), 083032.

Miao, H., Danilishin, S., Müller-Ebhardt, H., Rehbein, H., Somiya, K., and Chen, Y. (2010b). Probing macroscopic quantum states with a sub-Heisenberg accuracy. *Physical Review A*, **81**(1), 012114.

Miao, H., Ma, Y., Zhao, C., and Chen, Y. (2015, Nov). Enhancing the bandwidth of gravitational-wave detectors with unstable optomechanical filters. *Physical Review Letters*, **115**, 211104.

Mizuno, J., Strain, K. A., Nelson, P. G., Chen, J. M., Schilling, R., Rüdiger, A., Winkler, W., and Danzmann, K. (1993). Resonant sideband extraction: a new configuration for interferometric gravitational wave detectors. *Physics Letters A*, **175**(5), 273–6.

Müller-Ebhardt, H., Rehbein, H., Li, C., Mino, Y., Somiya, K., Schnabel, R., Danzmann, K., and Chen, Y. (2009). Quantum-state preparation and macroscopic entanglement in gravitational-wave detectors. *Physical Review A*, **80**(4), 043802.

Müller-Ebhardt, H., Rehbein, H., Schnabel, R., Danzmann, K., and Chen, Y. (2008, Jan). Entanglement of macroscopic test masses and the standard quantum limit in laser interferometry. *Physical Review Letters*, **100**, 013601.

Nation, P. D., Johansson, J. R., Blencowe, M. P., and Nori, F. (2012). Colloquium: Stimulating uncertainty: Amplifying the quantum vacuum with superconducting circuits. *Reviews of Modern Physics*, **84**(1), 1.

Nimmrichter, S., Hornberger, K., and Hammerer, K. (2014). Optomechanical sensing of spontaneous wave-function collapse. *Physical Review Letters*, **113**(2), 020405.

Ozawa, M. (2003, Apr). Universally valid reformulation of the Heisenberg uncertainty principle on noise and disturbance in measurement. *Physical Review A*, **67**, 042105.

Peano, V., Schwefel, H. G. L., Marquardt, Ch., and Marquardt, F. (2015, Dec). Intracavity squeezing can enhance quantum-limited optomechanical position detection through deamplification. *Physical Review Letters*, **115**, 243603.

Penrose, R. and Jorgensen, P. E. T. (2006). The road to reality: A complete guide to the laws of the universe. *The Mathematical Intelligencer*, **28**(3), 59–61.

Polchinski, J. (1991). Weinberg's nonlinear quantum mechanics and the Einstein-Podolsky-Rosen paradox. *Physical Review Letters*, **66**(4), 397.

Purdue, P. and Chen, Y. (2002, Dec). Practical speed meter designs for quantum nondemolition gravitational-wave interferometers. *Physical Review D*, **66**, 122004.

Reznik, B. and Aharonov, Y. (1995). Time-symmetric formulation of quantum mechanics. *Physical Review A*, **52**(4), 2538.

Saulson, P. R. (1994). *Fundamentals of interferometric gravitational wave detectors*. Volume 7. World Scientific, Singapore.

Schnabel, Roman (2015). Einstein-Podolsky-Rosen–entangled motion of two massive objects. *Physical Review A*, **92**(1), 012126.

Schumaker, B. L. and Caves, C. M. (1985). New formalism for two-photon quantum optics. ii. mathematical foundation and compact notation. *Physical Review A*, **31**(5), 3093.

Stamp, P. C. E. (2015). Rationale for a correlated worldline theory of quantum gravity. *New Journal of Physics*, **17**(6), 065017.

Suh, J., Weinstein, A. J., Lei, C. U., Wollman, E. E., Steinke, S. K., Meystre, P., Clerk, A. A., and Schwab, K. C. (2014). Mechanically detecting and avoiding the quantum fluctuations of a microwave field. *Science*, **344**(6189), 1262–5.

Tsang, M., Wiseman, H. M., and Caves, C. M. (2011). Fundamental quantum limit to waveform estimation. *Physical Review Letters*, **106**(9), 090401.

Unruh, W. G. (1976). Notes on black-hole evaporation. *Physical Review D*, **14**(4), 870.

Unruh, W. G. (1983). Quantum noise in the interferometer detector. In *Quantum Optics, Experimental Gravity, and Measurement Theory*, pp. 647–660. Springer.

Vanner, M. R., Hofer, J., Cole, G. D., and Aspelmeyer, M. (2013). Cooling-by-measurement and mechanical state tomography via pulsed optomechanics. *Nature Communications*, **4**.

Vyatchanin, S. P. and Zubova, E. A. (1995). Quantum variation measurement of a force. *Physics Letters A*, **201**(4), 269–74.

Wicht, A., Danzmann, K., Fleischhauer, M., Scully, M., Müller, G., and Rinkleff, R.-H. (1997). White-light cavities, atomic phase coherence, and gravitational wave detectors. *Optics Communications*, **134**(1), 431–9.

Wise, S., Quetschke, V., Deshpande, A. J., Mueller, G., Reitze, D. H., Tanner, D. B., Whiting, B. F., Chen, Y., Tünnermann, A., Kley, E. et al. (2005). Phase effects in the diffraction of light: beyond the grating equation. *Physical Review Letters*, **95**(1), 013901.

Wiseman, H. M. and Milburn, G. J. (2010). *Quantum Measurement and Control*. Cambridge University Press.

Wu, L.-A., Kimble, H. J., Hall, J. L., and Wu, H. (1986). Generation of squeezed states by parametric down conversion. *Physical Review Letters*, **57**(20), 2520.

Yang, H., Miao, H., Lee, D.-S., Helou, B., and Chen, Y. (2013). Macroscopic quantum mechanics in a classical spacetime. *Physical Review Letters*, **110**(17), 170401.

Zhou, M., Zhou, Z., and Shahriar, S. M. (2014). Quantum noise limits in white-light-cavity-enhanced gravitational wave detectors. *arXiv preprint arXiv:1410.6877*.

5

Optomechanics and Quantum Measurement

Aashish A. Clerk

Institute for Molecular Engineering, University of Chicago, Illinois, USA, and Department of Physics, McGill University, Montréal, Québec, Canada

Aashish Clerk

Aashish A. Clerk, *Optomechanics and Quantum Measurement* In: *Quantum Optomechanics and Nanomechanics.*
Edited by: Pierre-Francois Cohadon, Jack Harris, Florian Marquardt, Leticia F. Cugliandolo, Oxford University Press (2020).
© Oxford University Press. DOI: 10.1093/oso/9780198828143.003.0005

Chapter Contents

5.1   Introduction

These notes summarize lectures given at the 2015 Les Houches School on Optomechanics. The first part of the notes give a quick review of the basic theory of quantum optomechanical systems, based largely on linearized Heisenberg–Langevin equations. The notes then focus on selected topics relating to quantum measurement and quantum optomechanics. Section 5.3 gives comprehensive discussion of the quantum limit on the added noise of a continuous position detector, following the quantum linear response approach. While much of this discussion can already be found in (Clerk, 2004; Clerk *et al.*, 2010), I provide a greater discussion here of the role of noise correlations, and how these can be achieved in an optomechanical cavity (by using squeezed input light, or by modifying the choice of measured output quadrature). Section 5.4 turns to a discussion of back-action evading measurements of a mechanical quadrature, discussing how this can be achieved in a two-tone driven cavity system. I also provide a quick introduction to the theory of conditional continuous quantum measurement, and use it to discuss how a back-action evading measurement can be used to produce conditional mechanical squeezed states.

5.2   Basic Quantum Cavity Optomechanics Theory

This section will present a 'quick and dirty' introduction to the basic theoretical language used to describe quantum optomechanical systems. More complete introductions to some of the topics covered here can be found in (Clerk *et al.*, 2010; Aspelmeyer *et al.*, 2014; Clerk and Marquardt, 2014)

5.2.1   Optomechanical Hamiltonian

We start by considering a standard optomechanical system, consisting of a single mode of a resonant electromagnetic cavity whose frequency ω_{cav} depends on the position x of a mechanical resonator. Both the mechanical mode and cavity mode are harmonic oscillators, and the Hamiltonian takes the form ($\hbar = 1$):

$$\hat{H}_{OM} = \omega_{cav}[\hat{x}]\hat{a}^\dagger\hat{a} + \omega_M\hat{b}^\dagger\hat{b} \tag{5.1}$$

where $\hat{a}$ is the annihilation operator for the cavity mode, $\hat{b}$ is the annihilation operator for the mechanical mode. Finally, as we are typically interested in small mechanical displacements, we can Taylor expand the dependence of ω_{cav} on x keeping just the first term. Writing the mechanical position in terms of creation and destruction operators, the optomechanical Hamiltonian takes the form:

$$\hat{H}_{OM} = \omega_{cav}\hat{a}^\dagger\hat{a} + \omega_M\hat{b}^\dagger\hat{b} + g\hat{a}^\dagger\hat{a}\left(\hat{b} + \hat{b}^\dagger\right) \equiv \hat{H}_0 + \hat{H}_{int} \tag{5.2}$$

where

$$g = \frac{d\omega_{\text{cav}}}{dx} x_{\text{ZPF}} = \frac{d\omega_{\text{cav}}}{dx} \sqrt{\frac{\hbar}{2m\omega_M}}. \tag{5.3}$$

It is worth noting a few basic features of this Hamiltonian:

- The position x of the mechanical resonator sets the cavity frequency; thus, if you drive the cavity with a monochromatic laser, the average cavity photon number $\langle \hat{a}^\dagger \hat{a} \rangle$ will also depend on the mechanical position.
- Anything that multiplies $\hat{x}$ in a Hamiltonian acts as a force on the mechanical resonator. Hence, the cavity photon number $\hat{n} = \hat{a}^\dagger \hat{a}$ is a force on the mechanical resonator (i.e. the radiation-pressure force).
- Without loss of generality, we have defined the optomechanical interaction with a plus sign in this section. Hence, a positive displacement $x > 0$ of the mechanics results in an increased cavity frequency. We also use g to denote the single-photon optomechanical coupling strength in this section (the same quantity which is denoted g_0 in other works, e.g. (Aspelmeyer *et al.*, 2014)).

A rigorous derivation of this Hamiltonian (for the specific case of a Fabry–Perot resonator with a moveable end mirror) was given in (Law, 1995). This derivation keeps all the resonant modes of the optical cavity, and shows how in principle one also obtains interaction terms where the mechanical resonator can mediate scattering between different optical modes, and also terms corresponding to the dynamical Casamir effect, where, e.g. , destruction of a phonon can result in the creation of a pair of photons. Such additional terms are of negligible importance in the standard situation where the mechanical frequency ω_M is much smaller than all optical frequency scales.

Returning to the basic Hamiltonian $\hat{H}_{\text{OM}}$ above, one sees immediately that the cavity photon number $\hat{n}$ commutes with $\hat{H}$ and is thus a conserved quantity. Thus, in the absence of any driving or coupling to dissipation, $\hat{H}_{\text{OM}}$ can easily be exactly diagonalized. For a fixed photon number, the optomechanical interaction corresponds to a static force on the mechanics, which simply shifts its equilibrium position an amount $\Delta x_n = -2(g/\omega_M) x_{\text{ZPF}}$. The eigenstates are just tensor products of states of fixed photon number with displaced harmonic oscillator eigenstates. This is conveniently described by making a polaron transformation, i.e. a $\hat{n}$-dependent displacement transformation of the mechanical resonator:

$$\hat{U} = \exp\left(-i\hat{p}\Delta x_n\right) = \exp\left(\frac{g}{\omega_M} \hat{n} \left(\hat{b} - \hat{b}^\dagger\right)\right) \tag{5.4}$$

The transformed Hamiltonian takes the form:

$$\hat{U}^\dagger \hat{H}_{\text{OM}} \hat{U} = \left(\omega_{\text{cav}} - \frac{g^2}{\omega_M}\right) \hat{a}^\dagger \hat{a} + \omega_M \hat{b}^\dagger \hat{b} - \frac{g^2}{\omega_m} \hat{a}^\dagger \hat{a}^\dagger \hat{a} \hat{a} \tag{5.5}$$

In this new frame, we see explicitly from the last term that the eigenenergies of the Hamiltonian have a nonlinear dependence on photon number: the mechanical resonator mediates an optical Kerr-type nonlinearity (or equivalently, photon–photon interaction). In the simplest picture, we simply combine the two features noted above: as ω_{cav} depends on x, and x depends on $\hat{n}$ (as it is a force), the cavity frequency will depend on $\hat{n}$. This yields the $\hat{n}^2$ term above. In a more quantum picture, note that to leading order, the interaction $\hat{H}_{int}$ creates or destroys a mechanical excitation (a phonon) with a matrix element proportional to $\hat{n}$. While such a process would not conserve energy, to second order we could have an energy-conserving process that involves a virtual state with a mechanical excitation. The amplitude of such a process would be proportional to $\hat{n}^2$ and inversely proportional to ω_M (i.e. the energy cost of the virtual state). This is yet another way to understand the last term in Eq. (5.5).

5.2.2   Dissipation and Noise

Next, we need to include the coupling of both the cavity and the mechanical resonator to their respective dissipative environments, as well any driving terms (e.g. a coherent laser drive of the cavity). We follows the standard input–output theory route to treat these effects (see, e.g. (Walls and Milburn, 2008; Clerk *et al.*, 2010)), where the cavity mode (mechanical mode) acquires an energy damping rate κ (γ). To obtain these effects, one needs to include terms in the Hamiltonian describing the dissipative baths and their coupling to the system:

$$\hat{H} = \hat{H}_{OM} + \hat{H}_{\kappa} + \hat{H}_{\gamma} \tag{5.6}$$

Consider first the cavity dissipation, described by $\hat{H}_{\kappa}$. This just describes a linear coupling between the cavity mode and extra-cavity photon modes, which are themselves just free bosons (lowering operators $\hat{b}_q$):

$$\hat{H}_{\kappa} = \sum_q \omega_q \hat{b}_q^\dagger \hat{b}_q - i\sqrt{\frac{\kappa}{2\pi\rho}} \sum_q \left(\hat{a}^\dagger \hat{b}_q - h.c. \right) \tag{5.7}$$

As is standard, we approximate the the bath density of states to be a frequency-independent constant (which is typically an excellent approximation, as we are only interested in a small range of bath frequencies centred around the cavity frequency ω_{cav}):

$$\sum_q \delta(\omega - \omega_q) = \rho \tag{5.8}$$

As described in, e.g. (Clerk *et al.*, 2010), one can now derive the effective Heisenberg–Langevin equation of motion for the cavity field. This involves first solving the Heisenberg equation of motion for the bath operators $\hat{b}_q$, and then substituting these into the Heisenberg equation of motion for the cavity mode lowering operator $\hat{a}$. One obtains:

$$\frac{d}{dt}\hat{a} = -i\left[\hat{a}, \hat{H}_{\text{OM}}\right] - \frac{\kappa}{2}\hat{a} - \sqrt{\kappa}\hat{a}_{\text{in}} \tag{5.9}$$

The second term on the RHS describes a simple linear damping of the cavity (resulting from photons leaking from the cavity to the bath), while the third term describes a driving of the cavity mode by noise emanating from the bath. Note that $\hat{a}_{\text{in}}$ has been normalized so that $\hat{a}_{\text{in}}^{\dagger}\hat{a}_{\text{in}}$ represents a photon number flux (i.e. $\hat{a}_{\text{in}}$ has units of $1/\sqrt{\text{time}}$).

Taking the bath to be in a thermal equilibrium state, one finds that the operator-valued input noise $\hat{a}_{\text{in}}(t)$ is Gaussian (i.e. fully characterized by two-point correlation functions). Further, for the typically small frequency scales of interest, it can also be approximated as being white noise. One obtains:

$$\left\langle \hat{a}_{\text{in}}(t)\hat{a}_{\text{in}}^{\dagger}(t') \right\rangle = \delta(t - t')\left(1 + \bar{n}_{\text{th}}^{\text{cav}}\right) \tag{5.10}$$

$$\left\langle \hat{a}_{\text{in}}^{\dagger}(t)\hat{a}_{\text{in}}(t') \right\rangle = \delta(t - t')\bar{n}_{\text{th}}^{\text{cav}} \tag{5.11}$$

Here, $\bar{n}_{\text{th}}^{\text{cav}}$ is a Bose–Einstein occupancy factor evaluated at the cavity frequency and at the bath temperature. Note that the input noise satisfies the canonical commutation relation

$$\left[\hat{a}_{\text{in}}(t), \hat{a}_{\text{in}}^{\dagger}(t')\right] = \delta(t - t') \tag{5.12}$$

One can treat the effects of mechanical dissipation in a completely analogous manner. The Heisenberg–Langevin equation of motion for the mechanical lowering operator takes the form

$$\frac{d}{dt}\hat{b} = -i\left[\hat{b}, \hat{H}_{\text{OM}}\right] - \frac{\gamma}{2}\hat{b} - \sqrt{\kappa}\hat{b}_{\text{in}} \tag{5.13}$$

The mechanical input noise has correlation functions analogous to those in Eqs. (5.10), (5.11), except that $\bar{n}_{\text{th}}^{\text{cav}}$ is replaced by $\bar{n}_{\text{th}}^{\text{M}}$, a Bose–Einstein factor evaluated at the mechanical frequency and mechanical bath temperature.

Finally, we note that the approximations we have made in treating the bath (i.e. taking it to have a constant density of states, and taking its noise correlation functions to be delta-correlated) correspond to treating it as a Markovian bath (a vanishing correlation time, absence of memory effects). The resulting Heisenberg–Langevin equations are completely local in time.

5.2.3   Driving and Output Field

To include a coherent driving of the cavity (e.g. by a laser), we simply allow the input field $\hat{a}_{\text{in}}$ to have an average value $\bar{a}_{\text{in}}(t)$. One can easily confirm that this is completely equivalent to having added an explicit linear driving term

$$\hat{H}_{\text{drive}} = -i\left(\bar{a}_{\text{in}}(t)\hat{a}^\dagger - h.c.\right) \tag{5.14}$$

to the Hamiltonian. In many cases, the cavity bath corresponds to modes in the waveguide (or transmission line) used to drive and measure the cavity. In this case, we are also interested in knowing what the field emitted by the cavity into the waveguide is. The amplitude of this outgoing field is described by the operator $\hat{a}_{\text{out}}(t)$. In the simple Markovian limit we are focusing on, this field is completely determined by the input–output relation:

$$\hat{a}_{\text{out}}(t) = \hat{a}_{\text{in}}(t) + \sqrt{\kappa}\hat{a}(t) \tag{5.15}$$

where the intracavity field $\hat{a}$ is determined by the Heisenberg–Langevin equation Eq. (5.9). Note that $\hat{a}(t)$ is driven by the input field $\hat{a}_{\text{in}}(t)$, and hence the two terms on the RHS here are not independent. When $\kappa = 0$ (i.e. no coupling between the wave guide and the cavity), the input–output equation just describes a perfect reflection of waves off the end of the waveguide. For non-zero κ, the first term describes incident waves that are immediately reflected from the cavity–waveguide boundary, whereas the second term describes waves emitted from the waveguide.

Consider the simple case where we have a monochromatic coherent driving of the cavity: $\bar{a}_{\text{in}}(t) = \alpha_{\text{in}}e^{-i\omega_L t}$. It is convenient to work in a rotating frame where this driving looks time-independent. This is achieved by using the unitary $\hat{U}(t) = \exp\left(i\omega_L\hat{a}^\dagger\hat{a}t\right)$ to transform to a new frame. In this new frame, the Hamiltonian is given by:[1]

$$\hat{H}' = \hat{U}(t)\hat{H}\hat{U}^\dagger(t) + i\left(\frac{d}{dt}\hat{U}(t)\right)\hat{U}^\dagger(t) \tag{5.16}$$

and the Heisenberg–Langevin equation takes the form:

$$\frac{d}{dt}\hat{a} = -i\left[\hat{a}, -\Delta\hat{a}^\dagger\hat{a} + \omega_M\hat{b}^\dagger\hat{b} + \hat{H}_{\text{int}}\right] - \frac{\kappa}{2}\hat{a} - \sqrt{\kappa}\alpha_{\text{in}} - \sqrt{\kappa}\hat{a}_{\text{in}} \tag{5.17}$$

where the drive detuning $\Delta \equiv \omega_L - \omega_{\text{cav}}$, and we have explicitly separated out the average value of $\hat{a}_{\text{in}}$. Note that the RHS has no explicit time-dependence.

In the case where there is no optomechanical coupling, one can easily solve the above equation for the stationary value of $\hat{a}$; one finds:

$$\langle\hat{a}\rangle = -\frac{\sqrt{\kappa}\alpha_{\text{in}}}{\frac{\kappa}{2} - i\Delta} \equiv \alpha \tag{5.18}$$

---

[1] One also needs in principle to shift the frequency of the bath oscillators $\hat{b}_q$ to make sure that in the new frame, the cavity–bath coupling remains time-independent. While this is normally innocuous, it does imply the presence of negative-frequency bath modes in the new frame. The implications of this for a driven optomechanical system are discussed in detail in (Lemonde and Clerk, 2015).

It thus follows from Eq. (5.15) that the average output field is given by:

$$\langle \hat{a}_{\text{out}} \rangle = -\frac{\frac{\kappa}{2} + i\Delta}{\frac{\kappa}{2} - i\Delta} \alpha_{\text{in}} \equiv e^{i\theta} \alpha_{\text{in}} \tag{5.19}$$

We thus recover the expected expression for the reflection phase θ.

5.2.4 Displacement Transformation

Let's now include the optomechanical interaction in Eq. (5.17). We expect again that the cavity drive will induce an average value for the cavity field $\hat{a}$, and will also induce an average value for the mechanical lowering operator $\hat{b}$ (as the average photon number of the cavity is a static force on the mechanical resonator, which will displace its equilibrium position). It is useful to make displacement transformations to separate out these classical mean values from the additional dynamics that arises due to the noise operator $\hat{a}_{\text{in}}$ and $\hat{b}_{\text{in}}$ (operators which encode both classical and quantum noise driving the system). We thus introduce displaced cavity and mechanical lowering operators $\hat{d}$ and $\hat{b}_{\text{new}}$ defined via:

$$\hat{a} = \bar{a}_{\text{cl}} + \hat{d} \qquad \hat{b} = \bar{b}_{\text{cl}} + \hat{b}_{\text{new}} \tag{5.20}$$

where $\bar{a}_{\text{cl}}, \bar{b}_{\text{cl}}$ are the classical average values for the cavity and mechanical mode operators. These are found by solving the classical, noise-free version of the Heisenberg–Langevin equations Eq. (5.9) and (5.13) (i.e. one replaces $\hat{a} \to \bar{a}_{\text{cl}}$, $\hat{b} \to \bar{b}_{\text{cl}}$ in the equations and drops all noise terms). These classical equations are nonlinear, and regimes exist where multiple classical solutions can be found (the well-known optomechanical bistability, see Sec. V.A of (Aspelmeyer *et al.*, 2014) for a discussion). We assume however that we work in a regime where there is a unique solution to the classical equations. As $\hat{d}$ and $\hat{b}_{\text{new}}$ encode all quantum effects in our system, it is common to refer to them as the quantum parts of the cavity and mechanical annihilation operators. We stress that they are standard canonical bosonic annihilation operators.

Once the classical amplitudes have been found, one returns to the full Heisenberg–Langevin equations, expressed now in terms of the operators $\hat{d}$ and $\hat{b}_{\text{new}}$. One finds that there are no purely constant terms on the RHS of these equations (i.e. linear driving terms). In particular, the coherent cavity drive α_{in} enters only through the classical displacements $\bar{a}_{\text{cl}}, \bar{b}_{\text{cl}}$. The resulting equations are equivalent to having started with a coherent Hamiltonian

$$\hat{H}_{\text{OM}} = -\Delta' \hat{d}^\dagger \hat{d} + \omega_M \hat{b}_{\text{new}}^\dagger \hat{b}_{\text{new}} + \hat{H}_{\text{int}} \tag{5.21}$$

$$\hat{H}_{\text{int}} = g \left(\bar{a}_{\text{cl}}^* \hat{d} + \bar{a}_{\text{cl}} \hat{d}^\dagger + \hat{d}^\dagger \hat{d} \right) \left(\hat{b}_{\text{new}}^\dagger + \hat{b}_{\text{new}} \right) \tag{5.22}$$

where the modified detuning $\Delta' = \Delta - g(\bar{b}_{\text{cl}} + \bar{b}_{\text{cl}}^*)$. We see that in this new displaced frame the mechanical mode only interacts with the fluctuating parts of the cavity photon number $\hat{n}$ (i.e. terms involving $\hat{d}, \hat{d}^\dagger$). Note that the static, 'classical' part of the photon number $|\bar{a}_{cl}|^2$ determines the classical mechanical displacement $\bar{b}_{cl}$, but does not appear explicitly in $\hat{H}_{OM}$. In what follows, we drop the subscript 'new' on $\hat{b}_{new}$ to keep things clear, and also replace Δ' by Δ.

5.2.5   Linearized regime of optomechanics

By strongly driving the cavity (large α_{in}), $\bar{a}_{cl}$ increases in magnitude. This in turn increases the fluctuations in the intracavity photon number, namely the term that is *linear* in the $\hat{d}$ operators. Correspondingly (as per Eq. (5.22)), the large drive enhances the quadratic terms in the optomechanical interaction Hamiltonian. It is thus common to introduce a drive-enhanced many-photon optomechanical interaction strength, defined as[2]

$$G \equiv g|\bar{a}_{cl}| = g\sqrt{\bar{n}_{cav}} \tag{5.23}$$

In almost all current experiments, the single photon strength g is too weak to directly play a role (i.e. $g \ll \kappa, \omega_M$), and appreciable optomechanical effects are only obtained when the cavity is strongly driven and G made large. In this regime where $G \gg g$, it a good approximation to only retain the drive-enhanced terms in Eq. (5.22). The physics is then described by the approximate quadratic Hamiltonian:

$$\hat{H}_{lin} = -\Delta\hat{d}^\dagger\hat{d} + \omega_M\hat{b}^\dagger\hat{b} + G(\hat{b}+\hat{b}^\dagger)(\hat{d}+\hat{d}^\dagger) \tag{5.24}$$

with corresponding Heisenberg–Langevin equations:

$$\frac{d}{dt}\hat{d} = (i\Delta - \kappa/2)\hat{d} - iG\left(\hat{b}+\hat{b}^\dagger\right) - \sqrt{\kappa}\,\hat{d}_{in} \tag{5.25}$$

$$\frac{d}{dt}\hat{b} = (-i\omega_M - \gamma/2)\hat{b} - iG\left(\hat{d}+\hat{d}^\dagger\right) - \sqrt{\gamma}\,\hat{b}_{in} \tag{5.26}$$

Note that we have made a gauge transformation on the cavity field to absorb the phase of the classical cavity field $\bar{a}_{cl}$. In this regime, the equations of motion for the cavity and mechanical lowering operators are purely linear, and thus it is often termed the linearized regime of optomechanics. The physics is just that of a system of two linearly coupled harmonic oscillators, albeit one with a great deal of tunability, and one where the dissipative rates and effective temperatures of the two oscillators can be very different. The effective frequency of the photonic oscillator can be tuned by changing

---

[2] It is also common notation to use g_0 (instead of g) to denote the single-photon optomechanical coupling strength, and use g (instead of G) to denote the many-photon coupling strength. We prefer the notation used here, both as it avoids having to use yet another subscript, and because the use of a capital letter more dramatically emphasizes the coupling enhancement by the drive.

the frequency of the cavity drive, and the strength of the interaction G can be tuned by changing the amplitude of the cavity drive.

To get a feel for the various interesting things that can be done with linearized optomechanics, we quickly sketch different interesting possibilities below.

Beam-splitter Hamiltonian

Consider Eq. (5.24) in the regime where $\Delta = -\omega_M$, implying that the cavity drive frequency is detuned to the red of the cavity resonance an amount exactly equal to the mechanical frequency (a so-called red-sideband drive). The two effective oscillators in Eq. (5.24) are then resonant. If we further assume that $\omega_M \gg \kappa$ and that $G \ll \omega_M$, the interaction terms that create or destroy a photon–phonon pair are sufficiently non-resonant to not play a large role in the dynamics. We can thus make a rotating wave approximation, keeping only the energy conserving interaction terms. The Hamiltonian then takes the simple form:

$$\hat{H}_{\text{lin}} = \omega_M \hat{a}^\dagger \hat{a} + \omega_M \hat{b}^\dagger \hat{b} + G(\hat{b}^\dagger \hat{a} + \hat{a}^\dagger \hat{b}) \tag{5.27}$$

This interaction converts photons to phonons and vice versa, and is known as a beam-splitter Hamiltonian or 'swap' Hamiltonian. It is at the heart of quantum state-transfer applications of optomechanics (see, e.g., (Hill *et al.*, 2012; Dong *et al.*, 2012; Andrews *et al.*, 2014)). It also is at the heart of optimal cavity-cooling schemes, where one uses the driven cavity to cool a thermal mechanical resonator to close to the ground state (Marquardt *et al.*, 2007; Wilson-Rae *et al.*, 2007). The above swap-Hamiltonian can be used to transfer mechanical excitations to the cavity mode, where they are quickly emitted (at a rate κ) to the cavity bath.

Entangling Hamiltonian

Consider Eq. (5.24) for the opposite detuning choice, where $\Delta = \omega_M$: the cavity drive frequency is now detuned to the blue of the cavity resonance an amount ω_M. If we again assume that the good-cavity condition $\omega_M \gg \kappa$ is fulfilled and that $G \ll \omega_M$, we can again make a rotating-wave approximation where we only retain energy-conserving terms. As now the cavity photons have an effective negative frequency, the energy-conserving terms correspond to creating or destroying pairs of excitations, and the Hamiltonian becomes:

$$\hat{H}_{\text{lin}} = -\omega_M \hat{a}^\dagger \hat{a} + \omega_M \hat{b}^\dagger \hat{b} + G(\hat{b}^\dagger \hat{a}^\dagger + \hat{a}\hat{b}) \tag{5.28}$$

The dynamics of this Hamiltonian creates photons and phonons in pairs, and leads to states with a high degree of correlations between the mechanics and light (i.e. the photon number and phonon number are almost perfectly correlated). Such states are known as 'two-mode squeezed states', and their correlations correspond to quantum entanglement. Such optomechanical entanglement has recently been measured. The above Hamiltonian also corresponds to the Hamiltonian of a non-degenerate parametric amplifier. It can be used for near quantum-limited amplification. It also can exhibit dynamical instabilities for sufficiently large values of G.

5.3 Quantum Limit on Continuous Position Detection

One of the key motivations for studying optomechanics is the possibility of using light to measure mechanical motion with a precision limited by the fundamental constraints imposed by quantum mechanics. In some cases, one can even devise schemes that transcend the quantum constraints that limit more conventional measurement strategies. In this section, we will review the most general and rigorous formulation of the quantum limit on continuous position detection for a generic linear-response detector. The general derivation will make the origin of this quantum limit clear, as well establish a set of requirements which must be met to achieve it. We will then apply this general formalism to the basic optomechanical cavity as introduced in the previous section. The presentation here is closely related to that developed in (Clerk, 2004) and (Clerk *et al.*, 2010). In contrast to those works, we provide a discussion of how squeezing can be used to generate the noise correlations needed to reach the quantum limit in the free mass limit.

5.3.1 General Problem: Minimizing Total Detector Added Noise

To motivate things, let's return to Eq. (5.19) for the reflection phase shift for a driven single-sided cavity, and give a classical discussion of how our cavity could function as a position detector. For a fixed cavity drive frequency, the reflection phase shift θ is a function of the cavity frequency ω_{cav}. With the optomechanical coupling, the instantaneous cavity frequency becomes a function of the mechanical position. Consider first the simple case where x is fixed to some unknown position x_0. The mechanical displacement shifts the cavity frequency by an amount gx_0/x_{ZPF}. For small displacements, the change in phase will be linear in x_0. In the optimal case where $\Delta = 0$, we have

$$\theta = \pi + \frac{4g}{\kappa}\frac{x_0}{x_{\text{ZPF}}} \equiv \theta_0 + \Delta\theta \tag{5.29}$$

(with $\theta_0 = \pi$) and the output field from the cavity will have the form

$$a_{\text{out}} = e^{i(\theta_0 + \Delta\theta)}a_{\text{in}} \simeq e^{i\theta_0}(1 + i\Delta\theta)a_{\text{in}} = -(1 + i\Delta\theta)a_{\text{in}} \tag{5.30}$$

where a_{in} is the amplitude of the incident cavity driving field. Information on the mechanical displacement x_0 is encoded in the change in reflection phase $\Delta\theta$, and will thus be optimally contained in the phase quadrature of the output field. Taking (without loss of generality) a_{in} to be real, this quadrature is defined as:

$$X_\phi(t) = -i\left(a_{\text{out}}(t) - a_{\text{out}}^*(t)\right) \tag{5.31}$$

Using the standard technique of homodyne interferometry, this output-field quadrature can be directly converted into a photocurrent $I(t) \propto X_\phi$ (see e.g. (Walls and Milburn, 2008; Clerk *et al.*, 2010)). This photocurrent will have intrinsic fluctuations (due to shot noise in the light), and hence will have the form:

$$I(t) = \lambda x_0 + \delta I_0(t) \qquad (5.32)$$

The first term is the 'signal' associated with the measurement, where $\lambda \propto g$ parameterizes the response of the photocurrent to changes in position. The second term represents the imprecision noise in the measurement, i.e. the intrinsic fluctuations in the photocurrent. At each instant t, $\delta I_0(t)$ is a random variable with zero mean. To determine x_0, one would need to integrate $I(t)$ over some finite time interval to resolve the signal above these intrinsic fluctuations. It is common to scale these intrinsic output fluctuations by the response coefficient λ, and think of this noise as an equivalent mechanical position noise $\delta x_{\mathrm{imp}}(t) \equiv \delta I_0(t)/\lambda$.

   If the mechanical resonator is now in motion, and ω_M is sufficiently small (i.e. much smaller than κ), then the cavity will be able to adiabatically follow the mechanical motion. Eq. (5.32) will then still hold, with the replacement $x_0 \to x_0(t)$. Our goal here will not be to be measure the instantaneous position of the mechanical resonator: we consider the standard situation where the mechanical resonator is in motion (oscillating!), and the optomechanical coupling is far too weak to measure position in a time short compared to the mechanical period. Instead, our goal will be to determine the quadrature amplitudes describing the oscillating motion, i.e. the amplitudes of the sine and cosine components of the motion:

$$x(t) = X(t)\cos(\Omega t) + Y(t)\sin(\Omega t). \qquad (5.33)$$

These amplitudes typically evolve on a timescale much longer than the mechanical period.

   One might think that in principle, one could increase the signal-to-noise ratio indefinitely by increasing λ (e.g. by increasing g or the power associated with the cavity drive). From a quantum point of view, we know the situation cannot be that simple, as the quadrature amplitudes X and Y become non-commuting, conjugate observables, and cannot be known simultaneously with arbitrary precision. More concretely, in the quantum case, we absolutely need to think about an additional kind of measurement noise: the disturbance of the mechanical position by the detector, otherwise known as back-action. Returning to the optomechanical Hamiltonian of Eq. (5.2), we see that fluctuations of cavity photon number act like a noisy force on the mechanical resonator. This will give rise to extra fluctuations in its position, $\delta x_{\mathrm{BA}}(t)$. Eq. (5.32) thus needs to be updated to have the form:

$$I(t) = \lambda\left[x_0(t) + \delta x_{\mathrm{BA}}(t)\right] + \delta I_0(t) \equiv \lambda\left[x_0(t) + \delta x_{\mathrm{add}}(t)\right] \qquad (5.34)$$

We have introduced here the total added noise of the measurement $\delta x_{\mathrm{add}}(t) = \delta x_{\mathrm{BA}}(t) + \delta I_0(t)/\lambda$.

The goal is now to make this total added noise as small as possible, and to understand if there are any fundamental limits to its size. Simply increasing the coupling strength or optical power no longer is a good strategy: in this limit, the second 'imprecision' noise term in $\delta x_{\text{add}}(t)$ will become negligible, but the first back-action term will be huge. Similarly, in the opposite limit where the coupling (or optical drive power) is extremely weak, back-action noise will be negligible, but the imprecision noise will be huge. Clearly, some trade-off between these two limits will be optimal. Further, one needs to consider whether these two kinds of noises (imprecision and back-action) can be correlated, and whether this is desirable.

Our goal in what follows will be to first review how one characterizes the magnitude of noise via a spectral density, and then to establish the rigorous quantum limit on how small one can make the added noise $\delta x_{\text{add}}(t)$.

5.3.2 Quantum Noise Spectral Densities: Some Essential Features

In this section, we give a compact (and no doubt highly incomplete) review of some basic properties of quantum noise spectral densities. We start, however, with the simpler case of spectral densities describing classical noise.

Classical noise basics

Consider a classical random signal $I(t)$. The signal is characterized by zero mean $\langle I(t) \rangle = 0$, and autocorrelation function

$$G_{II}(t, t') = \langle I(t)I(t') \rangle. \tag{5.35}$$

The autocorrelation function is analogous to a covariance matrix: for $t = t'$, it tells us the variance of the fluctuations of $I(t)$, whereas for $t \neq t'$, it tells us if and how fluctuations of $I(t)$ are correlated with those at $I(t')$. Some crucial concepts regarding noise are:

- *Stationary noise.* The statistical properties of the fluctuations are time-translation invariant, and hence $G_{II}(t, t') = G_{II}(t - t')$.
- *Gaussian fluctuations.* The noise is fully characterized by its autocorrelation function; there are no higher-order cumulants.
- *Correlation time.* This timescale τ_c governs the decay of $G_{II}(t)$: $I(t)$ and $I(t')$ are uncorrelated (i.e. $G_{II}(t - t') \to 0$) when $|t - t'| \gg \tau_c$.

For stationary noise, it is often most useful to think about the fluctuations in the frequency domain. In the same way that $I(t)$ is a Gaussian random variable with zero mean, so is its Fourier transform, which we define as:

$$I_T[\omega] = \frac{1}{\sqrt{T}} \int_{-T/2}^{+T/2} dt\, e^{i\omega t} I(t), \tag{5.36}$$

where T is the sampling time. In the limit $T \gg \tau_c$ the integral is a sum of a large number $N \approx \frac{T}{\tau_c}$ of random uncorrelated terms. We can think of the value of the integral as the end point of a random walk in the complex plane which starts at the origin. Because the distance travelled will scale with $\sqrt{T}$, our choice of normalization makes the statistical properties of $I[\omega]$ independent of the sampling time T (for sufficiently large T). Notice that $I_T[\omega]$ has the peculiar units of $[I]\sqrt{\text{secs}}$ which is usually denoted $[I]/\sqrt{\text{Hz}}$.

The spectral density of the noise (or power spectrum) $S_{II}[\omega]$ answers the question 'how big is the noise at frequency ω?'. It is simply the variance of $I_T(\omega)$ in the large-time limit:

$$S_{II}[\omega] \equiv \lim_{T \to \infty} \langle |I_T[\omega]|^2 \rangle = \lim_{T \to \infty} \langle I_T[\omega] I_T[-\omega] \rangle. \tag{5.37}$$

A reasonably straightforward manipulation (known as the Wiener–Khinchin theorem) tells us that the spectral density is equal to the Fourier transform of the autocorrelation function

$$S_{II}[\omega] = \int_{-\infty}^{+\infty} dt \, e^{i\omega t} \, G_{II}(t). \tag{5.38}$$

We stress that Eq. (5.37) provides a simple intuitive understanding of what a spectral density represents, whereas in theoretical calculations one almost always starts with the expression in Eq. (5.38). We also stress that since the autocorrelation function $G_{II}(t)$ is real, $S_{II}[\omega] = S_{II}[-\omega]$. This is of course in keeping with Eq. (5.36), which tells us that negative and positive frequency components of the noise are related by complex conjugation, and hence necessarily have the same magnitude.

Definition of quantum noise spectral densities

In formulating quantum noise, one turns from a noisy classical signal $I(t)$ to a Heisenberg-picture Hermitian operator $\hat{I}(t)$. Similar to our noisy classical signal, one needs to think about measurements of $\hat{I}(t)$ statistically. One can thus introduce a quantum-noise spectral density which completely mimics the classical definition, e.g.:

$$S_{xx}[\omega] = \int_{-\infty}^{+\infty} dt \, e^{i\omega t} \langle \hat{x}(t)\hat{x}(0) \rangle. \tag{5.39}$$

We have simply inserted the quantum autocorrelation function in the classical definition. The expectation value is the quantum statistical average with respect to the noisy system's density matrix; we assume that this is time-independent, which then also gives us an autocorrelation function which is time-translational invariant.

What makes quantum noise so quantum? There are at least three answers to this question:

- *Zero-point motion.* While a classical system at zero-temperature has no noise, quantum mechanically there are still fluctuations, i.e. $S_{xx}[\omega]$ need not be zero.

- *Frequency asymmetry.* Quantum mechanically, $\hat{x}(t)$ and $\hat{x}(t')$ need not commute when $t \neq t'$. As a result the autocorrelation function $\langle \hat{x}(t)\hat{x}(t') \rangle$ can be complex, and $S_{xx}[\omega]$ need not equal $S_{xx}[-\omega]$. This of course can never happen for a classical noise spectral density.

- *Heisenberg constraints.* For any system that can act as a detector or amplifier, there are fundamental quantum constraints that bound its noise. These constraints have their origin in the uncertainty principle, and have no classical counterpart.

Noise asymmetry and fluctuation dissipation theorem

The asymmetry in frequency of quantum noise spectral densities is a topic that is discussed in great detail in (Clerk *et al.*, 2010). In short, this asymmetry directly reflects the asymmetry in the noisy system's ability to absorb versus emit energy. This aspect of quantum noise spectral densities provides an extremely useful route to understanding optomechanical damping effects. While this asymmetry will not be the main focus of our discussion here, it is useful to consider a simple but instructive example which helps demystify two-sided quantum noise spectral densities. Consider a harmonic oscillator that is coupled weakly to a noise force produced by a second quantum system. This force is described by an operator $\hat{F}$, and the coupling to the harmonic oscillator is

$$\hat{H}_{\text{int}} = -\hat{x}\hat{F}. \tag{5.40}$$

Of course, the basic optomechanical Hamiltonian is an example of such a coupling, where $\hat{F} \propto \hat{n}$, the intracavity photon number.

Classically, including this force in Newton's equation yields a Langevin equation:

$$M\ddot{x} = -M\omega_{\text{M}}^2 x - M\gamma_{\text{cl}}\dot{x} + F_{\text{cl}}(t). \tag{5.41}$$

In addition to the noisy force, we have included a damping term (rate γ_{cl}). This will prevent the oscillator from being infinitely heated by the noise source; we can think of it as describing the *average* value of the force exerted on the oscillator by the noise source, which is now playing the role of a dissipative bath. If this bath is in thermal equilibrium at temperature T, we also expect the oscillator to equilibrate to the same temperature. This implies that the heating effect of $F_{\text{cl}}(t)$ must be precisely balanced by the energy-loss effect of the damping force. More explicitly, one can use Eq. (5.41) to derive an equation for the average energy of the oscillator $\langle E \rangle$. As we are assuming a weak coupling between the bath and the oscillator, we can take $\gamma_{\text{cl}} \ll \omega_{\text{M}}$, and hence find

$$\frac{d}{dt}\langle E \rangle = -\gamma_{\text{cl}}\langle E \rangle + \frac{S_{FF}[\omega_{\text{M}}]}{2M}. \tag{5.42}$$

where $S_{FF}[\omega]$ is the classical noise spectral density associated with $F_{\text{cl}}(t)$ (c.f. Eq. (5.37)).

Insisting that the stationary value of $\langle E \rangle$ obey equipartition then leads directly to the classical fluctuation dissipation relation:

$$S_{FF}[\omega_M] = 2M\gamma_{cl}k_B T. \tag{5.43}$$

Let's now look at our problem quantum mechanically. Writing $\hat{x}$ in terms of ladder operators, we see that $\hat{H}_{int}$ will cause transitions between different oscillator Fock states. Treating $\hat{H}_{int}$ in perturbation theory, we thus derive Fermi Golden rule transition rates $\Gamma_{n\pm1,n}$ for transitions from the n to the $n\pm1$ Fock state. As shown explicitly in Appendix B of (Clerk *et al.*, 2010), these rates can be directly tied to the quantum noise spectral density associated with $\hat{F}$. One finds:

$$\Gamma_{n+1,n} = (n+1)\frac{x_{ZPF}^2}{\hbar^2}S_{FF}[-\omega_M] \equiv (n+1)\Gamma_\uparrow, \tag{5.44}$$

$$\Gamma_{n-1,n} = (n)\frac{x_{ZPF}^2}{\hbar^2}S_{FF}[\omega_M] \equiv n\Gamma_\downarrow. \tag{5.45}$$

Transitions where the noise source absorbs energy are set by the negative frequency part of the noise spectral density, while emission is set by the positive frequency part.

We can now write a simple master equation for the probability $p_n(t)$ that the oscillator is in the nth Fock state:

$$\frac{d}{dt}p_n = \left[n\Gamma_\uparrow p_{n-1} + (n+1)\Gamma_\downarrow p_{n+1}\right] - \left[n\Gamma_\downarrow + (n+1)\Gamma_\uparrow\right]p_n. \tag{5.46}$$

We can then connect this quantum picture to our classical Langevin equation by using Eq. (5.46) to derive an equation for the average oscillator energy $\langle E \rangle$. One obtains

$$\frac{d}{dt}\langle E \rangle = -\gamma\langle E \rangle + \frac{\bar{S}_{FF}[\omega_M]}{2M}, \tag{5.47}$$

where:

$$\gamma = \frac{x_{ZPF}^2}{\hbar^2}(S_{FF}[\omega_M] - S_{FF}[-\omega_M]), \tag{5.48}$$

$$\bar{S}_{FF}[\Omega] = \frac{S_{FF}[\omega_M] + S_{FF}[-\omega_M]}{2}. \tag{5.49}$$

We see that the quantum equation for the average energy, Eq. (5.47), has an identical form to the classical equation (Eq. (5.42)), which gives us a simple way to connect our quantum noise spectral density to quantities in the classical theory:

- The symmetrized quantum noise spectral density $\bar{S}_{FF}[\Omega]$ defined in Eq. (5.49) plays the same role as the classical noise spectral density $S_{FF}[\Omega]$: it heats the oscillator the same way a classical stochastic force would.

- The asymmetric-in-frequency part of the quantum noise spectral density $S_{FF}[\Omega]$ is directly related to the damping rate γ in the classical theory. The asymmetry between absorption and emission events leads to a net energy flow between the oscillator and the noise source, analogous to what one obtains from a classical viscous damping force.

We thus see that there is a direct connection to a classical noise spectral density, and moreover the 'extra information' in the asymmetry of a quantum noise spectral density also corresponds to a seemingly distinct classical quantity, a damping rate. This latter connection is not so surprising. The asymmetry of the quantum noise is a direct consequence of (here) $[\hat{F}(t), \hat{F}(t')] \neq 0$. However, this same non-commutation causes the average value of $\langle \hat{F} \rangle$ to change in response to $\hat{x}(t)$ via the interaction Hamiltonian of Eq. (5.40). Using standard quantum linear response (i.e. first-order time-dependent perturbation theory, see e.g. Ch. 6 of (Bruus and Flensberg, 2004)), one finds

$$\delta \langle \hat{F}(t) \rangle = \int_{-\infty}^{\infty} dt' \, \chi_{FF}(t - t') \langle \hat{x}(t') \rangle, \qquad (5.50)$$

where the force–force susceptibility is given by the Kubo formula:

$$\chi_{FF}(t) \equiv \frac{-i}{\hbar} \theta(t) \left\langle \left[\hat{F}(t), \hat{F}(0) \right] \right\rangle. \qquad (5.51)$$

From the classical Langevin equation Eq. (5.41), we see that part of $\langle \hat{F}(t) \rangle$ which is in phase with $\dot{x}$ is the damping force. This leads to the definition

$$\gamma = \frac{1}{M\Omega} \left(-\text{Im} \, \chi_{FF}[\Omega] \right). \qquad (5.52)$$

An explicit calculation shows that the above definition is identical to Eq. (5.48), which expresses γ in terms of the noise asymmetry. Note that in the language of many-body Green functions, $-\text{Im} \, \chi_{FF}$ is referred to as a spectral function, whereas the symmetrized noise $\bar{S}_{FF}[\omega]$ is known (up to a constant) as the 'Keldysh' Green function.

Quantum fluctuation-dissipation theorem and notion of effective temperature

Consider the case where the quantum system producing the noise $\hat{F}$ is in thermal equilibrium at temperature T. For weak coupling, we expect that the stationary value of $\langle E \rangle$ as given by Eq. (5.47) should match the thermal equilibrium value $\hbar \omega_M (1/2 + n_B[\omega_M])$. Insisting that this be the case forces a relation between the damping γ (which is set by the asymmetry of the noise, c.f. Eq. (5.48)) and symmetrized noise $\bar{S}_{FF}[\omega_M]$ which is nothing more than the quantum version of the fluctuation-dissipation theorem:

$$\bar{S}_{FF}[\omega_{\mathrm{M}}] = M\gamma[\omega_{\mathrm{M}}]\hbar\omega_{\mathrm{M}}\coth\left(\frac{\hbar\omega_{\mathrm{M}}}{2k_{\mathrm{B}}T}\right) = M\gamma[\omega_{\mathrm{M}}]\hbar\omega_{\mathrm{M}}\left(1 + 2n_B[\omega_{\mathrm{M}}]\right). \quad (5.53)$$

For $k_{\mathrm{B}}T \gg \hbar\omega_{\mathrm{M}}$ this reproduces the classical result of Eq. (5.43), whereas in the opposite limit, it describes zero-point noise. We stress that Eq. (5.53) can also be proved directly using nothing more than the fact that the system producing the noise has a thermal-equilibrium density matrix.

What happens if our noise source is *not* in thermal equilibrium? In that case, it is useful to use Eq. (5.53) to *define* an effective temperature $T_{\mathrm{eff}}[\Omega]$ from the ratio of the symmetrized noise and damping. Re-writing things in terms of the quantum noise spectral density, one finds

$$k_{\mathrm{B}}T_{\mathrm{eff}}[\Omega] \equiv \hbar\Omega\left[\ln\left(\frac{S_{FF}[\Omega]}{S_{FF}[-\Omega]}\right)\right]^{-1}. \quad (5.54)$$

The effective temperature at a given frequency Ω characterizes the asymmetry between absorption and emission rates of energy $\hbar\Omega$; a large temperature indicates that these rates are almost equal, whereas a small temperature indicates that emission by the noise source is greatly suppressed compared to absorption by the source. Away from thermal equilibrium, there is no guarantee that the ratio on the RHS will be frequency-independent, and hence T_{eff} will generally have a frequency dependence.

5.3.3 Heisenberg Inequality on Detector Quantum Noise

Generic two-port linear response detector

Having discussed two of the ways quantum noise spectral densities differ from their classical counterparts (zero-point noise, frequency asymmetry), we now turn to the third distinguishing feature: there are purely quantum constraints on the noise properties of any system capable of acting as a detector or amplifier. We will be interested in the generic two-port detector sketched in Fig. 5.1. The detector has an input port characterized by an operator $\hat{F}$: this is the detector quantity which couples to the system we wish to measure. Similarly, the output port is characterized by an operator $\hat{I}$: this is the detector quantity that we will readout to learn about the system coupled to the input. In our optomechanical

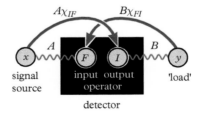

Fig. 5.1 *Schematic of a generic linear response detector.*

system, $\hat{F}$ would be the cavity photon number $\hat{n}$, while $\hat{I}$ would be proportional to the phase quadrature of the output field of the cavity (c.f. Eq. (5.31)).

We will be interested almost exclusively in detector-signal couplings weak enough that one can use linear-response to describe how $\hat{I}$ changes in response to the signal. For example, if we couple an input signal $\hat{x}$ to our detector via an interaction Hamiltonian

$$\hat{H}_{\text{int}} = \hat{x} \cdot \hat{F}, \tag{5.55}$$

linear response tells us that the change in the detector output will be given by:

$$\delta \langle \hat{I}(t) \rangle = \int_{-\infty}^{\infty} dt' \chi_{IF}(t - t') \langle \hat{x}(t') \rangle, \tag{5.56}$$

$$\chi_{IF}(t) = -\frac{i}{\hbar} \theta(t) \Big\langle [\hat{I}(t), \hat{F}(0)] \Big\rangle. \tag{5.57}$$

This is completely analogous to the way we discussed damping, c.f. Eq. (5.52). As is standard in linear-response, the expectation value in Eq. (5.57) is with respect to the state of the system (signal plus detector) at zero coupling (i.e. $\hat{H}_{\text{int}} = 0$). Also, without loss of generality, we will assume that both $\langle \hat{I} \rangle$ and $\langle \hat{F} \rangle$ are zero in the absence of any coupling to the input signal.

Even on a classical level, any noise in the input and output ports will limit our ability to make measurements with the detector. Quantum mechanically, we have seen that it is the symmetrized quantum spectral densities that play a role analogous to classical noise spectral densities. We will thus be interested in the quantities $\bar{S}_{II}[\omega]$ and $\bar{S}_{FF}[\omega]$. Given our interest in weak detector-signal couplings, it will be sufficient to characterize the detector noise at zero-coupling to the detector.

In addition to $\bar{S}_{II}, \bar{S}_{FF}$, we will also have to contend with the fact that the noise in $\hat{I}$ and $\hat{F}$ may be correlated. Classically, we would describe such correlations via a correlation spectral density $S_{IF}[\omega]$:

$$S_{IF}[\omega] \equiv \lim_{T \to \infty} \langle I_T[\omega] \, (F_T[\omega])^* \rangle = \int_{-\infty}^{\infty} dt \, \langle I(t) F(0) \rangle e^{i\omega t}, \tag{5.58}$$

where the Fourier transforms $I_T[\omega]$ and $F_T[\omega]$ are defined analogously to Eq. (5.36). Not surprisingly, such classical correlations correspond to a symmetrized quantum noise spectral density

$$\bar{S}_{IF}[\omega] \equiv \frac{1}{2} \int_{-\infty}^{\infty} dt \, \langle \{\hat{I}(t), \hat{F}(0)\} \rangle e^{i\omega t}. \tag{5.59}$$

Note that the classical correlation density $S_{IF}[\omega]$ is generally complex, and is only guaranteed to be real at $\omega = 0$; the same is true of $\bar{S}_{IF}[\omega]$.

Finally, we normally are only concerned about how large the output noise is compared to the magnitude of the 'amplified' input signal at the output (i.e. Eq. (5.56)). It is thus common to think of the output noise at a given frequency $\delta I_T[\omega]$ as an equivalent fluctuation of the signal $\delta z_{\text{imp}}[\omega] \equiv \delta I_T[\omega]/\chi_{IF}[\omega]$. We thus define the imprecision noise spectral density and imprecision-back-action correlation density as:

$$\bar{S}_{zz}[\omega] \equiv \frac{\bar{S}_{II}[\omega]}{|\chi_{IF}[\omega]|^2}, \quad \bar{S}_{zF}[\omega] \equiv \frac{\bar{S}_{IF}[\omega]}{\chi_{IF}[\omega]}. \tag{5.60}$$

Motivation and derivation of noise constraint

We can now ask what sort of constraints exist on the detector noise. In almost all relevant cases, our detector will be some sort of driven quantum system, and hence will not be in thermal equilibrium. As a result, any meaningful constraint should not rely on having a thermal equilibrium state. Classically, all we can say is that the correlations in the noise cannot be bigger than the noise itself. This constraint takes the form of a a Schwartz inequality, yielding

$$\mathcal{S}_{zz}[\omega]\mathcal{S}_{FF}[\omega] \geq |\mathcal{S}_{zF}[\omega]|^2. \tag{5.61}$$

Equality here implies a perfect correlation, i.e. $I_T[\omega] \propto F_T[\omega]$.

Quantum mechanically, additional constraints will emerge. Heuristically, this can be expected by making an analogy to the example of the Heisenberg microscope. In that example, one finds that there is a trade-off between the imprecision of the measurement (i.e. the position resolution) and the back-action of the measurement (i.e. the momentum kick delivered to the particle). In our detector, noise in $\hat{I}$ will correspond to the imprecision of the measurement (i.e. the bigger this noise, the harder it will be to resolve the signal described by Eq. (5.56)). Similarly, noise in $\hat{F}$ is the back-action: as we already saw, by virtue of the detector–signal coupling, $\hat{F}$ acts as a noisy force on the measured quantity $\hat{z}$. We thus might naturally expect a bound on the product of $\bar{S}_{zz}\bar{S}_{FF}$.

Alternatively, we see from Eq. (5.57) that for our detector to have any response at all, $\hat{I}(t)$ and $\hat{F}(t')$ cannot commute for all times. Quantum mechanically, we know that uncertainty relations apply any time we have non-commuting observables; here things are somewhat different, as the non-commutation is between Heisenberg-picture operators at different times. Nonetheless, we can still use the standard derivation of an uncertainty relation to obtain a useful constraint. Recall that for two non-commuting observables $\hat{A}$ and $\hat{B}$, the full Heisenberg inequality is (see, e.g. (Gottfried, 1966))

$$(\Delta A)^2 (\Delta B)^2 \geq \frac{1}{4}\left\langle\left\{\hat{A},\hat{B}\right\}\right\rangle^2 + \frac{1}{4}\left|\left\langle\left[\hat{A},\hat{B}\right]\right\rangle\right|^2. \tag{5.62}$$

Here we have assumed $\langle\hat{A}\rangle = \langle\hat{B}\rangle = 0$. We now take $\hat{A}$ and $\hat{B}$ to be cosine-transforms of $\hat{I}$ and $\hat{F}$, respectively, over a finite time interval T:

$$\hat{A} \equiv \sqrt{\frac{2}{T}}\int_{-T/2}^{T/2} dt\,\cos(\omega t + \delta)\,\hat{I}(t), \quad \hat{B} \equiv \sqrt{\frac{2}{T}}\int_{-T/2}^{T/2} dt\,\cos(\omega t)\,\hat{F}(t). \tag{5.63}$$

Note that we have phase-shifted the transform of $\hat{I}$ relative to that of $\hat{F}$ by a phase δ. In the limit $T \to \infty$ we find

$$\bar{S}_{zz}[\omega]\bar{S}_{FF}[\omega] \geq \left[\mathrm{Re}\left(e^{i\delta}\bar{S}_{zF}[\omega]\right)\right]^2 + \frac{\hbar^2}{4}\left[\mathrm{Re}\ e^{i\delta}\left(1 - \frac{(\chi_{FI}[\omega])^*}{\chi_{IF}[\omega]}\right)\right]^2. \qquad (5.64)$$

We have introduced here a new susceptibility $\chi_{FI}[\omega]$, which describes the reverse response coefficient or reverse gain of our detector. This is the response coefficient relevant if we used our detector in reverse: couple the input signal $\hat{z}$ to $\hat{I}$, and see how $\langle\hat{F}\rangle$ changes. A linear response relation analogous to Eq. (5.56) would then apply, with $F \leftrightarrow I$ everywhere. For the optomechanical system we are most interested in, this reverse response coefficient vanishes: coupling to the output field from the cavity cannot change the intracavity photon number (and hence $\hat{F}$). We thus take $\chi_{FI}[\omega] = 0$ in what follows (see (Clerk *et al.*, 2010) for further discussion on the role of a non-zero $\chi_{FI}[\omega]$).

If we now maximize the RHS of Eq. (5.64) over all values of δ, we are left with the optimal bound

$$\bar{S}_{zz}[\omega]\bar{S}_{FF}[\omega] - \left|\bar{S}_{zF}[\omega]\right|^2 \geq \frac{\hbar^2}{4}\left(1 + \Delta\left[\frac{2\bar{S}_{zF}[\omega]}{\hbar}\right]\right), \qquad (5.65)$$

where

$$\Delta[y] = \frac{\left|1+y^2\right| - \left(1 + |y|^2\right)}{2}, \qquad (5.66)$$

Note that for any complex number y, $1 + \Delta[y] > 0$. Related noise constraints on linear-response detectors are presented in (Braginsky and Khalili, 1996) and (Averin, 2003).

We see that applying the uncertainty principle to our detector has given us a rigorous constraint on the detector's noise which is stronger than the simple classical bound of Eq. (5.61) on its correlations. For simplicity, consider first the $\omega \to 0$ limit, where all noise spectral densities and susceptibilities are real, and hence the term involving $\Delta[y]$ vanishes. The extra quantum term on the RHS of Eq. (5.64) then implies:

- The product of the imprecision noise $\bar{S}_{zz}$ and back-action noise $\bar{S}_{FF}$ cannot be zero. The magnitude of both kinds of fluctuations must be non-zero.

- Moreover, these fluctuations cannot be perfectly correlated with one another: we cannot have $\left(\bar{S}_{zF}\right)^2 = \bar{S}_{zz}\bar{S}_{FF}$.

The presence of these extra quantum constraints on noise will lead directly (and rigorously!) to the fundamental quantum limits on continuous position detection. As we will see, reaching this quantum limit requires one to use a detector which has 'ideal' quantum noise (i.e. noise spectral densities for which the inequality of Eq. (5.65) becomes an equality).

5.3.4   Power Gain and the Large Gain Limit

Before finally turning to defining and deriving the quantum limit on the added noise, there is one more crucial aspect of the detector to address: the notion of 'power gain'. We are interested in detectors that turn the motion $x(t)$ of the mechanical resonator into a 'large' signal in the output of the detector, a signal so large that we do not need to worry about how this detector output is then read out. To be able to say that our detector truly amplifies the motion of the oscillator, it is not sufficient to simply say the response function χ_{IF} must be large (note that χ_{IF} is not dimensionless!). Instead, true amplification requires that the *power* delivered by the detector to a following amplifier be much larger than the power drawn by the detector at its input—i.e., the detector must have a dimensionless power gain $G_P[\omega]$ much larger than one. If the power gain was not large, we would need to worry about the next stage in the amplification of our signal, and how much noise is added in that process. Having a large power gain means that by the time our signal reaches the following amplifier, it is so large that the added noise of this following amplifier is unimportant.

The power gain of our position detector can be defined by imagining a situation where one couples a second auxiliary oscillator to the output of the detector, such that $I(t)$ acts as a driving force on this oscillator. The power gain is then defined as the power delivered to this auxiliary oscillator, divided by the power drawn from the measured mechanical resonator coupled to the detector output (optimized over properties of the auxiliary oscillator). The calculation is presented in Appendix A, and the result is the simple expression:

$$G_P[\omega] \equiv \max \left[\frac{P_{\text{out}}}{P_{\text{in}}} \right] = \frac{|\chi_{IF}[\omega]|^2}{4 \left(\text{Im}\, \chi_{FF}[\omega] \right) \left(\text{Im}\, \chi_{II}[\omega] \right)} \tag{5.67}$$

The susceptibility $\chi_{II}[\omega]$ is defined analogously to Eq. (5.51). Note that if there is no additional back-action damping of the measured oscillator by the detector, then Im χ_{FF} vanishes, and the power gain is strictly infinite. As we will see, this the case for an optomechanical cavity driven on resonance.

Having a large power gain also implies that we can treat the detector quantities $\hat{I}(t)$ and $\hat{F}(t)$ as being effectively 'classical'. A large power gain over some relevant frequency range implies that the imaginary parts of χ_{II} and χ_{FF} are negligible over this range. From Eqs. (5.48) and (5.52), this implies that the quantum noise spectral densities $S_{II}[\omega]$, $S_{FF}[\omega]$ are to a good approximation *symmetric* over these frequencies, just like a classical noise spectral density. This in turn implies that one can effectively treat $[\hat{I}(t), \hat{I}(t')] = [\hat{F}(t), \hat{F}(t')] = 0$ (i.e. these operators commute at different times, just like a classical noisy function of time). We note that in many discussions of linear quantum measurements, the fact that the detector input and output quantities commute with themselves at different times is taken as a starting assumption in calculations (see, e.g. (Braginsky and Khalili, 1992; Khalili *et al.*, 2012), as well as Chapter 4 of this volume).

Requiring both the quantum noise inequality in Eq. (5.65) to be saturated at frequency ω as well as a large power gain (i.e. $G_P[\omega] \gg 1$) leads to some important additional constraints on the detector, as derived in Appendix I of (Clerk *et al.*, 2010):

- $(2/\hbar)\mathrm{Im}\,\bar{S}_{zF}[\omega]$ is small like $1/\sqrt{G_P[\omega]}$.
- The detector's effective temperature must be much larger than $\hbar\omega$; one finds

$$k_\mathrm{B}\,T_\mathrm{eff}[\omega] \sim \sqrt{G_P[\omega]}\,\hbar\omega. \tag{5.68}$$

Conversely, it is the largeness of the detector's effective temperature that allows it to have a large power gain.

5.3.5   Defining the Quantum Limit

Having now understood the proper way to discuss the 'size' of noise, as well as the existence of quantum constraints on noise, we can now return to the question posed at the start of this section: how small can we make the added noise $\delta x_\mathrm{add}(t)$ (c.f. Eq. (5.34)) of a generic linear-response position detector? We assume that, like in our optomechanical setup, the detector couples to mechanical position via a Hamiltonian:

$$\hat{H}_\mathrm{int} = \hat{x} \cdot \left(A\hat{F}\right) \tag{5.69}$$

Note that unlike the generic system-detector interaction in Eq. (5.55), we have included a coupling strength in the definition of the system operator $\hat{F}$. We assume that A is weak enough that the detector output responds linearly to changes in position.

Added noise spectral density

We start by returning to the heuristic classical expression in Eq. (5.34) for the detector output current $I(t)$ and added noise $\delta x_\mathrm{add}(t)$, and Fourier transform these expression. From linear response theory, we know that the response coefficient λ in that expression should be replaced by $\lambda \to A\chi_{IF}[\omega]$, where the frequency dependence parameterizes that the detector output will not respond instantaneously to changes at the input.

$$I[\omega] = A\chi_{IF}[\omega]\,(x[\omega] + \delta x_\mathrm{add}[\omega]) \tag{5.70}$$

$$\delta x_\mathrm{add}[\omega] = \delta x_\mathrm{BA}[\omega] + \frac{\delta I_0[\omega]}{A\chi_{IF}[\omega]} \tag{5.71}$$

The first back-action term is just the mechanical response to the back-action force fluctuations:

$$\delta x_\mathrm{BA}[\omega] = A\chi_{xx}[\omega]\delta F[\omega], \tag{5.72}$$

where $\chi_{xx}[\omega]$ is the oscillator's force susceptibility,[3] and is given by

$$M\chi_{xx}[\omega] = \left(\omega^2 - \omega_M^2 + i\omega\gamma_0\right)^{-1}. \tag{5.73}$$

To state the quantum limit on position detection, we first define the total measured position fluctuations $x_{\text{meas}}[\omega]$ as simply the total detector output $I[\omega]$ referred back to the oscillator:

$$x_{\text{meas}}[\omega] = I_{\text{tot}}[\omega]/(A\chi_{IF}[\omega]). \tag{5.74}$$

If there was *no* added noise, and further, if the oscillator was in thermal equilibrium at temperature T, the spectral density describing the fluctuations $\delta x_{\text{meas}}(t)$ would simply be the equilibrium fluctuations of the oscillator, as given by the fluctuation-dissipation theorem:

$$\bar{S}_{xx}^{\text{meas}}[\omega] = \bar{S}_{xx}^{\text{eq}}[\omega, T] = \hbar \coth\left(\frac{\hbar\omega}{2k_B T}\right)[-\text{Im } \chi_{xx}[\omega]] \tag{5.75}$$

$$= \frac{x_{\text{ZPF}}^2(1 + 2n_B)}{2} \sum_{\sigma=\pm} \frac{\gamma_0}{(\omega - \sigma\Omega)^2 + (\gamma_0/2)^2}. \tag{5.76}$$

Here, γ_0 is the intrinsic damping rate of the oscillator, which we have assumed to be $\ll \Omega$.

Including the added noise, and for the moment ignoring the possibility of any additional oscillator damping due to the coupling to the detector, the above result becomes

$$\bar{S}_{xx}^{\text{meas}}[\omega] = \bar{S}_{xx}^{\text{eq}}[\omega, T] + \bar{S}_{xx}^{\text{add}}[\omega] \tag{5.77}$$

where the last term is the spectral density of the added noise (both back-action and imprecision noise).

We can now, finally, state the quantum limit on continuous position detection: at each frequency ω, we must have

$$\bar{S}_{xx}^{\text{add}}[\omega] \geq \bar{S}_{xx}^{\text{eq}}[\omega, T = 0]. \tag{5.78}$$

The spectral density of the added noise cannot be made arbitrarily small: at each frequency, it must be at least as large as the corresponding zero-point noise. Note that we do not call this the 'standard' quantum limit. As we will discuss later, what is usually termed the standard quantum limit (e.g. in the gravitational wave detection community)

---

[3] Strictly speaking, with our definitions the force susceptibility is $-\chi_{xx}[\omega]$, and the force driving the mechanics is $-\hat{F}$. This is because we took the interaction Hamiltonian to be $\hat{H}_{\text{int}} = +\hat{x}\hat{F}$ instead of the more physical $\hat{H}_{\text{int}} = -\hat{x}\hat{F}$.

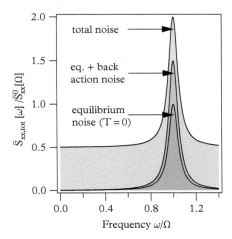

Fig. 5.2 *Spectral density of a mechanical resonator's position fluctuations as measured by a quantum limited position detector. The mechanical resonator's intrinsic fluctuations (for the case of a zero-temperature oscillator) are shown in blue. The back-action of the position detector will effectively heat the mechanical resonator, increasing the area under the Lorentzian; this is shown as the orange area. Finally, there is also imprecision noise, the fluctuations in the detector output that would be present even without any coupling to the mechanical resonator. These typically give frequency independent noise, giving a flat background. The ideal case where the quantum limit is reached on the added noise at mechanical resonance is the yellow area.*

only coincides with Eq. (5.78) exactly at resonance ($\omega = \omega_M$), and for other frequencies does not represent any kind of true quantum bound. The various contributions to the added noise spectral density are shown in Fig. 5.2.

Finally, the above result can be refined to include situations where the coupling to the detector also changes the mechanical damping (in addition to driving it with extra fluctuations). Letting γ_{BA} denote this extra damping, the added noise is now defined via

$$\bar{S}_{xx}^{meas}[\omega] = \frac{\gamma_0}{\gamma_{BA} + \gamma_0} \bar{S}_{xx}^{eq}[\omega, T] + \bar{S}_{xx}^{add}[\omega], \qquad (5.79)$$

where the susceptibility χ_{xx} now involves the total damping of the oscillator, i.e.:

$$M\chi_{xx}[\omega] = \left(\omega^2 - \omega_M^2 + i\omega(\gamma + \gamma_{BA})\right)^{-1}. \qquad (5.80)$$

With this definition, the quantum limit on the added noise is unchanged from the limit stated in Eq. (5.78).

A possible correlation-based loophole?

Our heuristic formulation of the quantum limit naturally leads to a possible concern. Even though quantum mechanics may require a position measurement to have a back-action (as position and momentum are conjugate quantities), couldn't this back-action noise be perfectly anti-correlated with the imprecision noise? If this were the case, the added noise $\delta x(t)$ (which is the sum of the two contributions, c.f. Eq. (5.71)) could be made to vanish.

One might hope that this sort of loophole would be explicitly forbidden by the quantum noise inequality of Eq. (5.65). However, this is not the case. Even in the ideal case of zero reverse gain, one can achieve a situation where back-action and imprecision are perfectly correlated at a given non-zero frequency ω. One needs the following:

- The correlator $\bar{S}_{IF}[\omega]$ should be purely imaginary; this implies that the part of $F(t)$ that is correlated with $I(t)$ is 90 degrees out of phase. Note that $\bar{S}_{IF}[\omega]$ can only be imaginary at non-zero frequencies.
- The magnitude of $\bar{S}_{IF}[\omega]$ should be larger than $\hbar/2$.

Under these circumstances, one can verify that there is no additional quantum constraint on the noise beyond what exists classically, and hence the perfect correlation condition of $\bar{S}_{FF}[\omega]\bar{S}_{II}[\omega] = |\bar{S}_{IF}[\omega]|^2$ is allowable. The $\pi/2$ phase of the back-action–imprecision correlations are precisely what is needed to make $\delta x_{\mathrm{add}}[\omega]$ vanish at the oscillator resonance, $\omega = \Omega$.

As might be expected, this seeming loophole is *not* a route to ideal, noise-free position detection free from quantum constraints. As already discussed, we need to be more careful in specifying what we want our detector to do. We aren't interested in just having the mechanical motion show up in the detector output $I(t)$, we want there to be amplification associated with this process—the mechanical signal should be 'bigger' at the output than it is at the input. It is only when we insist on amplification that there are quantum constraints on added noise; a passive transducer need not add any noise. On a heuristic level, one could view amplification as an effective expansion of the phase-space of the oscillator. Such a pure expansion is of course forbidden by Liouville's theorem, which tells us that volume in phase-space is conserved. The way out is to introduce additional degrees of freedom, such that for these degrees of freedom phase-space contracts. Quantum mechanically such degrees of freedom necessarily have noise associated with them (at the very least, zero-point noise); this then is the source of the limit on added noise.

The requirement that our detector produces a large signal is that the power gain (as defined in section 5.3.4) should be much larger than one. In that case, back-action–imprecision correlations must be purely real, and the possibility of zero added noise (due to perfect correlations) is excluded.

5.3.6   Derivation of the Quantum Limit

We now turn to a rigorous proof of the quantum limit on the added noise given in Eq. (5.78). From the classical-looking Eq. (5.71), we expect that the symmetrized quantum noise spectral density describing the added noise will be given by

$$\bar{S}_{xx,\text{add}}[\omega] = \frac{\bar{S}_{II}}{|\chi_{IF}|^2 A^2} + A^2 |\chi_{xx}|^2 \bar{S}_{FF} + \frac{2\text{Re}\left[\chi_{IF}^*(\chi_{xx})^* \bar{S}_{IF}\right]}{|\chi_{IF}|^2} \tag{5.81}$$

$$= \frac{\bar{S}_{zz}}{A^2} + A^2 |\chi_{xx}|^2 \bar{S}_{FF} + 2\text{Re}\left[(\chi_{xx})^* \bar{S}_{zF}\right]. \tag{5.82}$$

In the second line, we have introduced the imprecision noise $\bar{S}_{zz}$ and imprecision back-action correlation $\bar{S}_{zF}$ as in Eq. (5.60). We have also omitted writing the explicit frequency dependence of the gain χ_{IF}, susceptibility χ_{xx}, and noise correlators; they should all be evaluated at the frequency ω. Finally, the oscillator susceptibility χ_{xx} here is given by Eq. (5.80), and includes the effects of back-action damping. While we have motivated this equation from a seemingly classical noise description, the full quantum theory also yields the same result: one simply calculates the detector output noise perturbatively in the coupling to the oscillator (Clerk, 2004).

The first step in determining the limit on the added noise is to consider its dependence on the coupling strength strength A. If we ignore for a moment the detector-dependent damping of the oscillator, there will be an optimal value of the coupling strength A which corresponds to a trade-off between imprecision noise and back-action (i.e. first and second terms in Eq. (5.81)). We would thus expect $\bar{S}_{xx,\text{add}}[\omega]$ to attain a minimum value at an optimal choice of coupling $A = A_{\text{opt}}$ where both these terms make equal contributions. Defining $\phi[\omega] = \arg \chi_{xx}[\omega]$, we thus have the bound

$$\bar{S}_{xx,\text{add}}[\omega] \geq 2|\chi_{xx}[\omega]| \left(\sqrt{\bar{S}_{zz}\bar{S}_{FF}} + \text{Re}\left[e^{-i\phi[\omega]}\bar{S}_{zF}\right] \right), \tag{5.83}$$

where the minimum value at frequency ω is achieved when

$$A_{\text{opt}}^2 = \sqrt{\frac{\bar{S}_{zz}[\omega]}{|\chi_{xx}[\omega]|^2 \bar{S}_{FF}[\omega]}}. \tag{5.84}$$

Using the inequality $X^2 + Y^2 \geq 2|XY|$ we see that this value serves as a lower bound on $\bar{S}_{xx,\text{add}}$ even in the presence of detector-dependent damping. In the case where the detector-dependent damping is negligible, the RHS of Eq. (5.83) is independent of A, and thus Eq. (5.84) can be satisfied by simply tuning the coupling strength A; in the more general case where there is detector-dependent damping, the RHS is also a function of A (through the response function $\chi_{xx}[\omega]$), and it may no longer be possible to achieve Eq. (5.84) by simply tuning A.

While Eq. (5.83) is certainly a bound on the added displacement noise $\bar{S}_{xx,\text{add}}[\omega]$, it does not in itself represent the quantum limit. Reaching the quantum limit requires more than simply balancing the detector back-action and intrinsic output noises (i.e. the first two terms in Eq. (5.81)); *one also needs a detector with 'quantum-ideal' noise properties, that is a detector which optimizes* Eq. (5.65). Using the quantum noise constraint of Eq. (5.65) to further bound $\bar{S}_{xx,\text{add}}[\omega]$, we obtain

$$\bar{S}_{xx,\text{add}}[\omega] \geq \quad \hbar|\chi_{xx}[\omega]| \left[\sqrt{\left(1 + \Delta\left[\frac{\bar{S}_{zF}}{\hbar/2}\right]\right) + \left|\frac{\bar{S}_{zF}}{\hbar/2}\right|^2} + \frac{\text{Re}\left[e^{-i\phi[\omega]}\bar{S}_{zF}\right]}{\hbar/2} \right], \quad (5.85)$$

where the function $\Delta[z]$ is defined in Eq. (5.66). The minimum value of $\bar{S}_{xx,\text{add}}[\omega]$ in Eq. (5.85) is now achieved when one has *both* an optimal coupling (i.e. Eq. (5.84)) *and* a quantum limited detector, that is one which satisfies Eq. (5.65) as an equality.

Next, we further specialize to the relevant case where the detector acts as a good amplifier, and has a power gain $G_P[\omega] \gg 1$ over all frequencies of interest. As discussed, this implies that the ratio $\bar{S}_{zF}$ is purely real up to small $1/G_P$ corrections (see Appendix I of (Clerk *et al.*, 2010) for more details). This in turn implies that $\Delta[2\bar{S}_{zF}/\hbar] = 0$; we thus have

$$\bar{S}_{xx,\text{add}}[\omega] \geq \hbar|\chi_{xx}[\omega]| \left[\sqrt{1 + \left(\frac{\bar{S}_{zF}}{\hbar/2}\right)^2} + \cos(\phi[\omega])\frac{\bar{S}_{zF}}{\hbar/2} \right]. \quad (5.86)$$

Finally, as there is no further constraint on $\bar{S}_{zF}$ (beyond the fact that it is real), we can minimize the expression over its value. The minimum $\bar{S}_{xx,\text{add}}[\omega]$ is achieved for a detector whose cross-correlator satisfies

$$\bar{S}_{zF}[\omega]\Big|_{\text{optimal}} = \frac{\hbar}{2}\cot\phi[\omega] = -\frac{\hbar}{2}\frac{\omega^2 - \omega_M^2}{\omega\gamma}, \quad (5.87)$$

with the minimum value of the added noise being given precisely by

$$\bar{S}_{xx,\text{add}}[\omega]\Big|_{\text{min}} = \hbar|\text{Im }\chi_{xx}[\omega]| = \lim_{T\to 0}\bar{S}_{xx,\text{eq}}[\omega, T], \quad (5.88)$$

in agreement with Eq. (5.78). Thus, in the limit of a large power gain, we have that *at each frequency, the minimum displacement noise added by the detector is precisely equal to the noise arising from a zero-temperature bath.*

5.3.7 Simple Limits and Discussion

We have thus provided a rigorous derivation of the quantum limit on the added noise of a continuous position detector which possesses a large gain. The derivation shows explicitly what is needed to reach the quantum limit, namely:

1. A detector with quantum limited noise properties, that is one which optimizes the inequality of Eq. (5.65).
2. A coupling A which satisfies Eq. (5.84).
3. A detector cross-correlator $\bar{S}_{IF}$ which satisfies Eq. (5.87).

It is worth stressing that Eq. (5.87) implies that it will *not* in general be possible to achieve the quantum limit simultaneously at all frequencies, as the needed amount of back-action–imprecision correlation varies strongly with frequency. We consider a few important limits below.

Quantum limit on added noise at resonance $\omega = \omega_M$

To reach the quantum limit on the added noise at the mechanical resonance, Eq. (5.87) tells us that $\bar{S}_{zF}[\omega]$ must be zero: back-action and imprecision noises should be completely uncorrelated. If this is the case, reaching the quantum limit simply involves tuning the coupling A to balance the contributions from back-action and imprecision to the added noise. As we will see, in our optomechanical cavity, this is equivalent to optimizing the choice of the driving power.

This remaining condition on the coupling (again, in the limit of a large power gain) may be written as

$$\frac{\gamma_{BA}[A_{opt}]}{\gamma_0 + \gamma_{BA}[A_{opt}]} = \frac{\hbar\Omega}{4k_B T_{eff}}. \tag{5.89}$$

As $\gamma_{BA}[A] \propto A^2$ is the back-action-induced damping of the oscillator (c.f. Eq. (5.48)), we thus have that *to achieve the quantum-limited value of $\bar{S}_{xx,add}[\Omega]$ with a large power gain, one needs the intrinsic damping of the oscillator to be much larger than the back-action damping.* The back-action damping must be small enough to compensate the large effective temperature of the detector; if the bath temperature satisfies $\hbar\Omega/k_B \ll T_{bath} \ll T_{eff}$, Eq. (5.89) implies that at the quantum limit, the temperature of the oscillator will be given by

$$T_{osc} \equiv \frac{\gamma_{BA} \cdot T_{eff} + \gamma_0 \cdot T_{bath}}{\gamma_{BA} + \gamma_0} \to \frac{\hbar\Omega}{4k_B} + T_{bath}. \tag{5.90}$$

Thus, at the quantum limit and for large T_{eff}, the detector raises the oscillator's temperature by $\hbar\Omega/4k_B$.[4] As expected, this additional heating is only *half* the zero-point energy; in contrast, the quantum-limited value of $\bar{S}_{xx,add}[\omega]$ corresponds to the full zero-point result, as it also includes the contribution of the intrinsic output noise of the detector.

Quantum limit on added noise in the free-mass limit $\omega \gg \omega_M$

In gravitational wave detection, one is usually interested in the added noise at frequencies far above resonance, where the mechanical dynamics are effectively like those of a free mass. In this case, Eq. (5.87) tells us that reaching the quantum limit on the added noise requires back-action imprecision correlations satisfying:

---

[4] If in contrast our oscillator was initially at zero temperature (i.e. $T_{bath} = 0$), one finds that the effect of the back-action (at the quantum limit and for $G_P \gg 1$) is to heat the oscillator to a temperature $\hbar\Omega/(k_B \ln 5) \simeq 0.62\hbar\Omega/k_B$.

$$\bar{S}_{zF}[\omega]\Big|_{\text{optimal}} \rightarrow -\frac{\hbar\omega}{2\gamma} \tag{5.91}$$

In the limit where $\omega \gg \gamma$, the correlations are huge, implying that the optimal situation is to have back-action and imprecision noises almost perfectly correlated. If one could achieve this, the quantum limited value of the added noise is given by:

$$\bar{S}_{xx,\text{add}}[\omega] \rightarrow \frac{\hbar\gamma}{m\omega^3} \tag{5.92}$$

In contrast to the full quantum limit above, one often discusses the 'standard quantum limit' in the gravitational wave detection community. This is the minimum added noise possible (in the free mass limit) if you use a detector with quantum-ideal noise (i.e. saturates the inequality of Eq. (5.65)) but which has $\bar{S}_{zF}[\omega] = 0$. In this case, the only optimization involves tuning the coupling to balance back-action and imprecision noises, and one finds:

$$\bar{S}_{xx,\text{add}}[\omega]\Big|_{\text{SQL}} = \hbar|\chi_{xx}[\omega]| \rightarrow \frac{\hbar}{m\omega^2} \tag{5.93}$$

One sees that this larger than the true quantum limit by a large factor ω/γ.

The take-home message here is that while reaching the true quantum limit for $\omega \gg \omega_{\text{M}}$ might be challenging, one can do much better than the 'standard' quantum limit by using a detector having back-action–imprecision correlations. In the next section, we will review how injecting squeezed light into an optomechanical cavity can achieve this goal.

5.3.8 Applications to an Optomechanical Cavity (With and Without Input Squeezing)

We now apply our general approach for formulating the quantum limit to the specific case of an optomechanical cavity detector, as introduced in the first section. We consider a cavity which is strongly driven on resonance (implying that the detuning $\Delta = 0$), and which can be treated using the linearized equations of motion introduced in section 5.3.1. Our starting point is thus Eq. (5.25) for the displaced cavity field, with $\Delta = 0$. We will further work in the regime where the mechanical frequency $\omega_{\text{M}} \ll \kappa$, and we are interested in $\hat{d}[\omega]$ at frequencies $\omega \ll \kappa$. To capture the behaviour at these frequencies, we can make the adiabatic approximation, and ignore the $(d/dt)\hat{d}$ term on the LHS of Eq. (5.25). Defining $A = G/x_{\text{ZPF}} = g\bar{a}_{\text{cl}}/x_{\text{ZPF}}$, we have:

$$\hat{d}(t) \simeq -\frac{2iA}{\kappa}\hat{x}(t) - \frac{2}{\sqrt{\kappa}}\hat{d}_{\text{in}}(t) \tag{5.94}$$

where the effective coupling strength A is given by:

$$A = \frac{G}{x_{\mathrm{ZPF}}} = \frac{g\bar{a}_{\mathrm{cl}}}{x_{\mathrm{ZPF}}} \tag{5.95}$$

We see that, as anticipated, the effective coupling strength is indeed dependent on the strength of the cavity driving field. Using the input–output relation of Eq. (5.15), we also find:

$$\hat{d}_{\mathrm{out}}(t) = \hat{d}_{\mathrm{in}}(t) + \sqrt{\kappa}\,\hat{d}(t) \simeq -\frac{2iA}{\sqrt{\kappa}}\hat{x}(t) - \hat{d}_{\mathrm{in}}(t) \tag{5.96}$$

We can use these results to calculate the needed noise properties: the fluctuations in the back-action force acting on the mechanics, and the fluctuations in the detector output quantity, the phase quadrature of the output light defined in Eq. (5.31). The back-action force operator corresponds to the fluctuating part of the intracavity photon number. Keeping only the drive-enhanced term in this operator, we have that:

$$A \cdot \hat{F} \equiv \hbar A\left(\hat{d} + \hat{d}^{\dagger}\right) \tag{5.97}$$

We need to understand the fluctuations of $\hat{F}$ in the absence of any optomechanical coupling; we can thus substitute in Eq. (5.94) at $A = 0$ to find:

$$\hat{F} = -\frac{2\hbar}{\sqrt{\kappa}}\left(\hat{d}_{\mathrm{in}}(t) + \hat{d}^{\dagger}_{\mathrm{in}}(t)\right) \tag{5.98}$$

Similarly, the output quantity is given by the phase quadrature of the cavity output field, Eq. (5.31):[5]

$$\hat{I} = i\left(d_{\mathrm{out}} - \hat{d}^{\dagger}_{\mathrm{out}}\right) \tag{5.99}$$

$$= \frac{4A}{\sqrt{\kappa}}\hat{x}(t) - i\left(\hat{d}_{\mathrm{in}} - \hat{d}^{\dagger}_{\mathrm{in}}\right) \equiv A\chi_{IF}\hat{x}(t) + \hat{I}_0 \tag{5.100}$$

We thus directly can read-off both the response coefficient $\chi_{IF} = 4/\sqrt{\kappa}$ of the detector (which is frequency independent as we focus on $\omega \ll \kappa$), and the intrinsic imprecision noise in the output $\hat{I}_0$.

We see that two orthogonal (and hence canonically conjugate) quadratures of the input noise $\hat{d}_{\mathrm{in}}$ entering the cavity determine the two kinds of relevant noise (back-action and imprecision). This is to be expected: it is fluctuations in the amplitude quadrature of

---

[5] There is in principle a proportionality constant between $\hat{I}$ and the phase quadrature of the output field; however, this constant plays no role in determining the detector added noise, so we set it to unity.

the incident drive that cause photon number fluctuations and hence back-action, while it is the phase quadrature fluctuations which give imprecision noise.

Vacuum noise input

Taking the input noise to be vacuum noise and using Eqs. (5.10), (5.11), it is straightforward to calculate the needed noise correlators:

$$\bar{S}_{FF}[\omega] = \frac{4\hbar^2}{\kappa}, \qquad \bar{S}_{II}[\omega] = 1 \qquad \bar{S}_{IF}[\omega] = 0 \qquad (5.101)$$

Note crucially that there are no correlations between back-action and imprecision noise, as they correspond to conjugate quadratures of the input vacuum noise. The vanishing of correlations can ultimately be traced back to the fact that averages like $\langle \hat{d}_{\mathrm{in}}(t)\hat{d}_{\mathrm{in}}(t') \rangle$ are always zero in the vacuum state.

It thus follows that:

$$\bar{S}_{zz}\bar{S}_{FF} \equiv \frac{\bar{S}_{II}\bar{S}_{FF}}{\chi_{IF}^2} = \frac{4\hbar^2/\kappa}{16/\kappa} = \frac{\hbar^2}{4} \qquad (5.102)$$

Our driven cavity thus optimizes the quantum noise inequality of Eq. (5.65), but has no back-action–imprecision correlations. From our general discussion, this implies that it is able to reach the quantum limit on the added noise exactly at the mechanical resonance, but away from resonance, misses the true quantum limit by a large amount.

In the case where one wishes to reach the quantum limit on resonance, it is interesting to ask what the optimal coupling strength is using Eq. (5.84). It is convenient to parameterize the coupling in terms of the cooperativity $\mathcal{C}$, defined as:

$$\mathcal{C} \equiv \frac{4G^2}{\kappa\gamma} \qquad (5.103)$$

The optimal coupling required to reach the quantum limit on resonance then becomes:

$$\mathcal{C}_{\mathrm{opt}} = \frac{1}{4} \qquad (5.104)$$

Noise correlations via 'variational' readout

As we have seen, reaching the quantum limit in the free mass limit (i.e. at frequencies much larger than ω_M) requires strong correlations between the back-action noise in $\hat{F}$ and the imprecision noise in $\hat{I}_0$, c.f. Eq. (5.91). These correlations are absent in the simplest scheme described above, as the back-action and imprecision noise operators correspond to conjugate quadratures of the input vacuum noise.

An extremely simple way to induce correlations between back-action and imprecision is to alter the choice of which quadrature of the cavity output field to measure. Suppose, instead of the choice in Eq. (5.99), we chose to measure the output quadrature:

$$\hat{I}_{\text{new}} \equiv i\left(e^{i\varphi}\hat{d}_{\text{out}} - e^{-i\varphi}\hat{d}^{\dagger}_{\text{out}}\right) = \cos\varphi\,\hat{I}_{\text{old}} - \sin\varphi\left(\frac{\hat{F}}{2\hbar/\sqrt{\kappa}}\right) \tag{5.105}$$

where the angle φ determines the particular choice of quadrature. $\varphi = 0$ corresponds to measuring the phase quadrature as before, and ensures that $\hat{I}$ has a maximal sensitivity to $\hat{x}$ (as $\hat{x}$ only appears in the imaginary part of $\hat{d}_{\text{out}}$, c.f. Eq. (5.96)). By taking $\varphi \neq 0$, we reduce the response coefficient χ_{IF}, but trivially induce back-action–imprecision correlations, as now the measured detector output *includes* explicitly the back-action fluctuations $\hat{F}$. For an arbitrary choice of φ, one finds that $\bar{S}_{FF}$ and $\bar{S}_{II}$ are unchanged from Eq. (5.101), but:

$$\chi_{IF} = \frac{4\cos\varphi}{\sqrt{\kappa}} \qquad \bar{S}_{IF}[\omega] = -\sin\varphi\,\frac{2\hbar}{\sqrt{\kappa}} \tag{5.106}$$

The detector still saturates the Heisenberg bound on its quantum noise:

$$\bar{S}_{zz}\bar{S}_{FF} - \bar{S}^{2}_{zF} \equiv \frac{\bar{S}_{II}\bar{S}_{FF} - \bar{S}^{2}_{IF}}{\chi^{2}_{IF}} = \frac{(4\hbar^{2}/\kappa)(1 - \sin^{2}\varphi)}{16\cos^{2}\varphi/\kappa} = \frac{\hbar^{2}}{4} \tag{5.107}$$

Thus, this strategy in principle allows one to reach the quantum limit on the added noise for frequencies away from mechanical resonance, where one requires a non-zero cross-correlator. In the free-mass limit $\omega \gg \omega_{\text{M}}$, the optimal value of $\bar{S}_{zF}$ is given by Eq. (5.91). Achieving this value requires:

$$\frac{\hbar}{2}\tan\varphi = \frac{\omega}{\gamma} \tag{5.108}$$

One sees that the required choice of φ is frequency-dependent. To achieve the optimal correlations over a finite range of frequencies, one could first apply a frequency-dependent rotation to the output field from the optomechanical cavity (using, e.g. a second detuned cavity), and then measure a fixed quadrature. Perhaps more troubling is the fact that the RHS of the above equation is typically extremely large, implying that $\varphi \to \pi/2$. In this limit, the measured output quantity is almost equivalent to the amplitude quadrature of the input noise incident on the cavity, and has almost no information on the state of the mechanics.

The above approach is known as the 'variational' readout strategy in the gravitational wave detection community (Vyatchanin and Zubova, 1995; Vyatchanin and Matsko, 1996), and is also discussed (in a slightly different way) in Chapter 4 of this volume. While this seems like a simple strategy for reaching the quantum limit away from mechanical resonance, in practice it is not a good strategy: the strong reduction in the size of the signal means that even though the intrinsic detector noise may be as small as required by quantum mechanics, other non-intrinsic sources of added noise will start to dominate. In short, reaching the quantum limit by throwing away signal strength is almost never a good strategy.

Noise correlations via squeezing

We would like to find a way to induce the needed noise correlations for reaching the quantum limit away from mechanical resonance, while at the same time not modifying the size of the measured signal in the output of our detector. The trick will be to modify the input noise driving the cavity. It is convenient to introduce canonical (Hermitian) quadratures of the input noise $\hat{d}_{in}(t)$ via:

$$\hat{d}_{in} = \frac{1}{\sqrt{2}} \left(\hat{X}_{in} + i \hat{Y}_{in} \right) \tag{5.109}$$

As we have seen, back-action is controlled by $\hat{X}_{in}$ and imprecision noise by $\hat{Y}_{in}$. A straightforward calculation shows that

$$\bar{S}_{zF} = \hbar \bar{S}_{Y_{in}, X_{in}} \tag{5.110}$$

Hence, back-action–imprecision noise correlations require a state where the fluctuations in the two quadratures of the input noise are strongly correlated. Further, we still require that the quantum noise inequality of Eq. (5.65) be saturated, even in the presence of strong correlations. The required input noise corresponds to a quantum squeezed state (see, e.g., (Gerry and Knight, 2005) for an extensive discussion). In phase-space, the Wigner function of such a state has elliptical iso-probability contours (as opposed to the circular contours of a vacuum state or thermal state). The elongated direction of the ellipse should be aligned so as to yield the desired large correlations. As the needed correlations should be negative, the 'squeezed' direction of the ellipse (direction with minimal fluctuations) corresponds to the quadrature:

$$\tilde{X}_{in} = \frac{1}{\sqrt{2}} \left(\hat{X}_{in} + \hat{Y}_{in} \right) \tag{5.111}$$

For a squeezed state where $\tilde{X}$ is the squeezed quadrature, we have:

$$\bar{S}_{\tilde{X}_{in} \tilde{X}_{in}} = \frac{1}{2} e^{-2r} \tag{5.112}$$

where $r \geq 0$ is the squeeze parameter. Reaching the quantum limit at a frequency $\omega \gg \omega_M$ with our cavity detector thus requires one to use an input squeezing with a squeezing magnitude

$$e^{-2r} = \frac{\gamma}{2\omega}. \tag{5.113}$$

The idea of using squeezing to generate strong back-action–imprecision correlations was discussed extensively in (Pace *et al.*, 1993) (albeit using a somewhat different formulation), and is also discussed in Chapter 4 of this volume. We stress that the use of squeezing is to generate correlations; it is *not* being used to make up for a lack of incident laser power (i.e. $\bar{a}_{cl}$ too small).

An alternate use of squeezing which *is* equivalent to tuning the incident optical power was discussed in the seminal work (Caves, 1981). Imagine one wants to reach the quantum limit at mechanical resonance, but cannot achieve the optical power required to balance the contributions from back-action and imprecision noises, i.e. achieve the condition in Eq. (5.104). Squeezing can help in this situation. If one squeezes the phase quadrature of the incident light $\hat{Y}_{in}$ (and thus necessarily amplifies the amplitude quadrature $\hat{X}_{in}$), the back-action noise is enhanced, and the imprecision noise reduced. Hence, input squeezing in this case is equivalent to boosting the magnitude of the cavity drive, i.e. increasing the cooperativity C. Squeezing in this case does not generate any back-action–imprecision correlations.

5.4 Back-action Evasion and Conditional Squeezing

In this section, we discuss a method for monitoring mechanical position that is not subject to any fundamental quantum limit. As the quantum limit of the previous section is indeed a true, unavoidable limit, the only way to do better is to somehow change the rules of the game. Here, this is accomplished by being more modest in what we choose to measure. Recall that a standard weak, continuous position measurement one attempts to measure both quadrature components of the mechanical motion. i.e. $X(t)$ and $Y(t)$ as defined in Eq. (5.114). Quantum mechanically these are conjugate, non-commuting observables; as a result, one cannot measure both quadratures perfectly, as the measurement of X perturbs Y (and vice versa). In this section, we will give up trying to have full knowledge of $x(t)$, and will instead attempt to monitor only one of the two quadrature components. As we will show, this is something that can be done with no fundamental limit coming from quantum mechanics. This opens the door to force sensing with no fundamental quantum limit, as well as the possibility of using the measurement to generate quantum squeezed states of the mechanical resonator. The classic references discussing such back-action evading single quadrature measurements are (Braginsky *et al.*, 1980), (Caves *et al.*, 1980) and (Bocko and Onofrio, 1996), while the full quantum theory (including the production of conditional quantum squeezed states) was treated in (Clerk *et al.*, 2008). Note that several recent experiments have implemented this scheme using microwave circuit realizations of optomechanics: (Hertzberg *et al.*, 2010; Suh *et al.*, 2014; Lecocq *et al.*, 2015).

5.4.1 Single Quadrature Measurements

Our goal is to measure a single mechanical quadrature. The canonically conjugate quadrature operators are defined in the Heisenberg picture (with respect to the mechanical Hamiltonian) as:

$$\hat{x}(t) = \left(\sqrt{2}x_{\text{ZPF}}\right)\hat{X}(t)\cos(\omega_M t) + \hat{Y}(t)\sin(\omega_M t). \tag{5.114}$$

with

$$\hat{X} = \frac{1}{\sqrt{2}}\left(\hat{b} + \hat{b}^\dagger\right), \qquad \hat{Y} = \frac{-i}{\sqrt{2}}\left(\hat{b} - \hat{b}^\dagger\right) \qquad (5.115)$$

We stress that in the Schrödinger picture, $\hat{X}(t)$ and $\hat{Y}(t)$ are explicitly time-dependent observables, and do not simply correspond to the position and momentum operators of the mechanical resonator. For example, in the Schrödinger picture, we have:

$$\hat{X}(t) = \frac{1}{\sqrt{2}}\left(\hat{b}e^{i\omega_M t} + \hat{b}^\dagger e^{-i\omega_M t}\right) = \frac{1}{x_{ZPF}}\left(\cos(\omega_M t)\,\hat{x} - \frac{\sin(\omega_M t)}{m\omega_M}\hat{p}\right) \qquad (5.116)$$

In the second equality, $\hat{x}$ ($\hat{p}$) is the standard Schrödinger-picture position (momentum) operator of the oscillator. For an undamped oscillator, both quadrature operators are constants of the motion. Note that the definition of the quadrature operator necessarily requires some external phase reference, or equivalently, some choice defining the zero of time. With the above definition, $\hat{X}(t=0)$ is proportional to the mechanical position, while $\hat{Y}(t=0)$ is proportional to the mechanical momentum.

The goal is to measure say $\hat{X}$, and have all the corresponding back-action of the measurement drive the unmeasured quantity $\hat{Y}$. As $\hat{X}$ is dynamically independent of $\hat{Y}$, the back-action will never come back to corrupt subsequent measurements of $\hat{X}$ at later times, and we can in principle make the measurement better and better by increasing the measurement strength. Such a measurement is known as a back-action evading (BAE) measurement. It is also an example of a quantum non-demolition measurement, as one is measuring an observable (i.e. $\hat{X}(t)$) that is a constant of the motion.

While the basic idea is clear, upon first glance implementation would seem to be a challenge. To measure $\hat{X}$, we just need to couple this operator to some input operator $\hat{F}$ of our detector. From Eq. (5.116), we see that this would require time-dependent couplings between the detector and *both* the mechanical position $\hat{x}$ and mechanical momentum $\hat{p}$. This would be extremely difficult to achieve (see, e.g., Chapter 4 discussing the difficulties of coupling a detector to mechanical momentum).

Luckily, there is a simple trick to let us turn a standard coupling between detector and mechanical position (like we have in optomechanics) into the kind of single-quadrature coupling we need. The trick has two parts:

- Start with a detector which couples to the mechanical position, but modulate the coupling strength in time at the mechanical frequency ω_M. Working in the Schrödinger picture, we want to modify the basic system–detector interaction in Eq. (5.69) to now have the form:

$$\hat{H}_{int} = \left(\bar{A}\cos\omega_M t\right)\hat{x} \cdot \hat{F} = \hat{X}(t)\left(1 + \cos 2\omega_M t\right) + \hat{Y}(t)\sin 2\omega_M t \qquad (5.117)$$

In the last line, we have re-expressed things in terms of the quadrature operators. Note that the coupling to $\hat{X}$ has a time-independent part, where the coupling to $\hat{Y}$ is strictly oscillating.

- Next, imagine we work with a 'slow' detector, one that cannot respond to perturbations occurring at frequencies $\sim 2\omega_M$. In that case, the time-dependent oscillating terms in the above equation will average away, and we will be left with a time-independent coupling between the detector force operator $\hat{F}$ and the $\hat{X}$ quadrature only.

5.4.2 Two-tone QND Scheme

As first discussed in (Braginsky *et al.*, 1980), the above 'modulated' coupling scheme for measuring a single mechanical quadrature can be achieved using a standard optomechanical setup if one works in the resolved sideband regime $\omega_M \gg \kappa$ and drives the cavity equally with coherent tones at both the red and blue mechanical sidebands, i.e. at $\omega_{cav} \pm \omega_M$. This is equivalent to amplitude modulating the laser drive on the cavity. We thus consider a cavity input field (i.e. laser drive) of the form:

$$\bar{a}_{in}(t) \sim e^{-i\omega_{cav}t} \sin(\omega_M t + \phi) \tag{5.118}$$

The phase ϕ of the amplitude modulation will directly determine the definition of the mechanical quadrature which couples to the cavity field; we take $\phi = 0$ for simplicity in what follows. As usual, this coherent-state drive will induce an average (time-dependent) amplitude in the cavity $\alpha(t)$ that can be found by solving the classical equations of motion. We will again work in a displaced frame, and thus write the cavity lowering operator as this classical amplitude plus a correction $\hat{d}$ which describes noise effects and the effects of the optomechanical coupling:

$$\hat{a} = \bar{a}_{cl}(t) + \hat{d}, \qquad \bar{a}_{cl}(t) = \alpha_0 e^{-i\omega_{cav}t} \cos\omega_M t \tag{5.119}$$

We can take α_0 to be real without loss of generality.

Linearizing the optomechanical interaction in the usual way yields:

$$\hat{H}_{int} = g\hat{a}^\dagger \hat{a} \frac{\hat{x}}{x_{ZPF}} \simeq \left[\frac{g}{x_{ZPF}}\alpha_0 \cos(\omega_M t)\right] \hat{x}\left(\hat{d}e^{i\omega_{cav}t} + h.c.\right) \tag{5.120}$$

$$\equiv A(t)\hat{x} \cdot \hat{F}(t) \tag{5.121}$$

We see that modulating the average cavity intensity naturally provides the modulated coupling we are after. Working in an interaction picture with respect to the free mechanical and cavity Hamiltonians $\hat{H}_0 = \omega_M \hat{b}^\dagger \hat{b} + \omega_{cav}\hat{d}^\dagger \hat{d}$ yields

$$\hat{H}_{int} = \frac{G}{\sqrt{2}}\left[\hat{X}(1 + \cos 2\omega_M t) + \hat{Y}\sin 2\omega_M t\right]\left(\hat{d} + \hat{d}^\dagger\right) \tag{5.122}$$

where we have defined the many-photon coupling $G = g\alpha_0$. Note that in our interaction picture, the mechanical quadrature operators have no explicit time-dependence, and are given by Eqs. (5.114).

We now make use of the resolved sideband, good-cavity condition $\omega_M \gg \kappa$, and the fact that we will be interested in coupling $G \ll \kappa$. In this limit the terms oscillating at frequency $2\omega_M$ will average to zero on the timescales relevant to the dynamics (which will all be much longer than $1/\kappa$). We can thus safely make the rotating-wave approximation (RWA), and drop the oscillating terms, resulting in a time-independent interaction Hamiltonian:

$$\hat{H}_{\text{int}} = G\hat{X} \cdot \left(\frac{\hat{d} + \hat{d}^\dagger}{\sqrt{2}} \right) \equiv G\hat{X} \cdot \hat{X}_{\text{cav}} \qquad (5.123)$$

We see that the mechanical $\hat{X}$ quadrature is coupled only to the corresponding X quadrature of the cavity. Without dissipation, both $\hat{X}$ and $\hat{X}_{\text{cav}}$ are constants of the motion. Similar to the case of standard position detection, as the mechanics only couples to the X cavity quadrature, information on its motion will only drive the conjugate cavity quadrature $\hat{Y} \sim -i\left(\hat{d} - \hat{d}^\dagger\right)$ (i.e. the optical phase quadrature). By measuring the output Y quadrature, we thus obtain a measurement of the mechanical $\hat{X}$ quadrature.

To see this explicitly, we first solve the Heisenberg–Langevin equation for $\hat{d}$:

$$\frac{d}{dt}\hat{d}(t) = -\frac{\kappa}{2}\hat{d}(t) - \frac{iG}{\sqrt{2}}\hat{X}(t) - \sqrt{\kappa}\hat{d}_{\text{in}}(t) \qquad (5.124)$$

Information on the mechanics will be at the cavity field near resonance, in a bandwidth $\sim \gamma \ll \kappa$. We thus only need to describe $\hat{d}$ at frequencies $\ll \kappa$, and solve this equation adiabatically, i.e. ignoring the d/dt term. This yields:

$$\hat{d}(t) = \frac{-2iG}{\sqrt{2}\kappa}\hat{X}(t) - \frac{2}{\sqrt{\kappa}}\hat{d}_{\text{in}}(t) \qquad (5.125)$$

Information on the mechanical X quadrature is, as expected, encoded solely in the imaginary part of $\hat{d}$, and hence in the cavity Y quadrature. One thus measures the Y quadrature of the output field. The measurement output (i.e. output homodyne current) is then:

$$\hat{I}(t) = \sqrt{2}\hat{Y}_{\text{cav,out}} \sim \sqrt{2\kappa}\,\hat{Y}_{\text{cav}} \sim \frac{2\sqrt{2}G}{\sqrt{\kappa}}\hat{X}(t) + \hat{\xi}(t) \qquad (5.126)$$

$$\equiv \frac{2\sqrt{2}G}{\sqrt{\kappa}}\left(\hat{X}(t) + \delta\hat{X}_{\text{imp}}(t)\right) \qquad (5.127)$$

where $\hat{\xi} = -i(\hat{d}_{\text{in}} - \hat{d}_{\text{in}}^\dagger)$ is the Y quadrature of the input noise driving the cavity, and is delta-correlated. In the last line, we have introduced the imprecision noise operator

$\delta\hat{X}_{\text{imp}}(t)$ in the usual way, by referring the intrinsic noise in $\hat{I}(t)$ back to the measured mechanical quadrature. Note the $\sqrt{2}$ factor in the definition of the homodyne current has just been included for convenience (the overall prefactor plays no role in our analysis).

Using this form, one easily finds that symmetrized spectral density of the imprecision noise is given by:

$$\bar{S}_{XX, \text{imp}}[\omega] = \frac{\kappa}{8G^2} \equiv \frac{1}{\tilde{k}} \tag{5.128}$$

The imprecision noise spectral density has the units of an inverse rate, and we have thus used it to define an effective measurement rate $\tilde{k}$.

The rate $\tilde{k}$ has a simple interpretation: it tells us how quickly the power signal-to-noise (SNR) ratio grows for a measurement where we try to resolve whether the measured mechanical quadrature has been displaced an amount $\sim x_{\text{ZPF}}$, i.e. is $X = 1$ or 0? A simple estimator would be to simply integrate the output current $I(t)$, i.e.

$$\hat{m}(t) \equiv \int_0^t dt' \hat{I}(t') \tag{5.129}$$

It is easy to then check that in the absence of mechanical dissipation (and to lowest order in the coupling to the detector)

$$\text{SNR} \equiv \frac{\text{signal power}}{\text{noise power}} = \frac{[\langle \hat{m}(t) \rangle_{X=1} - \langle \hat{m}(t) \rangle_{X=0}]^2}{\langle\langle \hat{m}^2(t) \rangle\rangle} = \frac{\left(\sqrt{\tilde{k}}t - 0\right)^2}{t} = \tilde{k}t \tag{5.130}$$

where the variance $\langle\langle \hat{m}^2(t) \rangle\rangle = \langle \hat{m}^2(t) \rangle - \langle \hat{m}(t) \rangle^2$.

Having understood the basics of the measurement, we can now ask about the effects of back-action. Consider first the ideal limit where $\kappa/\omega_{\text{M}} \to 0$, and hence the cavity has strictly no coupling to the mechanical $\hat{Y}$ quadrature. In this case $\hat{X}$ commutes with the interaction Hamiltonian, implying that it is completely unaffected by the coupling. By solving the mechanical Heisenberg–Langevin equation, one finds that the fluctuations are the same as when $G = 0$, i.e.

$$\bar{S}_{XX}[\omega] \equiv \frac{1}{2}\int_{-\infty}^{\infty} dt\, e^{i\omega t} \left\langle \{\hat{X}(t), \hat{X}(0)\} \right\rangle = \frac{\gamma/2}{\omega^2 + (\gamma/2)^2}\left(1 + 2\bar{n}_{\text{th}}^M\right) \tag{5.131}$$

Again, this is exactly what we would have without any coupling to the cavity.

In contrast, the mechanical Y quadrature is driven by the fluctuations in the cavity $\hat{X}_{\text{cav}}$ operator. This back-action does not change the damping rate of the quadrature, but does heat it an amount corresponding to $\bar{n}_{\text{BA}}$ quanta. One finds:

$$\bar{S}_{YY}[\omega] = \frac{\gamma/2}{\omega^2 + (\gamma/2)^2}\left(1 + 2\bar{n}_{\text{th}}^M + 2\bar{n}_{\text{BA}}\right), \qquad \bar{n}_{\text{BA}} = \frac{2G^2}{\kappa\gamma} \equiv \mathcal{C} \tag{5.132}$$

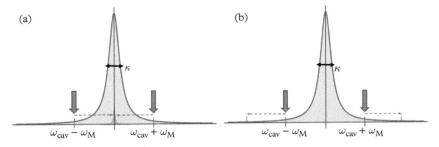

Fig. 5.3 *(a) Cavity density of states and drive frequencies for the two-tone BAE measurement scheme; the frequencies of the classical drives are marked with large solid arrows. Photons arriving at the cavity resonance could have been generated either by anti-Stokes scattering from the red-detuned drive tone, or Stokes scattering from the blue-detuned drive tone. As the amplitudes of these drives are equal, the cavity amplitude at resonance effects a measurement of the mechanical X quadrature. One needs to be in the good cavity limit, where the mechanical frequency ω_M is much larger than the cavity linewidth κ. (b) The mechanical motion can also generate weak amplitudes at frequencies $\omega_{cav} \pm 2\omega_M$ via non-resonant Raman processes. As the density of states for such processes is small, their amplitude is weak. These processes effectively measure the mechanical Y quadrature, and thus causing a small back-action heating of X.*

We have introduced here again the optomechanical cooperativity $\mathcal{C}$, which in this context, can be viewed as the ratio of the measurement rate to the intrinsic mechanical damping rate.

What if we include the effects of κ/ω_M? The oscillating terms in Eq. (5.122) are now not completely negligible. Heuristically, they give rise to scattering processes where incident drive photons are scattered to frequencies $\omega_{cav} \pm 2\omega_M$ (see Fig. 5.3 (b)). Such processes have a very small amplitude due to the very small cavity density of states at these frequencies. They do, however, contain information on the mechanical Y quadrature, and thus result in heating of the mechanical X quadrature. A careful calculation (see (Clerk, Marquardt and Jacobs, 2008)) finds that this additional heating results can be captured by making the replacement:

$$\bar{n}_{th}^M \to \bar{n}_{th}^M + \frac{1}{32}\left(\frac{\kappa}{\omega_M}\right)^2 \tag{5.133}$$

in Eqs. (5.131),(5.132) for the quadrature noise spectral densities.

5.4.3 Conditional Squeezing: Heuristic Description

We have seen that a perfect back-action evading measurement results in a heating of the unmeasured, conjugate mechanical quadrature Y, while the measured mechanical quadrature X is completely unaffected by the measurement, c.f. Eq. (5.131). This is in keeping with the fact that $\hat{X}$ commutes with the optomechanical coupling Hamiltonian,

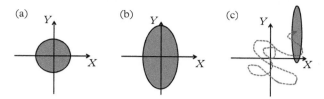

Fig. 5.4 *Mechanical state associated with various aspects of the BAE measurement, depicted as probability density in the phase-space associated with the mechanical quadrature amplitudes X and Y. (a) Mechanical state without the measurement: a thermal state with equal uncertainty in the X and Y quadratures. (b) Unconditional mechanical state when the measurement is on. The measured X quadrature is unaffected by the measurement, whereas the back-action of the measurement heats the Y quadrature and thus increases its uncertainty. (c) Conditional mechanical state. If we use the information in the measurement record $I(t)$ associated with a particular run of the experiment, we see that the mechanical resonator is in a squeezed state, where the X quadrature uncertainty has been greatly reduced compared to its original value. However, the mean values of X and Y (i.e. centre of the ellipse) undergo a random walk which is completely correlated with the measurement record. If we average over these fluctuations (e.g. discard the measurement record), we recover the picture in panel (b).*

and is thus completely unaffected by the light field used for the measurement. We thus expect the mechanical state (pictured as a phase-space distribution) to be modified as shown in Fig. 5.4(b). Note that the total entropy of the state appears to have been increased by the measurement.

While this seems simple enough, it would seem to contradict another aspect of our treatment. Namely, we showed that the added noise of the measurement (which has no back-action contribution) could be made arbitrarily small by increasing the measurement strength (i.e. G), meaning that one could make an essentially perfect measurement of X. This seems to imply that the measurement will greatly reduce the uncertainty in the X quadrature, in stark contrast to what is depicted in Fig. 5.4(b).

There is of course no contradiction here. To fully describe the reduction in X uncertainty occurring during a measurement, we need to understand what happens during a particular run of the experiment: given a particular history of the measurement output $I(t)$, what is the state of the mechanical resonator? We will show that there is a particularly simple picture for this 'conditional' mechanical state, as is shown schematically in Fig. 5.4(c). Once transients have died away, the conditional mechanical state will have greatly reduced X quadrature uncertainty—it will be a squeezed state. However, the mean position of this state in phase-space will undergo a random walk, a random walk which is completely correlated with the seemingly random fluctuations of the measurement record $I(t)$. If we have access to the measurement record, we can follow these fluctuations, and thus to us, this random motion does not represent a true uncertainty.

In contrast, for an observer who does not have access to $I(t)$, the fluctuations in the mean position of the mechanical squeezed state are just another uncertainty in the mechanical state. For such an observer, the mechanical resonator is best described by the unconditional state, where we average over all possible measurement outcomes, and thus include the random walk done by the conditional squeezed state means in the state uncertainty. Doing so, we return to the picture in Fig. 5.4(b): the fluctuations in the mechanical resonator X quadrature to an observer who does not have access to the measurement record are the same as they were without the measurement.

In what follows, we will develop in more detail the theoretical tools required to describe such a conditional evolution and the conditional mechanical state. We will then apply this formalism to our optomechanical back-action evading measurement to understand how one can generate squeezed mechanical states from the measurement.

5.4.4   Stochastic Master Equation Description of a Conditional Measurement

In this section, we will give a quick and dirty 'derivation' of the stochastic master equation approach that describes the conditional evolution of the mechanical state during an ideal version of our back-action evading measurement. Our approach follows closely the extremely pedagogical treatment in (Jacobs and Steck, 2006), which itself is related to the derivation in (Caves and Milburn, 1987). The derivation, interpretation and use of such stochastic master equations are treated extensively in recent textbooks (Wiseman and Milburn, 2014; Jacobs, 2014).

Discretized measurement record

We start by writing the measurement output (Eq. (5.127)) as

$$\hat{I}(t) = \sqrt{\bar{k}}\hat{X}(t) + \hat{\xi}(t) \tag{5.134}$$

As we already saw in the discussion surrounding Eq. (5.130), if we only look at $I(t)$ over a very short time interval, we get almost no information on the state of the mechanical resonator: the noise $\xi(t)$ in the measurement record completely dominates the contribution from the signal $X(t)$. Useful information is thus only acquired gradually over time. To describe our measurement, we want to quantitatively describe the small information gain that occurs over a given short time interval, and importantly, also describe the corresponding change in the mechanical state.

We start by discretizing time into finite intervals of width Δt and define $t_j \equiv j\Delta t$; we will eventually take the $\Delta t \to 0$ limit. During each of these finite intervals, we use the information in $I(t)$ to estimate the state of the mechanics. This is done analogously to Eq. (5.130). The estimate of X derived from $I(t)$ in the interval (t_{j-1}, t_j) is described by the operator $\hat{X}_{t_j}$, defined as

$$\hat{X}_{t_j} \equiv \frac{1}{\sqrt{\bar{k}}\Delta t} \int_{t_{j-1}}^{t_j} dt' \hat{I}(t'). \tag{5.135}$$

For simplicity, we start by assuming that the instantaneous mechanical resonator state is pure and described by the wavefunction $|\psi(t)\rangle$, and assume that it has no intrinsic dynamics—we ignore its intrinsic dissipation and any external forces. We will also assume that any back-action disturbance occurs instantaneously at the end of each interval (t_{j-1}, t_j), and hence $|\psi(t)\rangle$ is constant over the interval. As we are interested in the $\Delta t \to 0$ limit, this approximation is not unreasonable. Using this fact and the fact that $\langle \hat{\xi}(t) \rangle = 0$, we have

$$\langle \hat{X}_{t_j} \rangle \simeq \langle \hat{X}(t) \rangle \equiv \langle \psi(t) | \hat{X} | \psi(t) \rangle \tag{5.136}$$

We can also calculate the variance of $\hat{X}_{t_j}$:

$$\left\langle \left(\left(\hat{X}_{t_j} \right)^2 \right) \right\rangle \equiv \left\langle \left(\hat{X}_{t_j} - \left\langle \hat{X}_{t_j} \right\rangle \right)^2 \right\rangle$$

$$\simeq \frac{1}{\left(\sqrt{\tilde{k}} \Delta t \right)^2} \int_{t_{j-1}}^{t_j} dt' \int_{t_{j-1}}^{t_j} dt'' \langle \hat{\xi}(t') \hat{\xi}(t'') \rangle \tag{5.137}$$

$$= \frac{1}{\tilde{k} \Delta t} \equiv \sigma \tag{5.138}$$

Note that we have ignored the contribution to the variance from the intrinsic uncertainty of $\hat{X}$ in the mechanical resonator state $|\psi(t)\rangle$. This is because the contribution we keep (which is solely from the imprecision noise of the measurement, $\xi(t)$) completely dominates the intrinsic noise in the limit of small Δt. This imprecision-noise contribution scales like $1/\Delta t$, while the intrinsic noise would tend to a constant.

We have thus discretized the measurement record $I(t)$ that would be obtained in a given experimental run into a set of discrete X-quadrature estimates X_{t_j}. These estimates can be viewed as an effective classical stochastic process, i.e.

$$X_{t_j} = \langle \hat{X}(t_j) \rangle + \frac{\Delta W_j}{\sqrt{\tilde{k}} \Delta t} \tag{5.139}$$

Here, the ΔW_j are random variables describing the fluctuations of each estimate due to the imprecision noise $\xi(t)$ in the measurement record. Equivalently, they represent the difference in the outcome of the measurement during the given time interval from the mean of $\hat{X}$ in the state $|\psi(t)\rangle$. As $\xi(t)$ is Gaussian white noise, it follows that the ΔW_j are each Gaussian random variables with zero mean, and further, are not correlated with one another. From Eq. (5.138), we have simply:

$$\overline{\Delta W_j} = 0 \qquad \overline{\Delta W_j \Delta W_{j'}} = \delta_{j,j'} \Delta t \tag{5.140}$$

The bar here represents an average over the classical stochastic process. We stress that the classical stochastic process defined by Eq. (5.139) has been constructed so that it yields the same statistics as the microscopic (quantum) theory describing our measurement.

State evolution

Having come up with a simple way to think about the measurement record produced in a single run of the experiment, we now return to the question of back-action: in a particular run of the experiment (described by a particular set of X_{t_j}), how will the state $|\psi(t)\rangle$ of the mechanical resonator change over time? Again, we are only considering here mechanical dynamics caused by the measurement, and are assuming that these disturbances happen instantaneously between each measurement interval (t_{j-1}, t_j). At the end of a given measurement interval, the experimentalist will have obtained a particular measurement result X_{t_j}. If the measurement was a simple, strong projective measurement, the mechanical state after the measurement would be completely determined by the outcome X_{t_j}; the mechanics would just be projected into the eigenstate of $\hat{X}$ corresponding to this outcome.

In our system, we are however very far from the strong measurement limit, as we only obtain a tiny bit of information on X during the given time interval. The final mechanical state will thus depend *both* on the initial mechanical state and the measurement outcome, and we can write:

$$|\psi(t_j + \Delta t)\rangle = \hat{M}(X_{t_j})|\psi(t_j)\rangle \qquad (5.141)$$

The operator $\hat{M}(X_{t_j})$ describes the outcome-dependent disturbance of the mechanical state by the measurement in a given time interval. It is a so-called Krauss operator, used in the POVM description of the kind of incomplete, weak measurements relevant here (see, e.g., (Jacobs and Steck, 2006) for a more complete discussion). We assume the ideal case where this state disturbance is as small as possible given the information gain of the measurement. As the statistics of our measurement outcomes are Gaussian distributed and dominated by imprecision noise, the corresponding Krauss operators also have a Gaussian form:

$$\hat{M}(x) \propto \exp\left[-\frac{1}{4\sigma^2}\left(x - \hat{X}\right)^2\right], \qquad (5.142)$$

where $\sigma = 1/(\tilde{k}\Delta t)$ is the variance of each discrete measurement (c.f. Eq. (5.138)), and we have dropped a normalization constant. Note that in extreme limit where $\sigma \to \infty$, $\hat{M}(x) \to 1$: there is no information obtained in the measurement, and hence no measurement-induced change in the mechanical state. In the opposite case $\sigma \to 0$, then $\hat{M}(x) \to |x\rangle\langle x|$, i.e. the Krauss operators become projectors onto eigenstates of $\hat{X}$, and we recover the standard description of a strong measurement.

Using Eq. (5.139), we can re-express the post-measurement mechanical state in terms of the random variables ΔW_j as:

$$|\psi(t_j + \Delta t)\rangle \propto \exp\left[-\frac{\tilde{k}\Delta t}{4}\hat{X}^2 + \left(\frac{\tilde{k}\Delta t}{2}\langle\hat{X}(t_j)\rangle + \sqrt{\frac{\tilde{k}}{4}}\Delta W_j\right)\hat{X}\right]|\psi(t_j)\rangle, \qquad (5.143)$$

where again we have dropped purely constant prefactors affecting the normalization of the state.

Next, as we are interested in the $\Delta t \to 0$ limit, all the terms in the argument of the above exponential will become small, and we can thus Taylor expand keeping terms to order Δt only. This implies that we need terms that are up to second order in ΔW_j, as this random variable has a variance Δt, and hence a typical size $\sqrt{\Delta t}$. We thus find:

$$|\psi(t_j + \Delta t)\rangle \propto \left[1 - \frac{\tilde{k}\Delta t}{4}\hat{X}^2 + \left(\frac{\tilde{k}\Delta t}{2}\langle \hat{X}(t_j)\rangle + \sqrt{\frac{\tilde{k}}{4}}\Delta W_j \right) \hat{X} + \frac{\tilde{k}}{8}\left(\Delta W_j \right)^2 \hat{X}^2 \right] |\psi(t_j)\rangle$$

(5.144)

Finally, we want to take the limit where the duration of our small time intervals tends to an infinitesimal, $\Delta t \to dt$, which will give us a stochastic differential equation for the evolution of the mechanical state. In this limit, we will follow standard convention, and label ΔW as dW, the so-called Wiener increment. In general, one would think that dW^2 is itself a random variable which fluctuates. In the $\Delta t \to 0$ limit, these fluctuations play no role, and we can rigorously replace dW^2 by its average value, dt (see, e.g., Ch. 3 of (Jacobs, 2010) for a detailed discussion). Using this, and also adding terms to ensure that our state remains normalized, we find:

$$d|\psi\rangle \equiv |\psi(t + dt)\rangle - |\psi(t)\rangle$$

$$= \left[-\frac{\tilde{k}}{8}\left(\hat{X} - \langle \hat{X}\rangle \right)^2 dt + \sqrt{\frac{\tilde{k}}{4}}\left(\hat{X} - \langle \hat{X}\rangle \right) dW \right] |\psi(t)\rangle$$

(5.145)

In the same limit, the measurement record can be represented as:

$$dI \equiv \int_{t-\Delta t}^{t} I(t')\,dt' \to \sqrt{\tilde{k}}\langle \hat{X}(t)\rangle dt + dW$$

(5.146)

Several comments are now in order:

- Eq. (5.145) and Eq. (5.146) are coupled stochastic differential equations. The random Wiener increment dW both dictates the evolution of the measurement record and the evolution of the state of the mechanical resonator. Formally, it tells us that the noise in the measurement record directly reflects changes in the mechanical state.

- Eq. (5.145) is nonlinear, as the expectation value $\langle \hat{X}\rangle$ on the RHS is of course itself a function of the mechanical state. It should be evaluated using the mechanical state at time t.

- It is useful to interpret the two terms Eq. (5.146) in a manner that connects to Baysian probabilities. The first term is what we 'expect' from our measurement, given our current knowledge of the mechanical state (as represented by $|\psi(t)\rangle$). The second term (dW) represents the 'surprise' of our measurement. Just as in Baysian probabilities, we should update our knowledge of the mechanical state based on this new information. This is exactly the role of the dW term on the RHS of Eq (5.145).

Finally, it is straightforward to calculate the corresponding equation of motion for the density matrix describing the mechanical state, using $\hat{\rho} = |\psi\rangle\langle\psi|$ and Eq. (5.145). As usual, one needs to retain terms to order ΔW^2, as in the $\Delta t \to 0$ limit, $\Delta W^2 \to dW^2 = dt$. One finds:

$$d\hat{\rho}\big|_{\text{meas}} = -\frac{\tilde{k}}{8}\left[\hat{X},\left[\hat{X},\hat{\rho}\right]\right]dt + \sqrt{\frac{\tilde{k}}{4}}\left(\hat{X}\hat{\rho} + \hat{\rho}\hat{X} - 2\langle\hat{X}\rangle\hat{\rho}\right)dW \qquad (5.147)$$

Again, the last term on the RHS $\propto dW$ describes how the evolution of mechanical state is correlated with the noise in the measurement record; equivalently, it tells us how our knowledge of the mechanical state (as encoded in $\hat{\rho}$) is updated by the surprise dW of the measurement. Note that if $\hat{\rho}$ is initially in a pure state, then Eq. (5.147) keeps it in a pure state at all times (for the simple reason that the evolution described by Eq. (5.145) also keeps the mechanical state pure). The evolution of the density matrix clearly depends on the particular history and form of the measurement record (through the dW) terms, and we call this the *conditional density matrix*. One could also ask what the state of the mechanical resonator is averaged over *all possible* measurement outcomes. Equivalently, if we don't have access to $I(t)$, how would we describe the mechanical state? Averaging over possible measurement records is equivalent to averaging over the dW, which is simple, as $dW = 0$. Hence, the *unconditional* density matrix evolves only under the first term on the RHS of Eq. (5.147). As can easily be checked, this term causes off-diagonal elements of the the density matrix in the $\hat{X}$-eigenstate basis to decay exponentially. This is just the expected unconditional back-action of the measurement, which is a heating of the unmeasured Y quadrature.

Note that while we have given a rather heuristic derivation of the conditional master equation for our system, it is possible to give a more microscopic description, one that starts with the photodetection involved in the homodyne measurement of the cavity output field; see (Clerk *et al.*, 2008) for details.

Finally, we also need to include terms which correspond to the measurement-independent evolution of the mechanical resonator. In our case, the mechanical quadratures have no free Hamiltonian evolution, but will be subject to the heating and damping by the intrinsic sources of mechanical dissipation. We have discussed how these can be described using the Heisenberg–Langevin formalism. In the Markovian limit of interest, they can equivalently be described by Linblad terms in a quantum master equation:

$$d\hat{\rho}\big|_{\text{diss}} = \gamma\left(1 + \bar{n}_{\text{th}}^M\right)\left[\hat{b}\hat{\rho}\hat{b}^{\dagger} - \frac{1}{2}\left\{\hat{b}^{\dagger}\hat{b},\hat{\rho}\right\}\right]dt + \gamma\left(\bar{n}_{\text{th}}^M\right)\left[\hat{b}^{\dagger}\hat{\rho}\hat{b} - \frac{1}{2}\left\{\hat{b}\hat{b}^{\dagger},\hat{\rho}\right\}\right]dt \quad (5.148)$$

These describe the addition and removal of quanta from the mechanical resonator by a thermal reservoir. The total evolution of the system in the presence of the measurement is then:

$$d\hat{\rho} = d\hat{\rho}\big|_{\text{diss}} + d\hat{\rho}\big|_{\text{meas}} \tag{5.149}$$

5.4.5 Conditional Back-Action Evading Measurement

We can now apply the general theory of the previous section to our ideal cavity optomechanical single quadrature measurement. We want to describe the conditional state of the mechanical resonator, i.e. its state during a particular run of the measurement, with a particular measurement record $I(t)$. As the RHS of the conditional master equation has no terms that involve more than two mechanical raising or lowering operators, it has the property that Gaussian states (e.g. a thermal state, the ground state) remain Gaussian under the evolution. We can thus reduce our conditional master equation to a set of ODEs for the means and variances of the Gaussian state. For the $\hat{X}$ quadrature, we thus need to know:

$$\bar{X}(t) \equiv \langle \hat{X}(t) \rangle_{\text{cond}} \qquad V_X(t) \equiv \langle \hat{X}^2(t) \rangle_{\text{cond}} - \left(\langle \hat{X}(t) \rangle_{\text{cond}} \right)^2 \tag{5.150}$$

where all averages are with respect to the conditional density matrix of the mechanical resonator.

One finds straightforwardly that the evolution of the variances and covariances are completely deterministic, i.e. they do not involve dW. For the X quadrature, one finds:

$$\frac{dV_X}{dt} = -\tilde{k}V_X^2 - \gamma \left(V_X - \bar{n}_{\text{th}}^M - \frac{1}{2} \right) \tag{5.151}$$

This has an extremely simple interpretation. Without the measurement $\tilde{k} = 0$, the intrinsic mechanical dissipation causes V_X to relax exponentially (rate γ) to its thermal equilibrium value. In contrast, with the measurement, the first term tries to relax V_X towards zero, i.e. prepare a squeezed state where the X quadrature variance is minimal.

Note that for our system, the measurement strength $\tilde{k} = 4\mathcal{C}\gamma$, where the cooperativity is defined in Eq. (5.132). For the large cooperativity limit, the stationary value of the variance is given by:

$$\frac{V_X}{V_{X,zpt}} \rightarrow \sqrt{\frac{1 + 2\bar{n}_{\text{th}}^M}{2\mathcal{C}}} \equiv \sqrt{\frac{1}{2\mathcal{C}_{\text{th}}}} \tag{5.152}$$

where in the last line, we have introduced the so-called thermal cooperativity. We thus see that achieving strong quantum squeezing requires $\mathcal{C}_{\text{th}} > 1/2$; this is similar to the condition required for ground state cavity cooling with a system in the good cavity limit.

We can also ask about how we recover the unconditional picture of the measurement, where the X quadrature is unaffected. Recall the picture in Fig. 5.4(c): the X quadrature is in a squeezed state whose mean fluctuates in a way that is correlated with the measurement record. If we don't have access to the measurement record, we should include these fluctuations of the mean in the uncertainty of X, and expect that this will offset the squeezing found before. To make this quantitative, we can calculate the equation of motion for the average of X in the conditional mechanical state from the conditional master equation. One finds:

$$d\bar{X}(t) = -\frac{\gamma}{2}\bar{X}dt + \sqrt{\tilde{k}}V_X(t)dW \tag{5.153}$$

The evolution of the mean of X is indeed fluctuating, in a way that is completely correlated with the measurement record (i.e. dW both determines $\bar{X}$ and the measurement record $I(t)$). If one averages over these fluctuations, one can explicitly check that:

$$V_X + \langle \bar{X}(t)^2\rangle = \frac{1}{2} + \bar{n}^M_{\text{th}} \tag{5.154}$$

i.e. one recovers the results of the unconditional theory, where the X quadrature variance is the thermal equilibrium value, the same as though there were no measurement.

What about the unmeasured Y quadrature, how does it evolve? Letting $C(t)$ denote the covariance $\frac{1}{2}\langle\{\hat{X}(t), \hat{Y}(t)\}\rangle$, we have again that the remaining variances and covariances also evolve deterministically:

$$\frac{dV_Y}{dt} = -\tilde{k}C^2 - \gamma\left(V_Y - \bar{n}^M_{\text{th}} - \frac{1}{2}\right) + \frac{\tilde{k}}{4} \tag{5.155}$$

$$\frac{dC}{dt} = -\tilde{k}V_X C - \gamma C \tag{5.156}$$

The last term on the RHS of Eq. (5.155) describes the expected heating of the measurement. The first term tells us that if there are correlations between the quadratures, then our measurement of X also has the effect of reducing the uncertainty in Y. Finally, the last equation tells us that under the measurement, any initial correlations between the quadratures will decay away.

Finally, we have considered so far the case of a perfect measurement. It is of course also important to understand what happens to the conditional squeezing generated by the measurement when things are not so perfect. One key imperfection is that the final measurement of the cavity output quadrature is not perfect. This could be due to losses (which replace signal by vacuum noise), or due to unwanted added noise in the final homodyne measurement (e.g. due to a following non-quantum-limited amplifier in a microwave frequency optomechanics experiment). In either case, the next effect is to reduce the size of the signal term in Eq. (5.127) while keeping the noise term the same, i.e. $\tilde{k} \to \tilde{k}\sqrt{\eta}$, where the efficiency $\eta \leq 1$. Including this imperfection, one finds that the conditioning of the X quadrature (first term in Eq. (5.154) is reduced by a factor of η, whereas the back-action heating of the Y quadrature is unchanged (last term in Eq. (5.155)). As a result, in the large cooperativity limit, the condition for quantum squeezing becomes more stringent:

$$\left(\mathcal{C}_{\mathrm{th}} \equiv \frac{2G^2}{\kappa\gamma\left(1 + 2\bar{n}_{\mathrm{th}}^M\right)} \right) \geq \frac{1}{2\eta} \tag{5.157}$$

5.4.6   Feedback to Create Unconditional Squeezing

As discussed in (Clerk *et al.*, 2008), one can convert the conditional squeezing described above into true squeezing: one needs to use information in the measurement record to apply an appropriate feedback force which suppresses the fluctuations in the conditional mean $\bar{X}(t)$. The feedback force should create an extra damping of the $\bar{X}(t)$; we thus need to apply a linear force which couples to the mechanical $\hat{Y}$ quadrature:

$$\hat{H}_{\mathrm{fb}} = -\alpha\frac{\gamma}{2}\bar{X}(t) \cdot \hat{Y} \tag{5.158}$$

Here, $\alpha(\gamma/2)$ is the strength of the applied feedback. It is easy to check that this feedback force increases the damping of $\bar{X}$ in Eq. (5.153) by an amount $\alpha\gamma/2$. For large α, this feedback-induced damping will suppress the $\bar{X}$ fluctuations, and the unconditional state variance of X will be the same as the conditional (squeezed) variance.

This might still seem mysterious: how does the experimentalist know what $\bar{X}(t)$ is? To answer this, we re-write Eq. (5.153) (including the feedback force) in terms of the measurement record $dI(t)$ defined in Eq. (5.146). We find:

$$\frac{d}{dt}\bar{X} = -\frac{\gamma}{2}(1+\alpha)\bar{X} + \sqrt{\tilde{k}}V_X\left(dI(t) - \sqrt{\tilde{k}}\bar{X}\right) \tag{5.159}$$

We can now solve this equation, expressing $\bar{X}(t)$ in terms of the measurement record at earlier times. Assuming V_X has achieved its stationary value, we have

$$\bar{X}(t) = \sqrt{\tilde{k}}V_X \int_{-\infty}^{t} dt'\, e^{-\Gamma(t-t')} dI(t') \tag{5.160}$$

$$\Gamma = \frac{\gamma}{2}(1+\alpha) + \tilde{k}V_X \tag{5.161}$$

Thus, the conditional mean $\bar{X}(t)$ (i.e. our best estimate for the value of the mechanical X quadrature) is determined by filtering the measurement record obtained at earlier times. The optimal filter is just exponentially decaying (i.e. the measurement record at recent times influences our estimate more than the record at earlier times), with a time constant $1/\Gamma$ that depends on both the measurement strength (through $\tilde{k}$) and on the feedback strength (through α). As has been discussed extensively, the optimal filter here coincides with the classical Kalman filter (see, e.g. (Jacobs, 2014)).

We now have a concrete prescription for how to get unconditional squeezing via measurement plus feedback. At each instant in time, one first constructs the optimal estimate of $\bar{X}(t)$ from the measurement record as per the above equation. Next, one uses this estimate to apply the appropriate linear feedback force which damps $\bar{X}$. More details on feedback-induced squeezing in this system are provided in (Clerk *et al.*, 2008).

5.4.7   Extensions of Back-action Evasion Techniques

More elaborate versions of back-action evasion measurements in quantum optomechanics are possible. In a system where two mechanical resonators are coupled to a single cavity, one can extend the two-tone driving approach to make a back-action free measurement of two commuting collective mechanical quadratures (Woolley and Clerk, 2013) (e.g. the sum of the X quadratures of mechanical resonator 1 and 2, and the difference of their Y quadratures). This allows the possibility of measuring both quadratures of an applied force with absolutely no quantum limit; it is intimately connected to ideas developed in (Tsang and Caves, 2010) and (Wasilewski *et al.*, 2010).

Another interesting possibility is to 'break' the QND measurement described in the previous sections by slightly imbalancing the amplitudes of the two cavity drive tones. This imbalance ruins the back-action evasion nature of the measurement, as the mechanical X quadrature no longer commutes with the Hamiltonian. However, the resulting measurement back-action on X can be harnessed to directly squeeze the mechanical resonator. As discussed extensively in (Kronwald *et al.*, 2013), this is an example of coherent feedback, where the driven cavity both 'measures' the mechanical X quadrature, and also applied exactly the correct feedback force needed to squeeze the mechanical resonator. This coherent feedback approach has the strong advantage of requiring the experimentalist neither to make a shot-noise limited measurement of the cavity output, nor to process a classical measurement record and generate an ideal feedback force. In a sense, the driven cavity does all the work. This scheme has been implemented in three recent experiments to achieve true quantum squeezing, where the X-quadrature mechanical uncertainty drops below the zero-point value (even though one starts from a thermal state) (Wollman *et al.*, 2015; Pirkkalainen *et al.*, 2015; Lecocq *et al.*, 2015). This general idea of coherent feedback can also be extended to the 'two-mode' back-action evasion scheme described above, thus providing a means for generating mechanical entanglement (Woolley and Clerk, 2014).

5.5   Appendix: Derivation of Power Gain Expression

To be able to say that our detector truly amplifies the motion of the oscillator, it is not sufficient to simply say the response function χ_{IF} must be large (note that χ_{IF} is not dimensionless!). Instead, true amplification requires that the *power* delivered by the detector to a following amplifier be much larger than the power drawn by the detector at its input—i.e., the detector must have a dimensionless power gain $G_P[\omega]$ much larger than one. If the power gain was not large, we would need to worry about the next stage in the amplification of our signal, and how much noise is added in that process. Having a large power gain means that by the time our signal reaches the following amplifier, it is so large that the added noise of this following amplifier is unimportant.

To make the above more precise, we start with the ideal case of no reverse gain, $\chi_{FI} = 0$. We will define the power gain $G_P[\omega]$ of our generic position detector in a way that is analogous to the power gain of a voltage amplifier. Imagine we drive the oscillator we are trying to measure (whose position is x) with a force $2F_D \cos \omega t$; this will cause the output of our detector $\langle \hat{I}(t) \rangle$ to also oscillate at frequency ω. To optimally detect this signal in the detector output, we further couple the detector output I to a second oscillator with natural frequency ω, mass M, and position y: there is a new coupling term in our Hamiltonian, $H'_{int} = B \hat{I} \cdot \hat{y}$, where B is a coupling strength. The oscillations in $\langle I(t) \rangle$ will now act as a driving force on the auxiliary oscillator y (see Fig 5.1). We can consider the auxiliary oscillator y as a 'load' we are trying to drive with the output of our detector.

To find the power gain, we need to consider both P_{out}, the power supplied to the output oscillator y from the detector, and P_{in}, the power fed into the input of the amplifier. Consider first P_{in}. This is simply the time-averaged power dissipation of the input oscillator x caused by the back-action damping $\gamma_{BA}[\omega]$. Using a bar to denote a time average, we have

$$P_{in} \equiv M\gamma_{BA}[\omega] \cdot \overline{\dot{x}^2} = M\gamma_{BA}[\omega]\omega^2|\chi_{xx}[\omega]|^2 F_D^2. \tag{5.162}$$

Note that the oscillator susceptibility $\chi_{xx}[\omega]$ includes the effects of γ_{BA}, c.f. Eq. (5.80).

Next, we need to consider the power supplied to the 'load' oscillator y at the detector output. This oscillator will have some intrinsic, detector-independent damping γ_{d}, as well as a back-action damping γ_{out}. In the same way that the back-action damping γ_{BA} of the input oscillator x is determined by the quantum noise in $\hat{F}$ (cf. Eq. (5.48)), the back-action damping of the load oscillator y is determined by the quantum noise in the output operator $\hat{I}$:

$$\begin{aligned}
\gamma_{out}[\omega] &= \frac{B^2}{M\omega}[-\mathrm{Im}\,\chi_{II}[\omega]] \\
&= \frac{B^2}{M\hbar\omega}\left[\frac{S_{II}[\omega] - S_{II}[-\omega]}{2}\right],
\end{aligned} \tag{5.163}$$

where χ_{II} is the linear-response susceptibility which determines how $\langle \hat{I} \rangle$ responds to a perturbation coupling to $\hat{I}$:

$$\chi_{II}[\omega] = -\frac{i}{\hbar}\int_0^\infty dt \left\langle \left[\hat{I}(t), \hat{I}(0)\right]\right\rangle e^{i\omega t}. \tag{5.164}$$

As the oscillator y is being driven on resonance, the relation between y and I is given by $y[\omega] = \chi_{yy}[\omega]I[\omega]$ with $\chi_{yy}[\omega] = -i[\omega M \gamma_{\rm out}[\omega]]^{-1}$. From conservation of energy, we have that the *net* power flow into the output oscillator from the detector is equal to the power dissipated out of the oscillator through the intrinsic damping $\gamma_{\rm d}$. We thus have

$$
\begin{aligned}
P_{\rm out} &\equiv M\gamma_{\rm d} \cdot \overline{\dot{y}^2} \\
&= M\gamma_{\rm d}\omega^2 |\chi_{yy}[\omega]|^2 \cdot |BA\chi_{IF}\chi_{xx}[\omega]F_D|^2 \\
&= \frac{1}{M} \frac{\gamma_{\rm d}}{(\gamma_{\rm d} + \gamma_{\rm out}[\omega])^2} \cdot |BA\chi_{IF}\chi_{xx}[\omega]F_D|^2.
\end{aligned}
\tag{5.165}
$$

Using the above definitions, we find that the ratio between $P_{\rm out}$ and $P_{\rm in}$ is independent of γ_0, but depends on $\gamma_{\rm ld}$:

$$
\frac{P_{\rm out}}{P_{\rm in}} = \frac{1}{M^2\omega^2} \frac{A^2 B^2 |\chi_{IF}[\omega]|^2}{\gamma_{\rm out}[\omega]\gamma_{\rm BA}[\omega]} \frac{\gamma_{\rm d}/\gamma_{\rm out}[\omega]}{(1 + \gamma_{\rm d}/\gamma_{\rm out}[\omega])^2}.
\tag{5.166}
$$

We now define the detector power gain $G_P[\omega]$ as the value of this ratio maximized over the choice of $\gamma_{\rm d}$. The maximum occurs for $\gamma_{\rm d} = \gamma_{\rm out}[\omega]$ (i.e. the load oscillator is 'matched' to the output of the detector), resulting in:

$$
\begin{aligned}
G_P[\omega] &\equiv \max\left[\frac{P_{\rm out}}{P_{\rm in}}\right] \\
&= \frac{1}{4M^2\omega^2} \frac{A^2 B^2 |\chi_{IF}|^2}{\gamma_{\rm out}\gamma_{\rm BA}} \\
&= \frac{|\chi_{IF}[\omega]|^2}{4\operatorname{Im}\chi_{FF}[\omega] \cdot \operatorname{Im}\chi_{II}[\omega]}
\end{aligned}
\tag{5.167}
$$

In the last line, we have used the relation between the damping rates $\gamma_{\rm BA}[\omega]$ and $\gamma_{\rm out}[\omega]$ and the linear-response susceptibilities $\chi_{FF}[\omega]$ and $\chi_{II}[\omega]$, c.f. Eq. (5.52). We thus find that the power gain is a simple dimensionless ratio formed by the three different response coefficients characterizing the detector, and is independent of the coupling constants A and B. As we will see, it is completely analogous to the power gain of a voltage amplifier, which is also determined by three parameters: the voltage gain, the input impedance and the output impedance.

Finally, we note that the above results can be generalized to include a non-zero detector reverse gain, χ_{FI}, see (Clerk *et al.*, 2010). In the case of a perfectly symmetric detector (i.e. $\chi_{FI} = \chi_{IF}^*$), one can show that the power gain is at most equal to one: true amplification is never possible in this case.

Bibliography

Andrews, R. W., Peterson, R. W., Purdy, T. P., Cicak, K., Simmonds, R. W., Regal, C. A., and Lehnert, K. W. (2014, March). Bidirectional and efficient conversion between microwave and optical light. *Nat. Phys.*, **10**(4), 321–6.

Aspelmeyer, M., Kippenberg, T. J., and Marquardt, F. (2014, Dec). Cavity optomechanics. *Rev. Mod. Phys.*, **86**, 1391–452.

Averin, D. (2003). Linear quantum measurements. In *Quantum Noise in Mesoscopic Systems* (ed. Y. Nazarov), pp. 205–28. Kluwer, Amsterdam.

Bocko, M. F. and Onofrio, R. (1996). On the measurement of a weak classical force coupled to a harmonic oscillator: experimental progress. *Rev. Mod. Phys.*, **68**, 755–99.

Braginsky, V. B. and Khalili, F. Y. (1992). *Quantum Measurement*. Cambridge University Press, Cambridge.

Braginsky, V. B. and Khalili, F. Y. (1996). Quantum nondemolition measurements: the route from toys to tools. *Rev. Mod. Phys.*, **68**, 1.

Braginsky, V. B., Vorontsov, Y. I., and Thorne, K. P. (1980). Quantum nondemolition measurements. *Science*, **209**, 547–57.

Bruus, H. and Flensberg, K. (2004). *Many-body quantum theory in condensed matter physics: a introduction*. Oxford University Press, Oxford.

Caves, C. M. and Milburn, G. J. (1987). Quantum-mechanical model for continuous position measurements. *Phys. Rev. A*, **36**(12), 5543–55.

Caves, C. M. (1981). Quantum-mechanical noise in an interferometer. *Phys. Rev. D*, **23**, 1693–708.

Caves, C. M., Thorne, K. S., Drever, R. W. P., Sandberg, V. D., and Zimmermann, M. (1980). On the measurement of a weak classical force coupled to a quantum-mechanical oscillator. i. issues of principle. *Rev. Mod. Phys.*, **52**, 341–92.

Clerk, A. A. (2004). Quantum-limited position detection and amplification: A linear response perspective. *Phys. Rev. B*, **70**, 245306.

Clerk, A. A., Devoret, M. H., Girvin, S. M., Marquardt, F., and Schoelkopf, R. J. (2010). Introduction to quantum noise, measurement and amplification. *Rev. Mod. Phys.*, **82**, 1155.

Clerk, A. A. and Marquardt, F. (2014). Basic Theory of Cavity Optomechanics. In *Cavity Optomechanics* (M. Aspelmeyer, T. J. Kippenberg, F. Marquardt, eds), pp. 5–23. Springer, Berlin.

Clerk, A. A., Marquardt, F., and Jacobs, K. (2008). Back-action evasion and squeezing of a mechanical resonator using a cavity detector. *New J. Phys.*, **10**, 095010.

Dong, C., Fiore, V., Kuzyk, M. C., and Wang, H. (2012, December). Optomechanical Dark Mode. *Science*, **338**(6114), 1609–13.

Gerry, C. and Knight, P. (2005). *Introductory Quantum Optics*. Cambridge University Press, Cambridge.

Gottfried, K. (1966). *Quantum Mechanics, Volume I: Fundamentals*. W. A. Benjamin, Inc.

Hertzberg, J. B., Rocheleau, T., Ndukum, T., Savva, M., Clerk, A. A., and Schwab, K. C. (2010). Back-action evading measurements of nanomechanical motion. *Nature Phys.*, **6**, 213.

Hill, J. T., Safavi-Naeini, A. H., Chan, J., and Painter, O. (2012, October). Coherent optical wavelength conversion via cavity optomechanics. *Nat. Comm.*, **3**, 1196–7.

Jacobs, K. (2010). *Stochastic Processes for Physicists*. Cambridge University Press, Cambridge.

Jacobs, K.(2014). *Quantum Measurement Theory and its Applications*. Cambridge University Press, Cambridge.

Jacobs, K. and Steck, D. A. (2006). A straightforward introduction to continuous quantum measurement. *Contemporary Physics*, **47**, 279.

Khalili, F., Miao, H., Yang, H., Safavi-Naeini, A. H., Painter, O., and Chen, Y. (2012, September). Quantum back-action in measurements of zero-point mechanical oscillations. *Phys. Rev. A*, **86**(3), 033840.

Kronwald, A., Marquardt, F., and Clerk, A. A. (2013, Dec). Arbitrarily large steady-state bosonic squeezing via dissipation. *Phys. Rev. A*, **88**, 063833.

Law, C. K. (1995, March). Interaction between a moving mirror and radiation pressure: A Hamiltonian formulation. *Phys. Rev. A*, **51**(3), 2537–41.

Lecocq, F., Clark, J. B., Simmonds, R. W., Aumentado, J., and Teufel, J. D. (2015, December). Quantum nondemolition measurement of a nonclassical state of a massive object. *Phys. Rev. X*, **5**(4), 041037.

Lemonde, M.-A. and Clerk, A. A. (2015, Mar). Real photons from vacuum fluctuations in optomechanics: The role of polariton interactions. *Phys. Rev. A*, **91**, 033836.

Marquardt, F., Chen, J. P., Clerk, A. A., and Girvin, S. M. (2007). Quantum theory of cavity-assisted sideband cooling of mechanical motion. *Phys. Rev. Let.*, **99**, 093902.

Pace, A. F., Collett, M. J., and Walls, D. F. (1993, April). Quantum limits in interferometric detection of gravitational-radiation. *Phys. Rev. A*, **47**(4), 3173–89.

Pirkkalainen, J.-M., Damskägg, E., Brandt, M., Massel, F., and Sillanpää, M. A. (2015, Dec). Squeezing of quantum noise of motion in a micromechanical resonator. *Phys. Rev. Lett.*, **115**, 243601.

Suh, J., Weinstein, A. J., Lei, C. U., Wollman, E. E., Steinke, S. K., Meystre, P., Clerk, A., and Schwab, K. C. (2014, June). Mechanically detecting and avoiding the quantum fluctuations of a microwave field. *Science*, **344**(6189), 1262–5.

Tsang, M. and Caves, C. M. (2010, Sep). Coherent quantum-noise cancellation for optomechanical sensors. *Phys. Rev. Lett.*, **105**, 123601.

Vyatchanin, S. P. and Zubova, E. A. (1995). Quantum variation measurement of a force. *Physics Letters A*, **201**(4), 269–74.

Vyatchanin, S. P. and Matsko, A. B. (1996). Quantum variational measurements of force and compensation of the nonlinear backaction in an interferometric displacement transducer. *JETP*, **83**(4), 690.

Walls, D. F. and Milburn, G. J. (2008). *Quantum Optics* (2nd edn). Springer, Berlin.

Wasilewski, W., Jensen, K., Krauter, H., Renema, J. J., Balabas, M. V., and Polzik, E. S. (2010, Mar). Quantum noise limited and entanglement-assisted magnetometry. *Phys. Rev. Lett.*, **104**, 133601.

Wilson-Rae, I., Nooshi, N., Zwerger, W., and Kippenberg, T. J. (2007). Theory of ground state cooling of a mechanical oscillator using dynamical back-action. *Phys. Rev. Lett.*, **99**, 093901.

Wiseman, H. M. and Milburn, G. J. (2014). *Quantum Measurement and Control*. Cambridge University Press, Cambridge.

Wollman, E. E., Lei, C. U., Weinstein, A. J., Suh, J., Kronwald, A., Marquardt, F., Clerk, A. A., and Schwab, K. C. (2015). Quantum squeezing of motion in a mechanical resonator. *Science*, **349**(6251), 952–5.

Woolley, M. J. and Clerk, A. A. (2013, Jun). Two-mode back-action-evading measurements in cavity optomechanics. *Phys. Rev. A*, **87**, 063846.

Woolley, M. J. and Clerk, A. A. (2014, Jun). Two-mode squeezed states in cavity optomechanics via engineering of a single reservoir. *Phys. Rev. A*, **89**, 063805.

6

Coupling Superconducting Qubits to Electromagnetic and Piezomechanical Resonators

Andrew N. Cleland

Institute for Molecular Engineering, University of Chicago,
Illinois, USA

Andrew Cleland

Andrew N. Cleland, *Coupling Superconducting Qubits to Electromagnetic and Piezomechanical Resonators* In: *Quantum Optomechanics and Nanomechanics*. Edited by: Pierre-Francois Cohadon, Jack Harris, Florian Marquardt, Leticia F. Cugliandolo, Oxford University Press (2020). © Oxford University Press. DOI: 10.1093/oso/9780198828143.003.0006

Chapter Contents

6.1   Coupling Qubits to Other Systems

Quantum bits (qubits) have been under intense development since the late 1990s, due to the discovery of a number of potential applications for engineered quantum systems to problems in computation, communication and other areas in which quantum behaviour may be able to achieve better performance than classical systems. The systems under development include atomic ions, cold atoms, photons, semiconducting quantum bits, electrons on helium, exotic topological materials and superconducting circuits. Here I focus on the latter, in part because superconducting circuits provide a straightforward path to scaling up to large numbers of qubits, have electrical impedances close to those of easy-to-implement wiring as well as standard cabling, and are flexible in terms of their application to a range of different problems. The primary challenge in such circuits is that the splitting ω of the lowest two energy levels, which serve as the ground ($|g\rangle$ or $|0\rangle$) and excited ($|e\rangle$ or $|1\rangle$) states for computational purposes, is in the microwave band, so that the effective environmental temperature needs to be much less than $\hbar\omega/k_B \sim 1$ K, meaning that refrigeration of such systems to millikelvin temperatures is a requirement. However, commercial dilution refrigerators that are cryogen-free are now readily available, so operating at such temperatures, while still a technical impediment, no longer requires continuous supervision and care of the cryostat.

Here I focus on the problem of coupling superconducting qubits to other systems, in particular to microwave frequency electromagnetic resonators as well as mechanical resonators. The simple coupling requirements for superconducting circuits makes them relatively ideal for this kind of coupling, and allow, in the case of electromagnetic resonators, highly sophisticated measurement and control of the quantum states of the resonator.

I begin by introducing the topics of piezoelectricity and its role in solid mechanics, then turn to a description of one flavour of superconducting qubit, the phase qubit. I then describe how the phase qubit can be used to control and measure a superconducting electromagnetic resonator, and conclude by describing how a phase qubit can also be used to control and measure a piezomechanical resonator.

6.2   Historical Notes on Piezoelectricity

Piezoelectricity was first discovered in 1880 by the Curie brothers, Paul-Jacques (at the time 24 years old) and Pierre (21 years old), when they discovered that a number of materials would generate measurable voltages when placed under mechanical stress. The materials in which they saw this effect included quartz, a very important material today for a number of piezoelectric applications, including for timing applications and thin-film deposition monitoring. They coined the term piezoelectricity by combining the Greek words "$\pi\iota\epsilon\zeta\iota\nu$" (*piezin*, to stress) and "$\eta\lambda\epsilon\kappa\tau\rho\omega\nu$" (*electron*, interestingly the Greek word for amber, which was commonly used for generating electrostatic charge). The year following this discovery, the mathematician Gabriel Lippman suggested that symmetry arguments require that applying a voltage to a piezoelectric material should generate strain, or displacement. That same year, the Curie brothers showed that this was indeed the case.

The Curie brothers used this discovery to then build a very sensitive electrometer, able to measure extremely small electric currents; the function of this electrometer was based on piezoelectricity. This device was to play a central role in later studies by Pierre Curie and his later wife Marie Curie on radioactivity, for which they would win the Nobel Prize in Physics in 1906. However, the Curies did not pursue further studies on piezoelectricity. This topic was instead left to be picked up by one of Pierre Curie's doctoral students, Paul Langevin.

Langevin (1872–1946), among his many accomplishments, is regarded as the inventor of sonar, which he developed in competition with parallel efforts by Ernest Rutherford working with Robert William Boyle in England. Langevin developed sonar as a result of his interest in practical applications of piezoelectricity; he invented a device that is called the Langevin sandwich transducer, comprising a plate of piezoelectric material sandwiched between two metal plates. This structure is typically operated in a transverse dilatational mode, where the thickness of the plate oscillates at a resonance frequency scaling with the thickness t of the overall structure and inversely with the speed of sound c_s. The lowest such resonance frequency is approximately at $f_0 = 2t/v$, corresponding to a half wavelength acoustic wave fitting into the thickness of the overall structure (note we are ignoring differences in the mechanical properties of the outer metal plates and the inner piezoelectric material). In section 6.6, we will work out in detail the mechanics of this type of system.

Langevin was able to use the sandwich transducer as the active element for generating and detecting sound waves in water, i.e. for what was later called sonar. Langevin obtained a patent from the British Patent Office in 1916 for this device. The concept is simple in description: the transducer, submerged in water, is first used to generate a short pulse of sound by applying to it a short oscillating voltage pulse, at the 'loaded' resonance frequency of the transducer surrounded by water. This generates a pressure wave in the surrounding water that propagates at the speed of sound. The response of the transducer is then monitored as the sound wave propagates some distance away, reflects off an object, and returns to the transducer, generating a voltage in the transducer as the reflected sound wave interacts with it. By measuring the time for the return pulse, both the presence and the distance of the object can be calculated. This was of great interest at the time that Langevin invented it, during the First World War, as the submarine had been recently invented and deployed as a military weapon, and detecting (and hopefully locating and disabling) submarines was of great interest.

The Langevin sandwich transducer is still commonly used today, for example as the central actuating element in laboratory ultrasonic cleaners.

6.3   A Quick Introduction to Solid Continuum Mechanics

To understand in more detail the operation of the Langevin sandwich transducer, I now introduce a bit of the theory of solid continuum mechanics. Note there is no accepted set of symbols for the different physical quantities involved in continuum mechanics, with several different sets in current use. Other descriptions of solid mechanics may therefore use other notation from the kind I am using here. Much of the material I present here can be found in much greater detail in any book on solid mechanics, in e.g. (Cleland, 2003).

Solid mechanics deals with the deformation of solid objects, rather than the displacement and orientation of the object as a whole. The two dynamic quantities in solid mechanics are *stress* and *strain*, which are closely related to force and displacement, respectively, from the Newtonian description of the mechanics of the motion of rigid objects. Stress, which I will represent with the symbol T, is force per unit area (thus one can think of it as a kind of pressure); the SI units of stress are N/m^2. Strain, which I will represent with the symbol S, is relative displacement per unit length—in other words, if one changes the length of an object from L to $L + \Delta L$, the strain is $S = \Delta L/L$. Strain is dimensionless.

One typically learns Hooke's law as the relation between force and extension for a spring, one of the very few examples of a deformable object covered in first-year physics. The generalized form of Hooke's law, which plays a central role in continuum mechanics, relates stress to strain by a quantity termed an *elastic modulus*; in the simplest form, stress is proportional through the Young's modulus E to the strain through

$$T = ES. \qquad (6.1)$$

Clearly the Young's modulus has units of N/m^2 or Pa. Most solid materials have Young's moduli in the range of hundreds of GPa.

A complication I have avoided so far is the fact that forces and displacements are of course vectors, and when one is dealing with continuum solids as here, these forces and displacements can be applied to different differential surfaces of the solid in different directions. Briefly, then, this twofold complication is neatly dealt with by representing the stress and strain by *tensors*, second-rank matrices with pre-defined transformations under rotations of the coordinate system. Hence the stress **T** is actually a 3×3 array of stress components T_{ij} ($i,j = 1,2,3$), and the strain **S** is a 3×3 array of strain components S_{ij} ($i,j = 1,2,3$). What this means is fairly straightforward: T_{ij} is the stress in the $\hat{x}_i$ Cartesian direction on a surface perpendicular to the Cartesian unit vector $\hat{x}_j$ (where $\hat{x}_1 = \hat{x}$, $\hat{x}_2 = \hat{y}$ and $\hat{x}_3 = \hat{z}$). The notation $\hat{x}$ means a unit vector, $\hat{x} = \vec{x}/|\vec{x}|$. Hence, T_{11} is the force per unit area in the direction $\hat{x}_1 = \hat{x}$ on a differential surface perpendicular to $\hat{x}_1$ (hence a *longitudinal* stress), while T_{21} is the force per unit area in the $\hat{x}_2 = \hat{y}$ direction on a surface perpendicular to $\hat{x}_1$ (hence a *shear* stress).

Strain is equally simple, defined in terms of the relative vector displacement $\vec{u}$ of a point originally at $\vec{r}$ ('originally' here usually meaning in the unstrained solid); the relative displacement $\vec{u}$ means that, after a strain has occurred, the point in the solid that was at $\vec{r}$ is now at $\vec{r} + \vec{u}$. Note of course that the relative displacement can vary with position in the solid, so $\vec{u}$ is a function of position $\vec{r}$, $\vec{u} = \vec{u}(\vec{r})$. However, the relative displacement $\vec{u}$ could have occurred due to a displacement or rotation of the solid as a whole (i.e. due to rigid body motion), and we do not want to include this in the definition of the strain. Bulk displacements and rotations are instead usually described by the displacement of the object's centre of mass, and by rotations of the object about that point. In order to remove such rigid body motion from the definition of strain, the mathematical definition of the ij component of strain S_{ij} is instead through the differential equation

$$S_{ij} = \frac{1}{2}\left(\frac{\partial u_i}{\partial r_j} + \frac{\partial u_j}{\partial r_i}\right). \tag{6.2}$$

The derivatives ensure that rigid body displacements are eliminated (for a rigid body displacement, the relative displacement $\vec{u}$ does not vary with position $\vec{r}$ in the solid, so its derivative is zero), and the sum with the interchanged indices eliminates rigid body rotations (as these are always antisymmetric under the interchange of coordinate indices, as can be seen with a bit of thought). A non-zero strain S_{11} is then due to an x_1-dependent relative displacement in the $\hat{x}_1$ direction, and is termed a *dilatation*, while a non-zero strain $S_{21} = S_{12}$ is due to a displacement in the $\hat{x}_2$ $(\hat{x}_1)$ direction as a function of x_1 (x_2), in other words $S_{21} (S_{12})$ describes a *shear strain*. All the strains S_{ii} are dilatational in nature, while the strains S_{ij} with $i \neq j$ are due to shear displacements.

As an example, if one holds a stretchable ruler of length L along the x_1 axis, fixes one end of the ruler, and pulls the other end out by a distance ΔL along x_1, the displacement in this situation is u_1 and is a function of position along the length of the ruler, with the relative displacement $u_1(x_1)$ in the $\hat{x}_1$ direction at the position x_1 along the original ruler given by $u_1(x_1) = x_1(\Delta L/L)$. The only non-zero strain in this situation (ignoring any concomitant reduction in the transverse dimensions of the ruler due to the stretching) is $S_{11} = \partial u_1/\partial x_1 = \Delta L/L$ and is independent of position x_1 along the ruler.

The definition of strain is clearly symmetric under interchange of indices, i.e. $S_{ij} = S_{ji}$, apparent in the symmetry of Eq. (6.2). It turns out that the stress is also symmetric under exchange of indices, $T_{ij} = T_{ji}$, although the reason for this is a little more subtle. If the stress were not symmetric in this way, so that $T_{ij} \neq T_{ji}$, then an infinitesimal cube of side dL of the solid would experience a rotational torque $(T_{ij} - T_{ji})\, dL^2/dL^3$, which diverges in the limit $dL \to 0$, so we must have $T_{ij} = T_{ji}$. Furthermore, if we again consider the same infinitesimal cube, the net longitudinal force $T_{ii}dL^2$ in the $\hat{x}_i$ direction perpendicular to $\hat{x}_i$ on the surface of the cube at $x_i + dL$ must be balanced by an equal and opposite force $-T_{ii}dL^2$ acting on the opposite face of the cube at x_i. This holds for any T_{ij}, in other words the stresses on one face are balanced by equal and opposite stresses on the opposite face. Otherwise, the differential cube would experience infinite translational acceleration in the limit $dL \to 0$.

6.3.1 Hooke's Law

Stress and strain are related, as discussed above, and in general the relation between stress and strain is assumed to be a linear one, as given by Hooke's law. In tensor form, the most general such linear relation is given by

$$T_{ij} = \sum_{k,l=1,2,3} c_{ijkl} S_{kl}, \tag{6.3}$$

where the $3^4 = 81$ constants c_{ijkl} are called the elastic moduli, with SI units of N/m^2. The intrinsic symmetries of T and S, i.e. the fact that these are symmetric tensors, forces many of the 81 elastic constants to be equal on one another, and it turns out that there are only 36 at most independent constants.

6.3.2 Six-vector notation

The symmetry of both the stress and strain tensors means that instead of having 9 independent stress and strain components, there are only 6 independent variables for each. It is traditional therefore to define a *six-vector* form for each of these tensors, where the 3×3 tensor is replaced by a six-element vector. The correspondence between the tensor stress (strain) T_{ij} (S_{ij}) and the six-vector stress (strain) T_i (S_i) is as follows (and yes, it is traditional to use the same symbols for both the tensor and six-vector forms):

$$T = \begin{pmatrix} T_1 & T_6 & T_5 \\ T_6 & T_2 & T_4 \\ T_5 & T_4 & T_3 \end{pmatrix}, \quad S = \begin{pmatrix} S_1 & S_6 & S_5 \\ S_6 & S_2 & S_4 \\ S_5 & S_4 & S_3 \end{pmatrix}. \tag{6.4}$$

Equivalently, one can write the six-vector form in terms of the tensor elements, as

$$T = (T_{11}, T_{22}, T_{33}, T_{23}, T_{13}, T_{12})^T, \tag{6.5}$$

and

$$S = (S_{11}, S_{22}, S_{33}, S_{23}, S_{13}, S_{12})^T. \tag{6.6}$$

The Hooke's law relation between the six-vector form of stress and strain is then written in the form

$$T_i = \sum_{j=1\ldots6} c_{ij} S_j, \tag{6.7}$$

using the same symbol for the elastic moduli as for the tensor relation, but now with only two indices; note that there are now 36 independent values for c_{ij}, so these can all be independent.

We will instead now make a strong simplifying assumption, which will reduce the number of elastic constants c_{ij} to just two: we will assume the materials we will work with are isotropic. It turns out that assuming complete isotropy reduces the elastic moduli to just c_{11} and c_{12}, where for convenience we also use $c_{44} = (c_{11} - c_{12})/2$, which is termed the shear modulus. The form of the full 6×6 elastic modulus matrix is then given by

$$\mathbf{c} = \begin{pmatrix} c_{11} & c_{12} & c_{12} & 0 & 0 & 0 \\ c_{12} & c_{11} & c_{12} & 0 & 0 & 0 \\ c_{12} & c_{12} & c_{11} & 0 & 0 & 0 \\ 0 & 0 & 0 & c_{44} & 0 & 0 \\ 0 & 0 & 0 & 0 & c_{44} & 0 \\ 0 & 0 & 0 & 0 & 0 & c_{44} \end{pmatrix}. \tag{6.8}$$

This very strong assumption of isotropy eliminates the possibility of accurately handling crystalline or other non-isotropic materials, but the assumption is usually not so bad for most materials.

It is common to use other elastic constants in place of c_{11} and c_{12}; two common choices are the Lamé constants λ and μ, defined as

$$\lambda = c_{12}$$
$$\mu = c_{44} = (c_{11} - c_{12})/2, \tag{6.9}$$

and the Young's modulus E and Poisson ratio ν, defined in terms of the Lamé constants as

$$E = \frac{\mu(3\lambda + 2\mu)}{\lambda + \mu}$$
$$\nu = \frac{\lambda}{2(\lambda + \mu)}. \tag{6.10}$$

6.4   Dynamical Equations for a Solid

We will now work out the dynamical equations for a continuum isotropic solid, using the six-vector notation for stress and strain as appropriate.

Consider a differential cube of side dL, with faces oriented with the Cartesian coordinate axes. Let's look at the force along the x axis on the two faces of the cube perpendicular to the x axis; this comprises the force F_x on the face at x (where we will use the sign convention that this is positive for a force in the $-x$ direction) and F_{x+dL} on the face at $x + dL$ (which is positive for a force in the $+x$ direction). The net force in the x direction from these two forces is then

$$F_{x,\text{net}} = F_{x+dL} - F_x = \frac{dF_x}{dx}\, dL. \tag{6.11}$$

In terms of the stress components, here reverting to the tensor form for notational convenience, we have $F_x = T_{11}(x)dL^2$ and $F_{x+dx} = T_{11}(x + dL)dL^2$, so we can write

$$F_{x,\text{net}} = \frac{dT_{11}}{dx}(dL)^3. \tag{6.12}$$

This net force must result in a linear acceleration along the x direction, i.e. this force determines the second time derivative of the relative displacement u_x:

$$\rho(dL)^3 \frac{d^2 u_x}{dt^2} = \frac{dT_{11}}{dx}(dL)^3, \tag{6.13}$$

where the mass of the differential cube is $\rho(dL)^3$. Hence we find

$$\rho \frac{d^2 u_x}{dt^2} = \frac{dT_{11}}{dx}. \tag{6.14}$$

With some thought, we can extend this to include the forces in the x direction from the stress components $T_{12} = T_{21}$ on the two faces perpendicular to y and from the stress components $T_{13} = T_{31}$ from the two faces perpendicular to z. This then gives the more general result

$$\rho \frac{d^2 u_x}{dt^2} = \sum_{j=1,2,3} \frac{dT_{1j}}{dx}. \tag{6.15}$$

We can further extend this to work out the acceleration in the y and z directions, which yields the simple result for all i

$$\rho \frac{d^2 u_i}{dt^2} = \sum_{j=1,2,3} \frac{dT_{ij}}{dx_i}. \tag{6.16}$$

This is frequently simplified using the straightforward notation

$$\rho \frac{d^2 u_i}{dt^2} = \sum_{j} \nabla_j T_{ij}. \tag{6.17}$$

We can now substitute the relation between stress and strain for an isotropic solid, Eq. (6.6), and the expression for strain in terms of the derivatives of the displacement, to come up with a closed-form dynamical equation for the relative displacement:

$$\rho \frac{d^2 \vec{u}}{dt^2} = (\lambda + \mu)\nabla(\nabla \cdot \vec{u}) + \mu \nabla^2 \vec{u}. \tag{6.18}$$

6.5 Piezoelectricity

We now wish to include a simple description of piezoelectricity into our description of continuum mechanics. This is nonetheless a significant complication to the equations of motion, as we now have to include the electromagnetic degrees of freedom into our formalism. A more complete description can be found in e.g. (Auld, 1990).

First, we remind the reader of the relation between the electric displacement $\vec{D}$ and the electric field $\vec{E}$ in a linear dielectric:

$$\vec{D} = \epsilon \cdot \vec{E}, \qquad (6.19)$$

where ϵ is the dielectric tensor. In subscript notation this is

$$D_i = \sum_{j=1,2,3} \epsilon_{ij} E_j. \qquad (6.20)$$

In a piezoelectric solid, this relation must be changed to include the generation of an electric displacement due to strain, i.e. due to the fact that strain will generate a dipole moment in the material. Similarly, the linear relation between stress and strain, in six-vector form written as

$$\vec{T} = \mathbf{c} \cdot \vec{S}, \qquad (6.21)$$

must be modified to include the fact that in a piezoelectric, an electric field will generate stress in the solid.

The two new coupled equations relating electric displacement, electric field, stress and strain can be written

$$\vec{D} = \epsilon \cdot \vec{E} + \mathbf{e}^T \cdot \vec{S}, \qquad (6.22)$$

and

$$\vec{T} = -\mathbf{e} \cdot \vec{E} + \mathbf{c} \cdot \vec{S}. \qquad (6.23)$$

Note that here the electric displacement $\vec{D}$ and the electric field $\vec{E}$ are conventional three-vectors, while the stress $\vec{T}$ and strain $\vec{S}$ are six-vectors. Hence the new piezoelectric matrix $\mathbf{e}$ in Eq. 6.22 is a 6×3 rectangular array of numbers (the transpose of this appearing in Eq. (6.22) of course turns this into a 3×6 array). The fact that the same matrix $\mathbf{e}$ appears in both Eqs. (6.22) and (6.23) reflects the symmetry predicted by Gabriel Lippman when he heard about the Curie brother's discovery of piezoelectricity.

In component form we can write these equations as

$$D_i = \sum_{j=1,2,3} \epsilon_{ij} E_j + \sum_{j=1...6} e_{ji} S_j, \quad (i=1,2,3) \qquad (6.24)$$

and

$$T_i = - \sum_{j=1,2,3} e_{ij}E_j + \sum_{j=1...6} c_{ij}S_j, \quad (j = 1...6).$$ (6.25)

To re-derive the equations of motion for the relative displacement $\vec{u}$ these equations must be combined with Newton's law for a continuum solid, i.e. combined with Eq. 6.17, reproduced here:

$$\rho \frac{d^2 u_i}{dt^2} = \sum_j \nabla_j T_{ij}.$$ (6.26)

This yields a quite complex set of equations that are not particularly illuminating, so I do not reproduce them here.

We turn instead to an example of how to apply these equations to a particular problem, the Langevin sandwich transducer mentioned in section 6.2.

6.6 One-dimensional Model of a Piezoelectric Dilatational Resonator: the Langevin Sandwich Transducer

Here we look at the electromechanics of a one-dimensional piezoelectric resonator. We will look at the dynamic equations for a model system as depicted in Fig. 6.1, where a slab of piezoelectric material of area A and thickness b (extending from $z = 0$ to $z = b$) has two massless, zero-thickness plates of metal on the top and bottom surfaces (in other words, plates that affect the system only in that they can be used to set a voltage on the top and bottom surfaces). This is the Langevin sandwich transducer mentioned in section 6.2. We will assume motion only along the z direction, i.e. no displacements along x or y, and only consider a material that has non-zero elastic coefficient c_{33}, piezoelectric coefficient e_{33}, and dielectric constant ϵ_{33}, with all other coefficients either irrelevant or set to zero.

Consider first a non-piezoelectric material. For dilatational vibration along the z-axis, the relevant strain component in six-vector notation is S_3 (this is S_{33} or S_{zz} in tensor notation). We will use $u(z,t)$ to represent the z-axis relative displacement, so $S_3 = \partial u / \partial z$. We take this as the only non-zero strain, and also assume the only non-zero stress is $T_3 = c_{33}S_3$, where c_{33} is the relevant elastic constant (we are thus implicitly assuming that the Poisson ratio ν is zero). Note in terms of the Lamé constants, $c_{33} = \lambda + 2\mu$, and setting the Poisson ratio ν to zero is equivalent to setting $\lambda = 0$, so $c_{33} = 2\mu$.

For a non-piezoelectric, isotropic solid, the relative displacement then satisfies the dynamic equation

$$\rho \frac{\partial^2 u}{\partial t^2} = \frac{\partial T_3}{\partial z} = c_{33} \frac{\partial S_3}{\partial z} = c_{33} \frac{\partial^2 u}{\partial z^2},$$ (6.27)

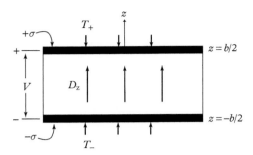

Fig. 6.1 *Schematic of the cross-section of a Langevin sandwich transducer, where the black rectangles are cross-sections through the top and bottom metal plates, and the clear rectangle in the centre is the piezoelectric slab. The direction of D_z corresponds to its algebraic positive value, but if the surface charge σ is positive, D_z will be negative.*

where ρ is the material density. Note this is a wave equation, with the sound velocity given by $v_s = \sqrt{c_{33}/\rho}$:

$$\frac{\partial^2 u}{\partial t^2} = v_s^2 \frac{\partial^2 u}{\partial z^2}. \tag{6.28}$$

For the simplest case of a non-piezoelectric slab of thickness b with free-free boundaries (meaning zero normal stress) at $z = 0$ and at $z = b$, we need to have $T_3(z = 0) = T_3(z = b) = 0$ so $\partial_z u|_{z=0, z=b} = 0$ (this still of course assumes the only non-zero elastic coefficient is c_{33}). The solutions to the wave equation with these boundary conditions are then $u_n(z,t) = \mathcal{U} \cos(k_n z) e^{i\omega_n t}$ with frequencies $\omega_n = v_s k_n = \sqrt{c_{33}/\rho}\, k_n$ and wave vectors $k_n = n\pi/b$, with $n = 1, 2, 3 \ldots$. The displacement amplitude is $\mathcal{U}$.

We now consider the piezoelectric slab shown in Fig. 6.1. We have metal surfaces on the top and bottom, on which we place equal and opposite oscillating charges with amplitude Q, or an equivalent surface charge density $\sigma = Q/A$; the charge will be assumed to oscillate at frequency ω, with the lower plate 180° out of phase with the upper plate (so its charge is always the opposite sign as that on the upper plate). We then apply the piezoelectric stress equations for the relevant z components:

$$\left. \begin{array}{rcl} D_z &=& \epsilon_{33} E_z + e_{33} S_3, \\ T_3 &=& -e_{33} E_z + c_{33} S_3. \end{array} \right\} \tag{6.29}$$

Here D_z and E_z are the z components of the electric displacement and electric field, ϵ_{33} is the relevant element of the dielectric tensor (in SI units, this would be $\epsilon_{33} = \epsilon_r \epsilon_0$), e_{33} is the relevant element in the piezoelectric modulus tensor, and T_3 is the relevant component of the stress six-vector ($T_3 = T_{zz}$).

Using Gauss' law in the interior of the slab, where there are no free charges, we have $\nabla \cdot \vec{D} = 0$. When restricted to z dependence only, this simplifies to $\partial_z D_z = 0$. Combining this with the dynamic relation $\rho \partial_t^2 u = \partial_z T_3$, we obtain from Eq. (6.29)

$$\left.\begin{aligned} \epsilon_{33}\frac{\partial E_z}{\partial z} &= -e_{33}\frac{\partial^2 u}{\partial z^2}, \\ \rho\frac{\partial^2 u}{\partial t^2} &= -e_{33}\frac{\partial E_z}{\partial z} + c_{33}\frac{\partial^2 u}{\partial z^2}, \end{aligned}\right\} \tag{6.30}$$

yielding the self-consistent dynamical equation

$$\rho\frac{\partial^2 u}{\partial t^2} = \left(c_{33} + \frac{e_{33}^2}{\epsilon_{33}}\right)\cdot\frac{\partial^2 u}{\partial z^2}, \tag{6.31}$$

This has exactly the same form as Eq. (6.28) but with a (slightly) modified effective stiffness $\bar{c}_{33} = c_{33} + e_{33}^2/\epsilon_{33}$. For a typical piezoelectric material the stiffness will change by at most a few per cent. The wave speed is $\bar{c}_s = \sqrt{\bar{c}_{33}/\rho}$. For reference, for the piezoelectric material AlN, the relevant coefficients are $e_{33} = 1.46$ C/m^2, $\epsilon_{33} = 10.7\epsilon_0$, and $c_{33} = 395$ GPa, with $e_{33}^2/\epsilon_{33}c_{33} = 0.057$.

Now we return to the basic piezoelectric relations (6.29), which we re-write with D_z and S_3 as the independent variables, appropriate because we will be fixing the free charge density σ on the surface of metal plates on the top and bottom of the piezoelectric slab (note we'll always assume that all variables oscillate at frequency ω, including the charge density σ, the displacement u, the stress T_3 and the electric field and electric displacement E_z and D_z). Then we have

$$\left.\begin{aligned} E_z &= \tfrac{1}{\epsilon_{33}}D_z - h_{33}S_3, \\ T_3 &= -h_{33}D_z + c_{33}S_3, \end{aligned}\right\} \tag{6.32}$$

using the definition $h_{33} = e_{33}/\epsilon_{33}$. As we have $\partial_z D_z = 0$ on the interior of the piezoelectric slab, and $D_z = -\sigma$ from Gauss' law on the top and bottom surfaces, we know that $D_z = -\sigma$ on the interior of the slab. Then

$$\left.\begin{aligned} E_z &= -\frac{\sigma}{\epsilon_{33}} - h_{33}S_3, \\ T_3 &= h_{33}\sigma + c_{33}S_3. \end{aligned}\right\} \tag{6.33}$$

We can now calculate the voltage across the two metal plates, from

$$V = -\int_0^b E_z(z)\,dz = \frac{\sigma}{\epsilon_{33}}b - h_{33}\int_0^b S_3\,dz. \tag{6.34}$$

We also have an oscillating current I due to the oscillating charge on the plates,

$$I = i\omega\sigma A, \tag{6.35}$$

assuming a uniform oscillation at frequency ω and a plate area A.

Again for the slab of thickness b with free-free boundaries at $z = \pm b/2$, zero stress boundaries gives $\partial_z u|_{z=\pm b/2} = 0$, and solutions $u_n(z,t) = \mathcal{U}\sin(k_n z)e^{i\omega_n t}$ for n odd, $u_n(z,t) = \mathcal{U}\cos(k_n z)e^{i\omega_n t}$ for n even, with $\omega_n = \sqrt{\bar{c}_{33}/\rho}\,k_n$ and $k_n = n\pi/b$.

More generally, allowing for a normal external stress $T_\pm$ on the top and bottom surfaces $z = \pm b/2$, a surface charge $\pm\sigma$ at $z = \pm b/2$, a voltage V across the slab, and displacements $u_\pm$ at the top and bottom surface, with all components oscillating at frequency ω, there is a matrix relation similar to (6.33),

$$
\left\{ \begin{array}{c} T_+ \\ T_- \\ V \end{array} \right\} = \mathcal{Z} \left\{ \begin{array}{c} u_+ \\ u_- \\ \sigma \end{array} \right\},
\tag{6.36}
$$

where $\mathcal{Z}$ is the constitutive matrix relating the two vectors.

These quantities are related to one another by (6.33), which can be written with S and D as the independent variables,

$$
\left. \begin{array}{l} E_z = \epsilon_{33}^{-1} D_z - h_{33} S_3, \\ T_3 = -h_{33} D_z + \bar{c}_{33} S_3. \end{array} \right\}
\tag{6.37}
$$

The external stress is $T_\pm = -T_3(z = \pm b/2)$, the surface charge is given by $\sigma = -D_z$, and the voltage is related to the electric field by $V = -\int E_z dz$. Note that as $\partial_z D_z = 0$, $D_z = -\sigma$ is constant in the slab.

The boundary conditions allow displacements $u(z) = (\mathcal{U}_s \sin kz + \mathcal{U}_c \cos kz)e^{i\omega t}$ with $\omega = \sqrt{\bar{c}_{33}/\rho}\,k \equiv \bar{c}_s k$. The displacements $u_\pm$ at $z = \pm b/2$ are related to the amplitudes by $U_s = (u_+ - u_-)/(2\sin(kb/2))$ and $U_c = (u_+ + u_-)/(2\cos(kb/2))$. The corresponding stress is $S_3 = \partial_z u = (k\mathcal{U}_s \cos kz - k\mathcal{U}_c \sin kz)e^{i\omega t}$, so the electric field is $E_z = -\epsilon_{33}^{-1}\sigma - h_{33}k(\mathcal{U}_s \cos kz - \mathcal{U}_c \sin kz)e^{i\omega t}$. The voltage is then $V = \epsilon_{33}^{-1}\sigma b + h_{33}(u_+ - u_-)$. We can thus write

$$
\left\{ \begin{array}{c} T_+ \\ T_- \\ V \end{array} \right\} = \left\{ \begin{array}{ccc} -kc_{33}\cot(kb) & kc_{33}\csc(kb) & -h_{33} \\ -kc_{33}\csc(kb) & kc_{33}\cot(kb) & -h_{33} \\ h_{33} & -h_{33} & b/\epsilon_{33} \end{array} \right\} \left\{ \begin{array}{c} u_+ \\ u_- \\ \sigma \end{array} \right\}.
\tag{6.38}
$$

We can now invert this matrix to get the displacements $u_\pm$ and charge σ in terms of the stress and voltage. For a stress-free system, $T_\pm = 0$, and we find the relation between charge and voltage,

$$
\sigma = (\epsilon_{33}/b)\frac{kb\cot(kb) + kb\csc(kb)}{-g + kb\cot(kb) + kb\csc(kb)}V,
\tag{6.39}
$$

with dimensionless coupling constant $g = 2h_{33}^2 \epsilon_{33}/c_{33}$. We also find that $u_+ = -u_-$, so $U_c = 0$ and only the sine part of the displacement is non-zero,

$$U_s = \frac{1}{h_{33}} \frac{g}{g - kb\cot(kb) - kb\csc(kb)} V. \tag{6.40}$$

The corresponding electrical admittance can be obtained from $I = Ad\sigma/dt = Y(\omega)V$, with

$$Y(\omega) = \frac{i\omega\epsilon_{33}A}{b} \frac{kb\cot(kb) + kb\csc(kb)}{kb\cot(kb) + kb\csc(kb) - g}. \tag{6.41}$$

If we set the piezoelectric coupling $g \to 0$, then we find $Y(\omega) = i\omega\epsilon_{33}A/b \equiv Y_C(\omega)$, corresponding to a capacitance $C = \epsilon_{33}A/b$ for a slab of area A and thickness b, as expected. We thus separate the admittance $Y(\omega) = Y_C(\omega) + Y_{\text{res}}(\omega)$, comprising the capacitive part $Y_C(\omega)$ and a part purely due to the piezoelectric response $Y_{\text{res}}(\omega)$, with

$$Y_{\text{res}}(\omega) = Y_C(\omega) \left[\frac{g}{kb\cot(kb) + kb\csc(kb) - g} \right] \tag{6.42}$$

$$= -Y_C(\omega) \frac{h_{33} U_s}{V}. \tag{6.43}$$

The resonator admittance $Y_{\text{res}}(\omega)$ has sharp resonances whenever $kb = b\omega_m/\bar{c}_s \equiv (2m+1)\pi - 2g/b\bar{c}_s\omega_m \approx (2m+1)\pi$. We focus only on the fundamental resonance $m = 0$, with $\omega_0 = \pi\bar{c}_s/b$. Near this resonance, with $x = b\omega/\bar{c}_s - \pi = \pi(\omega/\omega_0 - 1) = \pi(\omega - \omega_0)/\omega_0$, we can write $\sin(kb) \approx -x$ and $\cos(kb) + 1 \approx x^2/2$. The denominator of $Y_{\text{res}}(\omega)$ is then $kb\cot(kb) + kb\csc(kb) - g \approx \pi(\frac{x^2/2-1}{-x} + \frac{1}{-x}) - g$ or approximately $\pi(-x/2) - g \approx -\pi x/2 = -(\pi^2/2)(\omega - \omega_0)/\omega_0$ and

$$Y_{\text{res}}(\omega) \approx -\frac{2g}{\pi^2} Y_C(\omega) \frac{\omega_0}{\omega - \omega_0} = -\frac{4g}{\pi^2} Y_C \frac{\omega_0^2}{\omega^2 - \omega_0^2}, \tag{6.44}$$

where we have written the relation as an undamped Lorentzian using $2\omega_0^2/(\omega^2 - \omega_0^2) \approx \omega_0/(\omega - \omega_0)$.

Phenomenological damping can be added in to give

$$Y_{\text{res}}(\omega) \approx -(4g/\pi^2) Y_C(\omega) \frac{\omega_0^2}{\omega^2 - \omega_0^2 + 2i\beta\omega}, \tag{6.45}$$

with quality factor $Q = 2\beta/\omega_0$. Defining the Lorentzian response function $\mathcal{R}(\omega)$,

$$\mathcal{R}(\omega) \equiv \frac{\omega_0^2}{\omega^2 - \omega_0^2 + 2i\beta\omega}, \tag{6.46}$$

the total current through the resonator (including the displacement current through the capacitor C) is

$$I_{\text{res}}(\omega) = Y(\omega)V(\omega) \tag{6.47}$$

$$= \left[i\omega C - (4g/\pi^2)i\omega C \mathcal{R}(\omega) \right] V(\omega). \tag{6.48}$$

By a similar argument we can write the total displacement amplitude U_s as

$$U_s = \frac{1}{h_{33}} \frac{4g}{\pi^2} \mathcal{R}(\omega)V(\omega), \tag{6.49}$$

and with $\dot{V} = i\omega V$, $\dot{U}_s = i\omega U_s$, we can write

$$I_{\text{res}}(t) = C\dot{V} - h_{33}C\dot{U}_s. \tag{6.50}$$

If we define the average strain as $U = U_s/b$, we can rewrite this relation as

$$I_{\text{res}}(t) = C\dot{V} - h_{33}bC\dot{U}. \tag{6.51}$$

We will return to this relation later, as it will allow us to calculate the coupling between a phase qubit and a Langevin sandwich transducer; we now turn to a discussion of the physics of the phase qubit.

6.7   The Phase Qubit: the Current-biased Josephson Junction

In this section, I introduce the superconducting quantum device known as the 'phase qubit' and describe how it can be used to perform quantum measurements on other (typically linear) systems. For an introduction to superconductivity and to Josephson junctions, good references are (Tinkham, 1996) and (Barone and Paterno, 1982).

The phase qubit, which is a dc current/dc flux biased Josephson junction, can most easily be understood as a nonlinear LC resonator, built from a current-biased Josephson junction. A Josephson junction is one of the fundamental active devices in superconductivity, and comprises a pure Josephson element, described below, with a parallel capacitance, the latter in part from the geometric capacitance associated with a lithographically defined, thin film Josephson junction, and possibly from an additional capacitance placed in parallel with the Josephson junction. In the traditional model for the current-biased Josephson junction, one also includes a parallel resistance R in the description, which is used to model loss in the junction and/or in the wires connected to it; here we assume this resistance is very large, so that we can treat loss as a small perturbation on an otherwise lossless circuit element. ehp

The Josephson junction itself comprises two pieces of superconducting metal in contact with one another through a very thin insulating tunnel barrier, through which electrons in the normal metal state, or Cooper pairs in the superconducting state, can tunnel at an appreciable rate. This device is best thought of for the role it plays in the phase qubit as a nonlinear inductance, which can be seen as arising from the two Josephson relations that describe the dynamics of the Josephson junction. A circuit schematic, and a cartoon of a Josephson junction, are shown in Fig. 6.2; the image appearing in Fig. 6.2(c) is adapted from (O'Connell, Hofheinz, Ansmann, Bialczak, Lenander, Lucero, Neeley, Sank, Wang, Weides, Wenner, Martinis and Cleland, 2010).

The first of the two Josephson relations is the so-called 'dc Josephson' relation, relating the current I_J through the Josephson element to the difference in the phases ϕ of the macroscopic wave functions describing the superconducting condensates on either side of the junction:

$$I_J = I_0 \sin \phi, \tag{6.52}$$

where I_0, the critical current of the junction, is determined by the area of the junction contact and the thickness of the tunnel barrier (typically a metal oxide) in that contact.

The second is the 'ac Josephson' relation, giving the voltage V across the junction in terms of the rate of change of the same phase ϕ that appears in the dc Josephson relation:

$$V = \frac{\hbar}{2e}\frac{d\phi}{dt} = \frac{\hbar}{2e}\dot{\phi}. \tag{6.53}$$

Note that the phase ϕ is only defined to within a factor of 2π.

There are two energy scales relevant to the Josephson junction, in addition to the superconducting gap energy Δ. These are the charging energy E_C,

$$E_C = \frac{(2e)^2}{2C}, \tag{6.54}$$

(a)   (b)   (c)

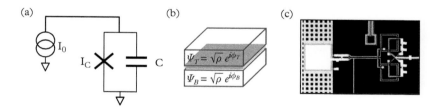

Fig. 6.2 *(a) Equivalent circuit for a current-biased Josephson junction, with a constant current I_B biasing a parallel circuit of a pure Josephson element (marked with a cross), with critical current I_0, and a capacitance C. (b) Physical construction of a Josephson junction, with top and bottom superconductors with macroscopic wave functions Ψ_T and Ψ_B, with equal amplitudes $\sqrt{\rho}$ but different macroscopic phases ϕ_T and ϕ_B. The Josephson phase ϕ is defined as $\phi = \phi_T - \phi_B$. The geometric capacitance C is due to the parallel metal plates spaced by the thin insulating layer separating the two superconductors. (c) Optical micrograph of a phase qubit, showing the Josephson junction (small triangle to left of centre), a parallel capacitance to the left, and a readout SQUID for measuring the qubit state to the right.*

which is the energy to put a charge equal to a Cooper pair on the capacitance C, and the Josephson energy $E_{\mathcal{J}}$,

$$E_{\mathcal{J}} = \frac{\hbar I_0}{2e}, \tag{6.55}$$

which is the coupling energy in the Hamiltonian linking the two superconducting wavefunctions through the Josephson junction.

For aluminum, the superconducting gap energy is $\Delta = 180 \ \mu\text{eV}$. For a typical phase qubit, the charging energy corresponding to a capacitance of about 1 pF is about $E_C \approx 0.3 \ \mu\text{eV}$, and the Josephson energy for a junction with a critical current of about 5 μA, is about $E_{\mathcal{J}} \approx 10$ meV.

We will now work out the dynamical equations for a current-biased Josephson junction. If we take the time derivative of the dc Josephson relation, we find

$$\dot{I}_{\mathcal{J}} = I_0 \cos(\phi) \, \dot{\phi}, \tag{6.56}$$

and substituting into the ac Josephson relation, we find

$$V = \frac{\hbar}{2e} \frac{1}{I_0 \cos(\phi)} \dot{I}_{\mathcal{J}}. \tag{6.57}$$

As long as ϕ does not vary too much, we can ignore the variation in $\cos(\phi)$ in the denominator of Eq. 6.57, and define an effective inductance $L_{\mathcal{J}}$ for the Josephson junction:

$$L_{\mathcal{J}} = \frac{\hbar}{2e} \frac{1}{I_0 \cos(\phi_0)}, \tag{6.58}$$

where ϕ_0 is the average value of ϕ over the dynamics we are interested in (i.e. we assume ϕ always remains close to ϕ_0).

We can then find the usual relation between voltage and current for an inductor:

$$V = L_{\mathcal{J}} \dot{I}_{\mathcal{J}}. \tag{6.59}$$

It is in this sense that a Josephson junction can be thought of as a nonlinear inductor: when ϕ remains near a constant value ϕ_0, the inductance is constant, but if something changes the average value of ϕ_0, then the inductance changes. Large dynamic changes in ϕ will correspond to different local values of the inductance at different values of ϕ.

In the phase qubit, the Josephson junction is in parallel with a capacitance C and the whole is biased with a dc current I_B. The dc current sets the average phase value ϕ_0 and thus the value of the inductance $L_{\mathcal{J}}$; however, changes in the quantum state of the qubit change the average phase value sufficiently (by a few per cent) that the inductance depends also on the state of the qubit, although to a smaller degree than the dependence on the bias current.

We can write the equations for the circuit as follows: for the capacitor C, we have the displacement current I_C,

$$I_C = C\dot{V}. \tag{6.60}$$

As the voltage V for the parallel circuit is the same across the capacitor and the junction, we can write the total bias current $I_B = I_{\mathcal{J}} + I_C = I_0 \sin\phi + C\dot{V}$. Defining the phase ϕ using Eq. (6.53), we can write this as

$$\frac{\hbar C}{2e}\ddot{\phi} + I_0 \sin\phi = I_B. \tag{6.61}$$

This can be converted into an equation that looks more like the classical equation of motion for a particle in a one-dimensional potential:

$$\frac{\hbar^2}{2E_C}\ddot{\phi} = F(\phi) = -\frac{d}{d\phi}E_{\mathcal{J}}(-s\phi - \cos\phi) = -\frac{dU(\phi)}{d\phi}, \tag{6.62}$$

where the effective force $F(\phi)$ is written as the ϕ derivative of a potential function $U(\phi)$ given by

$$U(\phi) = -E_{\mathcal{J}}(\cos\phi + s\phi), \tag{6.63}$$

and where we define the dimensionless bias

$$s = \frac{I_B}{I_0}. \tag{6.64}$$

We assume here that the bias current I_B is always less than the critical current I_0, so $|s| < 1$. We can now define an effective mass M for the particle with coordinate ϕ (note ϕ is dimensionless),

$$M = \frac{\hbar^2}{2E_C}. \tag{6.65}$$

Note that M actually has units of mass×length^2 due to the dimensionless nature of ϕ. We can also define the kinetic energy for the phase particle,

$$T = \frac{1}{2}M\dot{\phi}^2. \tag{6.66}$$

Equation (6.62) is the equation of motion for a particle with mass M, coordinate ϕ, moving in the 'tilted washboard' potential $U(\phi)$. A plot of the potential function is

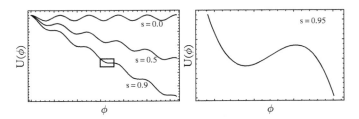

Fig. 6.3 *Left: Washboard potential $U(\phi)$ for different bias currents $s = I_B/I_0$; outlined box shows region for plot to right. Right: Washboard potential near a local minimum for $s = 0.95$.*

shown in Fig. 6.3. The overall tilt of the washboard potential is set by the bias current $s = I_B/I_0$; for $|s| < 1$, there are local minima in the potential, in which the phase particle can be trapped, while for $|s| > 1$ there are no minima, and the particle just runs in the $+\phi$ $(-\phi)$ direction for positive (negative) bias. In Fig. 6.3 we display as a detail plot a local minimum in $U(\phi)$ for $s = 0.95$.

For our purposes, we only consider situations where the phase particle is trapped in one of these local minima. The oscillation frequency for small motion is then determined by the curvature at the bottom of the well, with oscillation (or *plasma*) frequency

$$\omega_p = \sqrt{\frac{U''(\phi)}{M}} = \frac{\sqrt{2E_C E_{\mathcal{J}}}}{\hbar}(1 - s^2)^{1/4}, \tag{6.67}$$

with the zero bias ($s = 0$) plasma frequency

$$\omega_{p,0} = \frac{\sqrt{2E_C E_{\mathcal{J}}}}{\hbar} = \sqrt{\frac{2eI_0}{\hbar C}}. \tag{6.68}$$

For typical phase qubits, this frequency is typically in the range of $\omega_{p,0}/2\pi \sim 10 - 30$ GHz. When biased to typical bias currents $s \sim 0.9 - 0.95$ (typical for a phase qubit), the plasma frequency is of order half this value, i.e. in the range of 5–15 GHz. The bias current tunability of the plasma frequency is a very powerful feature of the phase qubit. In Fig. 6.4 we display the dependence of the plasma frequency on bias current.

6.8 Simple Quantization for the Current-biased Josephson Junction

There is a simple approach for quantizing the motion of the phase of the Josephson junction. First we define the Lagrangian from the classical kinetic and potential energies,

$$\mathcal{L} = T - U = \frac{1}{2}M\dot{\phi}^2 - U(\phi). \tag{6.69}$$

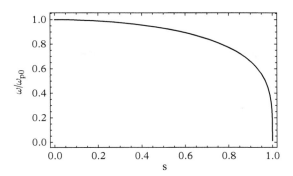

Fig. 6.4 *Plasma frequency ω/ω_{p0} as a function of bias current $s = I_B/I_0$.*

We then have the canonical momentum

$$p = \frac{\partial \mathcal{L}}{\partial \dot{\phi}} = M\dot{\phi}. \tag{6.70}$$

Note that we can identify the momentum with the voltage across the junction, $p = \hbar(eV/E_c)$, so the momentum is proportional to the charge on the capacitor C. The capacitor charge is thus conjugate to the junction phase, and in the quantum treatment, the two variables are non-commuting and are limited by a Heisenberg-like uncertainty relation.

The classical Hamiltonian is then

$$\mathcal{H} = \frac{1}{2M}p^2 + U = \frac{4E_C}{\hbar^2}p^2 + U. \tag{6.71}$$

We now follow the standard quantization procedure by setting $\hat{p} \rightarrow -i\hbar(d/d\phi)$, such that $[\phi, p] = i\hbar$. The Hamiltonian operator in position (ϕ) space is then

$$H = -E_C\frac{d}{d\phi^2} + U(\phi). \tag{6.72}$$

We then can solve the position-based, time-independent Schrödinger equation in the form

$$-E_C\frac{d^2\psi}{d\phi^2} + U(\phi)\psi = E_n\psi. \tag{6.73}$$

This is most easily solved for bias currents s close to, but less than, unity. For these values of bias, the local potential minimum in $U(\phi)$ can be very well approximated as a harmonic oscillator potential with a small cubic correction, where the frequency of the harmonic oscillator for small amplitude motion is given by the plasma frequency

ω_p, Eq. (6.67). The eigenstates of the Hamiltonian are then approximately the evenly spaced harmonic oscillator levels in the local minimum, with energies $E_n \approx (n+1/2)\hbar\omega_p$, but the cubic correction to the potential gives rise to a reasonably significant dispersive correction, such that the ground state $|g\rangle$ and excited state $|e\rangle$ are split by $\omega_{ge} \approx \omega_p$, but the third level (traditionally written as $|f\rangle$) is split from the excited state by approximately $\omega_{ef} \approx 0.95\omega_{ge}$. This energy-level dependent splitting means that the phase qubit can truly serve as a two-level system, as only microwave signals close to the energy-level splitting can cause oscillations in the occupation probabilities. The dispersive shift of ω_{ef} compared to ω_{ge} is sufficient to isolate each pair of energy levels from the other levels. Thus the experimentalist can selectively address the $|g\rangle - |e\rangle$ transition or the $|e\rangle - |f\rangle$ transition, giving rise to complete quantum control over this electronic, tunable qubit.

The dipole moments of the energy levels are slightly different from the harmonic oscillator dipole moments, but are well approximated by $x_{n+1,n} = \langle n+1|\phi|n\rangle \approx 0.04(n+1)^{1/2}$ for typical values of E_C, E_J and s and for n small; for the simple harmonic oscillator, these matrix elements are

$$\langle n+1|x|n\rangle = \sqrt{\frac{\hbar}{2M\omega_p}}\sqrt{n+1} = \left(\frac{E_C}{2E_J\sqrt{1-s^2}}\right)^{1/4}\sqrt{n+1}. \qquad (6.74)$$

There is also a small residual dipole coupling $\langle n+2|\phi|n\rangle$ roughly 30–50 times smaller than this (note in the simple harmonic oscillator, there is zero dipole coupling between the levels n and $n+2$).

6.9 Coupling to the Phase Qubit

The way the phase qubit interacts with its environment is through the bias current I_B. Slow variations in the bias current changes the energy level spacing, and thus causes changes in the relative phases of the energy levels, in addition to the usual time evolution $e^{-i\Delta Et/\hbar}$ for an energy difference ΔE. Variations in the bias current at a frequency close to the energy difference $\Delta E/\hbar$, however, cause transitions between the energy levels, with a constant tone generating Rabi oscillations with unit amplitude when the frequency matches the energy level difference exactly, with a Rabi oscillation rate that scales with the amplitude of the current modulation.

To capture this physics, we assume the bias current has a time dependence

$$I_B(t) = I_0 + \Delta I(t). \qquad (6.75)$$

We will further assume the time dependence is close to harmonic, and in resonance with the qubit $|g\rangle - |e\rangle$ transition frequency ω_{ge}, with

$$\Delta I(t) = I_c(t)\cos(\omega_{ge}t) + I_s(t)\sin(\omega_{ge}t), \qquad (6.76)$$

with slowly varying cosine and sine quadrature amplitudes $I_c(t)$ and $I_s(t)$. We include this variation in an approximate fashion by including it in the potential U which now becomes time-dependent,

$$U(t) = U_0 + \Delta U(t), \tag{6.77}$$

where U_0 is the potential including the constant bias current I_0, and the time-varying potential $\Delta U(t)$ is given by

$$\Delta U(t) = E_J \frac{\Delta I(t)}{I_C} \phi = \frac{\hbar \Delta I(t)}{2e} \phi. \tag{6.78}$$

For small current amplitude $\Delta I(t) \ll I_C$, the resulting effect on the quantum state can be treated by first-order perturbation theory, with the Hamiltonian $\hat{H} = \hat{H}_0 + \delta\hat{H}(t) = \hat{H}_0 + \Delta U(t)$. Using the lowest two unperturbed eigenstates of $\hat{H}_0$, $|g\rangle$ and $|e\rangle$ with splitting $\hbar\omega_{ge} = E_e - E_g$, we can express the perturbation in matrix form as

$$\delta\hat{H} = \begin{pmatrix} \langle g|\Delta U|g\rangle & \langle g|\Delta U|e\rangle \\ \langle e|\Delta U|g\rangle & \langle e|\Delta U|e\rangle \end{pmatrix} = \begin{pmatrix} \langle g|\phi|g\rangle & \langle g|\phi|e\rangle \\ \langle e|\phi|g\rangle & \langle e|\phi|e\rangle \end{pmatrix} \frac{\hbar \Delta I(t)}{2e}. \tag{6.79}$$

If we approximate the eigenstates as simple harmonic oscillator eigenstates, with $|g\rangle \approx |0\rangle$ and $|e\rangle \approx |1\rangle$, the diagonal matrix elements are zero, so we can write

$$\delta\hat{H} \approx \begin{pmatrix} 0 & \langle g|\phi|e\rangle \\ \langle e|\phi|g\rangle & 0 \end{pmatrix} \frac{\hbar \Delta I(t)}{2e}. \tag{6.80}$$

Writing this in the interaction picture, this is

$$\delta\hat{H}_{\text{int}} \approx \begin{pmatrix} 0 & \langle g|\phi|e\rangle e^{-i\omega_{ge}t} \\ \langle e|\phi|g\rangle e^{+i\omega_{ge}t} & 0 \end{pmatrix} \frac{\hbar \Delta I(t)}{2e}. \tag{6.81}$$

We can expand the cosine and sine dependence of $\Delta I(t)$,

$$\begin{aligned} \Delta I(t) &= I_c(t)\cos(\omega_{ge}t) + I_s(t)\sin(\omega_{ge}t) \\ &= \frac{I_c(t)}{2}\left(e^{i\omega_{ge}t} + e^{-i\omega_{ge}t}\right) + \frac{I_s(t)}{2i}\left(e^{i\omega_{ge}t} - e^{-i\omega_{ge}t}\right). \end{aligned} \tag{6.82}$$

Then the interaction picture perturbation is

$$\delta\hat{H}_{\text{int}} \approx \frac{\hbar}{2e} \begin{pmatrix} 0 & \frac{I_c}{2} + \frac{I_s}{2i} + \frac{I_c}{2}e^{-2i\omega_{ge}t} - \frac{I_s}{2i}e^{-2\omega_{ge}t} \\ \frac{I_c}{2} - \frac{I_s}{2i} + \frac{I_c}{2}e^{+2i\omega_{ge}t} + \frac{I_s}{2i}e^{+2\omega_{ge}t} & 0 \end{pmatrix} \langle g|\phi|e\rangle. \tag{6.83}$$

We now use the *rotating wave approximation* and throw away all the rapidly varying terms, leaving us with

$$\delta\hat{H}_{\text{RWA}} \approx \frac{\hbar}{2e} \begin{pmatrix} 0 & \frac{I_c}{2} + \frac{I_s}{2i} \\ \frac{I_c}{2} - \frac{I_s}{2i} & 0 \end{pmatrix} \langle g|\phi|e\rangle. \tag{6.84}$$

We can write this in terms of the Pauli matrices $\hat{\sigma}_x = \begin{pmatrix} 0 & 1 \\ 1 & 0 \end{pmatrix}$ and $\hat{\sigma}_y = \begin{pmatrix} 0 & -i \\ i & 0 \end{pmatrix}$,

$$\delta\hat{H}_{\text{RWA}} \approx \frac{\hbar}{2e} \langle g|\phi|e\rangle \left(\frac{I_c(t)}{2}\hat{\sigma}_x + \frac{I_s(t)}{2i}\hat{\sigma}_y \right). \tag{6.85}$$

Hence we see that $I_c(t)$ controls rotations about the $\hat{x}$ axis, and $I_s(t)$ controls rotations about $\hat{y}$. These amplitude envelopes thus afford control over the qubit state. For example, a continuous cosine tone with $I_c(t) = I_{rf}$ exactly at the ω_{ge} splitting frequency, with $I_s(t) = 0$, gives rise to a rotation (Rabi flop) between $|g\rangle$ and $|e\rangle$ at a rate $\hbar\Omega = \frac{\hbar}{2e}\langle g|\phi|e\rangle I_{rf}$ with unit amplitude. If the tone is applied slightly off resonance, differing from ω_{ge} by $\Delta\omega$, then the Rabi flop rate is increased to $\hbar\Omega = \sqrt{(\frac{\hbar}{2e}\langle g|\phi|e\rangle I_{rf})^2 + (\hbar\Delta\omega_{ge})^2}$, but with less than unit amplitude. An exactly analogous response occurs if instead the amplitude is in the sine quadrature.

6.10   Coupling a Qubit to an Electrical Resonator

Here we look at the interesting physics associated with coupling a qubit resonantly to an electrical resonator. This relates primarily to an experiment described in (Hofheinz, Weig, Ansmann, Bialczak, Lucero, Neeley, O'Connell, Wang, Martinis and Cleland, 2008) and (Hofheinz, Wang, Ansmann, Bialczak, Lucero, Neeley, O'Connell, Sank, Wenner, Martinis and Cleland, 2009). The electrical resonator was a half-wavelength coplanar resonator, meaning a length of coplanar waveguide that is effectively open-terminated (by small capacitors) at either end, so that the resonant electrical modes correspond to waveguide resonances at $\lambda/2$, $3\lambda/2$, etc. We will only look at the fundamental half-wave resonance; in that case, the resonator can be approximated quite accurately as a lumped LC resonator. We first work out the equivalent circuit, for the half-wave resonator with a small coupling capacitor C_c at one end of the resonator.

6.10.1   *LC* Approximation to a Half-wave Resonator

We assume the waveguide has length ℓ, so that the maximum voltage along the length of the resonator (at the time that the current is zero) is given by

$$V(x) = V_{\text{max}} \cos(\pi x/\ell) \tag{6.86}$$

where x runs from 0 to ℓ. At this point in the oscillation, the energy is all in the electrical field, with

$$\mathcal{E} = \int_0^\ell \frac{1}{2} C' V(x)^2 = \frac{1}{4} C' \ell V_{\mathrm{max}}^2. \tag{6.87}$$

Here C' is the capacitance per unit length of the waveguide; $C'\ell$ is its total geometric capacitance.

We want the equivalent LC lumped resonator to have the same frequency as the coplanar waveguide, and we also want the coupling capacitor to give the same effective coupling (to the qubit via radiation through this capacitor) in both systems. The latter condition means that we want the same voltage across the effective capacitance C_{eff} as at the end of the resonator, where we will place the capacitor, when the two resonators have the same energy $\mathcal{E}$. This latter condition means that we need to have

$$\mathcal{E} = \frac{1}{4} C' \ell V_{\mathrm{max}}^2 = \frac{1}{2} C_{\mathrm{eff}} V_{\mathrm{max}}^2. \tag{6.88}$$

This clearly implies that the effective capacitance is related to the total capacitance of the waveguide by

$$C_{\mathrm{eff}} = \frac{1}{2} \ell C'. \tag{6.89}$$

To obtain the same resonance frequency, we set the effective inductance L_{eff} using

$$\frac{1}{\sqrt{L_{\mathrm{eff}} C_{\mathrm{eff}}}} = \omega_{\mathrm{res}} = \frac{\pi v_\phi}{\ell}, \tag{6.90}$$

the second expression coming from the 1/2-wavelength nature of the waveguide, where $v_\phi = 1/\sqrt{L'C'}$ is the speed of light in the waveguide, with inductance per unit length L'. This yields the effective inductance

$$L_{\mathrm{eff}} = \frac{2}{\pi^2} \ell L'. \tag{6.91}$$

6.10.2 Coupling a Qubit to a Resonator

We consider the combined circuit of a qubit in parallel with an LC equivalent lumped resonator (i.e. we replace the coplanar waveguide resonator with its equivalent lumped circuit).[1] With charge Q on the resonator capacitance C_{eff}, and phase ϕ for the junction, the full equations of motion for the combined circuit are

$$\left(\frac{\hbar}{2e}\right)^2 (C + C_c)\ddot{\phi} + E_J \sin\phi = \frac{\hbar}{2e}\frac{C_c}{C_{\text{res}}}\dot{Q} \tag{6.92}$$

for the qubit side of the circuit, and

$$L_{\text{res}}\left(1 + \frac{C_c}{C_{\text{res}}}\right)\ddot{Q} + \frac{1}{C_{\text{res}}}Q = \frac{\hbar}{2e}L_{\text{res}}C_c\frac{d^3\phi}{dt^3} \tag{6.93}$$

for the resonator side of the circuit. Note the coupling term on the right side of (6.93) involves the 3rd time derivative of the junction phase.

We now define the resonator phase φ using the equation

$$\varphi(t) = \frac{2e}{\hbar C_{\text{res}}}\int_{-\infty}^{t} Q(t')\,dt', \tag{6.94}$$

which is completely analogous to the voltage–phase relation for a Josephson junction given that the resonator voltage $V = Q/C_{\text{res}}$. Now we use $Q(t) = (\hbar C_{\text{res}}/2e)\dot{\varphi}$, and integrate (6.93) over time. This leads to the pair of coupled equations

$$\left(\frac{\hbar}{2e}\right)^2 (C + C_c)\ddot{\phi} + E_J \sin\phi = \left(\frac{\hbar}{2e}\right)^2 C_c\ddot{\varphi} \tag{6.95}$$

and

$$\left(\frac{\hbar}{2e}\right)^2 (C_{\text{res}} + C_c)\ddot{\varphi} + \frac{1}{L_{\text{res}}}\left(\frac{\hbar}{2e}\right)^2 \varphi = \left(\frac{\hbar}{2e}\right)^2 C_c\ddot{\phi} + \text{const.} \tag{6.96}$$

Note the symmetry in the coupling terms in these two equations.

We can now write a Lagrangian that generates both of these dynamical equations:

$$\mathcal{L} = \frac{1}{2}\frac{\hbar^2}{4e^2}\tilde{C}\dot{\phi}^2 + E_J \cos\phi + \frac{1}{2}\frac{\hbar^2}{4e^2}\tilde{C}_{\text{res}}\dot{\varphi}^2 - \frac{1}{2}\frac{\hbar^2}{4e^2 L_{\text{res}}}\varphi^2 - \frac{\hbar^2}{4e^2}C_c\dot{\phi}\dot{\varphi}, \tag{6.97}$$

where we use renormalized capacitances $\tilde{C} = C + C_c$ and $\tilde{C}_{\text{res}} = C_{\text{res}} + C_c$.

We can now work out the canonical momenta,

$$p_\phi = \frac{\hbar^2}{4e^2}\tilde{C}\dot{\phi} - \frac{\hbar^2}{4e^2}C_c\dot{\varphi}, \tag{6.98}$$

---

[1] Much of this calculation was adapted from M. Geller, and is otherwise unpublished.

and

$$p_\varphi = \frac{\hbar^2}{4e^2}\tilde{C}_{\text{res}}\dot{\varphi} - \frac{\hbar^2}{4e^2}C_c\dot{\phi}. \tag{6.99}$$

The Hamiltonian is then

$$H = H_\phi + H_\varphi + H_{\text{int}}, \tag{6.100}$$

where

$$H_\phi = \frac{E_C}{\hbar^2}p_\phi^2 - E_{\mathcal{J}}\cos\phi, \tag{6.101}$$

$$H_\varphi = \frac{2e^2}{\tilde{C}_{\text{res}}\hbar^2}p_\varphi^2 + \frac{\hbar^2}{4e^2}\frac{\varphi^2}{L_{\text{res}}}, \tag{6.102}$$

and

$$H_{\text{int}} = \frac{4e^2}{\hbar^2}\frac{C_c}{\tilde{C}\tilde{C}_{\text{res}} - C_c^2}p_\phi p_\varphi \approx \frac{4e^2}{\hbar^2}\frac{C_c}{CC_{\text{res}}}p_\phi p_\varphi, \tag{6.103}$$

where the approximation holds in the usual limit where $C_c \ll C_{\text{res}}, C$.

We can now identify the momentum operators in terms of the Pauli raising and lowering operators, with

$$p_\phi = \frac{\hbar^2}{2E_C}\omega_{eg}\langle g|\phi|e\rangle(\hat{\sigma}_+ - \hat{\sigma}_-), \tag{6.104}$$

and p_φ in terms of the lowering and raising operators for the resonator a and $a^\dagger$,

$$p_\varphi = \frac{\hbar}{\sqrt{2}}\sqrt{\frac{\hbar}{4e^2}C_{\text{res}}\omega_{\text{res}}}i(\hat{a}^\dagger - \hat{a}). \tag{6.105}$$

This allows us to write the interaction Hamiltonian as

$$H_{\text{int}} = \hbar g(\hat{\sigma}_+ - \hat{\sigma}_-)(\hat{a}^\dagger - \hat{a}) \tag{6.106}$$

with

$$g = \frac{1}{2}\frac{C_c}{C_{\text{res}}}\frac{\sqrt{2}\langle g|\phi|e\rangle}{\ell_\varphi}\omega_{ge}, \tag{6.107}$$

with $\ell_\varphi = \sqrt{\frac{4e^2}{\hbar\omega_{res}C_{res}}}$. In the harmonic limit, the second fraction in (6.107) goes to unity, and $g \approx (C_c/2C_{res})\omega_{ge}$, scaling with the size of the coupling capacitor.

By then applying the rotating wave approximation, all the 'energy non-conserving' (i.e. rapidly rotating) terms in (6.106) drop out and we are left with the Jaynes–Cummings Hamiltonian (Jaynes and Cummings 1963),

$$H_{int} = \hbar g \left(\hat{\sigma}_+ \hat{a} + \hat{\sigma}_- \hat{a}^\dagger \right). \tag{6.108}$$

When the qubit frequency ω_{ge} is far detuned from the resonator frequency ω_{res}, there is no energy exchange between the qubit and the resonator. If the qubit frequency is tuned close to the resonator frequency, however, the Jaynes–Cummings interaction leads to swaps of energy between the two.

If, for example, we prepare the qubit in $|g\rangle$ and the resonator in an n photon state $|n\rangle$, and the qubit is brought into resonance with the resonator, the two will exchange one quantum of energy at the rate $\Omega_n = 2\sqrt{n}g/\hbar$, with the qubit oscillating between $|g\rangle$ and $|e\rangle$ while the resonator oscillates between $|n\rangle$ and $|n-1\rangle$; the $\sqrt{n}$ multiplier comes from the action of the raising and lowering operators.

If the qubit is instead detuned from the resonator by Δ, the oscillation rate becomes $\sqrt{\Omega_n^2 + \Delta^2}$, the energy swaps are less than unity in amplitude, with $|g, n\rangle$ swapping to $|e, n-1\rangle$ with amplitude $\Omega_n^2/(\Omega_n^2 + \Delta^2)$. This dependence is shown in Fig.

This physics allows us to use the qubit to both pump photons into the resonator and measure the resonator state. Starting with the system in $|g0\rangle$ (i.e. qubit in its ground state with zero photons in the resonator), which occurs naturally by waiting while the system is kept at a temperature $k_B T \ll \hbar\omega_{res}, \hbar\omega_{ge}$, we detune the qubit from the resonator and then apply a carefully calibrated microwave current tone to the qubit, at the qubit's resonance frequency. As per the perturbation theory worked out earlier, where we can ignore the effect of the resonator due to the large detuning, this causes an X or Y rotation of the qubit which can be calibrated to complete a full swap from $|g\rangle \rightarrow |e\rangle$, leaving the system in $|e0\rangle$. We then tune the qubit into resonance with the resonator, and through the Jaynes–Cummings interaction allow the excitation to swap fully from the qubit to the resonator, which takes a time $h/4g$, leaving the system in $|g1\rangle$. By then detuning the qubit from the resonator, the interaction is effectively turned off, leaving the photon trapped in the resonator. We can now continue the process, injecting another excitation in the detuned qubit using a microwave pulse, putting the system into $|e1\rangle$. Tuning the qubit into resonance with the resonator again, but for a shorter time $h/4\sqrt{2}g$, we can swap this second excitation into the resonator, putting the system in $|g2\rangle$. This process can be continued in principle indefinitely, with the nth quantum swapping in a time $1/\sqrt{n}$ faster than the first excitation; the limit is set by the energy lifetime of the photons in the resonator, which is $T_1 = Q/\omega_{res}$ for the first photon if the resonator has quality factor Q, but T_1/n for the nth photon state.

To then measure the resonator state, we take the system (nominally in $|g, n\rangle$ and tune the qubit into resonance with the resonator for a time τ, allowing the two systems to interact and swap one quantum, so the qubit and resonator oscillator between $|g, n\rangle$ and

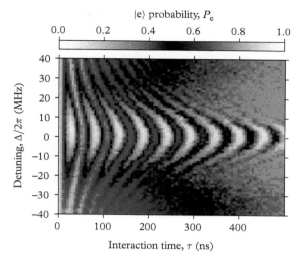

Fig. 6.5 *Rabi swaps between a phase qubit and an electromagnetic resonator, where the vertical axis is qubit tuning with respect to the fixed resonator frequency, the horizontal axis is the interaction time τ, and the colour code is the probability P_e of measuring the qubit in the excited state. The slowest oscillations, with the largest amplitudes, occur when the qubit is tuned into resonance with the resonator. Adapted with permission from (Hofheinz, Wang, Ansmann, Bialczak, Lucero, Neeley, O'Connell, Sank, Wenner, Martinis and Cleland, 2009).*

$|e, n - 1\rangle$. After this time, the qubit is detuned from the resonator and its state measured, resulting in a projective measurement outcome of $|g\rangle$ or $|e\rangle$ (see the literature for details on how this measurement process is performed for the phase qubit). This measurement is destructive, so the system is then allowed to relax to the ground state (by waiting), the state of interest recreated, and the measurement repeated, typically about 3,000 times to get sufficient statistics for each interaction time τ. The time τ is then varied over a range from zero to a few times h/g, allowing the measurement to capture a statistically significant number of Rabi swaps between the qubit and resonator.

The simplest analysis of the resulting oscillations of the qubit excited state probability P_e is to Fourier transform this time dependence. For a pure initial state $|g, n\rangle$, there will be a single Fourier peak in this transform at the Rabi swap frequency Ω_n. In practice the state is not pure, but is a mixture of states that comprise mostly $|g, n\rangle$, but with an admixture of $|g, n - 1\rangle$, etc., resulting in a Fourier transform with its primary peak at Ω_n but with some width associated with the small but non-negligible population of lower n states. In Fig. 6.6, we display both the time-dependent qubit–resonator interaction as reflected in the qubit $P_e(\tau)$ and its Fourier transform.

I have outlined the process for generating the Fock states $|g, n\rangle$, nominally pure states with n photons in the resonator, starting from the ground state $|g0\rangle$, and the process for then measuring these states. However, due to early work by Law and Eberly, with an interest in generating *arbitrary* states of harmonic motion for ions in an ion trap, it turns

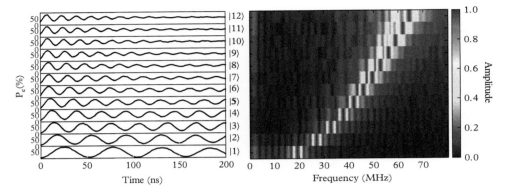

Fig. 6.6 *Fock states in a resonator. After programming the state $|n\rangle$ into the resonator, the qubit is used to measure the resonator, with the left plot showing the measured excited state probability $P_e(\tau)$ as a function of the qubit–resonator interaction time τ after creating the resonator state. Clear Rabi swap oscillations are visible, increasing in frequency with $|n\rangle$. Right plot shows Fourier transform of these oscillations, showing the clear $\sqrt{n}$ dependence of the swap rate expected from the Jaynes–Cummings model. Adapted with permission from (Hofheinz, Weig, Ansmann, Bialczak, Lucero, Neeley, O'Connell, Wang, Martinis and Cleland, 2008).*

out that it is possible to synthesize *arbitrary* photon states in the resonator, of the type

$$|\psi\rangle = |g\rangle \times \left(\sum_n a_n |n\rangle \right), \tag{6.109}$$

with the qubit in the ground state $|g\rangle$, disentangled from the resonator, and the complex amplitudes a_n for the resonator $|n\rangle$ states programmed to specific, intentional values. There is a limit on the largest n that can be occupied in this way, limited mostly by the resonator lifetime T_1, and of course some limit on the precision with which the amplitudes a_n are programmed, limited by the degree of control achievable over the qubit.

The way this is done is worked out in detail by (Law and Eberly, 1996), with the implementation for the phase qubit appearing in (Hofheinz, Wang, Ansmann, Bialczak, Lucero, Neeley, O'Connell, Sank, Wenner, Martinis and Cleland, 2009). Briefly, by programming the qubit into a specified superposition of the form $a|g\rangle + b|e\rangle$ and performing a partial swap into the resonator, arbitrary states can be generated in the resonator; this, however, requires some careful computation, as when the resonator is no longer in its ground state, there are energy (amplitude) swaps in both directions, and these must be accounted for in working out the detailed sequence. The scheme is best worked out by figuring out, given some desired final state of the form (6.109), how to *empty* the photons from the resonator in a sequence that empties the highest photon number state completely on each step, and simultaneously changes, in a completely calculable way, the amplitudes of the lower photon number states.

A sequence that empties each highest photon number state in turn is relatively easy to calculate, and terminates with the resonator and qubit in their ground states. As this is a reversible process, by actually applying this sequence in the opposite order, one can (in principle) take the qubit and resonator from an initial ground state to the desired initial state.

Once a state of the form (6.109 has been generated, the resonator state can then be measured, using the qubit as above to perform the measurement, with the resulting qubit excited state probability $P_e(\tau)$ allowing extraction of the resonator state probabilities P_n. For these more complex resonator states, one can show that the probability $P_e(\tau)$ of measuring the qubit in the excited state at time τ is given by

$$P_e(\tau) = \frac{1}{2}\left(1 - \sum_n P_n \cos(\Omega_n \tau)\right), \tag{6.110}$$

where we assume the qubit is initially in its ground state with unit probability, and P_n is the probability the resonator is in the n photon state. A maximum likelihood analysis is performed, using the measured $P_e(\tau)$, to obtain the most likely set of probabilities P_n, constrained by the maximum number of photons in the resonator, and by the condition that the resonator probabilities sum to unity. In Fig. 6.7 we display for two different states $|1\rangle + |3\rangle$ and $|1\rangle + i|3\rangle$ the probability $P_e(\tau)$ as the qubit and resonator undergo Rabi swaps, which due to the more complex resonator state do not have a simple sinusoidal dependence. An analysis of the probabilities P_n as in Eq. 6.110 yields the histograms on the right.

A complete analysis of the resonator state, allowing extraction not just of the probabilities P_n but the actual state amplitudes a_n, can be performed by adding to this measurement analysis a sequence where an external microwave source, capacitively

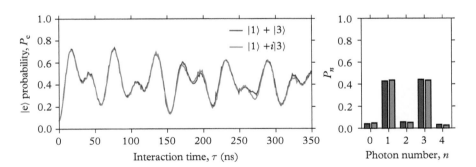

Fig. 6.7 *Left: Rabi swaps between a qubit and a resonator, the latter in either the $|1\rangle + |3\rangle$ (left) and the $|1\rangle + i|3\rangle$ state (centre), showing the incomplete swaps with a more complex time dependence. Right: An analysis of the type given in Eq. 6.110 yields the displayed histograms. Adapted with permission from (Hofheinz, Wang, Ansmann, Bialczak, Lucero, Neeley, O'Connell, Sank, Wenner, Martinis and Cleland, 2009).*

coupled to the resonator, is used to inject a Gaussian coherent state pulse into the resonator. This shifts the state of the resonator in resonator phase-space by an amount and phase determined by the amplitude and phase (represented by the complex value α) of the coherent pulse, and the probabilities P_n of the resulting resonator state as a function of this vector shift are used to calculate the expectation value of the parity Π of the resonator state,

$$\langle \Pi \rangle = \sum_n (-1)^n P_n. \tag{6.111}$$

A map of this function in resonator phase-space, multiplied by the factor $2/\pi$, comprises the *Wigner function* $W(\alpha)$ of the resonator,

$$W(\alpha) = \frac{2}{\pi} \langle \psi | D^\dagger(-\alpha) \Pi D(-\alpha) | \psi \rangle. \tag{6.112}$$

The Wigner quasiprobability distribution thus generated ((Banaszek and Wódkiewicz, 1996), (Banaszek, Radzewicz, Wódkiewicz and Krasiński, 1999)) is equivalent to the full density matrix representation of the resonator state, up to the maximum value n in the measurement sequence. An example of a Wigner distribution (theoretical and experimental) is shown in Fig. 6.8.

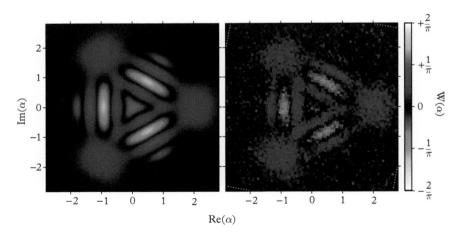

Fig. 6.8 *Wigner tomograms measured for the superposition state of a resonator comprising* $|\alpha\rangle +$ $|\alpha e^{2\pi i/3}\rangle + |\alpha e^{4\pi i/3}\rangle$, *an equal superposition of three coherent states with amplitude* $\alpha = 2$ *distributed evenly about* 2π *in resonator phase-space. Left shows calculated tomogram, right shows experimentally synthesized and measured tomogram. Adapted with permission from (Hofheinz, Wang, Ansmann, Bialczak, Lucero, Neeley, O'Connell, Sank, Wenner, Martinis and Cleland, 2009).*

6.11   Coupling a Qubit to a Mechanical Resonator

In the previous section, we worked through the physics for coupling a superconducting phase qubit to an electromagnetic resonator in the form of a half-wave coplanar waveguide resonator. Here we look at a closely related problem, namely coupling a phase qubit to a *mechanical* resonator, in the form of a Langevin sandwich transducer made from a plate of piezoelectric material with metal sheets on the top and bottom surfaces of this plate, designed so that the structure has its fundamental dilatational resonance at a frequency in the few GHz range, allowing us to tune the qubit into frequency resonance with the mechanical system. The circuit has the phase qubit electrically in parallel with the resonator, as for the qubit coupled to the electromagnetic resonator.

First we note that in section **6.6**, we worked out the current that flows into the capacitor-like metal plate structure,

$$I_{\text{res}} = C_{\text{res}} \left(\dot{V} - h_{33} b \dot{U} \right), \tag{6.113}$$

where C_{res} is the geometric capacitance of the structure, and $C_{\text{res}} \dot{V}$ is just the standard displacement current through a capacitor whose voltage is changing in time. The second term involves the piezoelectric interaction, with h_{33} the relevant piezoelectric coefficient, b the thickness of the plate, and $\dot{U}$ is the rate of change of the average strain in the transducer.

The capacitive part of the current can be absorbed into the identical expression for the capacitive response of the qubit itself, by simply increasing the qubit capacitance by C_{res} due to the parallel electrical arrangement of these two capacitances. This leaves the classical expression for the current due to the piezoelectric response,

$$I'_{\text{res}} = -h_{33} b C_{\text{res}} \dot{U}. \tag{6.114}$$

We would like to find the quantum mechanical expression corresponding to this expression, i.e. we want to quantize the motion (strain) of the mechanical resonator.

The displacement of the resonator is described by the classical relative displacement $\vec{u}(\vec{r}, t) = u(z, t)\hat{z}$, where we only consider motion perpendicular to the plane of the resonator, and assume the displacement does not vary with lateral position in the resonator. The local kinetic energy density associated with this motion is given by $\frac{1}{2}\rho(\dot{\vec{u}})^2 = \frac{1}{2}\rho\dot{u}(z,t)^2$. The local strain energy (equivalent to the potential energy associated with displacing a Hooke's law spring) is $\frac{1}{2}c_{33}S^2$, where c_{33} is the relevant elastic constant and $S = S_{33}$ the local strain associated with the relative displacement $\vec{u}$. We can then write the Hamiltonian for the solid as a whole,

$$H = \int_V \left(\frac{1}{2}\rho\dot{u}^2 + \frac{1}{2}c_{33}S^2 \right) dV, \tag{6.115}$$

integrated over the volume V of the resonator.

In order to quantize the motion, we must define the momentum density $\Pi(z,t) = \rho\dot{u}(z,t)$, using which we can write the Hamiltonian as

$$H = \int_V \left(\frac{1}{2\rho}\Pi^2 + \frac{1}{2}c_{33}S^2 \right) dV. \tag{6.116}$$

The momentum density and the strain have a certain mode shape for motion in a single resonant mode of the structure, leaving only the amplitude and phase of the motion in that mode as dynamical variables; this is analogous to the voltage mode shape for the coplanar waveguide resonator discussed in section 6.10 above. For the fundamental mode we are considering here, we have for the relative displacement in the z direction

$$u(z,t) = u_0(t)\sin(\pi z/b), \tag{6.117}$$

with dynamical time-dependent amplitude $u_0(t)$ and mode shape given by the sinusoidal function. The amplitude is assumed to oscillate at frequency ω_{res}. We define a mode shape function $f(z)$ by

$$f(z) = \sqrt{\frac{2}{bA}}\cos(\pi z/b), \tag{6.118}$$

where A is the cross-sectional area of the volume $V = bA$. This function is defined so that its modulus squared volume integral gives unity:

$$\int_v |f(z)|^2\, dV = 1. \tag{6.119}$$

We now write the strain $S = \partial u/\partial z$ in terms of the mode function $f(z)$, a geometric scale factor, and a dynamical displacement amplitude variable $x_0(t)$ that has units of length:

$$S(z,t) = \frac{\pi}{b}\sqrt{bA}f(z)x_0(t). \tag{6.120}$$

For the momentum density, we define a second mode shape function $g(z)$

$$g(z) = \sqrt{\frac{2}{bA}}\sin(\pi z/b), \tag{6.121}$$

which has the same modulus squared volume integral of unity. Using this, we write the momentum density in terms of a dynamical momentum amplitude variable $p_0(t)$ as

$$\Pi(z,t) = \frac{1}{\sqrt{bA}}g(z)p_0(t). \tag{6.122}$$

Note that p_0 has units of momentum.

For motion only in the fundamental mode described by the mode shape functions $f(z)$ and $g(z)$, we can re-write the Hamiltonian as

$$
\begin{aligned}
H &= \int_V dV \left(\frac{\Pi^2}{2\rho} + \frac{c_{33}S^2}{2} \right) \\
&= \int_V dV \left(\frac{1}{2\rho} \frac{1}{bA} g(z)^2 p_0^2 + \frac{c_{33}}{2} \frac{\pi^2}{b^2} bAf(z)^2 x_0^2 \right) \\
&= \frac{1}{2\rho bA} p_0^2 + \frac{1}{2} c_{33} \frac{\pi^2}{b^2} bA x_0^2 \\
&= \frac{p_0^2}{2M_{\text{res}}} + \frac{1}{2} M_{\text{res}} \omega_{\text{res}}^2 x_0^2,
\end{aligned}
\tag{6.123}
$$

where $M_{\text{res}} = \rho bA = \rho V$ is the mass of the resonator, and $\omega_{\text{res}} = (\pi/b)\sqrt{c_{33}/\rho}$. The last line in (6.123) is identical to the expression for the Hamiltonian of a classical harmonic oscillator moving in one dimension, completely analogous to our replacing the coplanar waveguide resonator operating in its fundamental mode by the equivalent LC simple harmonic oscillator.

We can then quantize the amplitudes p_0 and x_0 using the standard methods, defining the operators for these quantities in the quantum treatment in terms of the raising and lowering operators $\hat{b}^\dagger$ and $\hat{b}$, where we use different lettering for these *phonon* operators from the raising and lowering operators $a^\dagger$ and a for photons. Hence we have

$$
\hat{x}_0 = \sqrt{\frac{\hbar}{2M_{\text{res}}\omega_{\text{res}}}} \left(\hat{a}^\dagger + \hat{a} \right),
\tag{6.124}
$$

and

$$
\hat{p}_0 = \sqrt{\frac{\hbar M_{\text{res}}\omega_{\text{res}}}{2}} \left(\hat{a}^\dagger - \hat{a} \right),
\tag{6.125}
$$

in terms of which we can write the Hamiltonian operator

$$
\hat{H} = \hbar\omega_{\text{res}} \left(\hat{a}^\dagger \hat{a} + \frac{1}{2} \right).
\tag{6.126}
$$

We can now work out the coupling term between a qubit and a resonator, coupled by the current flowing between the two, (6.113), reproduced here:

$$
I_{\text{res}} = C_{\text{res}} \left(\dot{V} - h_{33} b \dot{U} \right),
\tag{6.127}
$$

The average strain U can be written in terms of the relative displacement at the top and bottom surface of the transducer,

$$U = \frac{1}{b} \left(u(b/2) - u(-b/2) \right). \tag{6.128}$$

The rate of change of the average strain is then

$$\dot{U} = \frac{1}{b} \left(\dot{u}(b/2) - \dot{u}(-b/2) \right). \tag{6.129}$$

This can be written in terms of the momentum density Π,

$$\begin{aligned}
\dot{U} &= \frac{1}{\rho b} \left(\Pi(b/2) - \Pi(-b/2) \right) \\
&= \frac{1}{\rho b} \frac{1}{\sqrt{bA}} \sqrt{\frac{2}{bA}} 2 p_0 \\
&= \frac{2\sqrt{2}}{M_{\text{res}} b} p_0.
\end{aligned} \tag{6.130}$$

We can now quantize this by using the operator representation for $\hat{p}_0$, so that

$$\hat{U} = \frac{2\sqrt{2}}{M_{\text{res}} b} i \sqrt{\frac{\hbar M_{\text{res}} \omega_{\text{res}}}{2}} (\hat{b}^\dagger - \hat{b}) = \frac{2i}{b} \sqrt{\frac{\hbar M_{\text{res}} \omega_{\text{res}}}{2}} (\hat{b}^\dagger - \hat{b}). \tag{6.131}$$

Hence we can write the interaction term as

$$\begin{aligned}
\hat{H}_{\text{int}} &= -E_J \frac{\hat{I}'_{\text{res}}}{I_0} \hat{\phi} \\
&= -\frac{\hbar}{2e} \hat{I}'_{\text{res}} \hat{\phi} \\
&= \frac{\hbar}{2e} h_{33} b C_{\text{res}} \hat{U} \hat{\phi}.
\end{aligned} \tag{6.132}$$

Replacing all the operators with raising and lowering operators, this can be written

$$\hat{H}_{\text{int}} = \frac{\hbar}{2e} h_{33} b C_{\text{res}} \frac{2i}{b} \sqrt{\frac{\hbar \omega_{\text{res}}}{M_{\text{res}}}} \left(\hat{b}^\dagger e^{-i\omega_{\text{res}}t} - \hat{b} e^{i\omega_{\text{res}}t} \right) x_{ge} \left(\hat{\sigma}_+ e^{-i\omega_{ge}t} + \hat{\sigma}_- e^{i\omega_{ge}t} \right), \tag{6.133}$$

where $x_{ge} = \langle g|\hat{\phi}|e\rangle$ is the dipole moment coupling the qubit ground and excited states.

We now assume we only interact on resonance, so $\omega_{\text{res}} = \omega_{ge}$, and use the rotating wave approximation to drop all non-cancelling time dependence, leaving us with

$$\hat{H}_{\text{int}} = i\frac{\hbar}{e}h_{33}\,C_{\text{res}}\sqrt{\frac{\hbar\omega_{\text{res}}}{M_{\text{res}}}}x_{ge}\left(\hat{\sigma}_{-}\hat{b}^{\dagger} - \hat{\sigma}_{+}\hat{b}\right). \qquad (6.134)$$

We define the coupling strength g by

$$g = \frac{h_{33}}{e}C_{\text{res}}\sqrt{\frac{\hbar\omega_{\text{res}}}{M_{\text{res}}}}x_{ge}, \qquad (6.135)$$

so that we can write the interaction Hamiltonian as

$$\hat{H}_{\text{int}} = i\hbar g\left(\hat{\sigma}_{-}\hat{b}^{\dagger} - \hat{\sigma}_{+}\hat{b}\right). \qquad (6.136)$$

While the Hamiltonian in Eq. 6.136 differs in some details from the Hamiltonian that describes the qubit–electromagnetic interaction, with a differing phase and sign, this difference is in fact difficult to discern experimentally, given that in the mechanics experiment we were only able to perform swaps and not e.g. tomographic measurements, that might otherwise have revealed these differences.

A significant challenge with mechanical resonators operating in the microwave band has been achieving high quality factors, in other words long lifetimes for energetic states. This is due to a combination of factors, including internal mechanical loss as well as radiated energy from the mechanical supports. Recent results have, however, shown that for single crystal resonators fabricated from silicon, that include carefully designed structures that minimize acoustic radiation through the supports, very high quality factors can be achieved. As a result, the sophistication of the experiments that have been completed to date remains fairly primitive, restricted to: tests that the resonator has indeed been cooled to its quantum ground state; injection of a single quantum of excitation (one phonon) into the resonator; Rabi swaps between the qubit and resonator of that single excitation; and the creation of a superposition state $|0\rangle + |1\rangle$ in the resonator. If higher quality factors can be achieved in similar mechanical resonators, much more sophisticated experiments could be pursued; this awaits some scientific advances.

In Fig. 6.9 we display the verification that a mechanical resonator coupled to a phase qubit was indeed in its ground state, obtained by placing the qubit and mechanical resonator into near-resonance for 1 μs, allowing the two to equilibrate, then measuring the state of the qubit. If the resonator remained in its excited state with any non-negligible probability, the qubit would be measured in its excited state with a non-negligible probability; instead, the qubit was found in its excited state at the typical background level of around 5% of the time, independent of the tuning of the qubit with respect to the resonator. This is consistent with the resonator being in its ground state more than 95% of the time.

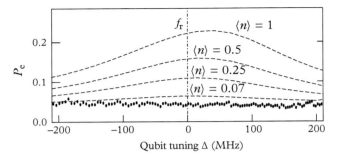

Fig. 6.9 *Verification that a mechanical resonator was indeed in its ground state when operated at an environmental temperature of around 25 mK. A phase qubit is strongly coupled to the resonator, each with a frequency near 6 GHz. The qubit was tuned close to the resonator frequency and left there for 1 μs, following which the qubit excited state probability P_e was measured. The qubit detuning with respect to the resonator frequency was then scanned over a range corresponding to about 5% of the resonator frequency. Experimentally the resonator does not excite the qubit from its ground state, indicating the resonator is always in its ground state. Adapted with permission from (O'Connell, Hofheinz, Ansmann, Bialczak, Lenander, Lucero, Neeley, Sank, Wang, Weides, Wenner, Martinis and Cleland, 2010).*

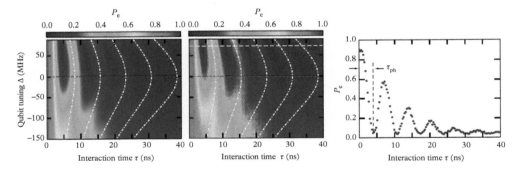

Fig. 6.10 *Observation of Rabi swaps between a qubit and a mechanical resonator. Left and centre shows the simulated and experimentally measured oscillations as a function of time and qubit detuning from resonator resonance. Right panel shows a cut through the experimental data for the largest amplitude swaps, showing the clear oscillations in qubit excited state probability P_e as the excitation is swapped to and from the mechanical resonator. Adapted with permission from (O'Connell, Hofheinz, Ansmann, Bialczak, Lenander, Lucero, Neeley, Sank, Wang, Weides, Wenner, Martinis and Cleland, 2010).*

In Fig. 6.10 we display a measurement in which a qubit is put in its excited state $|e\rangle$, then brought into resonance with a mechanical resonator and their interaction monitored by periodically measuring the qubit state. The oscillations in the qubit excited state probability P_e demonstrate that the quantum of energy is being swapped into the mechanical resonator, where it exists as a phonon, and then getting swapped back into the qubit, at a rate that is equal to the independently measured coupling g between the qubit and the resonator. The rapid decay in the amplitude of the oscillations is due to the short lifetime of the excitation in the resonator.

Bibliography

Auld, B. A. (1990). *Acoustic Waves and Fields in Solids*. John Wiley and Sons, New York.

Banaszek, K., Radzewicz, C., Wódkiewicz, K., and Krasiński, J. S. (1999). Direct measurement of the Wigner function by photon counting. *Phys. Rev. A*, **60**, 674–7.

Banaszek, K. and Wódkiewicz, K. (1996). Direct probing of quantum phase space by photon counting. *Phys. Rev. Lett.*, **76**, 4344–7.

Barone, A. and Paterno, G. (1982). *Physics and Applications of the Josephson Effect*. John Wiley and Sons, New York.

Cleland, A. N. (2003). *Foundations of Nanomechanics: From Solid-State Theory to Device Applications*. Springer, Berlin Heidelberg.

Hofheinz, M., Wang, H., Ansmann, M., Bialczak, R. C., Lucero, E., Neeley, M., O'Connell, A. D., Sank, D., Wenner, J., Martinis, J. M., and Cleland, A. N. (2009). Synthesizing arbitrary quantum states in a superconducting resonator. *Nature*, **459**, 546–9.

Hofheinz, M., Weig, E. M., Ansmann, M., Bialczak, R. C., Lucero, E., Neeley, M., O'Connell, A. D., Wang, H., Martinis, J. M., and Cleland, A. N. (2008). Generation of Fock states in a superconducting quantum circuit. *Nature*, **454**, 310–14.

Jaynes, E. T. and Cummings, F. W. (1963). Comparison of quantum and semiclassical radiation theories with application to the beam maser. *Proc. IEEE*, **51**, 89–109.

Law, C. K. and Eberly, J. H. (1996). Arbitrary control of a quantum electromagnetic field. *Phys. Rev. Lett.*, **76**, 1055–8.

O'Connell, A. D., Hofheinz, M., Ansmann, M., Bialczak, R. C., Lenander, M., Lucero, E., Neeley, M., Sank, D., Wang, H., Weides, M., Wenner, J., Martinis, J. M., and Cleland, A. N. (2010). Quantum ground state and single-phonon control of a mechanical resonator. *Nature*, **464**, 697–703.

Tinkham, Michael (1996). *Introduction to Superconductivity*. McGraw-Hill, New York.

7

Spin-coupled Mechanical Systems

Ania Bleszynski Jayich

University of California Santa Barbara
Department of Physics
United States of America

Ania Bleszynski Jayich

Ania Bleszynski Jayich, *Spin-coupled Mechanical Systems* In: *Quantum Optomechanics and Nanomechanics.* Edited by:
Pierre-Francois Cohadon, Jack Harris, Florian Marquardt, Leticia F. Cugliandolo, Oxford University Press (2020).
© Oxford University Press. DOI: 10.1093/oso/9780198828143.003.0007

Chapter Contents

7.1   Introduction

7.1.1   Motivation

A hybrid system composed of a spin coupled to a mechanical degree of freedom is interesting to study for several reasons: fundamental and applied, for enhancing the functionality of the spin and of the mechanical system, and in the quantum and semiclassical regimes. First I will give a brief overview of some of the goals of spin-coupled mechanical systems.

A major challenge in the field of quantum optomechanics is to realize a nonlinearity at the single phonon level in the quest to generate and detect nonclassical states of motion. To date classical and linear tools have been almost exclusively used in optomechanical experiments, restricting most optomechanical experiments to the manipulation of Gaussian states. A spin, an inherently nonlinear system that can be readily experimentally observed in the quantum regime, offers an elegant path towards achieving mechanical nonlinearities and hence the generation and detection of nonclassical states of motion. Therefore realizing sufficiently strong spin-mechanical coupling in the quantum regime is currently a hotly pursued goal.

In a hybrid quantum system, which utilizes and interfaces disparate quantum elements and capitalizes on their distinct advantages, mechanical excitations in the quantum regime constitute a useful element. A hybrid quantum processor could, for instance, be made up of nuclear spins or atoms to store quantum information for long periods of time, solid-state electronic spins that can be readily manipulated with microwave photons, and optical photons that provide a robust long-distance quantum bus. Phonons, relative newcomers to the field of quantum technologies, offer another way of realizing long range interactions between quantum bits (qubits), a major challenge in the field of quantum information processing with solid-state qubits, and therefore could facilitate scalable quantum information processing. Resonant phonons can play the role of microwave cavity photons in the analogous field of microwave cavity QED, as discussed in two recent proposals (Habraken, Stannigel, Lukin, Zoller and Rabl, 2012; Schuetz, Kessler, Giedke, Vandersypen, Lukin and Cirac, 2015). A proposed phonon-cavity QED schematic is shown in Fig. 7.1. Localized phonons inside a high-Q cavity interact strongly with a qubit, allowing exchange of quantum information. Phononic waveguides, weakly coupled to the cavity, transport quantum information between remote qubit-cavity nodes. The Hamiltonian describing the interaction at each node is given by:

$$H_{node} = \frac{\omega_q}{2}\sigma^z + \omega_c a^\dagger a + g(\sigma^+ a + \sigma^- a^\dagger) \tag{7.1}$$

where ω_q defines the transition frequency of the qubit, σ is the Pauli spin operator for the qubit, ω_c is the frequency of the mechanical mode, a is the bosonic annihilation operator for the mechanical mode, and g is the single-phonon coupling strength between

(a)

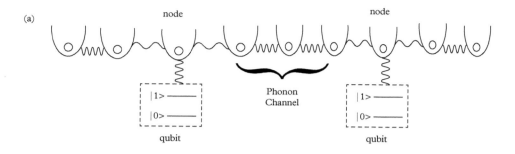

(b)

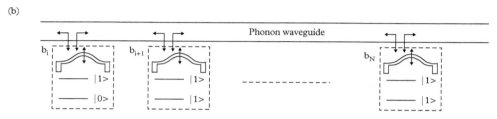

Fig. 7.1 *a) Schematic of a phononic quantum network with localized qubits interacting with a localized phonon mode, which is weakly coupled to a phonon channel or waveguide, as in (Habraken, Stannigel, Lukin, Zoller and Rabl, 2012). b) Similar schematic representation of a phononic quantum network with each node consisting of a qubit coupled (with strength $\lambda(t)$) to a localized mechanical mode, b_i. A global phononic quantum bus (waveguide) generates long distance interactions between remote qubits.*

the mechanical mode and the qubit. The so-called beam-splitter interaction between the qubit and the resonator describes resonant exchange of single quanta between the two systems, which can be used to entangle the two systems, for instance. The coupling g can take on a variety of forms as will be discussed and detailed in these lecture notes.

An important advantage of a phononic quantum bus, such as the one in Fig. 7.1, is its ability to couple dissimilar qubits, such as superconducting qubits, atoms, and photons, and to transduce quantum information between them. This flexibility arises because mechanical resonators can be made sensitive to a wide variety of fields, *e.g.* strain, electric, and magnetic fields. Phonons can thus be used as flying qubits themselves or as transducers between stationary qubits and photonic flying qubits, which have the advantage of long-distance propagation and long coherence times.

Other applications of spin-coupled mechanical systems include mechanical detection (Rugar, Budakian, Mamin and Chui, 2004) and control (Kolkowitz, Bleszynski Jayich, Unterreithmeier, Bennett, Rabl, Harris and Lukin, 2012; Yeo, de Assis, Gloppe, Dupont-Ferrier, Verlot, Malik, Dupuy, Claudon, Gérard, Auffeves, Nogues, Seidelin,

Poizat, Arcizet and Richard, 2013; Montinaro, Wust, Munsch, Fontana, Russo-Averchi, Heiss, Fontcuberta i Morral, Warburton and Poggio, 2014; Barfuss, Teissier, Neu, Nunnenkamp and Maletinsky, 2015; Ovartchaiyapong, Lee, Myers and Jayich, 2014; MacQuarrie, Gosavi, Moehle, Jungwirth, Bhave and Fuchs, 2015) of spin states and spin dynamics. In the pioneering magnetic resonance force microscopy (MRFM) experiments of Dan Rugar *et al.* (Rugar, Budakian, Mamin and Chui, 2004), a magnetically functionalized cantilever was used to sense the magnetic field associated with a single electron spin. Building on these results, MRFM experiments continue with the aim of detecting and imaging single nuclear spins to image ultimately, e.g., protein structure. On the control side, mechanical control offers several advantages including local control and access to spin transitions that are forbidden by magnetic dipole selection rules in standard microwave magnetic field control. In these lectures, I will use the term 'spin' for a generic 2-level system, thus encompassing, e.g., distinct orbital levels in atomic-like systems connected by optical dipole transitions. Mechanically induced strain control of optical transitions (both polarization and frequency) is also powerful, in particular for photon-mediated qubit entanglement.

Spin systems, in particular highly coherent ones, can make excellent sensors of magnetic, electric, thermal and strain fields. As a final example of the utility of spin-coupled mechanical systems, ensembles of spins interacting with a common phonon mode can be prepared in a spin-squeezed state, as proposed for NV centres in diamond(Bennett, Yao, Otterbach, Zoller, Rabl and Lukin, 2013) thus offering sensitivity beyond the standard quantum limit.

An interesting side note: until recently, spins and mechanics were generally considered mutually detrimental to each other's functionality. To be more explicit, fluctuating two-level systems (spins) are often blamed for the ultimate limit to achieving high mechanical quality factors (Pohl, Liu and Thompson, 2002; Remus, Blencowe and Tanaka, 2009; O'Connell, Hofheinz, Ansmann, Bialczak, Lenander, Lucero, Neeley, Sank, Wang, Weides, Wenner, Martinis and Cleland, 2010). Likewise, the energy relaxation time (T_1) and dephasing time (T_2)of a solid-state spin is often ultimately limited by interaction with a thermal bath of phonons. It is only recently that with increased experimental control over mechanics and spins (reaching the *single quantum* level) and their coupling, we can imagine the two systems complementing and enhancing each other's functionality.

Some promising spin systems that have reached the level of maturity to be fruitfully coupled to mechanical systems include nitrogen vacancy (NV) centres in diamond, defects/donors in Si and SiO_2, superconducting qubits, semiconducting quantum dots, and atoms. In several of these systems, mechanical coupling has been demonstrated(Metcalfe, Carr, Muller, Solomon and Lawall, 2010; LaHaye, Suh, Echternach, Schwab and Roukes, 2009; O'Connell, Hofheinz, Ansmann, Bialczak, Lenander, Lucero, Neeley, Sank, Wang, Weides, Wenner, Martinis and Cleland, 2010; Okazaki, Mahboob, Onomitsu, Sasaki and Yamaguchi, 2016) and in others mechanical coupling could add to their functionality, such as providing single addressability of spin-defects in Si.

7.1.2 Organization

In section 7.2, I will introduce the NV centre, its energy level structure and its sensitivity to a variety of fields. Included is a discussion of the NV's promising qualities that make it exciting for hybrid quantum technologies and in particular integration with mechanical oscillators. Simple group theory arguments will be introduced to explain the NV's energy level structure as well as its coupling to strain. Section 7.3 discusses different NV centre-mechanical coupling mechanisms, specifically magnetic- and strain-based ones. I will present realistic values for the coupling strength and protocols for coupling, as well as technical challenges to achieving a functional NV-mechanical coupled system, in particular in the quantum regime.

7.2 Nitrogen Vacancy Centres in Diamond

In these notes, we will focus on the NV centre in diamond as a model quantum two-level system to which mechanical systems can be coupled.

7.2.1 Background

The NV centre is a point-like defect in the diamond lattice consisting of a substitutional nitrogen atom adjacent to a missing carbon atom, a vacancy (Fig. 7.2). It forms an atom-like system with many promising quantum properties.

   The NV is particularly promising because of its long ($\sim$ ms) spin coherence time that persists up to room temperature, the ease of individual addressability with optical detection methods, facile optical and microwave-based quantum control on the single spin level, and high sensitivity to a variety of fields. As an artificial atom, it marries the benefits of solid-state qubit implementations with atomic physics: it is easy to 'trap', patternable, compatible with on-chip electronic and optical devices, stable and scalable,

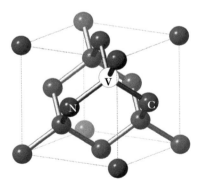

Fig. 7.2 *Atomic structure of the nitrogen vacancy centre in diamond. Carbon (C), Nitrogen (N), and Vacancy (V) are labelled.*

while having long coherence times, stable optical transitions and an atomically small size. It is in fact quite remarkable that the NV centre's attractive atomic-like features (in particular long coherence time) exist in a solid-state environment, where phonons and proximal atoms, defects, etc. often contribute to decoherence and smearing-out of microwave and optical transitions. However, because of diamond's high Debye temperature, small spin-orbit coupling, predominantly nuclear-spin-free background (^{12}C), and wide band gap (coupled with the fact that the NV defect level lies deep within the band gap, thus protecting it from excitations to the valence/conduction band), the NV centre maintains strong immunity from the lattice. However, important departures from atomic properties do exist and they must be addressed. For instance, local inhomogeneities in the lattice (arising from, e.g., strain or built-in electric fields) introduce variability from one NV to another and also time-varying properties, such as spectral diffusion of the optical lines. These issues make entanglement schemes, for instance, particularly challenging but as will be shown, mechanical control of optical transitions provides a path to alleviate these issues.

Simplified structure

The NV centre's energy level structure, simplistically shown in Fig. 7.3, consists of two functional 2-level systems ($\{|g\rangle, |e\rangle\}$ and $\{|\downarrow\rangle, |\uparrow\rangle\}$) and hence allows us to study mechanical coupling to both an optical dipole transition ($|g\rangle \leftrightarrow |e\rangle$) and a microwave magnetic dipole transition ($|\downarrow\rangle \leftrightarrow |\uparrow\rangle$).

The ($\{|\downarrow\rangle, |\uparrow\rangle\}$) two-level system is a 'real' spin in that the two states differ in angular momentum quantum number and hence couple to magnetic fields. We will refer to this manifold as the ground state spin of the NV. The energy splitting of these two levels is in the microwave regime and hence the ground state spin is amenable to microwave magnetic coherent manipulation. The $\{|g\rangle, |e\rangle\}$ levels are two different orbital levels of the NV centre, commonly referred to as the ground and excited states, and their energy splitting is on the order of PHz (637 nm). Note: the ground state is in fact a three-level system, a spin triplet, and the $|e\rangle$ state actually consists of six states, but we avoid these subtleties for now and go into more detail later. Both the $\{|g\rangle, |e\rangle\}$ and $\{|\downarrow\rangle, |\uparrow\rangle\}$

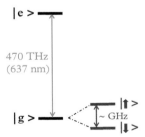

Fig. 7.3 *Simplified electronic energy level structure of the NV centre in diamond. The $\{|g\rangle, |e\rangle\}$ manifold and the $\{|\downarrow\rangle, |\uparrow\rangle\}$ manifold form two distinct two-level systems, with transitions in the optical and microwave domains, respectively.*

can be and will be referred to as spin systems, as they can be treated as isolated two-level systems with characteristic quantum decoherence and energy relaxation times, and coherent quantum manipulation of the two-state system is possible via external fields. There are important differences, however, (notably in their sensitivity to external fields) and the nature of these two different NV spin systems will be described at length in these notes.

NV formation

NV centres can be formed in a variety of different ways. They occur naturally or by intentional introduction of nitrogen and vacancies into the lattice. Nitrogen ion implantation is the most common way to introduce nitrogen at predetermined densities and depths. Vacancies form concurrently (and inevitably) with the implantation process and subsequent annealing at $T > 850°$ induces vacancy diffusion and NV formation. Another recently developed NV formation technique is that of nitrogen δ-doping during chemical vapour deposition (CVD) growth of diamond (Myers, Das, Dartiailh, Ohno, Awschalom and Bleszynski Jayich, 2014), a gentle, bottom-up technique with nm-scale depth control of nitrogen. Vacancies can then be formed via e.g. electron irradiation or carbon implantation and again, annealing induces NV formation. When coupling NV centres to mechanical oscillators, it is often desirable to position NV centres deterministically with nanometre-scale precision, for instance inside the high-strain region of a diamond nanomechanical resonator. A variety of NV-patterning techniques have been developed using patterned N formation e.g. N ion implantation through masks (Toyli, Weis, Fuchs, Schenkel and Awschalom, 2010)), and patterned vacancy formation (e.g. TEM-based electron irradiation (McLellan, Myers, Kraemer, Ohno, Awschalom and Bleszynski Jayich, 2016) and ^{12}C implantation through masks (Ohno, Joseph Heremans, de las Casas, Myers, Aleman, Bleszynski Jayich and Awschalom, 2014). TEM-based electron irradiation has the distinct advantage that it does not require a mask and hence is compatible with already fabricated mechanical structures. Furthermore, one can simultaneously image a structure and form vacancies in the high-strain region.

Major accomplishments with NV centres

Significant research activity in the last 10 years has focused on NV centres in a wide variety of applications. Major accomplishments include:

- Entanglement between NV spin centres located 3 m apart (Bernien, Hensen, Pfaff, Koolstra, Blok, Robledo, Taminiau, Markham, Twitchen, Childress and Hanson, 2013)

- Quantum memories at room temperature with 1-second coherence time (Maurer, Kucsko, Latta, Jiang, Yao, Bennett, Pastawski, Hunger, Chisholm, Markham, Twitchen, Cirac and Lukin, 2012)

- Loophole free Bell's inequality tests (Hensen, Bernien, Dréau, Reiserer, Kalb, Blok, Ruitenberg, Vermeulen, Schouten, Abellán, Amaya, Pruneri, Mitchell, Markham, Twitchen, Elkouss, Wehner, Taminiau and Hanson, 2015)

- NV-based imaging of single electron spins (Grinolds, Hong, Maletinsky, Luan, Lukin, Walsworth and Yacoby, 2013), nanoscale volumes of nuclear spins (Staudacher, Shi, Pezzagna, Meijer, Du, Meriles, Reinhard and Wrachtrup, 2013; Mamin, Kim, Sherwood, Rettner, Ohno, Awschalom and Rugar, 2013), superconducting vortices (Pelliccione, Jenkins, Ovartchaiyapong, Reetz, Emmanouilidou, Ni and Bleszynski Jayich, 2016; Thiel, Rohner, Ganzhorn, Appel, Neu, Müller, Kleiner, Koelle and Maletinsky, 2016), and nano magnets (Tetienne, Hingant, Martínez, Rohart, Thiaville, Diez, Garcia, Adam, Kim, Roch, Miron, Gaudin, Vila, Ocker, Ravelosona and Jacques, 2015)

Outstanding challenges facing NV centres include:

- deteriorated optical and spin properties near surfaces
- truly 1-nm scale deterministic position control
- spectral stability of the zero phonon line and broad phonon sideband.

7.2.2   NV structure

The NV centre that is the predominant focus of current experiments is the negatively charged NV^-, and we will use the label 'NV' for that setting as is commonly done in the literature. In its negatively charged state, the NV consists of 6 valence electrons that occupy 2 non-degenerate symmetric orbitals (a_1, a_1') and 2 degenerate orbitals (e_x, e_y) (Loubser and van Wyk, 1978). The single-electron orbitals (a_1, a_1', e_x, e_y) are constructed from linear combinations of the dangling bonds of the 3 carbon atoms and 1 nitrogen atom surrounding the vacancy: $\sigma_1, \sigma_2, \sigma_3, \sigma_N$, shown in Fig. 7.4.

$$a_1 = \alpha(\sigma_1 + \sigma_2 + \sigma_3) + \beta\sigma_N \tag{7.2}$$
$$a_1' = \beta'(\sigma_1 + \sigma_2 + \sigma_3) + \alpha'\sigma_N \tag{7.3}$$

Fig. 7.4 *Schematic of the four dangling bonds $(\sigma_1, \sigma_2, \sigma_3, \sigma_N)$ from which the molecular orbitals of the NV centre are constructed. Nitrogen is blue, vacancy is white, and carbon is grey.*

$$e_x = \frac{1}{\sqrt{6}}(2\sigma_1 - \sigma_2 - \sigma_3) \qquad (7.4)$$

$$e_y = \frac{1}{\sqrt{2}}(\sigma_2 - \sigma_3) \qquad (7.5)$$

These linear combinations are derived from group theory considerations, and the ordering of the orbitals relative to the valence and conduction bands of diamond, shown in Fig. 7.5, can be obtained by considering the electron–ion Coulomb interaction. The ground state configuration of the NV corresponds to four electrons in the totally symmetric (a_1, a_1') orbitals and two electrons in the degenerate (e_x, e_y) orbitals. In the excited state of the NV, one electron is excited from the a_1 orbital to either the e_x or e_y orbitals. In general, we are only concerned with the a_1, e_x, e_y orbitals as the relevant dynamics involve transitions between these states.

Short group theory primer for NV centres

In this short section, I introduce some terminology and symmetry concepts from group theory, specifically applied to the NV centre. The symmetry properties of the NV centre provide insight into the nature of its electronic states, and in particular these concepts will serve as a useful tool when considering the strain-mediated interaction of the NV centre with a mechanical oscillator.

The NV centre exhibits C_{3v} symmetry, i.e. it belongs to the C_{3v} symmetry point group. The C_{3v} group consists of six elements, $\{I, C_3, C_3^{-1}, R_1, R_2, R_3\}$: symmetry operations under which the NV centre is invariant. (C_3, C_3^{-1}) refer to $(2\pi/3, 4\pi/3)$ rotations around the NV axis, (R_1, R_2, R_3) refer to reflections about a plane containing the NV axis and any of its nearest-neighbour carbon atoms, and I is the identity operator. We can represent any of these six elements by matrices defined in a particular basis. One particularly important representation, termed the irreducible representation (IR), is the block diagonal representation of the the group elements. The three IRs for the NV centre are termed A_1, A_2, and E. A_1 is a one-dimensional IR (1×1 matrix equal to 1) that is fully symmetric under all the C_{3v} symmetry operations.

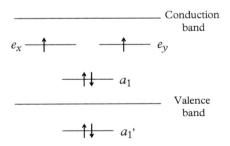

Fig. 7.5 *NV molecular orbitals and their ordering with respect to diamond's valence and conduction bands. Six electrons occupy these orbitals (three from the carbon dangling bonds, two from the nitrogen dangling bond, and one from a nearby donor, presumably a substitutional nitrogen).*

It is often useful to write functions, or operators, in terms of operators that transform according to particular IRs. For example, the position operator $\hat{z}$ is an operator that transforms as A_1, as it is unchanged under all C_{3v} operations. Other examples of 'A_1-symmetric' functions are $\hat{x}^2 + \hat{y}^2$ and $\hat{z}^2$. And what about $\hat{x}$ and $\hat{y}$? These are not 'A_1-symmetric' functions as they do not remain unchanged under C_{3v} operations, but rather they transform into each other and hence are basis functions for the two-dimensional IR E. Other examples of functions that transform as the E IR include the pairs $(\hat{x}^2 - \hat{y}^2, xy)$ and (yz, xz).

One can further show that the a_1 and a_1' orbitals are 'A_1-symmetric' in that they transform into themselves under all C_{3v} operations and the e_x, e_y are 'E-symmetric'.

Later on in our discussion of the interaction of strain with the NV centre, we will invoke these symmetry considerations again, and for instance, write the elements of the stress tensor as entities that transform as A_1 and E (we will not need to invoke the A_2 IR).

To simplify things, we will move from the six-electron picture of the NV centre to a two-hole picture, and only consider the a_1, e_x, e_y states that lie within the bandgap of diamond. The orbital diagrams in the electron and hole pictures are shown in Fig. 7.6, where spin has been indicated. The spin triplet (pictured) is favoured due to minimizing Coulomb interactions with an antisymmetric orbital wave function. The (spin triplet, orbital singlet) ground state thus consists of three states and the (spin triplet, orbital doublet) excited state consists of six states. The fine structure of the ground and excited states is determined by considering the effects of spin–spin interactions, spin–orbit interactions, strain, and electric and magnetic fields. We will not go through the derivations in detail, but rather state the relevant Hamiltonians in the following section.

7.2.3 The fine structure of the ground and excited states of the NV centre

The ground state

The interaction of magnetic fields, electric fields, and strain with the NV centre is important to understand when coupling NV centres to mechanical resonators, as the coupling will be engineered through these types of field interactions.

The Hamiltonian for the spin triplet ground state of the NV center is:

$$H_{gs} = (D_{gs} + d_{\parallel}\Pi_{\parallel})S_z^2 + \gamma\vec{B}\cdot\hat{S} - d_{\perp}[\Pi_x(S_xS_y + S_yS_x) + \Pi_y(S_x^2 - S_y^2)] \tag{7.6}$$

The terms in the Hamiltonian are defined below:

- D_{gs}, the zero-field splitting, is an internal effect that arises from spin–spin interactions. In the absence of any applied external fields, the NV centre's ground state triplet levels are split by $D_{gs} = 2.87$ GHz (between the $|\pm1\rangle$ and $|0\rangle$ triplet levels).
- Π is the combined effect of electric (E) and strain (σ) fields: $\Pi_{\parallel} = E_{\parallel} + \sigma_{\parallel}$, $\Pi_x = E_x + \sigma_x$, $\Pi_y = E_y + \sigma_y$. $E_{\parallel}$ refers to the electric field along the NV axis and E_x, E_y lie

in the plane perpendicular to the NV axis. Of course strain is a tensor and electric field is a vector and hence treating σ and E on equal footing is not, strictly speaking, correct. We will come back to this point in later sections, and address the full tensorial nature of strain and show that it can be mapped onto an effective electric field. At this point, however, we use this simple analogy and we define $\sigma_{\parallel}$ to be the strain 'along' the NV axis, made up of components that are A_1-symmetric, such as σ_{zz} and $\sigma_{xx} + \sigma_{yy}$, and we define $\sigma_{x,y}$ to be 'transverse' strain, made up of components that are E-symmetric, i.e. those that do not preserve the symmetry of the NV centre.

- $\Pi_{\parallel}$ effectively shifts D_{gs} by $(d_{\parallel,E}E_{\parallel} + d_{\parallel,\sigma}\sigma_{\parallel})$, where $d_{\parallel,E}$ and $d_{\parallel,\sigma}$ are the parallel electric field and strain susceptibilities, respectively. As $\Pi_{\parallel}$ commutes with both the D_{gs} and B_z terms in the Hamiltonian, it preserves the $|0,-1,+1\rangle$ eigenstates.

- $\Pi_{\perp}$ mixes the $|-1,+1\rangle$ eigenstates and splits them.

- The experimentally measured values of the ground-state strain and electric field susceptibilities are (Van Oort and Glasbeek, 1990; Ovartchaiyapong, Lee, Myers and Jayich, 2014; Teissier, Barfuss, Appel, Neu and Maletinsky, 2014):

  - $d_{\parallel,E} = 0.35\ Hz(V/cm)^{-1}$
  - $d_{\perp,E} = 17\ Hz(V/cm)^{-1}$
  - $d_{\parallel,\sigma} = 13\ GHz/strain$
  - $d_{\perp,\sigma} = 20\ GHz/strain$

- $\gamma\vec{B}\cdot\hat{S}$ is the Zeeman splitting that splits $|+1\rangle$ and $|-1\rangle$ by 2.8 MHz/G. Since $B_z S_z$ commutes with $D_{gs}S_z^2$, a magnetic field along the NV axis preserves the

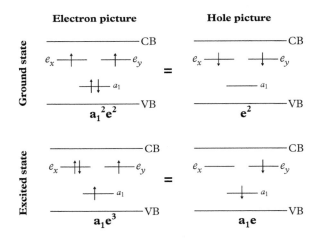

Fig. 7.6 *The electron and hole pictures of the ground and excited states of the NV centre, showing the spin occupation of the single-electron levels.*

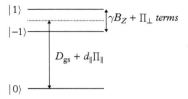

Fig. 7.7 *Energy levels of the NV spin triplet ground state.*

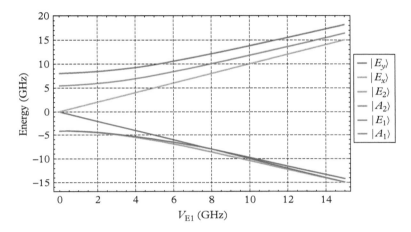

Fig. 7.8 *Energy levels of the NV excited state and their evolution with applied strain of E_1 symmetry. The zero of the y-axis is set to be the energy of the unstrained $\left|E_x, E_y\right\rangle$ states.*

$|0, -1, +1\rangle$ eigenstates. A nonzero $B_\perp$ mixes the $|0, -1, +1\rangle$ eigenstates but as long as $B_\perp << D_{gs}$, the mixing is insignificant.

The ground state 'spin' can be defined as any two of the three triplet levels: $|0, 1\rangle$, $|0, -1\rangle$, or $|-1, 1\rangle$. Because microwave magnetic fields couple $|0\rangle \leftrightarrow |1\rangle$ and $|0\rangle \leftrightarrow |-1\rangle$, the former two spin choices are most commonly used. However, because microwave strain fields can drive coherent transitions between $|-1\rangle \leftrightarrow |1\rangle$, strain manipulation via mechanical resonators opens up the possibility of a useful and accessible $|-1, 1\rangle$ qubit.

The excited state

The spin triplet, orbital doublet excited state consists of six states, conventionally labelled A_1, A_2, E_x, E_y, E_1 and E_2 (where the nomenclature relates to their symmetry). The orbital components of these states are made up of superpositions of the a_1, e_x and e_y orbitals (see (Doherty, Manson, Delaney and Hollenberg, 2011) for details). The E_x and E_y states have spin projection $m_s = 0$ and hence couple via optical dipole transitions to the $m_s = 0$ ground state level. The other four states have $m_s = \pm 1$ spin projection and hence couple via optical dipole transitions to the $m_s = \pm 1$ ground state levels. The ordering of the six excited state levels, their splittings, and their evolution with strain are pictured in Fig. 7.8. In the absence of any applied external fields, spin–spin and spin–orbit interactions split the levels. Strain (or electric fields) further split the levels. Magnetic fields are neglected here because their effect is much weaker than that of strain and electric fields in the excited state. For example, the susceptibility of the $E_{x,y}$ levels to strain is 1 PHz/strain, which is ~ 5 orders of magnitude larger than in the ground state. The $E_{x,y}$ levels are particularly sensitive to strain because of their orbital degeneracy.

An experimental note: due to phonon broadening at higher temperatures, the excited state structure is only resolvable at low (≤ 10 K) temperatures. Photoluminescence excitation spectroscopy is commonly used to probe the level shifts.

Finally, I note that the richness of the excited state structure, its sensitivity to strain, and the intimate relationship between orbital and spin states, allow for novel ways of quantum control of both the NV and of the mechanics.

7.3 Coupling Mechanics and Spins

Now that we have introduced the basics of the NV centre's level structure and its susceptibilities to different fields, let's examine the possible ways to couple mechanical

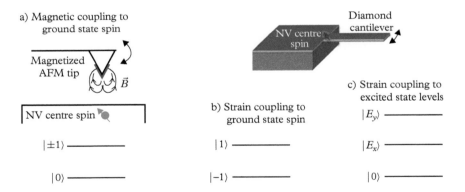

Fig. 7.9 *Schematics of various coupling modalities for NV-mechanical coupling. a) Magnetic coupling. A magnetized atomic force microscope (AFM) tip is brought into close proximity to an NV centre residing near the surface of a bulk diamond. As the cantilever oscillates, the NV experiences, and senses, an oscillating magnetic field. The spin qubit energy levels, $|+1\rangle$ and $|-1\rangle$, are modulated through the Zeeman coupling. b)–c) Strain coupling. An NV centre located inside a diamond mechanical oscillator experiences an oscillating strain field as the cantilever moves. In b), the strain modulates the ground state spin $|-1\rangle \rightarrow |+1\rangle$ transition. In c) the strain modulates the $|E_x, E_y\rangle$ excited state energies.*

motion to spin, using the NV centre as our model spin system. We have several different choices of 'spin' system as well as several different choices of field coupling type. In the following, I will dwell on three specific NV-mechanics coupling strategies, illustrated in Fig. 7.9.

For each of the spin-mechanical coupling scenarios in Fig. 7.9, it is important to consider the relevant figures of merit, namely 1) the coupling strength g between the spin and mechanics, 2) the relevant coherence decay rates of the spin Γ_{NV} and the mechanical system Γ_{mech}, and 3) the experimental difficulties/challenges associated with each of the strategies.

- The coupling strength g is defined as the energy shift of the target spin state per quantum of motion of the mechanical oscillator. For magnetic coupling in Fig. 7.9a, for instance, $g = \gamma a_0 \nabla B$, where a_0 is the zero-point amplitude of the mechanical oscillator, γ is the gyromagnetic ratio of the spin, and $\vec{\nabla} B$ is the gradient of the cantilever-induced magnetic field at the spin location. For the strain coupling in Fig. 7.9b–c, $g = d_\perp \sigma_{\perp,zpm}$ where $\sigma_{\perp,zpm}$ is the strain induced at the spin location for a cantilever displacement of a_0. Thus, one maximizes g through large zero-point motion amplitudes and through either large magnetic field gradients or wise choice of mechanical modes (strain induced by mechanical modes will be discussed in section 7.3.2).

- When considering coupling to the ground state spin manifold of the NV, $\Gamma_{NV} = 1/T_2$ (T_2 is the spin coherence time), which for NV centres has been shown to reach as low as $1\ Hz$ at low temperatures. When considering coupling to the optical dipole transitions, as in Fig. 7.9c, Γ_{NV} is the inverse linewidth of the optical transition, which can be as small as $\sim$20 MHz.

- For 'strong coupling', it is required that $g > \Gamma_{NV}, \Gamma_{mech}$ and for the 'high cooperativity regime', it is required that $g^2 > \Gamma_{NV}\Gamma_{mech}$.

- The ground state spin of the NV centre has relatively low Γ_{NV}, high magnetic susceptibility, and low strain susceptibility. The excited state of the NV centre has relatively large Γ_{NV} and high strain susceptibility. These factors must be weighed quantitatively, and furthermore one has to consider the effects of the coupling modality on both Γ_{NV} and Γ_{mech} (e.g. how does the close proximity of a magnetized tip affect Γ_{NV}?

I will now move on to study each of the examples in Fig. 7.9 in more detail.

7.3.1 Magnetic coupling between an NV centre spin and a mechanical oscillator

The Hamiltonian that describes the system of Fig. 7.9a is a sum of the NV centre Hamiltonian H_{NV}, the mechanical oscillator Hamiltonian H_{osc}, and the Hamiltonian describing the interaction between the two systems H_{int}:

$$H = H_{NV,gs} + H_{osc} + H_{int}$$

$$H_{NV,gs} = D_{gs} S_z^2$$

$$H_{osc} = \hbar \omega_m a^\dagger a \tag{7.7}$$

$$H_{int} = \gamma \vec{B}_{AC} \cdot \hat{S}.$$

where ω_m is the angular frequency of the mechanical oscillator, $a^\dagger (a)$ are the creation (annihilation) operators for the phonon mode of the cantilever, and $\vec{B}_{AC} \sim a + a^\dagger$ is the magnitude of the AC magnetic field produced by the oscillating cantilever.

The interaction Hamiltonian can be written as

$$H_{int} = \hbar \gamma \nabla B a_0 (a + a^\dagger) \sigma_z$$
$$= \hbar \lambda (a + a^\dagger) \sigma_z \tag{7.8}$$

where σ_z is the z-component of Pauli spin-1 operator and the coupling strength λ, which gives the vibrational shift per mechanical energy quantum, is

$$\lambda = \gamma a_0 \nabla B. \tag{7.9}$$

Let's now input some reasonable experimental numbers and see how λ compares to the relevant system decoherence rates: $\Gamma_{NV} \sim 1 - 10 \, \text{kHz}$ and $\Gamma_{mech} = \frac{k_B T}{\hbar Q} = 10$ kHz ($T = 100 \, \text{mK}$, mechanical quality factor $Q = 2 \cdot 10^5$). Note that Γ_{mech} is the true mechanical decoherence rate, i.e. the rate for the mechanical oscillator's phonon number to change by one, rather than the classical damping rate, $\Gamma_m = \frac{\omega}{Q}$, the rate for the mechanical oscillator to 'ring-down'. State-of-the-art magnetic field gradients are $\sim 6 \cdot 10^6 \, T/m$ (Mamin, Rettner, Sherwood, Gao and Rugar, 2012). The zero point motion amplitude $a_0 = \sqrt{\frac{\hbar}{2m\omega_m}}$ depends on the resonator geometry and assuming a singly clamped cantilever beam, $a_0 \sim \sqrt{\frac{l}{wt^2}}$ where l, w, t are the length, width and thickness of the beam. Hence an optimized design calls for long, narrow and thin cantilevers. Assuming $l = 20\mu m$, $w = 200 nm$, and $t = 100 nm$, $a_0 = 1.8 \cdot 10^{-13} m$, and therefore $\lambda = 28 \, \text{kHz} > \Gamma_{NV}, \Gamma_{mech}$. For these parameters, then, it is indeed possible to reach both the strong coupling regime and the high cooperativity regime.

It is important to caution, however, that the presence of a magnetic tipped cantilever in close proximity to the NV centre and the diamond surface introduces important experimental considerations. Thermomagnetic noise in the magnet produces fluctuating magnetic fields at the NV centre location, which may cause decoherence (i.e. increase in Γ_{NV}). This effect has been observed in several experiments and approaches to mitigate the problem include engineering single-domain magnets and using soft magnets in high magnetic fields, where the domains are highly aligned. Another important consideration is the effect of a proximal surface on the cantilever Q. There is ample evidence that bringing a high-Q mechanical resonator near a surface significantly reduces its Q, due to noncontact friction that arises from surface-induced forces fluctuating near ω_m (Stipe, Mamin, Stowe, Kenny and Rugar, 2001), whose origin is not well understood. This surface-induced friction has been the primary limitation to magnetic resonance force microscopy (MRFM), and the most successful mitigation strategy to date has been reducing the size of the cantilever cross-sectional area.

Engineering a Jaynes–Cummings interaction between the spin and the mechanical resonator

In this section, I will derive a Jaynes–Cummings interaction between an NV centre spin and a magnetized mechanical resonator (Rabl, Cappellaro, Dutt, Jiang, Maze and Lukin, 2009). This type of interaction allows for strong coherent coupling between the spin and the mechanics, which can enable spin-driven cooling of the resonator down to its quantum ground state as well as arbitrary quantum state preparation of the mechanical resonator.

Working in the NV ground state spin manifold, we apply a static magnetic field (to split the $|1\rangle$ and $|-\rangle 1$ states) and simultaneously drive two microwave magnetic fields to address the $|0\rangle \rightarrow |1\rangle$ and $|0\rangle \rightarrow |-\rangle 1$ transitions, as shown in Fig. 7.10. Writing the resulting Hamiltonian in a frame rotating with the microwave frequencies, Ω_- and Ω_+:

$$H_{NV} = -\hbar\Delta_- \, |-1\rangle\langle-1| - \hbar\Delta_+ \, |+1\rangle\langle+1|$$
$$+ \frac{\hbar\Omega_-}{2}(|0\rangle\langle-1| + |-1\rangle\langle0|)$$
$$+ \frac{\hbar\Omega_+}{2}(|0\rangle\langle+1| + |+1\rangle\langle0|) \tag{7.10}$$

Note: the values Ω_- and Ω_+ are NOT the frequencies of the microwave drives, but rather indicate the amplitudes of each microwave drive, $\Omega_\pm = \gamma B_{MW,\pm}$, where $B_{MW,\pm}$ is the strength of the microwave magnetic field addressing the two transitions. Now if we assume $\Omega_- = \Omega_+ = \Omega$ and $\Delta_- = \Delta_+ = \Delta$, we can write

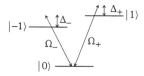

Fig. 7.10 *The NV ground state spin manifold in the presence of two microwave driving fields, detuned from the $|0\rangle \rightarrow |1\rangle$ and $|0\rangle \rightarrow |-\rangle 1$ transitions.*

$$H_{NV} = \sum_{i=-1,1} -\hbar\Delta \, |i\rangle \langle i| + \frac{\hbar\Omega}{2}(|0\rangle \langle i| + |i\rangle \langle 0|) \tag{7.11}$$

Notice that H_{NV} couples the state $|0\rangle$ to a 'bright' superposition of $|+1\rangle$ and $|-1\rangle$: $|b\rangle = \frac{|+1\rangle + |-1\rangle}{\sqrt{2}}$ and there is no coupling to the 'dark' state $|d\rangle = \frac{|+1\rangle - |-1\rangle}{1}\sqrt{2}$. Feel free to check this on your own by calculating: $H_{NV}|d\rangle = -\hbar\Delta \, |d\rangle$ and $H_{NV}|0\rangle = \frac{\hbar\Omega}{\sqrt{2]}}|b\rangle$.

Furthermore, one can find the eigenstates and eigenvalues of H_{NV} to be

$$
\begin{array}{ll}
|d\rangle & \omega_d = -\Delta \\
|e\rangle = \cos\theta \, |b\rangle + \sin\theta \, |0\rangle & \\
|g\rangle = \cos\theta \, |0\rangle - \sin\theta \, |b\rangle & \omega_{e,g} = \dfrac{-\Delta \pm \sqrt{\Delta^2 + 2\Omega^2}}{2}
\end{array}
\tag{7.12}
$$

where $tan(2\theta) = \frac{-\sqrt{2}\Omega}{\Delta}$. Again, feel free to derive these on your own.

When $\Delta < 0$, $|g\rangle$ is the lowest energy state. We can now write the whole system Hamiltonian (including mechanics) in the $|g\rangle, |d\rangle, |e\rangle$ basis to find (note: do this on your own):

$$
\begin{aligned}
H_s = {}& \hbar\omega_m a^\dagger a + \hbar\omega_{eg}|e\rangle\langle e| + \omega_{dg}|d\rangle\langle d| \\
& + \hbar(\lambda_g|g\rangle\langle d| + \lambda_e|d\rangle\langle e| + H.C.)(a + a^\dagger)
\end{aligned}
\tag{7.13}
$$

where $\lambda_g = -\lambda\sin\theta$ and $\lambda_e = \lambda\cos\theta$. The convenient aspect of this dressed basis is that we can tune ω_{dg} to match ω_m by adjusting Δ and Ω, as shown in Fig. 7.11.

With $\omega_m = \omega_{dg}$ and ω_m far detuned from ω_{ed}, we can reduce the three-level system shown in Fig. 7.11 to an effective two-level system. Entering into the interaction picture,[1] and making the rotating wave approximation, we arrive at the Jaynes–Cummings

Fig. 7.11 *The NV ground state spin eigenstates, dressed by a microwave magnetic field. Matching ω_m to ω_{dg} yields a Jaynes–Cummings Hamiltonian.*

---

[1] Operators gain time dependence in the interaction picture: $a^\dagger \to a^\dagger e^{i\omega_m t}, a \to a e^{-i\omega_m t}, \sigma_+ \to \sigma_+ e^{i\omega_m t}, \sigma_- \to \sigma_- e^{-i\omega_m t}$

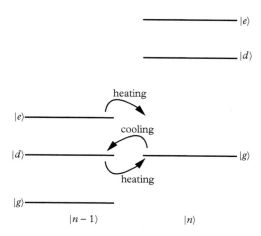

Fig. 7.12 *Schematic of NV spin-driven cooling of a mechanical resonator.*

Hamiltonian:

$$
\begin{aligned}
H_{int}^{RWA} &= \hbar(\lambda_g \,|g\rangle\,\langle d|\,a^\dagger + \lambda_g^*\,|d\rangle\,\langle g|\,a) \\
&= \frac{\hbar\lambda_g}{2}(a\sigma^+ + a^\dagger\sigma^-)
\end{aligned}
\tag{7.14}
$$

The Jaynes–Cummings Hamiltonian in 7.14 describes the coherent transfer of quantum states between spin and resonator modes, the basic ingredient for generating and detecting nonclassical states of mechanical motion. As an example, this interaction can be used for cooling the mechanical resonator. Optical pumping of the NV centre allows us to cool the NV centre spin very efficiently, i.e. polarize it into a highly non equilibrium $|0\rangle$ state, and hence with strong spin-mechanical coupling, we can controllably dissipate mechanical energy by optically pumping the spin state. The process is schematically shown in Fig. 7.12. The Jaynes–Cummings Hamiltonian describes coherent oscillations between $|n\rangle\,|g\rangle$ and $|n-1\rangle\,|d\rangle$ where $|n\rangle$ describes the number of phonons in the resonator. After a coherent $|n\rangle\,|g\rangle \rightarrow |n-1\rangle\,|d\rangle$ exchange, $|d\rangle$ gets quickly optically pumped to the excited state, from where the NV decays predominantly to the $|0\rangle$ ground state. Since microwaves do not couple $|0\rangle$ to $|d\rangle$, there will be no population in $|d\rangle$, thus suppressing the heating process $|n-1\rangle\,|d\rangle \rightarrow |n\rangle\,|g\rangle$. Another possible heating process $(|n-1\rangle\,|e\rangle \rightarrow |n\rangle\,|d\rangle$ is also suppressed as the process is detuned from resonance by $\omega_{dg} - \omega_{ed}$.

Spin decoherence protection scheme

To achieve the high magnetic field gradients necessary for the strong coupling or high cooperativity regime using the geometry of Fig. 7.9a, the spin must be very close to the magnetic tip, and hence the surface of the diamond sample. Achieving long T_2's near the diamond surface and near a large magnet is a nontrivial challenge. An advantage of the dressed-state approach just presented for spin-mechanical coupling allows for a clever scheme to extend T_2. The main idea is the following: the bare spin states of the NV centre $|\pm 1\rangle$, and hence the $|0\rangle \rightarrow |\pm 1\rangle$ spin splitting, couple linearly to magnetic fields, and hence to magnetic field fluctuations, a primary source of decoherence for NV centres. The eigenstates in the dressed-state $(|e, g, d\rangle)$ basis, however, are to first order insensitive to magnetic field fluctuations. One can check this by noting that the diagonal elements of the operator $\hat{S}_z \hat{B}_z$ are all zero in the dressed-state basis, e.g. $\langle d | \sigma_z B_z | d \rangle = 0$ etc. So if decoherence is due to fluctuating magnetic fields, then T_2 will be greatly increased in this scheme. But, if fluctuating strain or electric fields (which give a $(\sigma^+ + \sigma^-)$ interaction) dominate decoherence, then the dressed-state eigenstates are in fact sensitive to this noise to first order.

7.3.2 Strain coupling between an NV centre spin and a mechanical oscillator

In this section, I will discuss strain-mediated coupling of a spin to a mechanical degree of freedom and specifically address the two schemes shown in Fig. 7.9b–c. Strain-coupled defect systems are very promising platforms for realizing hybrid spin–phonon systems. Symmetry concepts are important in understanding spin–strain coupling because in a solid, symmetries are broken from the free atom case in a way that depends on the lattice and the nature of the defect. The techniques we describe are translatable to other defects with different symmetries.

Strain in a singly or doubly clamped beam

First we will calculate what strain-mediated coupling strengths λ can be realistically achieved. For that we will need to calculate the strain in a mechanical beam as a function of beam parameters, and use our results to design a structure to maximize λ. And, as always, we have to be mindful of the effect on the cantilever's Q and the NV's T_2. In this case, T_2 will likely be limited by proximity to the surface and nano-fabrication induced damage. Cantilever Q's are generally not easily predictable but we will at least want to minimize known sources of dissipation, such as clamping loss.

Lets start with a simple mechanical structure: a singly or doubly clamped beam undergoing flexural motion as shown in Fig. 7.13. We will apply Euler–Bernoulli (EB) theory, a simple theory that works well for long, thin beams and small amplitudes of motion. An important assumption of the EB theory is that the plane cross-sections of the beam remain perpendicular to the neutral axis of the strained beam. The neutral axis is the locus of points at which the strain is zero.

Under the assumptions of EB theory, the dynamic equation of motion of the beam is given by

$$EI\frac{\partial^2 U(z,t)}{\partial^4 z} = -\rho A \frac{\partial^2 U(z,t)}{\partial^2 t} \tag{7.15}$$

where $U(z,t)$ is the displacement of the beam in the y-direction, E is the elastic modulus, ρ is the mass density, $A = w * t$ is the cross-sectional area of the beam, and I is the moment of inertia of the beam's cross-section: $I = \frac{wt^3}{12}$ (Cleland, 2013).

Assuming time-harmonic motion, we can separate the spatial- and time-dependent parts of the solution, writing $U_n(z,t) = u_n(z,t)e^{-i\omega_n t}$, where the subscript n indicates the flexural mode.

We can solve for $u_n(z)$ by imposing appropriate boundary conditions (the displacement and its derivative are zero at the clamped end and the bending moment (torque) and shearing force (rate of change of force) at the free end are zero):

$$u_n(0) = u_n'(0) = u_n''(L) = u_n'''(L) = 0 \tag{7.16}$$

where L is the length of the beam. This yields the solution

$$u_n(z) = a_n \left[\cos(\beta_n z) - \cosh(\beta_n z)\right] + b_n \left[\sin(\beta_n z) - \sinh(\beta_n z)\right] \tag{7.17}$$

where β_n satisfies the equation $\cos(\beta_n L)\cosh(\beta_n L) + 1 = 0$, thus giving $\beta_1 L = 1.875, \beta_2 L = 4.69$, etc. The mode shape of the 1st, 2nd, and 3rd modes are shown in Fig. 7.14.

It is important to normalize the amplitude of the modes. This normalization is particularly important when we want to calculate the zero-point motional amplitude and zero-point strain. To normalize to the zero-point motional amplitude, we can set $u_1(L) = x_0 = \sqrt{\frac{\hbar}{2m_{eff}\omega_1}}$ where the effective mass $m_{eff} = \frac{1}{4}Ltw\rho$ and x_0 is the amplitude of zero-point motion. However, this approach only works when one knows the effective mass of a particular mechanical resonator geometry, and this becomes nontrivial for geometries more complicated than beams, etc. In general, a better normalization method is integrating the strain energy density $(\frac{E}{2})(\frac{du}{dz})^2$ over the volume of the cantilever and equating it to one-half the zero-point energy, $\frac{1}{4}\hbar\omega_m$.

For a doubly clamped beam, the above analysis is almost identical though with different boundary conditions (which yields different β_n's) and a different effective mass.

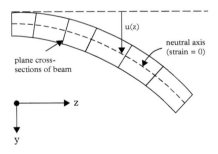

Fig. 7.13 *Singly clamped cantilever beam, showing the neutral axis and plane cross-sections that do not deform significantly under the Euler–Bernoulli approximation.*

Now that we have the mode shapes and their normalized amplitudes, we can calculate the strain ϵ profile along the beam using the relation

$$\epsilon(z) = d_0 \frac{\partial^2 u(z)}{\partial z^2} \tag{7.18}$$

where d_0 is the distance from the neutral axis, i.e. $d_0 = 0$ corresponds to a location on the neutral axis and $d_0 = \frac{t}{2}$ corresponds to a position on the top surface of the beam.

For a doubly clamped beam, which gives slightly higher strain per zero-point motion than a singly clamped beam of similar dimensions, the strain profile along the beam axis due to zero-point motion for flexural modes is given by

$$\epsilon_n(z) = 0.63 \frac{\beta_n^2 x_0}{l^2} (t/2 - d) \left[\cos(\beta_n z/l) + \cosh(\beta_n z/l) \right] - c_n \left([sin(\beta_n z/l) + \sinh(\beta_n z/l)] \right) \tag{7.19}$$

where d is the depth of the NV centre, $d = \frac{t}{2} - d_0$. As can be seen in Eq. 7.19, the strain depends linearly on x_0, and for a doubly clamped (and singly clamped) beam, $x_0 \sim (\frac{L}{t^2 w})^{.5}$. The strain is maximized at the base of the beam ($z = 0$) and at the surface of the beam ($d = 0$). At this maximum strain location,

$$\epsilon_{n,max} \sim \frac{1}{L^{3/2} w^{1/2}}. \tag{7.20}$$

Therefore, to maximize spin–strain coupling in a beam geometry, one wants to design the beam to be short and narrow. However, it is important to consider how

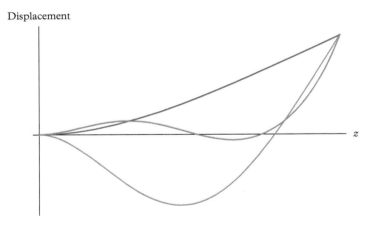

Displacement

z

Fig. 7.14 *Flexural mode shapes of a singly clamped cantilever beam. First mode is blue, second mode is orange, and third mode is green.*

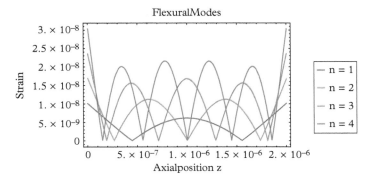

Fig. 7.15 *Zero-point motion strain associated with first four flexural modes of a doubly clamped cantilever beam with dimensions (2 μm, 100 nm, 50 nm).*

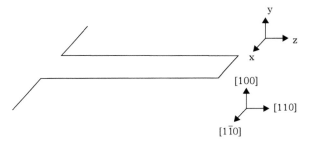

Fig. 7.16 *Schematic of singly clamped beam with crystallographic orientations and coordinate frame indicated.*

mechanical Q's scale with beam geometry, and unfortunately, mechanical dissipation due to clamping losses $\Gamma_{mech,clamping} \sim \frac{w}{L^5}$ (Judge, Photiadis, Vignola, Houston and Jarzynski, 2007). Therefore one has to balance the high strain with reduced Q's that short beams give.

For realistic beam dimensions of (2 μm, 100 nm, 50nm), the strain profiles for zero-point motion of the first four flexural modes are shown in Fig. 7.15, where the NV is assumed to reside at the surface of the beam. Note: the higher order modes have higher zero-point motion strain.

Effect of motionally induced strain on the NV centre

The simple equations for motionally induced strain that we derived in the previous section were for the strain along the beam axis. Now let's calculate the whole strain tensor in the NV centre's coordinate frame, which will be crucial for understanding the strain effect on the NV energy levels. The cantilever reference frame is shown in Fig. 7.16.

In the cantilever frame, flexural bending corresponds to a stress σ solely along the z-axis, $\sigma_{zz} \neq 0$ and all other elements of stress tensor are zero. The corresponding strain tensor, in the cantilever basis, is given by

$$\Xi = \begin{vmatrix} -\nu\epsilon(z) & 0 & 0 \\ 0 & -\nu\epsilon(z) & 0 \\ 0 & 0 & \epsilon(z) \end{vmatrix} \tag{7.21}$$

where $\nu = 0.11$ is the Poisson ratio of diamond and ϵ is the strain due to resonator flexing. We need to transform this strain tensor into the reference frame of the NV centre (of which there are four possible orientations), where z will lie along the NV's z-axis and x will lie along a projection of the carbon bond in the plane perpendicular to the NV axis. The strain tensor in the NV coordinate frame Ξ_{NV} can be obtained through the following transformation

$$\Xi_{NV} = Q\Xi Q^T \tag{7.22}$$

where Q is the rotation matrix that takes you from the cantilever reference frame to one of the four possible NV centre reference frames. For an NV oriented along the [111] direction, $Q_{111} = R_z(180°)R_Y(54.5°)R_x(90°)$, where Q_{111} is the rotation matrix that takes you from [110] $\rightarrow$ [111], $R_i(\theta)$ is a rotation by θ along axis i, and [110] is Miller index notation.

For NV centres oriented along the four possible directions,

$$\Xi_{111,\bar{1}\bar{1}1} = \begin{vmatrix} (0.33 - 0.67\nu)\epsilon(z) & 0 & -0.47(1+\nu)\epsilon(z) \\ 0 & -\nu\epsilon(z) & 0 \\ -0.47(1+\nu)\epsilon(z) & 0 & (0.67 - 0.33\nu)\epsilon(z) \end{vmatrix} \tag{7.23}$$

$$\Xi_{\bar{1}1\bar{1},1\bar{1}1} = \begin{vmatrix} \epsilon(z) & 0 & 0 \\ 0 & -\nu\epsilon(z) & 0 \\ 0 & 0 & -\nu\epsilon(z) \end{vmatrix} \tag{7.24}$$

These strain tensors highlight the fact that NV centres of different orientations will experience different strain effects under the same motion of the cantilever.

We will now determine how the elements of the strain tensor affect the NV energy levels. To do so, we express the strain in terms of matrices that transform according to the irreducible representations of the C_{3v} point group.

$$H_{strain} = V_{A_1'}(\epsilon_{xx} + \epsilon_{yy}) + V_{A_1}\epsilon_{zz}$$
$$+ V_x(\epsilon_{yy} - \epsilon_{xx}) + V_x'(\epsilon_{xz} + \epsilon_{zx})$$
$$+ V_y(\epsilon_{xy} + \epsilon_{yx}) + V_y'(\epsilon_{yz} + \epsilon_{zy}) \tag{7.25}$$

where $\epsilon_{\alpha\beta}$ is an element of the 3×3 strain tensor Ξ. The V operators on the first line of Eq. 7.25 transform as A_1, the operators on the second and third lines transform as E.

To write the Hamiltonian in Eq. 7.25 in terms of the relevant ground state and excited state two-electron wave functions requires some more mathematics, which will not be detailed here, but we state the answers, with λ_i the strain-coupling coefficients. In the excited state $|E_x, E_y\rangle$ basis we get

$$
\begin{aligned}
H_{strain} = {}& (\lambda_1 \epsilon_{zz} + \lambda_2 (\epsilon_{xx} + \epsilon_{yy}))(|E_x\rangle \langle E_x| + |E_y\rangle \langle E_y|) \\
& - (\lambda_3 (\epsilon_{yy} - \epsilon_{xx}) + \lambda_4 (\epsilon_{xz} + \epsilon_{zx}))(|E_x\rangle \langle E_x| - |E_y\rangle \langle E_y|) \\
& - (\lambda_3 (\epsilon_{xy} + \epsilon_{yx}) + \lambda_4 (\epsilon_{yz} + \epsilon_{zy}))(|E_x\rangle \langle E_y| + |E_y\rangle \langle E_x|)
\end{aligned}
\tag{7.26}
$$

The terms in the first line of Eq. 7.26 shift $|E_x\rangle$ and $|E_y\rangle$ together. The terms in the second line split $|E_x\rangle$ and $|E_y\rangle$ and the terms in the third line mix $|E_x\rangle$ and $|E_y\rangle$. Analogously, in the ground state $|0, +1, -1\rangle$ basis we get

$$
\begin{aligned}
H_{strain} = {}& (\lambda_1 \epsilon_{zz} + \lambda_2 (\epsilon_{xx} + \epsilon_{yy})) S_z^2 \\
& - (\lambda_3 (\epsilon_{yy} - \epsilon_{xx}) + \lambda_4 (\epsilon_{xz} + \epsilon_{zx}))(S_x^2 - S_y^2) \\
& - (\lambda_3 (\epsilon_{xy} + \epsilon_{yx}) + \lambda_4 (\epsilon_{yz} + \epsilon_{zy}))(S_x S_y + S_y S_x)
\end{aligned}
\tag{7.27}
$$

Looking back at Eq. 7.6, we demonstrate a mathematical equivalence between strain and electric field (i.e. which electric field would have to be applied to yield the same effect on the NV):

$$
d_\| E_z = \lambda_1 \epsilon_{zz} + \lambda_2 (\epsilon_{xx} + \epsilon_{yy})
\tag{7.28}
$$

$$
d_x E_x = \lambda_3 (\epsilon_{yy} - \epsilon_{xx}) + \lambda_4 (\epsilon_{xz} + \epsilon_{zx})
\tag{7.29}
$$

$$
d_y E_y = \lambda_3 (\epsilon_{xy} + \epsilon_{yx}) + \lambda_4 (\epsilon_{yz} + \epsilon_{zy})
\tag{7.30}
$$

Strain engineering a Jaynes–Cummings interaction between the spin and the mechanical resonator

Having established the relationship between strain and electric field, lets go back to the simple electric field treatment of Eq. 7.6. And let us only consider the interaction part of the Hamiltonian and the perpendicular field coupling $d_\perp$.

$$
H_{int,strain} = -d_\perp [E_x(S_x S_y + S_y S_x) + E_y(S_x^2 - S_y^2)]
\tag{7.31}
$$

Defining $S_\pm = S_x \pm i S_y$ and doing some algebra, we get

$$
H_{int,strain} = -d_\perp [\frac{E_x}{4i}(2S_+^2 - 2S_-^2) + \frac{E_y}{4}(2S_+^2 + 2S_-^2)]
\tag{7.32}
$$

Next we define $a = E_y - i E_x$ and $a^\dagger = E_y + i E_x$ as the boson creation/annihilation operators for the mechanical mode, where E_x and E_y are properly normalized so that

$[a, a^\dagger] = 1$, and we arrive at

$$H_{int,strain} = \frac{d_\perp E_0}{2}(|+1\rangle\langle-1|a + |-1\rangle\langle+1|a^\dagger) \qquad (7.33)$$

where E_0 is the effective electric field magnitude associated with zero-point motion of the mechanical mode. Note that the Hamiltonian in Eq. 7.33 couples the $|+1\rangle$ and $|-1\rangle$ states. By applying a magnetic field that splits $|+1\rangle$ and $|-1\rangle$ by Δ_B and tuning the magnetic field such that $\hbar\omega_m = \Delta_B$, we can eliminate the $|0\rangle$ state from the picture, achieving an effective two-level system. Thus we can convert the $SO(3)$ Pauli matrices to $SU(2)$ spin-1/2 operators and write the Hamiltonian for the spin-mechanical system as

$$H = \Delta_B \sigma^z + \omega_m a^\dagger a + g(\sigma^+ a + \sigma^- a^\dagger) \qquad (7.34)$$

Again, we have arrived at a Jaynes–Cummings interaction between the spin and the mechanical mode, where the interaction is mediated by strain.

Strain coupling strengths

We now compare the NV ground and excited state strain-coupling parameters to the mechanical decay rate and spin decay rate. A doubly clamped beam with dimensions $l = 2\,\mu m$, $w = 100\,nm$, and $t = 50\,nm$ achieves a ground-state perpendicular strain coupling strength g_{gs} of 400 Hz. At $T = 100\,mK$, $Q = 10^6$, and $T_2 = 100\,ms$, $\Gamma_{NV} \sim 10\,Hz$, $\Gamma_{mech} = \frac{k_B T}{\hbar Q} \sim 1\,kHz$, and hence the high cooperativity regime can be reached: $\eta = g_{gs}^2/(\Gamma_{NV}\Gamma_{mech}) \sim 5$. Strong coupling is difficult to achieve because the small value of strain coupling in the ground state.

In the excited state for the same cantilever parameters as above and $\Gamma_{NV} \sim 100\,MHz$, we achieve an excited-state coupling strength of g_{es} of $\sim 20\,MHz$. Again strong coupling is difficult to achieve because of the rapid decoherence rate of the NV centre in the excited state, but high cooperativity $\eta > 10^3$ can be reached, and in fact $\eta > 1$ can be reached for much less stringent requirements on the mechanical Q(10^5) and temperature (4 K). Note: for these calculations, we use the experimentally measured strain-coupling parameters to the excited state E_x and E_y levels. These strain-coupling parameters have been measured on NV ensembles in the high-strain regime in bulk diamond samples (Davies and Hamer, 1976) as well as recently in the low-strain regime using micromechanical resonators coupled to single NV centres (Lee, Lee, Ovartchaiyapong, Minguzzi, Maze and Bleszynski Jayich, 2016).

Acknowledgement

These lectures are based on work done in collaboration with many students, postdocs and colleagues, in particular Kenny Lee and Donghun Lee.

Bibliography

Barfuss, A., Teissier, J., Neu, E., Nunnenkamp, A., and Maletinsky, P. (2015, August). Strong mechanical driving of a single electron spin. *Nature Physics*, **11**, 820–4.

Bennett, S., Yao, N., Otterbach, J., Zoller, P., Rabl, P., and Lukin, M. (2013, April). Phonon-induced spin-spin interactions in diamond nanostructures: application to spin squeezing. *Physical Review Letters*, **110**(15), 156402.

Bernien, H., Hensen, B., Pfaff, W., Koolstra, G., Blok, M. S., Robledo, L., Taminiau, T. H., Markham, M., Twitchen, D. J., Childress, L., and Hanson, R. (2013, April). Heralded entanglement between solid-state qubits separated by three metres. *Nature*, **497**(7447), 86–90.

Cleland, A.N. (2013). *Foundations of Nanomechanics: From Solid-State Theory to Device Applications*. Advanced Texts in Physics. Springer Berlin Heidelberg.

Davies, G. and Hamer, M. F. (1976, February). Optical studies of the 1.945 eV vibronic band in diamond. *Proceedings of the Royal Society A: Mathematical, Physical and Engineering Sciences*, **348**(1653), 285–98.

Doherty, M. W., Manson, N. B., Delaney, P., and Hollenberg, L. C. L. (2011, February). The negatively charged nitrogen-vacancy centre in diamond: the electronic solution. *New Journal of Physics*, **13**(2), 025019.

Grinolds, M. S., Hong, S., Maletinsky, P., Luan, L., Lukin, M. D., Walsworth, R. L., and Yacoby, A. (2013, February). Nanoscale magnetic imaging of a single electron spin under ambient conditions. *Nature Physics*, **9**(4), 215–19.

Habraken, S. J. M., Stannigel, K., Lukin, M. D., Zoller, P., and Rabl, P. (2012, November). Continuous mode cooling and phonon routers for phononic quantum networks. *New Journal of Physics*, **14**(11), 115004.

Hensen, B., Bernien, H., Dréau, A. E., Reiserer, A., Kalb, N., Blok, M. S., Ruitenberg, J., Vermeulen, R. F. L., Schouten, R. N., Abellán, C., Amaya, W., Pruneri, V., Mitchell, M. W., Markham, M., Twitchen, D. J., Elkouss, D., Wehner, S., Taminiau, T. H., and Hanson, R. (2015, Oct). Loophole-free bell inequality violation using electron spins separated by 1.3 kilometres. *Nature*, **526**(7575), 682–6.

Judge, J. A., Photiadis, D. M., Vignola, J. F., Houston, B. H., and Jarzynski, J. (2007). Attachment loss of micromechanical and nanomechanical resonators in the limits of thick and thin support structures. *Journal of Applied Physics*, **101**(1), 013521.

Kolkowitz, S., Bleszynski Jayich, A. C., Unterreithmeier, Q. P., Bennett, S. D., Rabl, P., Harris, J. G. E., and Lukin, M. D. (2012, March). Coherent sensing of a mechanical resonator with a single-spin qubit. *Science*, **335**(6076), 1603–6.

LaHaye, M. D., Suh, J., Echternach, P. M., Schwab, K. C., and Roukes, M. L. (2009, 06). Nanomechanical measurements of a superconducting qubit. *Nature*, **459**(7249), 960–4.

Lee, K. W., Lee, D., Ovartchaiyapong, P., Minguzzi, J., Maze, J. R., and Bleszynski Jayich, A. C. (2016, March). Strain coupling of a mechanical resonator to a single quantum emitter in diamond. Physical Review Applied, **6**, 034005.

Loubser, J. H. N. and van Wyk, J. A. (1978). Electron spin resonance in the study of diamond. *Reports on Progress in Physics*, **41**, 1201.

MacQuarrie, E. R., Gosavi, T. A., Moehle, A. M., Jungwirth, N. R., Bhave, S. A., and Fuchs, G. D. (2015). Coherent control of a nitrogen-vacancy center spin ensemble with a diamond mechanical resonator. *Optica*, **2**(3), 233–8.

Mamin, H. J., Kim, M., Sherwood, M. H., Rettner, C. T., Ohno, K., Awschalom, D. D., and Rugar, D. (2013, January). Nanoscale nuclear magnetic resonance with a nitrogen-vacancy spin sensor. *Science*, **339**(6119), 557–60.

Mamin, H. J., Rettner, C. T., Sherwood, M. H., Gao, L., and Rugar, D. (2012). High field-gradient dysprosium tips for magnetic resonance force microscopy. *Applied Physics Letters*, **100**(1), 013102.

Maurer, P. C., Kucsko, G., Latta, C., Jiang, L., Yao, N. Y., Bennett, S. D., Pastawski, F., Hunger, D., Chisholm, N., Markham, M., Twitchen, D. J., Cirac, J. I., and Lukin, M. D. (2012). Room-temperature quantum bit memory exceeding one second. *Science*, **336**(6086), 1283–1286.

McLellan, C. A, Myers, B. A., Kraemer, S., Ohno, K., Awschalom, D. D., and Bleszynski Jayich, A. C. (2016, March). Patterned formation of highly coherent nitrogen-vacancy centers using a focused electron irradiation technique. *Nano Letters*, **16**, 2450–2454.

Metcalfe, M., Carr, S. M., Muller, A., Solomon, G. S., and Lawall, J. (2010, 07). Resolved sideband emission of InAs/GaAs quantum dots strained by surface acoustic waves. *Physical Review Letters*, **105**(3), 037401.

Montinaro, M., Wust, G., Munsch, M., Fontana, Y., Russo-Averchi, E., Heiss, M., Fontcuberta i Morral, A., Warburton, R. J., and Poggio, M. (2014, August). Quantum dot opto-mechanics in a fully self-assembled nanowire. *Nano Letters*, **14**(8), 4454–60.

Myers, B. A., Das, A., Dartiailh, M. C., Ohno, K., Awschalom, D. D., and Bleszynski-Jayich, A. C. (2014, July). Probing surface noise with depth-calibrated spins in diamond. *Physical Review Letters*, **113**(2), 027602.

O'Connell, A. D., Hofheinz, M., Ansmann, M., Bialczak, R. C., Lenander, M., Lucero, E., Neeley, M., Sank, D., Wang, H., Weides, M., Wenner, J., Martinis, J. M., and Cleland, A. N. (2010, March). Quantum ground state and single-phonon control of a mechanical resonator. *Nature*, **464**, 697.

Ohno, K., Joseph Heremans, F, de las Casas, C. F., Myers, B. A., Aleman, B. J, Bleszynski Jayich, A. C., and Awschalom, D. D. (2014, August). Three-dimensional localization of spins in diamond using 12C implantation. *Applied Physics Letters*, **105**(5), 052406.

Okazaki, Y., Mahboob, I., Onomitsu, K., Sasaki, S., and Yamaguchi, H. (2016, 04). Gate-controlled electromechanical backaction induced by a quantum dot. *Nat Commun*, **7**, 11132.

Ovartchaiyapong, P., Lee, K. W., Myers, B. A., and Bleszynski Jayich, A. C. (2014, July). Dynamic strain-mediated coupling of a single diamond spin to a mechanical resonator. *Nature Communications*, **5**, 4429.

Pelliccione, M., Jenkins, A., Ovartchaiyapong, P., Reetz, C., Emmanouilidou, E., Ni, N., and Bleszynski Jayich, A. C. (2016, 05). Scanned probe imaging of nanoscale magnetism at cryogenic temperatures with a single-spin quantum sensor. *Nat Nano*, **11**, 700–5.

Pohl, R. O., Liu, X., and Thompson, E. (2002, 10). Low-temperature thermal conductivity and acoustic attenuation in amorphous solids. *Reviews of Modern Physics*, **74**(4), 991–1013.

Rabl, P., Cappellaro, P., Dutt, M., Jiang, L., Maze, J., and Lukin, M. (2009, January). Strong magnetic coupling between an electronic spin qubit and a mechanical resonator. *Physical Review B*, **79**(4).

Remus, L. G., Blencowe, M. P., and Tanaka, Y. (2009, 11). Damping and decoherence of a nanomechanical resonator due to a few two-level systems. *Physical Review B*, **80**(17), 174103.

Rugar, D., Budakian, R., Mamin, H. J., and Chui, B. W. (2004). Single spin detection by magnetic resonance force microscopy. *Nature*, **430**(6997), 329–32.

Schuetz, M. J. A., Kessler, E. M., Giedke, G., Vandersypen, L. M. K., Lukin, M. D., and Cirac, J. I. (2015, September). Universal quantum transducers based on surface acoustic waves. *Physical Review X*, **5**(3), 031031.

Staudacher, T., Shi, F., Pezzagna, S., Meijer, J., Du, J., Meriles, C. A., Reinhard, F., and Wrachtrup, J. (2013, January). Nuclear magnetic resonance spectroscopy on a (5-nanometer)3 sample volume. *Science*, **339**(6119), 561–3.

Stipe, B., Mamin, H., Stowe, T., Kenny, T., and Rugar, D. (2001, August). Noncontact friction and force fluctuations between closely spaced bodies. *Physical Review Letters*, 87(9), 096801.

Teissier, J., Barfuss, A., Appel, P., Neu, E., and Maletinsky, P. (2014, July). Strain coupling of a nitrogen-vacancy center spin to a diamond mechanical oscillator. *Physical Review Letters*, **113**(2), 020503.

Tetienne, J. P., Hingant, T., Martínez, L. J., Rohart, S., Thiaville, A., Herrera Diez, L., Garcia, K., Adam, J. P., Kim, J. V., Roch, J. F., Miron, I. M., Gaudin, G., Vila, L., Ocker, B., Ravelosona, D, and Jacques, V (2015, April). The nature of domain walls in ultrathin ferromagnets revealed by scanning nanomagnetometry. *Nature Communications*, **6**, 6733.

Thiel, L., Rohner, D., Ganzhorn, M., Appel, P., Neu, E., Müller, B., Kleiner, R., Koelle, D., and Maletinsky, P. (2016, 05). Quantitative nanoscale vortex imaging using a cryogenic quantum magnetometer. *Nature Nanotechnology*, **11**, 677–81.

Toyli, D. M., Weis, C. D., Fuchs, G. D., Schenkel, T., and Awschalom, D. D. (2010, 08). Chip-scale nanofabrication of single spins and spin arrays in diamond. *Nano Letters*, **10**(8), 3168–72.

Van Oort, E. and Glasbeek, M. (1990). Electric-field-induced modulation of spin echoes of NV centers in diamond. *Chemical Physics Letters*, **168**(6), 529–32.

Yeo, I., de Assis, P.-L., Gloppe, A., Dupont-Ferrier, E., Verlot, P., Malik, N. S., Dupuy, E., Claudon, J., Gérard, J.-M., Auffeves, A., Nogues, G., Seidelin, S., Poizat, J.-Ph., Arcizet, O., and Richard, M. (2013, December). Strain-mediated coupling in a quantum dot–mechanical oscillator hybrid system. *Nature Nanotechnology*, **9**(2), 106–10.

8

Dynamic and Multimode Electromechanics

Konrad W. Lehnert

JILA, University of Colorado and the National Institute of Standards and Technology, Boulder, USA

Konrad W. Lehnert, *Dynamic and Multimode Electromechanics* In: *Quantum Optomechanics and Nanomechanics.*
Edited by: Pierre-Francois Cohadon, Jack Harris, Florian Marquardt, Leticia F. Cugliandolo, Oxford University Press (2020).
© Oxford University Press. DOI: 10.1093/oso/9780198828143.003.0008

Chapter Contents

8.1   Introduction

These notes discuss the phenomena of electromechanical devices in the quantum regime. This topic is closely related to a topic that has come to be known as "cavity optomechanics" Marquardt and Girvin (2009). Both cavity optomechanics and quantum electromechanics have their roots in the effort to detect gravitational waves. In building ultrasensitive sensors of mechanical strain, physicists had to confront most directly quantum noise, backaction, and the ways in which quantum mechanics can limit measurement resolution Caves *et al.* (1980). It was in gravitational wave detection that quantum measurement first became an engineering challenge. As such, most applications of cavity optomechanics and its electrical cousin are associated with sensing and ultrasensitive measurement.

In contrast, I will develop these notes for an emerging application of quantum electromechanics, namely signal processing. Signal processing applications are a natural consequence of shrinking the mechanical elements from the metre-scale resonators used in gravitational wave detectors to the micron scale, where quantum effects are more evident. Indeed, for classical information processing micro-electromechanical systems (MEMS) are a crucial technology; they form the compact, low-power and high-performance filters, clocks and delay lines necessary for modern wireless communication.

The advent of quantum information processing, particularly with superconducting circuits, means that there is now a need for analogue signal processing functions that operate at microwave frequencies and in the quantum regime. A simply expressed requirement for a device to operate in the quantum regime is that it be able to perform some useful signal processing function without losing or gaining even one signal photon to or from its environment.

Electromechanical devices have entered the quantum regime. They can store (Palomaki, Harlow, Teufel, Simmonds and Lehnert, 2013*a*), amplify (Palomaki, Teufel, Simmonds and Lehnert, 2013*b*), squeeze (Wollman, Lei, Weinstein, Suh, Kronwald, Marquardt, Clerk and Schwab, 2015; Pirkkalainen, Damskägg, Brandt, Massel and Sillanpää, 2015), entangle (Palomaki, Teufel, Simmonds and Lehnert, 2013*b*), temporally shape (Andrews, Peterson, Purdy, Cicak, Simmonds, Regal and Lehnert, 2014), and frequency convert microwave signals (Andrews, Peterson, Purdy, Cicak, Simmonds, Regal and Lehnert, 2014; Lecocq, Clark, Simmonds, Aumentado and Teufel, 2016). Although Josephson junction circuits can also perform these functions, an electromechanical circuit exploits a distinct nonlinearity that is interesting as it is subject to different limitations than the Josephson junction.

8.1.1   The Electromechanical Nonlinearity

The basic phenomenon underlying quantum MEMS devices is that electrical forces can alter the dimensions of a circuit element, thereby altering its electrical properties. Through this effect any capacitor or inductor is electrically nonlinear. For machines that use electrical energy to do mechanical work (e.g. motors and relays) this effect is engineered to be large. However, in signal processing applications the mechanical compliance of circuit elements is usually small enough to ignore.

To gain intuition let's consider a simple example: a parallel plate capacitor with plate area A and separation $d + x$, with d the static separation with no charge on the capacitor. The dielectric between the plates acts as a spring with constant k_s holding the plates apart. By imposing a charge separation q onto the two plates, the spring must compress under the attraction of the plates. Balancing the mechanical and electrical forces on the plates leads to the following conclusion; the voltage V across the capacitor is a nonlinear function of the imposed charge as

$$V = \frac{q}{C_0} \left(1 - \frac{q^2}{C_0 d^2 k_s} \right), \tag{8.1}$$

where $C_0 = \epsilon A / d$ and ϵ is the dimensionful dielectric constant. The second term in the parentheses gives the scale of the nonlinear correction. It is simply the ratio of electrical energy that would be stored in the infinitely rigid capacitor $q^2 / (2 C_0)$ compared to the mechanical work $(k_s d^2 / 2)$ required to fully compress $(x = -d)$ the plates. The scale for this nonlinearity can be expressed as the maximum possible (critical) voltage that the electromechanical capacitor can support

$$V_{\text{crit}} = \frac{2}{9} \cdot d \cdot \sqrt{\frac{3 k_s}{C_0}} = \frac{2}{9} \cdot d \cdot \sqrt{\frac{3 Y}{\epsilon}}, \tag{8.2}$$

where Y is the Young's modulus of the material between the capacitor plates. For a capacitor with $d = 1$ μm and sapphire dielectric, $V_{\text{crit}} \approx 10^5$ V. Not only is this voltage large compared to the ~ 1 V scales used in typical classical signal processing applications, it is enormous compared to the ~ 3 μV $= \omega_0 \sqrt{\hbar Z_0 / 2}$ scales associated with the voltage of a single microwave photon in a $\omega_0 \approx 2\pi \times 10$ GHz resonant circuit, where $Z_0 = 50$ Ω.

From this simple example it is clear that an electromechanical nonlinearity is not naturally relevant at the quantum scale. Quantum electromechanical devices must be carefully engineered to greatly enhance this nonlinearity. In particular, an electromechanical capacitor should have d and k_s as small as possible. Furthermore, it should be the dominant capacitive element in a high quality factor Q inductor-capacitor (LC) resonant circuit because a $1/Q$ fractional change in capacitance is a large effect in such a circuit. Similarly, the effect of the small electrostatic forces can be enhanced by ensuring that the mechanical element is part of a high quality mechanical oscillator. Indeed, these considerations are evident in most of the quantum electromechanical structures. Ultimately the nonlinear coupling between electrical energy and mechanical motion yields parametric processes, where a parameter of the circuit's equations of motion (EOMs) becomes a dynamical variable linked to the oscillator's motion.

8.2   Finding Electromechanical Equations of Motion

The types of electromechanical circuits I endeavour to describe consist of at least one LC resonance, weakly coupled to a transmission line, and with some unavoidable dissipation in both the circuit and the mechanical oscillator. Figure 8.1 shows an electromechanical device of the type described in reference (Andrews, Reed, Cicak, Teufel and Lehnert, 2015). Because it comprises three reactive elements, the circuit is described exactly by EOMs that have some uninteresting complexity.

Classical equations of motion for a simple electromechanical circuit

To avoid this uninteresting complexity, I will instead consider a simpler circuit (Fig. 8.2) that exhibits interesting electromechanical phenomena, but with fewer poles than the circuit in Fig. 8.1. I begin by choosing the charge on the capacitor plate q, and the plate separation x, as the dynamical variables. I can then write the total energy of the system

(a)   (b)

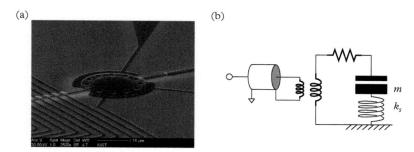

Fig. 8.1 *Image and schematic of an electromechanical circuit. a) The scanning electron microscope image shows primarily the mechanically compliant capacitor of an electromechanical device. The parallel plate capacitor is the central disc in the image, with a portion of the resonant spiral inductor visible on the left edge. b) A schematic of the device depicts a compliant and mechanically resonant capacitor plate (depicted as a mass m on a spring of constant k_s), electrically connected to an inductor and inductively coupled to a transmission line. The mutual inductance can be chosen to make the rate of energy lost to the transmission line as small as desired. An explicit resistor within the circuit models internal energy absorption.*

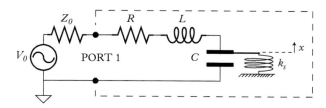

Fig. 8.2 *A model electromechanical circuit. The capacitor is assumed to be a mechanically compliant element whose capacitance depends on a coordinate x. That coordinate has a mass m; it experiences both a spring force $-k_s x$ and a dissipative force $-\gamma_0 \dot{x}$. In addition, it experiences a force F_{ext} exerted by some external agency. The dashed box separates portions of the circuit that will later be considered as part of the electromechanical resonant circuit (inside the box) and part of the external circuitry (outside the box). Dividing the resistor into two pieces R and Z_0 allows me to separately model energy absorbed in the circuit and energy emitted into a transmission line of characteristic impedance Z_0.*

in terms of $x, q, \dot{x}, \dot{q}$, and thus form the electromechanical Lagrangian

$$\mathcal{L} = \frac{1}{2}\left[L\dot{q}^2 + m\dot{x}^2 - q^2/C(x) - k_s x^2\right] + qV + F_{ext}x \tag{8.3}$$

The resulting EOMs are

$$L\ddot{q} = V - \frac{q}{C(x)} - \dot{q}(R + Z_0) \tag{8.4}$$

$$m\ddot{x} = F_{ext} - k_s x + \frac{q^2}{2C^2}\frac{\partial C}{\partial x} - \gamma_0 m\dot{x}. \tag{8.5}$$

In these expressions, I have introduced the dissipative forces 'by hand'. In the right hand side of Eq. 8.5, the third term is the electrostatic force, which has arisen naturally from the Lagrangian formulation.

The additional reactive elements used in most electromechanical devices exist to better isolate the electrical resonator from the transmission line and achieve a higher Q resonance than is possible in Fig. 8.2 for which $Q < \sqrt{L/C}/Z_0$. (Because Maxwell's equations determine both $\sqrt{L/C}$ and Z_0, the natural scale for both quantities is $\sqrt{\mu_0/\epsilon_0}$ (Pozar, 1998).) This practical limitation is concealed by writing the classical EOMs in a form where the coefficients have dimensions of (1/s) to some power

$$\ddot{q} + \dot{q}(\kappa_I + \kappa_E) + q\omega_{cir}^2(x) = \frac{V}{L} \tag{8.6}$$

$$\ddot{x} + \gamma_0 m\dot{x} + x\Omega^2 = \frac{F_{ext}}{m} + \frac{q^2}{2mC^2}\frac{\partial C}{\partial x},$$

with $\Omega^2 = k_s/m$ and $\omega_{cir}^2(x) = 1/LC(x)$, $R/L = \kappa_I$, $Z_0/L = \kappa_E$. For frequencies near one resonance of a more complex high-Q circuit, it is usually possible to identify the internal κ_I and external κ_E energy dissipation rates and thus regard Fig. 8.2 as an effective circuit that accurately models electromechanical behaviour for frequencies close to $\omega_{cir}(0)$.

Even for Fig. 8.2, it is a helpful simplification to make a further, single-pole approximation. Indeed one of the silly but confusing aspects encountered when relating optomechanical theory to electromechanics is the mathematically exact and universally agreed upon meaning of electrical circuit symbols in comparison to the less precisely defined schematic symbols of optomechanical systems. Consequently optomechanical EOMs are often written immediately in a single-pole approximation, while standard electrical circuit analysis yields more complex EOMs described by more poles. To illustrate the point in a simple and familiar case, I find the solution to the electromechanical EOMs in the uncoupled limit, $\partial\omega_r/\partial x = 0$

$$q(t) = \text{Re}\left[A e^{-\left(\frac{\kappa_I + \kappa_E}{2} - j\omega_{\text{cir}}' t \right)} \right]$$

$$\omega_{\text{cir}}' = \omega_{\text{cir}} \sqrt{\left(1 - \frac{\kappa_I + \kappa_E}{2\omega_{\text{cir}}} \right)}.$$

Note that the *field* $q(t)$ decays at rate $(\kappa_I + \kappa_E)/2$, but the energy decays at rate $\kappa \equiv (\kappa_I + \kappa_E)$. Similarly, I write the circuit susceptibility [$V(t) = \text{Re}(V_0 \exp(+j\omega t))$ and $q(t) = \text{Re}(q_0 \exp(+j\omega t))$]

$$\frac{q_0 L}{V_0} \equiv \chi_{\text{cir}} = \frac{1}{\omega_{\text{cir}}^2 - \omega^2 + j\kappa\omega}. \tag{8.7}$$

The form of Eq. 8.7 tempts me to define ω_{cir} as the resonance frequency, but this choice is irksome when approximating this susceptibility as a single Lorentzian. Instead, by writing a partial fraction expansion of this rational function, I express the susceptibility as the sum of two Lorentzians

$$\chi_{\text{cir}} = \frac{1}{2\omega_{\text{cir}}'^2} \left[\frac{1}{1 - z + j/(2Q)} + \frac{1}{1 + z - j/(2Q)} \right], \tag{8.8}$$

with $z = \omega/\omega_{\text{cir}}'$, $Q = \omega_{\text{cir}}'/\kappa$. In the limit $(z - 1) \ll 1$ and $Q \gg 1$, the zeroth-order single-pole approximation simply evaluates the second term in Eq. 8.8 at $z = 1$, and thus ω_{cir}' is the more natural definition of the resonance frequency. In fact, the constant term is usually ignored as it predominantly contributes a phase shift in the susceptibility, which is a difficult quantity to distinguish from other phase shifts in an experiment. Clearly, the uncoupled susceptibility of the mechanical oscillator would have the same form but with $\omega_{\text{cir}} \to \Omega$ and $\kappa \to \gamma_0$.

Quantized equations of motion

I could work with classical EOMs in this circuit language and find solutions using classical perturbation theory. This is the approach I followed in 2011 (Devoret, Huard, Schoelkopf and Cugliandolo, 2014). Let's follow the quantum path instead. There are virtues to the quantum path. Namely it builds in the single-pole approximation automatically, makes use of the powerful and widely known quantum optics formalism and is relevant to the ambitions of the field! Nevertheless, in the limit in which electromechanics currently operates the classical and quantum EOMs are usually in very close correspondence (Devoret, Huard, Schoelkopf and Cugliandolo, 2014).

I begin by transforming the Lagrangian of Eq. 8.3 to a classical Hamiltonian $\mathcal{H} = (1/2)[p^2/m + k_s x^2 + \Phi^2/L + q^2/C(x)]$, with $p = m\dot{x}$ and $\Phi = L\dot{q}$, and where I have dropped the qV and xF_{ext} terms. I will reintroduce applied voltages and forces when I add dissipation to the EOMs. The two harmonic oscillators in the Hamiltonian are coupled through the x dependence of C. Expanding this last term to first order in x yields

$$\frac{q^2}{2C_0}\left(1 - x\frac{1}{C_0}\left[\frac{\partial C}{\partial x}\right]_{x=0}\right).$$

The quantized version of this Hamiltonian is

$$\frac{\hat{H}}{\hbar} = \omega_{\text{cir}}\left(\hat{a}^\dagger\hat{a} + \frac{1}{2}\right) + \Omega\left(\hat{b}^\dagger\hat{b} + \frac{1}{2}\right) + g_0(\hat{b}^\dagger + \hat{b})\hat{a}^\dagger\hat{a} \tag{8.9}$$

$$+ \frac{g_0}{2}(\hat{a}\hat{a} + \hat{a}^\dagger\hat{a}^\dagger + 1)(\hat{b}^\dagger + \hat{b}),$$

with the following lengthy set of nested definitions:

$$Z_{\text{cir}} = \sqrt{\frac{L}{C}}, \quad Z_{\text{m}} = \sqrt{k_s m}, \quad x_{\text{zpm}} = \sqrt{\frac{\hbar}{2Z_{\text{m}}}},$$

$$\hat{a} = \hat{q}\sqrt{\frac{Z_{\text{cir}}}{2\hbar}} + i\hat{\Phi}\sqrt{\frac{1}{2\hbar Z_{\text{cir}}}},$$

$$\hat{b} = \hat{x}\sqrt{\frac{Z_{\text{m}}}{2\hbar}} + i\hat{p}\sqrt{\frac{1}{2\hbar Z_{\text{m}}}},$$

$$g_0 = G x_{\text{zpm}}, \quad G = -\omega_{\text{cir}}\left(\frac{1}{2C_0}\frac{\partial C}{\partial x}\right)_{x=0}.$$

The canonical commutation relations $[\hat{x}, \hat{p}] = [\hat{q}, \hat{\Phi}] = i\hbar$ imply the usual bosonic relations for the harmonic oscillator ladder operators $[\hat{a}, \hat{a}^\dagger] = [\hat{b}, \hat{b}^\dagger] = 1$.

Now that the model Hamiltonian has been reduced to a few parameters, I can use physical arguments to guide approximations. In order for the resonant circuit to be in its quantum ground state by refrigeration alone $\omega_{\text{cir}} \gg k_B T/\hbar$ should be a microwave frequency as electrical circuits can be cooled to a temperature $T \approx 10$ mK. In contrast, Ω is a much smaller frequency, because the flexible element of the capacitor must be large enough to have an appreciable capacitance in a microwave resonator ($\omega_{\text{cir}} C \sim \sqrt{\mu_0/\epsilon_0}$); therefore, Ω is much smaller than ω_{cir} essentially because the speed of sound is much less than the speed of light. For such low frequency mechanical oscillators built from top down lithography (Fig. 8.1), x_{zpm} is about 10^{-15} m, thus the electromechanical coupling $g_0 \sim \omega_{\text{cir}}(x_{\text{zpm}}/d)$ is the smallest rate in the Hamiltonian, about 10^7 times smaller than ω_{cir}. Electromechanical circuits have the hierarchy $\omega_{\text{cir}} \gg \Omega \gg g_0$. Consequently, we drop the term in the second line of Eq. 8.9; terms like $\hat{a}\hat{a}$ and $\hat{a}^\dagger\hat{a}^\dagger$ create electrical forces at $2\omega_{\text{cir}}$ where the mechanical oscillator will have vanishing susceptibility. The constant term is a static force created by the circuit's vacuum fluctuations—a kind of single-mode Casimir force—an intriguing notion, but not something that will influence the electromechanical dynamics.

To find the dynamics, we work in the Heisenberg representation to find the EOMs for the operators $\hat{a}$ and $\hat{b}$, $\dot{a} = i\omega_{\text{cir}}a - ig_0(b^\dagger + b)a$ and $\dot{b} = i\omega_{\text{cir}}b - ig_0a^\dagger a$. Conjugate these

EOMs for the $a^\dagger$ and $b^\dagger$ EOMs.[1] (Note that the 'hats' have been suppressed to simplify the notation.) From the EOMs it is clear that the electrical resonance frequency depends on the mechanical displacement, and that there is a force on the mechanical oscillator proportional to the number of microwave photons stored in the resonator.

To complete the model, I add to these EOMs terms that describe applied forces, applied voltages, dissipation and noise. The field of quantum optics has developed a formalism for introducing these effects; if the open quantum system meets certain conditions, it can be described using so-called input–output (IO) theory by implementing the following prescription:

1. To each source of loss define a port and associate input and output fields with these ports.

2. Add terms to the Hamiltonian to model the decay of the system state from the Hamiltonian operators to the output field operators.

3. Insist that the IO fields couple to the Hamiltonian operators and to each other in a manner consistent with the fluctuation dissipation theorem and unitary evolution.

This procedure yields the Heisenberg–Langevin EOMs.

$$\dot{a} = \left(-i\omega_{\text{cir}} - \frac{\kappa}{2}\right)a - ig_0(b^\dagger + b)a + \sqrt{\kappa_{\text{E}}}a_{\text{inE}} + \sqrt{\kappa_{\text{I}}}a_{\text{inI}} \qquad (8.10)$$

$$\dot{b} = \left(-i\Omega - \frac{\gamma_0}{2}\right)b - ig_0 a^\dagger a + \sqrt{\gamma_0}b_{\text{in}}$$

The output relations complete the quantum optics model by determining the fields leaving the electromechanical circuit as

$$a_{\text{outE}} = \sqrt{\kappa_{\text{E}}}a(t) - a_{\text{inE}}(t) \qquad (8.11)$$

$$a_{\text{outI}} = \sqrt{\kappa_{\text{I}}}a(t) - a_{\text{inI}}(t)$$

$$b_{\text{out}} = \sqrt{\gamma_0}b(t) - b_{\text{in}}(t).$$

Derivation of the Heisenberg–Langevin EOMs and the detailed definitions of the field operators in terms of bath modes are well described in the literature(Clerk, Devoret, Girvin, Marquardt and Schoelkopf, 2010). Briefly, the field operators obey continuous time commutation relations e.g. $[a_{\text{inE}}(t), a^\dagger_{\text{inE}}(t')] = \delta(t - t')$. They have dimensions of $\sqrt{[\text{number/time}]}$, in the same sense that $a^\dagger a$ is a number of photons. The fact that they do not commute leads directly to the quantum fluctuations of these bosonic modes.

---

[1] Electrical engineers and physicists use different conventions for the Fourier transform. For quantum electrical circuits this conundrum has been resolved by reserving j for $F[\omega] = \int dt f(t)e^{-j\omega t}$ and i for $F'[\omega] = \int dt f(t)e^{i\omega t}$; hence, $j = -i$.

Within a semiclassical limit they can be regarded as variables with a random component that take on new values at each instant in time. Although these modes must have (at least) quantum fluctuations, as any harmonic oscillator they can be prepared in large coherent states. Indeed I will model voltage and currents propagating in transmission lines using these operators, where the physical voltage or current wave incident on the electromechanical circuit is proportional to $a_{\mathrm{inE}}(t) \pm a_{\mathrm{inE}}^{\dagger}(t)$. As always there are many different normalization and sign conventions, but most conflicts of convention can be discovered by comparing the expectation values of these equations to their classical counterparts.

Now that I regard some of the input and output modes as waves in a transmission line, I can introduce a transmission line into the circuit model, as shown in Fig. 8.3. The IO modes a_{inE} and a_{outE} are special because they are experimentally accessible. I can inject microwave fields by choosing the state of the mode associated with a particular a_{inE}, and measuring the state associated with a particular a_{outE}.

Linearized equations of motion

Assume I drive a circuit with a single tone so that

$$a_{\mathrm{inE}}(t) = (\alpha_{\mathrm{in}} + d_{\mathrm{in}}(t))e^{-i\omega_{\mathrm{d}}t}$$

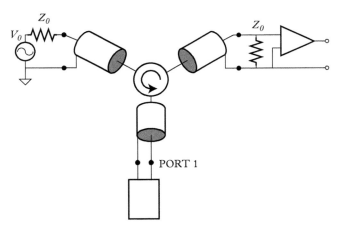

Fig. 8.3 *A microwave network for separating the excitation and response of a circuit. The square element represents the one-port network enclosed by the dashed box in Fig. 8.2. The circulator in the centre of the figure separates incident and reflected waves from each other at port 1 of the RLC circuit. The incident waves come from the source (of source impedance Z_0) while reflected waves are directed to the voltage amplifier of input impedance Z_0. If the cables also have a wave impedance Z_0, incident and reflected waves only couple at port 1. With this microwave network connected to port 1 instead of the simpler external circuit, the dynamics of the RLC circuit are unchanged.*

and seek solutions of the form

$$a(t) = (\alpha + d(t))e^{-i\omega_d t}$$

and

$$b(t) \rightarrow (\bar{b} + b(t)),$$

where α and $\bar{b}$ are the static solutions and $d(t)$ and $b(t)$ are small fluctuations about the static solutions. The static solutions are:

$$\bar{b} + \bar{b}^\dagger = \left(\frac{2g_0 |\alpha|^2}{\Omega \left(1 + \left(\frac{\gamma_0}{2\Omega} \right)^2 \right)} \right) \tag{8.12}$$

$$\alpha = \frac{\alpha_{in} \sqrt{\kappa_E}}{-i\Delta + \kappa/2}, \tag{8.13}$$

where $\Delta = \omega_d - \omega_{cir} - g_0(\bar{b} + \bar{b}^\dagger)$.

There are two things to note about these equations. First, the static deflection of the mechanical oscillator has a strange, seemingly unphysical, dependence on γ_0. It's just a consequence of failing to redefine the resonance frequency $\Omega \rightarrow \Omega'$ when adding dissipative terms to form the Heisenberg–Langevin EOMs, a minor point but one that is rarely noted in the literature. Second, α has an implicit α dependence through the definition of Δ. It is simple to solve for α more explicitly and identify a kind of 'Kerr' effect where the circuit resonates at a lower frequency the harder it is driven. This effect can lead to bifurcation, parametric amplification, squeezing and instability, but it is a relatively small effect in electromechanical circuits.

In a frame rotating at ω_d and to linear order in d the EOMs describing a small deviation from these static solutions are:

$$\dot{d} = (i\Delta - \kappa/2)d - ig_0\alpha(b^\dagger + b) + \sqrt{\kappa_E}d_{in} + \sqrt{\kappa_I}a_{inI}$$

$$\dot{b} = (-i\Omega - \gamma_0/2)b - ig_0(\alpha d^\dagger + \alpha^* d) + \sqrt{\gamma_0}b_{in}, \tag{8.14}$$

(c.f. Marquardt, Chen, Clerk and Girvin, 2007). The EOMs are now linear in the fields and the weak electromechanical coupling g_0 is enhanced by the amplitude of the coherent drive giving an effective electromechanical coupling $g = \alpha g_0$. Had we kept the aa and $a^\dagger a^\dagger$ terms in the original Hamiltonian, we would now drop these terms in a rotating wave approximation, as they oscillate at $2\omega_d$. The output relations complete the linearized EOMs. As long as we interpret the output fields as rotating at ω_d, e.g. $a_{outI}(t) \rightarrow a_{outI}(t)e^{-i\omega_d t}$, these equations are unchanged from Eqs. 8.11. However, we renamed the fields for the external port in the rotating frame; hence,

$$d_{\text{out}} = \sqrt{\kappa_E} d - d_{\text{in}}.$$

8.3   Dynamical Electromechanics

At this point I could flee for the familiar comfort of the frequency domain, as I eventually do in section 8.5. I know as a mathematical fact that this domain is equivalent to the time domain, and it is easy to work with. Let me offer two cultural and emotional reasons not to transform to the frequency domain at this stage. First, it will prejudice you to thinking about steady state effects. Second, the internal Hamiltonian degrees of freedom will disappear from view, leaving you with susceptibilities that relate input modes to output modes. This formulation hides the possibility of using mechanical oscillators as dynamical signal processing elements such as memories (Palomaki, Harlow, Teufel, Simmonds and Lehnert, 2013*a*) and pulsed entanglement generators (Palomaki, Teufel, Simmonds and Lehnert, 2013*b*).

8.3.1   Time-dependent Electromechanical Coupling

Notice that in electromechanics the coupling between electricity and motion can be time-dependent. If $\alpha_{\text{in}}(t)$ is varied $g(t)$ will vary. This effect is a key virtue of parametric coupling; the linear EOMs themselves are easily changed in time. And if $g(t)$ is indeed time-dependent, the frequency domain is no longer so natural.

For these linear but time-dependent EOMs (Eqs. 8.14), I will consider a few cases that can easily be solved analytically. In particular, I can consider cases where $g(t)$ is piecewise constant. Rather than leave the detuning unspecified, greater clarity is provided by selecting the detuning from natural choices evident in the EOMs: $\Delta = \pm\Omega$ if $\kappa \ll \Omega$ or $\Delta = 0$ if $\kappa > \Omega$. As a an example, let's choose $\Delta = -\Omega$, which for energy considerations alone we expect to swap or exchange information between electrical and mechanical modes.

I begin by transforming into a rotating frame of Eqs. 8.14, which is a rotating frame of a rotating frame, $d(t) \to d(t)e^{-i\Omega t}$ and $b(t) \to b(t)e^{-i\Omega t}$. Substituting into the linearized EOMs with $\Delta = -\Omega$, $\kappa_I = 0$, and dropping terms oscillating around 2Ω in the rotating wave approximation ($\kappa \ll 4\Omega$) gives,

$$\dot{d} = -(\kappa/2)d - igb + \sqrt{\kappa_E} d_{\text{in}} \tag{8.15}$$
$$\dot{b} = -(\gamma_0/2)b - ig^* d + \sqrt{\gamma_0} b_{\text{in}}$$

In the absence of dissipation, the Hamiltonian which would give these EOMs is $H/\hbar = g b^\dagger d + g^* b d^\dagger$. These EOMs indeed describe an exchange between mechanical and electrical modes, sometimes called a beam-splitter operation.

Strong coupling limit

If I first solve in the strong coupling limit (Teufel, Li, Allman, Cicak, Sirois, Whittaker and Simmonds, 2011*b*) of $g \gg \max[\kappa, \gamma_0]$ the EOMs are particularly simple: $\dot{d} = -igb$ and $\dot{b} = -ig^*d$. Choosing a piecewise constant coupling as $\alpha(t) = \alpha_0 \theta(t) \theta(T - t)$ these equations have solutions for $0 < t \leq T$

$$d(t) = d(0)\cos(|\alpha_0|g_0 t) - ib(0)\sin(|\alpha_0|g_0 t)e^{i\psi} \tag{8.16}$$
$$b(t) = b(0)\cos(|\alpha_0|g_0 t) + id(0)\sin(|\alpha_0|g_0 t)e^{-i\psi},$$

where $\psi = \arg\alpha_0$. If $|\alpha_0|g_0 T = \pi/2$, the resulting operation is an 'i-swap gate' with $d(T) = -ib(0)$ and $b(T) = id(0)$. The κ and γ terms introduce decay and the $a_{\mathrm{inE}}\sqrt{\kappa}$ and $b_{\mathrm{in}}\sqrt{\gamma_0}$ terms introduce decoherence to this otherwise unitary operation. Current experimental parameters are $g < 1.5$ MHz and $\kappa > 100$ kHz (Teufel, Li, Allman, Cicak, Sirois, Whittaker and Simmonds, 2011*b*). As such, it is a fairly low-fidelity swap.

Weak coupling limit

Returning to Eqs. 8.15, I investigate the weak coupling limit, $\kappa/2 \gg g \gg \gamma_0/2$. In this limit, the circuit's coherent dynamics are slow compared to it's decay $\dot{d} \ll (\kappa/2)d(t)$. Thus, I 'adiabatically eliminate' the circuit dynamics by approximating $\dot{d} \approx 0$. (See Lugiato *et al.*, 1984 for a discussion of the validity of this approximation.) This procedure leaves a single first order differential equation to work with:

$$\dot{b} = -(1/2)[\gamma_0 - \Gamma(t)]b(t) - i\sqrt{\Gamma(t)}e^{-i\psi}d_{\mathrm{in}}(t) + \sqrt{\gamma_0}b_{\mathrm{in}}(t), \tag{8.17}$$

where $\Gamma = 4|g|^2/\kappa$. Furthermore, the state of the circuit is no longer an independent degree of freedom, but rather determined by the instantaneous value of $b(t)$ and $d_{\mathrm{in}}(t)$ as

$$d(t) = (i\sqrt{\Gamma}e^{i\psi}/\kappa)b(t) + 2d_{\mathrm{in}}(t). \tag{8.18}$$

Even with a time-dependent coefficient, the solution to a single linear first order differential equation can be written in a simple analytical form. Nonetheless, I will again consider a piecewise constant $\alpha(t) = \alpha_0 \theta(t)$, implying that g and Γ are similarly piecewise constant. The solution to the differential equation for $t > 0$ is

$$b(t) = b(0)h(t) + [-i\sqrt{\Gamma}e^{i\psi}h \star d_{\mathrm{in}}(t) + \sqrt{\gamma_0}h \star b_{\mathrm{in}}(t)],$$

where $h(t) = e^{-(\gamma_0 + \Gamma)t/2}$ is the impulse response of Eq. 8.17 and the convolution integral is $h \star f(t) = \int_0^t dt' \, h(t - t')f(t')$.

What happens when I turn on the electromechanical coupling suddenly? The output field is

$$d_{\mathrm{out}}(t) = d_{\mathrm{in}} - i\sqrt{\Gamma}b(0)h(t)e^{i\psi} - \Gamma\left(h \star d_{\mathrm{in}}(t) + ih \star b_{\mathrm{in}}(t)\sqrt{\frac{\gamma_0}{\Gamma}}e^{-i\psi}\right).$$

A pulse emerges from the port of the electromechanical circuit with a decaying exponential envelope and characterized by a complex amplitude which is the state of the electrical circuit at $t = 0$ plus some input noise from d_{in} and b_{in}. What sort of filter best determines $b(0)$ with the least perturbation from the noise terms? With intuition from optimal Wiener filtering (Wiener, 1949), let's try $d_{meas} = \sqrt{\Gamma + \gamma_0} \int_0^\infty dt\, d_{out}(t) h(t)$. After some algebra, the result is

$$d_{meas} = -ib(0)e^{i\psi}\sqrt{\frac{c_p}{c_p + 1}} + \sqrt{\frac{1}{c_p + 1}} \int_0^\infty dt\, h(t)\sqrt{\gamma_0}\left[d_{in}(t) + ie^{-i\psi}\sqrt{\frac{1}{c_p}}b_{in}(t) \right].$$

In the limit of large cooperativity $c_p = \Gamma/\gamma_0$, the filtered output field is just the state of the mechanical oscillator at $t = 0$. In electromechanical systems, it is usually the case that the mechanical oscillator is coupled to an environment for which the temperature $T \gg \hbar\Omega/k_B$. As such b_{in} will be dominated by thermal rather than quantum noise and to reach the quantum regime requires $c_p \gg k_B T/(\hbar\Omega)$. Even so, electromechanical experiments are capable of operating deeply in the large cooperativity limit (Teufel, Donner, Li, Harlow, Allman, Cicak, Sirois, Whittaker, Lehnert and Simmonds, 2011a).

Rather than convert the state of the mechanical oscillator into a propagating microwave pulse, I can imagine running this process backward in time to catch a propagating mode incident on the electromechanical circuit. The time reversal is achieved by choosing $\alpha(t) = \alpha_0\theta(-t)$ (Palomaki, Harlow, Teufel, Simmonds and Lehnert, 2013a). With this choice the state of the mechanical oscillator at $t < 0$ is

$$b(t) = -i\sqrt{\Gamma}e^{i\psi}h \star d_{in}(t) + \sqrt{\gamma_0}h \star b_{in}(t), \tag{8.19}$$

but where the convolution integral is $h \star f(t) = \int_{-\infty}^t dt'\, h(t - t')f(t')$. Let's define a particular input mode operator $D = \sqrt{\Gamma + \gamma_0} \int_{-\infty}^0 dt\, h(-t)d_{in}(t)$ and mechanical noise operator $B = \sqrt{\Gamma + \gamma_0} \int_{-\infty}^0 dt\, h(-t)b_{in}(t)$. The mechanical oscillator state at $t = 0$ is easily written in terms of these modes

$$b(0) = -ie^{-i\psi}\sqrt{\frac{c_p}{c_p + 1}}D + \sqrt{\frac{1}{c_p + 1}}B. \tag{8.20}$$

In the high cooperativity limit, the mechanical oscillator acquires at $t = 0$ the state that was contained in D. The information is trapped in the mechanical oscillator when the coupling turns off $\alpha = 0 \rightarrow \Gamma = 0$.

What happens to the microwave output mode while an input mode is trapped? To understand this, imagine I programmed a microwave function generator to create a particular exponentially rising pulse, thus preparing mode D in a coherent state and determining the ensemble average of $d_{in}(t)$ to be $\bar{d}_{in}(t) = Mh(-t)\theta(-t)$. The ensemble average output is

$$\bar{d}_{\text{out}}(t) = \Gamma \int_{-\infty}^{0} [h(t - t')\bar{d}_{\text{in}}(t)\,dt'] - \bar{d}_{\text{in}}(t) = \bar{d}_{\text{in}}(t)\frac{1}{c_p + 1}.$$

In the high cooperativity limit, none of the input mode is reflected.

8.3.2 Signal Processing with Electromechanics

What if I wanted to 'catch' a more general temporal mode than the rising exponential pulse, could I do that? Not perfectly for an arbitrary temporal mode, even in the high cooperativity limit, but there is a well defined notion of an optimum $\Gamma(t)$ and it can be found analytically in the weak coupling limit where adiabatic elimination is valid (Harlow, 2013). The capture efficiency approaches unity if the maximum value of Γ is much greater than the spectral width of the signal to be captured.

The ability to turn on and off the coupling has enabled a kind of mode conversion. In this sense one can catch a microwave pulse with a particular temporal envelope running through a transmission line, converting the information it encoded into mechanical motion. At a later time one can convert the mechanical state back into a microwave pulse that is delayed with respect to, and has a different temporal envelope than, the original pulse. If one can also tune the microwave resonator while information is stored in the mechanical oscillator, it is even possible to emit a pulse with a very different frequency than the captured pulse (Andrews, Reed, Cicak, Teufel and Lehnert, 2015). To transfer a state between modes that propagate in two different transmission lines requires a circuit with two electromagnetic resonances (Lecocq, Clark, Simmonds, Aumentado and Teufel, 2016).

8.4 State Transfer between the Microwave and Optical Domains

Figure 8.4 shows a device with two microwave resonators coupled to a single mechanical oscillator and capable of transferring a microwave state from one transmission line to

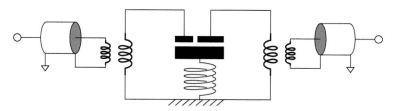

Fig. 8.4 *A schematic of a two mode converter. Two LC circuits, each with different resonance frequency, are modulated by the motion of a single mechanical element. Electromechanical coupling enables noiseless conversion of information between the two modes in the high cooperativity limit. A mechanically based quantum electro-optoconverter would operate on the same principle but with one LC circuit replaced with an optomechanical cavity.*

the other. But given the strong analogy between optomechanics and electromechanics, it's clear that one of these two electromagnetic resonators can be optical rather than electrical. If one can reach the regime of sufficiently large cooperativity and tame optical and electrical losses, such a device provides a quantum link between the optical and electrical domains. From the perspective of the analysis I need only say that I have two electromagnetic modes with resonances split by much more than Ω, yielding EOMs

$$\dot{a}(t) = (i\Delta_c - \kappa_c/2)a(t) - ig_c(b(t) + b^\dagger(t)) + \sqrt{\kappa_{c,E}}a_{inE}(t) + \sqrt{\kappa_{c,I}}a_{inI}(t) \qquad (8.21)$$

$$\dot{c}(t) = (i\Delta_o - \kappa_o/2)c(t) - ig_o(b(t) + b^\dagger(t)) + \sqrt{\kappa_{o,I}}c_{inI}(t) + \sqrt{\kappa_{o,E}}c_{inE}(t)$$

$$\dot{b}(t) = (-i\Omega - \gamma_0/2)b(t) - i(g_c a^\dagger(t) + g_c^* a(t))$$
$$- i(g_o c^\dagger(t) + g_o^* c(t)) + \sqrt{\gamma_0}b_{in}(t),$$

where the subscript 'c' indicates parameters associated with the electrical circuit and the subscript 'o' to indicate parameters associated with the optical cavity. The field amplitudes $a(t)$ and $c(t)$ are the fluctuations about, and in the rotating frame of, the microwave and optical pumps, respectively. The pumps populate the microwave circuit and optical cavity with mean field amplitudes α and $\bar{c}$. The mean field amplitude controls the strength of the optomechanical coupling: $g_c = G_c x_{zpf}\alpha$ and $g_o = G_o x_{zpf}\bar{c}$.

I will follow the same path with these EOMs as for the single cavity case. In particular, I assume that microwave and optical pump fields are optimally detuned, with $\Delta_c = \Delta_o = -\Omega$, and in the resolved sideband limit for both cavities. Finally, I transform to a frame rotating at Ω, with $a(t) \to a(t)e^{-i\Omega t}$, $b(t) \to b(t)e^{-i\Omega t}$, $c(t) \to c(t)e^{-i\Omega t}$ and drop terms rotating at 2Ω. The resulting EOMs are:

$$\dot{a}(t) = (-\kappa_c/2)a(t) - ig_c b(t) + \sqrt{\kappa_{c,E}}a_{inE}(t) + \sqrt{\kappa_{c,I}}a_{inI}(t) \qquad (8.22)$$

$$\dot{c}(t) = (-\kappa_o/2)c(t) - ig_o b(t) + \sqrt{\kappa_{o,I}}c_{inI}(t) + \sqrt{\kappa_{o,E}}c_{inE}(t)$$

$$\dot{b}(t) = (-\gamma_0/2)b(t) - i(g_c^* a(t) + g_o^* c(t)) + \sqrt{\gamma_0}b_{in}(t),$$

The last approximation is again to adiabatically eliminate both the optical cavity and microwave circuit fields as $\dot{a} \approx 0$ and $\dot{c} \approx 0$:

$$a(t) = -\frac{2ig_c}{\kappa_c}b(t) + \frac{2}{\kappa_c}(\sqrt{\kappa_{c,E}}a_{inE}(t) + \sqrt{\kappa_{c,I}}a_{inI}(t)) \qquad (8.23)$$

$$c(t) = -\frac{2ig_o}{\kappa_o}b(t) + \frac{2}{\kappa_o}(\sqrt{\kappa_{o,I}}c_{inI}(t) + \sqrt{\kappa_{o,E}}c_{inE}(t))$$

$$\dot{b}(t) = -\frac{1}{2}\left(\gamma_0 + \frac{4|g_c|^2}{\kappa_c} + \frac{4|g_o|^2}{\kappa_o}\right)b(t) - ig_c^*\frac{2}{\kappa_c}\left(\sqrt{\kappa_{c,E}}a_{inE}(t) + \sqrt{\kappa_{c,I}}a_{inI}(t)\right)$$

$$- ig_o^*\frac{2}{\kappa_o}\left(\sqrt{\kappa_{o,I}}c_{inI}(t) + \sqrt{\kappa_{o,E}}c_{inE}(t)\right) + \sqrt{\gamma_0}b_{in}(t).$$

Equations 8.23 and their associated output relations are simply solved in the frequency domain yielding a susceptibility matrix. The elements of most interest are the optical response to an electrical input S_{21} and the reciprocal electrical response to an optical input S_{12}.

$$S_{21}(\omega) = \frac{a_{\text{out}}(\omega)}{b_{\text{in}}(\omega)} = e^{i\phi} \times \frac{\sqrt{\Gamma_{\text{o}}\Gamma_{\text{c}}}}{-i(\omega - \omega_{\text{m}}) + (\Gamma_{\text{o}} + \Gamma_{\text{c}} + \gamma_{\text{m}})/2} \times \sqrt{\eta_{\text{o}}\eta_{\text{c}}} \qquad (8.24)$$

where $\phi = \text{Arg}(-g_{\text{o}}g_{\text{c}}^*)$, $\Gamma_{\text{o}} = 4|g_{\text{o}}|^2/\kappa_{\text{o}}$ and $\Gamma_{\text{c}} = 4|g_{\text{c}}|^2/\kappa_{\text{c}}$, and $\eta_{\text{c}} = \kappa_{\text{c,E}}/\kappa_{\text{c}}$ and $\eta_{\text{o}} = \kappa_{\text{o,E}}/\kappa_{\text{o}}$. Eq. 8.24 is in the frame of the pumps so that $b_{\text{in}}(\omega)$ and $a_{\text{out}}(\omega)$ are a frequency ω above the microwave and optical pumps, respectively. In this analysis, upconversion and downconversion efficiency are identical, $|S_{21}(\omega)| = |S_{12}(\omega)|$, meaning that frequency conversion is bidirectional(Andrews, Peterson, Purdy, Cicak, Simmonds, Regal and Lehnert, 2014). The conversion efficiency can approach unity if three conditions are met: 1.) the electrical and optical cavity modes must decay primarily to the propagating modes ($\kappa_{\text{c,E}} \gg \kappa_{\text{c,I}}$ and $\kappa_{\text{o,E}} \gg \kappa_{\text{o,I}}$); 2.) the electromechanical and optomechanical conversion rates must be matched $\Gamma_{\text{o}} = \Gamma_{\text{c}}$; and 3.) both rates must be large enough to be in the high cooperativity limit $\Gamma_{\text{o}} \gg \gamma_{\text{m}}$ and $\Gamma_{\text{c}} \gg \gamma_{\text{m}}$. As for the single mode case, quantum state preserving transfer requires a larger cooperativity $\Gamma_{\text{c}} \gg \gamma_{\text{m}}[k_B T/(\hbar\omega_M)]$.

8.5   State-space Models

Working with electromechanical systems usually entails solving systems of linear differential equations with additional IO relations. Engineers have developed a nice systematic way of handling this task; they work in a set of conventions and definitions known collectively as state-space models (Zadeh and Desoer, 1963). As originally conceived, state-space models describe classical systems. Although the EOMs Eqs. 8.14 describe the evolution of quantum degrees of freedom, they are also linear. Furthermore electromechanical experiments mostly use linear detectors with the input modes prepared in Gaussian quantum states. In this limit, quantum and classical mechanics are in perfect correspondence, as long as I insist that input modes carry classical Gaussian noise with variance chosen to mimic quantum fluctuations (Bartlett, Sanders, Braunstein and Nemoto, 2002; Tsang, 2013; Safavi-Naeini, Chan, Hill, Gröblacher, Miao, Chen, Aspelmeyer and Painter, 2013). Indeed all of the perplexities of back-action, quantum-limited measurement (Clerk, 2004), and sideband asymmetry (Weinstein, Lei, Wollman, Suh, Metelmann, Clerk and Schwab, 2014) are captured in this semiclassical picture. As such, it is possible to use classical state-space models to describe electromechanical systems even if they operate in the quantum regime of large cooperativity.

Why adopt state-space models?

1. They are a systematic way of solving linear EOMs. In particular, you can make some conventions static.

2. They are easy to automate. Both Mathematica and Matlab implement SS models.

3. They provide fast heuristics. 'Is my model unitary?' That's easy to check.

4. It is straightforward to add more elements to an existing model. There are simple rules for combining SS-models in series, in parallel, and for feeding the outputs back to the inputs to form larger more complex networks (Gough and James, 2009).

5. They are the language of modern control theory for both classical and quantum (James, Nurdin and Petersen, 2008) systems, with many powerful theorems written in this language. For example, stability, controllability and observability criteria are expressed as compact mathematical relationships of the primitive elements.

A linear state-space model is defined by four matrices that encode the EOMs and IO relations as

$$\dot{\vec{X}}(t) = \mathbf{A}\vec{X} + \mathbf{B}\vec{u} \tag{8.25}$$

$$\dot{\vec{y}}(t) = \mathbf{C}\vec{X} + \mathbf{D}\vec{u}. \tag{8.26}$$

The definitions are:

- $\vec{X}$ is a vector of n 'internal' or Hamiltonian variables
- $\vec{u}$ is a vector of m input fields
- $\vec{y}$ is a vector of m output fields
- $\mathbf{A}$ is an $n \times n$ matrix that models the dynamics of the internal variables
- $\mathbf{B}$ is an $n \times m$ matrix that models the dependence of the internal variables on input fields
- $\mathbf{C}$ is an $m \times n$ matrix that models the dependence of the output fields on the internal variables
- $\mathbf{D}$ is an $m \times m$ matrix that models the direct scattering of input modes to output modes.

Let's work an example to get a sense of the utility of this method. For Eqs. 8.14 the vectors are:

$$\vec{X}(t) = \begin{pmatrix} a \\ b \\ a^\dagger \\ b^\dagger \end{pmatrix}, \quad \vec{u}(t) = \begin{pmatrix} d_{\text{in}} \\ a_{\text{inI}} \\ b_{\text{in}} \\ d_{\text{in}}^\dagger \\ a_{\text{inI}}^\dagger \\ b_{\text{in}}^\dagger \end{pmatrix}, \quad \vec{y}(t) = \begin{pmatrix} d_{\text{out}} \\ a_{\text{outI}} \\ b_{\text{out}} \\ d_{\text{in}}^\dagger \\ a_{\text{outI}}^\dagger \\ b_{\text{out}}^\dagger \end{pmatrix}$$

and the matrices are:

$$
\mathbf{A} = \begin{pmatrix}
i\Delta - \kappa/2 & -ig & 0 & -ig \\
-ig^* & -i\Omega - \gamma_0/2 & -ig & 0 \\
0 & ig* & -i\Delta - \kappa/2 & ig^* \\
ig^* & 0 & ig & i\Omega - \gamma_0/2
\end{pmatrix},
$$

$$
\mathbf{B} = \begin{pmatrix}
\sqrt{\kappa_E} & \sqrt{\kappa_I} & 0 & 0 & 0 & 0 \\
0 & 0 & \sqrt{\gamma_0} & 0 & 0 & 0 \\
0 & 0 & 0 & \sqrt{\kappa_E} & \sqrt{\kappa_I} & 0 \\
0 & 0 & 0 & 0 & 0 & \sqrt{\gamma_0}
\end{pmatrix}
$$

$$
\mathbf{C} = \mathbf{B}^T, \quad \mathbf{D} = -\mathbf{I}_6.
$$

These definitions completely capture the quantum optics model developed in section 8.2.

This formulation is well adapted to working in the time domain, both analytically and numerically. Nevertheless, it's also easy to go to the Laplace or Fourier domain and find the 6×6 susceptibility matrix $\mathbf{\Xi}(\omega)$ that describes how input fields scatter to output fields $\vec{y} = \mathbf{\Xi}(\omega)\vec{u}$. Formally the matrix is

$$
\mathbf{\Xi}(\omega) = \mathbf{C}[i\omega\mathbf{I} - \mathbf{A}]^{-1}\mathbf{B} + \mathbf{D}.
$$

One command in Mathematica finds this matrix in all its symbolic glory.

It's not much more difficult to find the 6×6 cross spectral density matrix of the output noise fields given the cross correlation matrix (the diagonal elements are the usual (self) spectral densities) as $\mathbf{S}(\omega) = \mathbf{\Xi}^* \mathbf{\Sigma} \mathbf{\Xi}^T$, where the spectral density $\mathbf{S}(\omega)$ of the output fields is (Clerk, Devoret, Girvin, Marquardt and Schoelkopf, 2010; Loudon, 2000)

$$
2\pi \mathbf{S}(\omega)\delta(\omega - \omega') = \left\langle [\vec{y}(\omega')]^\dagger \vec{y}^T(\omega) \right\rangle = \mathbf{\Xi}^*(\omega') \left\langle [\vec{u}(\omega')]^\dagger \vec{u}^T(\omega) \right\rangle \mathbf{\Xi}^T(\omega) \tag{8.27}
$$

More compactly, $\mathbf{S}(\omega) = \mathbf{\Xi}^*(\omega)\mathbf{\Sigma}\mathbf{\Xi}^T(\omega)$ where $\mathbf{\Sigma}(2\pi\delta(\omega - \omega')) = \langle [\vec{u}(\omega')]^\dagger \vec{u}^T(\omega) \rangle$. Unpacking the notation shows that these definitions compactly reproduce the definitions in (Clerk, Devoret, Girvin, Marquardt and Schoelkopf, 2010), but note that the Hermitian conjugation is applied element by element to the input and output vectors without transposing the vectors themselves. Finally, to relate the elements of this spectral density matrix to the outcome of a linear measurement, which does not distinguish emission and absorption of quanta, one must use the symmetrized form of spectral density (Clerk, Devoret, Girvin, Marquardt and Schoelkopf, 2010). For example, the microwave noise measured at the external port is $N_{1,1} = ([\mathbf{S}(\omega)]_{1,1} + [\mathbf{S}(-\omega)]_{4,4})/2$.

The correlation matrix $\mathbf{\Sigma}$ describes the noise properties of the input modes. If these modes are driven by noise from thermal environments—that is, black-body radiation or Nyquist–Johnson noise—the elements of $\mathbf{\Sigma}$ have a simple form (Clerk, Devoret, Girvin, Marquardt and Schoelkopf, 2010). For example, $\langle [\hat{a}_{in}(\omega')]^\dagger \hat{a}_{in}(\omega) \rangle = 2\pi\delta(\omega - \omega')n_{in}$, where n_{in} is the Bose–Einstein occupation factor for the transmission line modes. Thus, the $(1,1)$ component of $\boldsymbol{\Sigma}$ will have the value $n_{in} = 1/(e^{\hbar\omega/k_B T} - 1)$ where T is the temperature of the modes in the transmission line. If the electromagnetic resonator has a narrow freqeuency response ($\kappa \ll \omega_0$), we can approximate $\omega \approx \omega_0$ in the expression for n_{in}. For $k_B T \gg \hbar\omega_0$, $n_{in} \approx k_B T/\hbar\omega_0$, which is the same as Johnson noise expressed in units of photons/sec·Hz. If all the ports of the optomechanical system are driven by thermal noise, then

$$
\boldsymbol{\Sigma} = \begin{bmatrix}
n_{in} & 0 & 0 & 0 & 0 & 0 \\
0 & n_0 & 0 & 0 & 0 & 0 \\
0 & 0 & n_m & 0 & 0 & 0 \\
0 & 0 & 0 & n_{in}+1 & 0 & 0 \\
0 & 0 & 0 & 0 & n_0+1 & 0 \\
0 & 0 & 0 & 0 & 0 & n_m+1
\end{bmatrix}. \tag{8.28}
$$

Almost all of the experimental phenomena of opto- and electromechanics can be described in this compact way of writing linear EOMs and their associated IO relations. Even for the simple electromechanical system considered here, the state-space approach already provides helpful automation of the analysis. For example, in Fig. 8.5 I plot the microwave output noise for several experimentally interesting cases that examine the

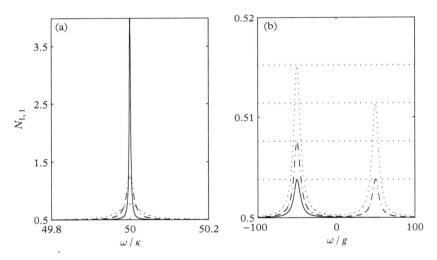

Fig. 8.5 *Microwave output spectral density of an electromechanical device. a) The microwave output noise shows steady-state mechanical cooling of a mechanical oscillator for $\Delta = -\Omega$, $\gamma_0/\kappa = 4 \times 10^{-4}$, $\kappa_I = 0$, $n_{in} = 0$, $n_m = 10$, and for $g/\kappa = 0.03$ (solid), $g/\kappa = 0.07$ (dashed dotted), $g/\kappa = 0.12$ (dotted). b) Sidebands in the microwave output noise associated with mechanical motion show one quantum of asymmetry for $\Delta = 0$, $\omega_M/g = 50$, $\gamma_0/g = 10$, $\kappa/g = 400$, $\kappa_I = 0$, and for $n_m = 0$ (solid), $n_m = 1$ (dashed dotted), $n_m = 3$ (dotted). Note that the upper sidebands are proportional to n_m and the lower sidebands to $n_m + 1$ (dotted horizontal lines).*

electromechanical model in very different limits. Although they illustrate different limits, the plots are created from the same symbolic state-space representation. In particular, I show the phenomena of both steady-state mechanical cooling in the resolved sideband limit with $\Delta = -\Omega$ and the asymmetry of the sidebands associated mechanical motion in the unresolved sideband limit with $\Delta = -\Omega$. (This latter effect is usually attributed to a quantum asymmetry between emission and absorption of individual mechanical quanta, but it can be modelled correctly with a classical state-space approach by introducing the quantum asymmetry through the form of Σ.) For more complex electromechanical and optomechanical systems, such as considered in section 8.4, this sort of automation becomes a necessity.

Bibliography

Andrews, R. W., Peterson, R. W., Purdy, T. P., Cicak, K., Simmonds, R. W., Regal, C. A., and Lehnert, K. W. (2014, March). Bidirectional and efficient conversion between microwave and optical light. *Nat. Phys.*, **10**(4), 321–6.

Andrews, R. W., Reed, A. P., Cicak, K., Teufel, J. D., and Lehnert, K. W. (2015). Quantum-enabled temporal and spectral mode conversion of microwave signals. *Nature Communications*, **6**, 10021.

Bartlett, S. D., Sanders, B. C., Braunstein, S. L., and Nemoto, K. (2002). Efficient classical simulation of continuous variable quantum information processes. *Phys. Rev. Lett.*, **88**(9), 097904.

Caves, C. M., Thorne, K. S., Drever, R. W. P., Sandberg, V. D., and Zimmermann, M. (1980, Apr). On the measurement of a weak classical force coupled to a quantum-mechanical oscillator. I. Issues of principle. *Rev. Mod. Phys.*, **52**, 341–92.

Clerk, A. A. (2004). Quantum-limited position detection and amplification: A linear response perspective. *Physical Review B*, **70**(24), 245306.

Clerk, A. A., Devoret, M. H., Girvin, S. M., Marquardt, F., and Schoelkopf, R. J. (2010, Apr). Introduction to quantum noise, measurement, and amplification. *Rev. Mod. Phys.*, **82**(2), 1155–208.

Devoret, M., Huard, B., Schoelkopf, R., and Cugliandolo, L. F. (eds) (2014). *Quantum Machines: Measurement and Control of Engineered Quantum Systems* (First edn)., Chapter 9, pp. 351–68. Les Houches Lecture Notes. Oxford University Press, Oxford.

Gough, J. and James, M. R. (2009). The series product and its application to quantum feedforward and feedback networks. *IEEE Transactions on Automatic Control*, **54**(11), 2530–44.

Harlow, J. W. (2013). *Microwave Electromechanics: Measuring and Manipulating the Quantum State of a Macroscopic Mechanical Oscillator*. Ph.D. thesis, University of Colorado. at Boulder.

James, M. R., Nurdin, H. I., and Petersen, I. R. (2008). H^{∞} control of linear quantum stochastic systems. *IEEE Transactions on Automatic Control*, **53**(8), 1787–803.

Lecocq, F., Clark, J. B., Simmonds, R. W., Aumentado, J., and Teufel, J. D. (2016). Mechanically mediated microwave frequency conversion in the quantum regime. *Phys. Rev. Lett.*, **116**, 043601.

Loudon, R. (2000). *The Quantum Theory of Light* (Third edn). Oxford University Press, New York.

Lugiato, L. A., Mandel, P., and Narducci, L. M. (1984, Mar). Adiabatic elimination in nonlinear dynamical systems. *Phys. Rev. A*, **29**, 1438–52.

Marquardt, F., Chen, J. P., Clerk, A. A., and Girvin, S. M. (2007, Aug). Quantum theory of cavity-assisted sideband cooling of mechanical motion. *Phys. Rev. Lett.*, **99**(9), 093902.

Marquardt, F. and Girvin, S. M. (2009, May). Optomechanics. *Physics*, **2**, 40.

Palomaki, T. A., Harlow, J. W., Teufel, J. D., Simmonds, R. W., and Lehnert, K. W. (2013*a*). Coherent state transfer between itinerant microwave fields and a mechanical oscillator. *Nature*, **495**(7440), 210–14.

Palomaki, T. A., Teufel, J. D., Simmonds, R. W., and Lehnert, K. W. (2013*b*). Entangling mechanical motion with microwave fields. *Science*, **342**(6159), 710–13.

Pirkkalainen, J.-M., Damskägg, E., Brandt, M., Massel, F., and Sillanpää, M. A. (2015, Dec). Squeezing of quantum noise of motion in a micromechanical resonator. *Phys. Rev. Lett.*, **115**, 243601.

Pozar, D. M. (1998). *Microwave Engineering* (Second edn). John Wiley and Sons, New York.

Safavi-Naeini, A. H, Chan, J., Hill, J. T, Gröblacher, S., Miao, H., Chen, Y., Aspelmeyer, M., and Painter, O. (2013). Laser noise in cavity-optomechanical cooling and thermometry. *New Journal of Physics*, **15**(3), 035007.

Teufel, J. D., Donner, T., Li, D., Harlow, J. W., Allman, M. S., Cicak, K., Sirois, A. J., Whittaker, J. D., Lehnert, K. W., and Simmonds, R. W. (2011*a*). Sideband cooling of micromechanical motion to the quantum ground state. *Nature*, **475**, 359–63.

Teufel, J. D., Li, D., Allman, M. S., Cicak, K., Sirois, A. J., Whittaker, J. D., and Simmonds, R. W. (2011*b*). Circuit cavity electromechanics in the strong-coupling regime. *Nature*, **471**(7337), 204–8.

Tsang, M. (2013). Testing quantum mechanics: a statistical approach. *Quantum Measurements and Quantum Metrology*, **1**, 84–109.

Weinstein, A. J., Lei, C. U., Wollman, E. E., Suh, J., Metelmann, A., Clerk, A. A., and Schwab, K. C. (2014). Observation and interpretation of motional sideband asymmetry in a quantum electromechanical device. *Phys. Rev. X*, **4**, 041003.

Wiener, N. (1949). *Extrapolation, Interpolation, and Smoothing of Stationary Time Series with Engineering Applications*. MIT Press.

Wollman, E. E., Lei, C. U., Weinstein, A. J., Suh, J., Kronwald, A., Marquardt, F., Clerk, A. A., and Schwab, K. C. (2015). Quantum squeezing of motion in a mechanical resonator. *Science*, **349**(6251), 952–5.

Zadeh, L. A., and Desoer, C. A. (1963). *Linear system theory: the state space approach*. McGraw-Hill series in system science. McGraw-Hill, New York.

9

Atom Optomechanics

Philipp Treutlein

University of Basel, Department of Physics,
Switzerland

Philipp Treutlein

Philipp Treutlein, *Atom Optomechanics* In: *Quantum Optomechanics and Nanomechanics*. Edited by: Pierre-Francois Cohadon,
Jack Harris, Florian Marquardt, Leticia F. Cugliandolo, Oxford University Press (2020). © Oxford University Press.
DOI: 10.1093/oso/9780198828143.003.0009

Chapter Contents

These lecture notes give an introduction to optomechanics with ultracold atoms, focusing in particular on hybrid systems where atoms are interfaced with micro- and nanofabricated mechanical structures.

I would like to thank K. Hammerer, C. Genes, K. Stannigel, B. Vogell, M. Wallquist and P. Zoller for the fruitful collaborations which laid the theoretical ground for our atom-optomechanics activities. I am equally grateful to the talented and motivated students who have contributed to our experiments: D. Hunger, S. Camerer, M. Korppi, A. Jöckel, M. Mader, T. Lauber, M. T. Rakher, A. Faber, T. Kampschulte, L. Beguin, T. Karg and G. Buser. In particular, I would like to thank T. Karg for helping with the preparation of these lecture notes.

9.1   Introduction

The mechanical effects of light on matter are at the heart of research in the fields of optomechanics and ultracold atoms. In optomechanics, a growing community of researchers is developing techniques for laser cooling, manipulation, and measurement of micro- and nanofabricated mechanical oscillators (Aspelmeyer *et al.*, 2012; Aspelmeyer *et al.*, 2014). An important goal is to control mechanical vibrations of a massive solid-state oscillator on the quantum level and to exploit this control for fundamental tests of quantum physics and applications in precision sensing and signal transduction.

For ultracold atoms, quantum control of mechanical vibrations is well established. The techniques of laser cooling and trapping developed since the 1980s allow one to prepare single atoms as well as large ensembles in the ground-state of a trap, to coherently manipulate their motion, and to detect their vibrations with quantum-limited precision (Chu, 1991; Adams and Riis, 1997; Chu, 2002; Weidemüller and Zimmermann, 2009). The coupling of ultracold atoms to the light field inside high-finesse cavities has been studied on the single-photon level in cavity quantum electrodynamics (Miller *et al.*, 2005; Tanji-Suzuki *et al.*, 2011). The availability of these techniques makes ultracold atoms attractive for optomechanics experiments deep in the quantum regime.

Atomic implementations of optomechanics give access to new regimes of optomechanical coupling, such as the 'granular' regime where the coupling of photons and phonons is significant at the level of single quanta (Stamper-Kurn, 2014), and connect optomechanics to research in many-body physics (Ritsch *et al.*, 2013). Besides their mechanical degrees of freedom, atoms have discrete internal levels that can be controlled with high fidelity. This adds new features to optomechanical systems such as two-level systems or collective spins that can be prepared in highly nonclassical states (Riedel *et al.*, 2010; Gross *et al.*, 2010; Vasilakis *et al.*, 2015; McConnell *et al.*, 2015). The new regimes and new functionality provided by atomic optomechanical systems are currently being explored in a number of experiments (Brennecke *et al.*, 2008; Murch *et al.*, 2008; Purdy *et al.*, 2010; Schleier-Smith *et al.*, 2011; Wolke *et al.*, 2012).

The close analogies between optomechanics and laser manipulation of atoms have also played a role in the development of hybrid mechanical-atomic systems (Hunger *et al.*, 2011; Treutlein *et al.*, 2014). In these systems, laser light is used to couple the vibrations of a solid-state mechanical oscillator to the vibrations or internal states of atoms in a trap. Hybrid mechanical-atomic systems provide new opportunities for quantum control. For example, atoms can enhance the optomechanical cooling of mechanical oscillators, which could enable ground-state cooling in regimes where purely optomechanical techniques fail (Genes *et al.*, 2011; Vogell *et al.*, 2013; Bennett *et al.*, 2014; Bariani *et al.*, 2014). By engineering strong coherent interactions between mechanical oscillator and atoms, non-classical atomic states could be swapped to the mechanical device, realizing non-classical states of mechanical motion (Hammerer *et al.*, 2009*b*; Vogell *et al.*, 2015). Coupling a mechanical oscillator to a spin oscillator with negative effective mass allows one to create Einstein–Podolsky–Rosen entanglement between the two systems, which could be exploited for remote sensing of mechanical vibrations with a precision beyond the standard quantum limit (Hammerer *et al.*, 2009*a*; Polzik and Hammerer, 2015). Exploiting the nonlinearities offered by interacting atomic systems, control of mechanical vibrations on the level of single phonons could be achieved (Carmele *et al.*, 2014). The rich toolbox for quantum control and measurement of atoms thus becomes available for control of solid-state mechanical devices.

These lecture notes give an introduction to optomechanics with ultracold atoms, with a particular focus on hybrid systems. In the first half, we derive the basic optomechanical interactions of atoms and light. Section 9.2 introduces essentials of atom trapping in far-detuned laser light. In section 9.3 we discuss the properties of trapped atoms as mechanical oscillators from an optomechanics point of view. Section 9.4 presents a very useful model to describe the optomechanical interactions of atoms and light, treating the atoms as polarizable particles. In section 9.5, we use this model to derive the optomechanical coupling of atoms and a cavity field and briefly review cavity optomechanics experiments with atoms in the quantum regime.

The second half of the chapter is devoted to hybrid mechanical-atomic systems. We start with an overview of different coupling mechanisms that are explored in recent experiments (section 9.6). In the following, we focus on light-mediated interactions and derive the long-distance coupling of a membrane to an ensemble of laser-cooled atoms (section 9.7). In section 9.8 we review experiments on sympathetic cooling of a membrane with cold atoms. The requirements and perspectives for mechanical quantum control are discussed in section 9.9. In section 9.10 we introduce the new possibilities that arise if the mechanical oscillator is coupled to the atomic internal state.

9.2 Optical Forces on Atoms

Laser light can exert forces on matter through radiation pressure. Harnessing these forces for laser cooling and trapping of atoms triggered a revolution in the field of atomic physics, which led to the observation of new quantum states of matter such as Bose–Einstein condensates, first implementations of quantum information processing tasks, and the development of precision measurement devices such as atomic fountain clocks and atom interferometers (Chu, 1991; Adams and Riis, 1997; Chu, 2002; Weidemüller and Zimmermann, 2009). One distinguishes two types of optical forces on atoms:

- The *optical dipole force* arises from absorption of photons by the atom followed by stimulated reemission into the laser field. This is a coherent process that results in a redistribution of photons between laser field modes and an associated momentum transfer to the atom. The optical dipole force is conservative and it is frequently used for atom trapping, but it also plays a role in sub-Doppler laser cooling.
- The *scattering force* has its origin in absorption of photons followed by spontaneous emission. This is an incoherent process whereby photons are taken from the incident laser field and scattered into vacuum field modes. The scattering force is dissipative and it is mostly used in laser cooling.

We will discuss basic properties of these forces for the case of an atom with an optical transition at frequency ω_0 and natural linewidth $\Gamma = 1/\tau$, where τ is the excited-state lifetime. As an example, consider the D2 line of ^{87}Rb with $\omega_0 = 2\pi \times 384$ THz, corresponding to a wavelength of 780 nm, and $\Gamma = 2\pi \times 6$ MHz. We focus on the experimentally relevant case where the driving laser field at frequency ω is far-detuned from the atomic transition, so that the detuning $\Delta = \omega - \omega_0$ satisfies $|\Delta| \gg \Gamma$ and the atomic transition is very far from being saturated. In this regime, the optical dipole force is much stronger than the scattering force, as we will see below. A very useful review of far-detuned optical dipole traps is given in (Grimm *et al.*, 2000), which forms the basis of the following discussion.

9.2.1 Optical Dipole Force and Photon Scattering Rate

We consider an atom driven by a classical laser field

$$\mathbf{E}(\mathbf{r}, t) = \tfrac{1}{2}\,\mathbf{e}\,\tilde{E}(\mathbf{r})\,e^{-i\omega t} + \text{c.c.} \tag{9.1}$$

The field induces an oscillating electric dipole moment in the atom,

$$\mathbf{p}(\mathbf{r}, t) = \tfrac{1}{2}\,\mathbf{e}\,\tilde{p}(\mathbf{r})\,e^{-i\omega t} + \text{c.c.} \tag{9.2}$$

with $\tilde{p} = \alpha\tilde{E}$, where $\alpha = \alpha(\omega)$ is the frequency-dependent, complex atomic polarizability. The interaction potential of this induced dipole in the driving laser field is

$$U_{\text{dip}}(\mathbf{r}) = -\frac{1}{2}\langle \mathbf{p}(\mathbf{r}, t) \cdot \mathbf{E}(\mathbf{r}, t) \rangle = -\frac{1}{2\epsilon_0 c}\text{Re}(\alpha)I(\mathbf{r}). \tag{9.3}$$

Here, $\langle \cdot \rangle$ denotes a time average over one oscillation period of the electric field and the electric field intensity is $I(\mathbf{r}) = \tfrac{1}{2}\epsilon_0 c|\tilde{E}(\mathbf{r})|^2$. The optical dipole force is given by the gradient of the potential

$$\mathbf{F}_{\text{dip}}(\mathbf{r}) = -\nabla U_{\text{dip}}(\mathbf{r}) = \frac{1}{2\epsilon_0 c}\Re(\alpha)\nabla I(\mathbf{r}). \tag{9.4}$$

The dipole force is a conservative force that depends on the real part $\Re(\alpha)$ of the atomic polarizability, representing the in-phase response of the atom.

The power absorbed by the oscillator from the driving field and re-emitted as dipole radiation into free space is given by

$$P_{\text{abs}} = \langle \dot{\mathbf{p}} \cdot \mathbf{E} \rangle = \frac{\omega}{\epsilon_0 c} \Im(\alpha) I. \tag{9.5}$$

It depends on the imaginary part $\Im(\alpha)$ of the atomic polarizability, representing the out-of-phase response of the atom. In a photon picture, the underlying process is absorption of photons followed by spontaneous emission. The corresponding photon scattering rate is

$$\Gamma_{\text{sc}}(\mathbf{r}) = \frac{P_{\text{abs}}}{\hbar \omega} = \frac{1}{\hbar \epsilon_0 c} \Im(\alpha) I(\mathbf{r}). \tag{9.6}$$

The above expressions for the optical dipole potential and the photon scattering rate hold for any polarizable neutral particle in an oscillating electric field, as long as a proper model for the polarizability α is used. For example, they also apply to experiments with levitated dielectric nanoparticles (Romero-Isart *et al.*, 2011).

9.2.2 Oscillator Model and Atomic Polarizability

There are different ways to describe the atom–light interaction and to derive the optical forces on an atom. If the atomic transition is far from being saturated, we can use the Lorentz oscillator model of the atom to obtain the atomic polarizability. In section 9.2.4, we will briefly discuss the connections to the two-level model.

In the Lorentz model, an electron of charge $-e$ and mass m_e is considered to be elastically bound to the atomic core, forming a simple harmonic oscillator of eigenfrequency ω_0. The oscillator is damped at a rate Γ_ω due to the power radiated by the accelerated charge. The equation of motion of the oscillator driven by the electric field is thus

$$\frac{d^2 x}{dt^2} + \Gamma_\omega \frac{dx}{dt} + \omega_0^2 x = -\frac{e}{m_e} E(t), \tag{9.7}$$

where x is the distance of the electron from the core. Transforming this equation to the frequency domain $[x(t) = \frac{1}{2}\tilde{x}e^{-i\omega t} + \text{c.c.}]$ yields

$$-\omega^2 \tilde{x} - i\omega \Gamma_\omega \tilde{x} + \omega_0^2 \tilde{x} = -\frac{e}{m_e} \tilde{E}. \tag{9.8}$$

The induced electric dipole moment of the atom is $\tilde{p} = -e\tilde{x}$. Using Eq. (9.8) and $\tilde{p} = \alpha \tilde{E}$ we obtain the complex atomic polarizability

$$\alpha(\omega) = \frac{e^2}{m_e} \cdot \frac{1}{\omega_0^2 - \omega^2 - i\omega\Gamma_\omega}. \tag{9.9}$$

The classical damping rate of an oscillating charge due to radiative energy loss is given by (Feynman *et al.*, 1964)

$$\Gamma_\omega = \frac{e^2\omega^2}{6\pi\epsilon_0 m_e c^3}. \tag{9.10}$$

Using Eq. (9.10) and $\Gamma \equiv \Gamma_{\omega_0} = \frac{\omega_0^2}{\omega^2}\Gamma_\omega$ we can rewrite the atomic polarizability as

$$\alpha(\omega) = \frac{6\pi\epsilon_0 c^3}{\omega_0^2} \cdot \frac{\Gamma}{\omega_0^2 - \omega^2 - i(\omega^3/\omega_0^2)\Gamma}. \tag{9.11}$$

For detunings $|\Delta| \ll \omega_0$ we can approximate $\omega_0^2 - \omega^2 \approx -2\omega_0\Delta$ and $\omega/\omega_0 \approx 1$. Within this rotating-wave approximation the atomic polarizability is

$$\alpha \simeq -\frac{3\pi\epsilon_0 c^3}{\omega_0^3} \cdot \frac{\Gamma}{\Delta + i\Gamma/2} = -\frac{3\pi\epsilon_0 c^3}{\omega_0^3} \cdot \frac{\Gamma}{\Delta} \cdot \frac{1 - i\frac{\Gamma}{2\Delta}}{1 + \left(\frac{\Gamma}{2\Delta}\right)^2}. \tag{9.12}$$

For $|\Delta| \gg \Gamma$ we can expand to second order in Γ/Δ and find

$$\alpha \simeq -\frac{3\pi\epsilon_0 c^3}{\omega_0^3} \cdot \frac{\Gamma}{\Delta} \cdot \left(1 - i\frac{\Gamma}{2\Delta}\right), \tag{9.13}$$

which is the atomic polarizability for far-detuned radiation and a single optical transition without substructure.

Inserting this result in Eqs. (9.3) and (9.6), we obtain the optical dipole potential and scattering rate for far-detuned radiation:

$$U_{\text{dip}}(\mathbf{r}) \simeq \frac{3\pi c^2}{2\omega_0^3} \cdot \frac{\Gamma}{\Delta} \cdot I(\mathbf{r}), \tag{9.14}$$

$$\Gamma_{\text{sc}}(\mathbf{r}) \simeq \frac{3\pi c^2}{2\hbar\omega_0^3} \cdot \left(\frac{\Gamma}{\Delta}\right)^2 \cdot I(\mathbf{r}). \tag{9.15}$$

The two quantities are related by

$$\hbar\Gamma_{\text{sc}} = \frac{\Gamma}{\Delta} U_{\text{dip}}, \tag{9.16}$$

a consequence of the relation between the dispersive and absorptive properties of the atom. From Eq. (9.14) we can read off the important property that the sign of the dipole potential depends on the sign of the detuning of the laser from the atomic resonance:

red detuning $(\Delta < 0)$ $\Rightarrow$ attractive potential
blue detuning $(\Delta > 0)$ $\Rightarrow$ repulsive potential

Moreover, since $U_{\mathrm{dip}} \sim I/\Delta$ but $\Gamma_{\mathrm{sc}} \sim I/\Delta^2$, one can make the scattering rate negligible by working at large detuning and high laser power. The resonant character of the atom–light interaction and the tunability it offers are one of the main differences between atom trapping and levitation of dielectric nanoparticles.

So far we have neglected the fine and hyperfine structure of the atom. For alkali atoms, Eq. (9.13) correctly describes the case where Δ is much larger than the fine structure splitting of the D1 and D2 lines. For detunings such that the fine structure is resolved, the two lines have to be treated separately. If the detuning is still much larger than the hyperfine splitting and the laser is linearly polarized, α takes the form Eq. (9.13) multiplied by the line strength factor 1/3 (2/3) for D1 (D2). For general laser polarization or a detuning that resolves the hyperfine structure, α depends on the ground-state hyperfine level in which the atom is prepared. Formulae for various relevant cases are given in (Grimm *et al.*, 2000).

9.2.3 Scattering Force

The scattering force on an atom is due to absorption of photons followed by spontaneous emission. It is thus connected to the scattering rate Γ_{sc}. Every absorption event goes along with a momentum kick $\hbar\mathbf{k}$ on the atom, where $\mathbf{k}$ is the wave vector of the laser mode from which the photons is absorbed. When the photon is spontaneously reemitted into the vacuum field mode $\mathbf{k}'$, the atom received another momentum kick of $-\hbar\mathbf{k}'$. Since spontaneous emission is a random process that occurs with equal probability into opposite directions, the mean momentum transfer due to the emission averages to zero over many absorption–emission cycles. Therefore, only the absorption processes contribute to the mean scattering force. For a single laser mode of wave vector $\mathbf{k}$, the mean scattering force is

$$\mathbf{F}_{\mathrm{sc}} = \hbar\mathbf{k}\,\Gamma_{\mathrm{sc}}. \tag{9.17}$$

Compare this to the dipole force $\mathbf{F}_{\mathrm{dip}} = -\nabla U_{\mathrm{dip}}$. For a tightly focused laser beam or a standing wave of light, the intensity and thus the dipole potential varies on the wavelength scale, $\nabla U_{\mathrm{dip}} \approx k U_{\mathrm{dip}}$. Using this and Eq. (9.16), we can estimate

$$\frac{F_{\mathrm{sc}}}{F_{\mathrm{dip}}} \sim \frac{\hbar k \Gamma_{\mathrm{sc}}}{k U_{\mathrm{dip}}} = \frac{\Gamma}{\Delta}. \tag{9.18}$$

While the scattering force is important for near-resonant light such as in laser cooling of atoms, it is much weaker than the optical dipole force in far-detuned optical traps with $|\Delta| \gg \Gamma$.

9.2.4   Two-level model of the atom

The results of the previous sections can also be obtained in a two-level model of the atom. In the electric dipole approximation, the atom–light interaction Hamiltonian is

$$\hat{V}(\mathbf{r}) = -\hat{\boldsymbol{\mu}} \cdot \mathbf{E}(\mathbf{r}), \tag{9.19}$$

where $\hat{\boldsymbol{\mu}} = -e\hat{\mathbf{x}}$ is the electric dipole operator and $\hat{\mathbf{x}}$ refers to the position of the electron in the atom.

In the limit of large detuning $|\Delta| \gg (\Gamma, |\Omega_R|)$ one can apply second-order perturbation theory to calculate the energy shift of the atomic ground state $|g\rangle$ due to the atom–light interaction (also called 'light shift' or 'AC Stark shift'):

$$U_{\mathrm{dip}}(\mathbf{r}) = \frac{\left| \langle e| \hat{V}(\mathbf{r}) |g\rangle \right|^2}{\hbar \Delta} = \frac{\hbar |\Omega_R(\mathbf{r})|^2}{4\Delta}. \tag{9.20}$$

Here, $\Omega_R = \frac{1}{\hbar} \langle e| \hat{\boldsymbol{\mu}} |g\rangle \cdot \tilde{\mathbf{E}}(\mathbf{r})$ is the Rabi frequency of the driving laser on the optical transition of the atom.

The atomic excited state radiatively decays at a rate Γ. In steady state, the excited state population of the laser-driven atom is $\frac{|\Omega_R|^2}{4\Delta^2}$. Hence the photon scattering rate is given by

$$\Gamma_{\mathrm{sc}} = \Gamma \frac{|\Omega_R|^2}{4\Delta^2}. \tag{9.21}$$

According to the Wigner–Weisskopf theory, the spontaneous decay rate is related to the dipole matrix element as

$$\Gamma = \frac{\omega_0^3}{3\pi \epsilon_0 \hbar c^3} \left| \langle e| \hat{\boldsymbol{\mu}} |g\rangle \right|^2. \tag{9.22}$$

This allows us to express the Rabi frequency as

$$|\Omega_R|^2 = \Gamma \frac{3\pi \epsilon_0 c^3}{\hbar \omega_0^3} |\tilde{E}|^2 = \Gamma \frac{6\pi c^2}{\hbar \omega_0^3} I \tag{9.23}$$

Inserting this expression in Eqs. (9.20) and (9.21) reproduces the expressions for the dipole potential Eq. (9.14) and scattering rate Eq. (9.15) obtained from the Lorentz oscillator model in the far-detuned limit.

9.3   Trapped Atoms as Mechanical Oscillators

An atom trapped in a far-detuned optical trap is a microscopic mechanical oscillator that can be prepared and manipulated deep in the quantum regime using the well-established techniques of atomic physics. A trapped ensemble of N atoms represents a collection of mechanical oscillators with similar frequency. The centre-of-mass mode of the ensemble behaves like a simple harmonic oscillator with the same frequency as a single atom but N times larger mass. In this chapter, we will discuss the properties of such atomic mechanical oscillators.

9.3.1   One-dimensional Optical Lattice

For concreteness, we consider an atom in a one-dimensional optical lattice potential (Bloch, 2005), see Fig. 9.1. The lattice is generated by interference of two counterpropagating laser beams of equal intensity I_0, wave vector k, and detuning Δ from atomic resonance. The resulting standing-wave intensity pattern is

$$I(x) = 4I_0 \cos^2(kx),$$
(9.24)

where I_0 is the single-beam intensity. The far-detuned optical dipole potential obtained from Eq. (9.14) is

$$U_{\text{dip}} = V_m \cos^2(kx)$$
(9.25)

with a modulation depth

$$V_m = \frac{3\pi c^2}{2\omega_0^3} \frac{\Gamma}{\Delta} 4I_0.$$
(9.26)

If the laser is red detuned, $\Delta < 0$, the potential is attractive and the atoms are trapped near the intensity maxima of the standing wave. Using standard techniques of laser cooling (see section 9.3.2), the atoms can be prepared with energies $\ll |V_m|$, so that they are confined near the bottom of the sinusoidal potential wells. In a harmonic approximation to the trap bottom ($kx \ll 1$),

$$U_{\text{dip}} \simeq V_m - V_m k^2 x^2 \stackrel{!}{=} V_m + \frac{1}{2} m\Omega_a^2 x^2,$$
(9.27)

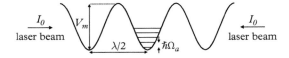

Fig. 9.1 *One-dimensional optical lattice potential created by interference of two laser beams.*

where m is the atomic mass and the trap frequency is defined as

$$\Omega_a = \sqrt{\frac{2|V_m|k^2}{m}}. \tag{9.28}$$

An ultracold atom trapped in the lattice represents a mechanical oscillator whose frequency can be adjusted via the laser intensity or the detuning, $\Omega_a \propto \sqrt{I_0/|\Delta|}$. This allows for fast *in situ* changes of Ω_a, e.g. to bring the atoms in resonance with other systems.

If an ensemble of N atoms is prepared in the lattice, each atom experiences an optical dipole potential. As long as the back-action of the atoms onto the light field is small (see section 9.4), each atom can be treated as an independent oscillator of frequency Ω_a. The centre-of-mass mode of the ensemble represents a mechanical oscillator with the same frequency Ω_a as a single atom, but the mass is increased to Nm. Deviations from this idealized picture arise e.g. from inhomogeneities across the ensemble due to the intensity profile of the trapping beams, which limit the mechanical quality factor of the centre-of-mass mode (see section 9.3.3).

In a red-detuned lattice ($\Delta < 0$), transverse confinement of the atoms is automatically provided by the transverse Gaussian laser profile. In the case of a blue-detuned lattice ($\Delta > 0$), the atoms are trapped near the intensity minima of the standing wave, but Eq. (9.28) for the lattice trap frequency still holds. In this case, the transverse confinement needs to be provided by another trapping beam. Conveniently, the beams creating the lattice usually need not be interferometrically stabilized. Fluctuations in the relative phase of the two beams only lead to a translation of the lattice potential but do not change its shape. Since these fluctuations usually are small and occur at frequencies far below Ω_a, the atoms can adiabatically follow the lattice position.

9.3.2 Ground-state Cooling of Atoms

With atoms in an optical lattice, mechanical frequencies up to a few MHz can be achieved (see also Table 9.1). To prepare the atoms in the ground state of the lattice potential, i.e. to reach $k_B T < \hbar\Omega_a$, microkelvin temperatures are required. Such low temperatures are routinely achieved using the techniques of atomic laser cooling (Adams and Riis, 1997). In the case of Rb atoms, simple optical molasses cooling provides temperatures $< 10 \ \mu$K

Table 9.1 *Parameters of rubidium atoms trapped in a one-dimensional optical lattice. P is the single-beam power and S the beam cross-sectional area so that the peak intensity is $I_0 = P/S$.*

P	S	$\Delta/2\pi$	$\Omega_a/2\pi$	$x_{a,zpf}$	Γ_{heat}
1 W	$(250 \ \mu m)^2$	10^5 GHz	50 kHz	34 nm	5 mHz
1 mW	$(250 \ \mu m)^2$	1 GHz	510 kHz	11 nm	5 kHz

in all three dimensions. Moreover, these cooling techniques have been shown to work well in the presence of optical lattice potentials (Winoto *et al.*, 1999). More advanced laser cooling techniques such as Raman sideband cooling (Kerman *et al.*, 2000; Treutlein *et al.*, 2001) have been used to achieve temperatures of a few hundred nanokelvin, preparing a large fraction of the atoms in the ground state of the lattice.

Since these laser cooling techniques cool each atom individually, all vibrational modes of an atomic ensemble are simultaneously cooled. Coupling between different vibrational modes of the ensemble, while providing a mechanism of mechanical decoherence, does not lead to heating because all modes are cold.

In ultracold atom experiments, the atoms are trapped under ultra-high vacuum conditions, with typical pressures in the range of 10^{-10} mbar and lower. Under these conditions, the atoms are very well isolated from the environment. The experimental apparatus itself is at room temperature, cryogenic cooling is not required. Decoherence of atomic motion mainly arises from the intrinsic fluctuations of the trapping potentials, which we discuss next.

9.3.3 Decoherence due to Photon Recoil Heating

So far, we have considered the mean optical forces experienced by the atom. They are the forces that remain in the limit of a classical polarizable object in a classical electromagnetic field. However, both the field as well as the induced atomic dipole show quantum fluctuations. These lead to fluctuations of the scattering force and the optical dipole force, which heat the trapped atoms and represent a fundamental source of decoherence in atom trapping.

In a red-detuned single-beam trap, the heating rate can be understood in terms of the spontaneous photon scattering at rate Γ_{sc} (Grimm *et al.*, 2000). Each scattered photon increases the energy of the atom by an amount proportional to the photon recoil energy $E_r = \hbar^2 k^2 / 2m$. In a standing-wave trap, as in Fig. 9.1, the situation is more subtle: both the spontaneous photon scattering as well as the quantum fluctuations of the dipole force contribute to the heating. The resulting heating power is (Gordon and Ashkin, 1980)

$$P_{\text{heat}} = E_r \cdot \Gamma_{\text{sc,max}} = E_r \frac{\Gamma V_m}{\hbar \Delta}, \tag{9.29}$$

where $\Gamma_{\text{sc,max}}$ is the maximum photon scattering rate in the lattice, i.e. the rate evaluated at the intensity maxima. Note that P_{heat} is independent of the position in the lattice and independent of the sign of the detuning, so that blue- and red-detuned lattices show the same heating rate.

The resulting phonon heating rate along the lattice direction, Γ_{heat}, limits the coherent manipulation of the atomic motion. For example, the coherence time of the vibrational ground state is limited by $1/\Gamma_{\text{heat}}$. Neglecting the three-dimensional character of the heating, we can estimate

$$\Gamma_{\text{heat}} = \frac{P_{\text{heat}}}{\hbar\Omega_a} = (kx_{\text{a,zpf}})^2 \, \Gamma_{\text{sc,max}}. \tag{9.30}$$

Here, $x_{\text{a,zpf}} = \sqrt{\hbar/2m\Omega_a}$ is the atomic zero-point motion. Note that for typical parameters, $kx_{\text{a,zpf}} = \sqrt{E_r/\hbar\Omega_a} \ll 1$ and therefore $\Gamma_{\text{heat}} \ll \Gamma_{\text{sc,max}}$. This means that many photons need to be scattered to change the vibrational quantum state of the atom by one phonon.

Table 9.1 shows parameters for atomic mechanical oscillators in a one-dimensional optical lattice. The heating rate Γ_{heat} is usually smaller than the typical cooling rates of a few kHz achievable with atomic laser cooling techniques. The atoms can therefore be prepared in the ground-state of the lattice potential using standard laser cooling techniques as discussed in section 9.3.2. In very far-detuned lattices, the heating rate is so small that the atomic motion can be coherently manipulated on timescales of milliseconds or even seconds.

In experiments with micro- and nanofabricated mechanical oscillators, the mechanical quality factor is a key parameter that determines also the thermal heating rate. In the case of trapped atoms, the situation is very different. In contrast to nanomechanical oscillators, which are clamped to a support, trapped atoms in an ultra-high vacuum chamber are nearly perfectly isolated. The quality factor $Q_a = \Omega_a/\Gamma_a$, where Γ_a is the mechanical linewidth, is typically limited by trap anharmonicities, by drifts in the trapping potential, or, in the case of atomic ensembles, by inhomogeneities across the ensemble. These mechanisms result in pure dephasing or in a coupling of the mechanical mode of interest to other modes of the ensemble that are also at microkelvin temperatures. Although the resulting Q_a in the range of 10–10^4 is relatively small, there is negligible heating associated with these damping or dephasing mechanisms. When comparing with other systems, it is therefore more meaningful to compare the decoherence rates rather than the quality factors.

9.4   Atoms as Optical Elements

In the previous chapters, we treated the optical potential as a static container that holds the atoms and is not affected by their presence. This approximation is very well satisfied in most optical lattice experiments, where the lattice light is detuned by tens or even hundreds of nanometres from the atomic transition. In optomechanics experiments, on the other hand, the back-action of the atoms onto the light field is of primary interest. This implies that one needs to work at moderate detunings of the order of GHz. In this section, we discuss a model of atom–light interactions that is very well suited to describing optomechanics experiments with atoms (Asboth *et al.*, 2008). It treats the atoms as polarizable objects that are trapped by the light but also influence the propagation of the trapping fields in a consistent manner.

9.4.1   Beam-splitter Model of Atoms in an Optical Lattice

Consider a one-dimensional optical lattice whose transverse size is much larger than the $\lambda/2$ spacing of the lattice potential wells. Ultracold atoms tightly trapped in the lattice form an array of disc-shaped atomic clouds whose thickness is much smaller than λ. We can model the atoms in a given potential well as an infinitesimally thin sheet of polarizable material, containing N atoms of polarizability α in a cross-sectional area S, see Fig. 9.2. We consider for the moment just a single disc of atoms at position x_0 and model the incoming and outgoing light fields as plane waves with identical polarization and complex amplitudes E_i as shown in Fig. 9.2a.

To determine the light field in the presence of the atoms, we have to solve the scalar Helmholtz equation

$$\left(\partial_x^2 + k^2\right) E(x) = -2k\zeta \, E(x)\delta(x - x_0),\tag{9.31}$$

where the dimensionless coupling constant

$$\zeta = k \cdot \frac{N}{S} \cdot \frac{\alpha}{2\epsilon_0}\tag{9.32}$$

is proportional to the areal density of the cloud polarizability. Integrating Eq. (9.31) over a small interval centred at x_0, we obtain the boundary conditions for the electric field at the position of the atoms

$$E(x_0^-) = E(x_0^+),$$
$$\partial_x E(x_0^-) = \partial_x E(x_0^+) + 2k\zeta \, E(x_0).\tag{9.33}$$

Fig. 9.2 *Atoms in one potential well of an optical lattice form a thin sheet of polarizable material that acts as a beam splitter for the light fields. (a) Incoming and outgoing electric fields modelled as plane waves with amplitudes E_i. (b) The atoms act as a beam splitter with reflection and transmission coefficients r and t, respectively. Here, $A = E_2 e^{-ikx_0}$, $B = E_0 e^{ikx_0}$, $C = E_1 e^{-ikx_0}$, and $D = E_3 e^{ikx_0}$ are the field amplitudes at the position of the atoms x_0.*

When evaluating these boundary conditions for the situation shown in Fig. 9.2a, it is useful to define the field amplitudes at the position of the atoms $A = E_2 e^{-ikx_0}$, $B = E_0 e^{ikx_0}$, $C = E_1 e^{-ikx_0}$, and $D = E_3 e^{ikx_0}$. We find that the thin sheet of atoms acts as a beam splitter with reflection and transmission coefficients

$$r = \frac{i\zeta}{1 - i\zeta}, \quad t = \frac{1}{1 - i\zeta}, \tag{9.34}$$

so that $\zeta = -ir/t$ and $A = rB + tC$ and $D = tB + rC$, see Fig. 9.2b.

These relations can also be expressed as a transfer matrix connecting the field amplitudes on the left of the atoms to those on the right,

$$\begin{pmatrix} A \\ B \end{pmatrix} = \begin{pmatrix} 1 + i\zeta & i\zeta \\ -i\zeta & 1 - i\zeta \end{pmatrix} \cdot \begin{pmatrix} C \\ D \end{pmatrix}. \tag{9.35}$$

Using the expression for α at large detuning from Eq. (9.13) we find

$$\zeta = -\frac{N\sigma}{2S} \cdot \frac{\Gamma}{2\Delta} \cdot \left(1 - i\frac{\Gamma}{2\Delta}\right), \tag{9.36}$$

where $\sigma = 3\lambda^2/(2\pi)$ is the resonant scattering cross-section of a single atom. The quantity $N\sigma/S$ is the resonant optical depth of the N atoms. This formalism now allows us to calculate the outgoing fields for given incoming fields and to determine the optical forces experienced by the atoms.

9.4.2 Optical Forces Experienced by the Atoms

The force F acting on the atomic cloud is determined by the rate at which momentum is extracted from the electromagnetic field. It can be expressed in terms of the power of the incoming and outgoing beams,

$$F = \frac{1}{c}(P_A + P_B - P_C - P_D), \tag{9.37}$$

reflecting the fact that a beam incident from the left (P_B) and being reflected to the left (P_A) results in a force to the right ($F > 0$), while a beam incident from the right (P_C) and being reflected to the right (P_D) leads to a force to the left ($F < 0$). Relating the power $P_A = I_A \cdot S$ to the intensity $I_A = \frac{1}{2}\epsilon_0 c |A|^2$ and similar for the other beams, we obtain

$$F = \frac{\epsilon_0}{2} \cdot S \cdot \left(|A|^2 + |B|^2 - |C|^2 - |D|^2\right). \tag{9.38}$$

The formalism described in the previous section allows us to express A and D in terms of the incoming fields $B = E_0 e^{ikx_0}$ and $C = E_1 e^{-ikx_0}$. Inserting the result in Eq. (9.38), we obtain the force as a function of the atomic position x_0 and the intensities of the incoming beams $I_0 \equiv I_B$ and $I_1 \equiv I_C$,

$$
\begin{aligned}
F(x_0) &= \frac{2S}{c}(I_0 - I_1)\,\frac{\Im(\zeta)}{|1 - i\zeta|^2} &&\Biggr\} = F_{\text{sc}} \\
&\quad -\frac{4S}{c}\sqrt{I_0 I_1}\,\frac{\Re(\zeta)}{|1 - i\zeta|^2}\sin(2kx_0 + \varphi) &&\Biggr\} = F_{\text{dip}} \\
&\quad +\frac{2S}{c}(I_0 - I_1)\,\frac{|\zeta|^2}{|1 - i\zeta|^2} &&\Biggr\} = F_{\text{refl}} \qquad (9.39)
\end{aligned}
$$

where $\varphi = \arg(E_0) - \arg(E_1)$ is the relative phase of the two incoming beams at $x = 0$. The first term can be identified with the scattering force F_{sc} due to spontaneous scattering of photons out of the laser beams. The second term corresponds to the dipole force F_{dip} due to stimulated redistribution of photons between the two laser beams. The third term F_{refl} arises from the incoherent reflection of light at the atomic cloud. It is of order $|\zeta|^2$ and insensitive to the phase of the two incoming laser beams. Equation (9.39) allows us to calculate the optical force even in the regime $|\zeta| \gtrsim 1$ where the light field is strongly perturbed by the presence of the atoms.

The standard expressions for the optical dipole force and scattering force are obtained in the limit $|\zeta| \ll 1$, corresponding to low atomic density or small $|\alpha|$, where the backaction of the atoms on the light field is small. In this limit, the term F_{refl} is negligible and to lowest order in ζ we recover the results of section 9.2,

$$
F(x_0) \simeq \underbrace{\frac{2S}{c}(I_0 - I_1)\Im(\zeta)}_{} - \underbrace{\frac{4S}{c}\sqrt{I_0 I_1}\,\Re(\zeta)\sin(2kx_0 + \varphi)}_{} \qquad (9.40)
$$

$$
= F_{\text{sc}} + F_{\text{dip}}.
$$

If we furthermore take the limit of large detuning $(\Gamma/|\Delta| \ll 1)$, only F_{dip} remains to lowest order.

9.4.3   Back-action of Atoms on the Light Field

The back-action of the atoms on the light field manifests itself in a modulation of the outgoing laser beams. For example, we find that the intensity I_A depends on the atomic position x_0,

$$
\begin{aligned}
I_A &= I_1\,\frac{1}{|1 - i\zeta|^2} &&\Biggr\} \text{ transmitted} \\
&\quad + I_0\,\frac{|\zeta|^2}{|1 - i\zeta|^2} &&\Biggr\} \text{ incoherent reflection} \qquad (9.41)
\end{aligned}
$$

$$- 2\sqrt{I_0 I_1} \; \frac{\Re(\zeta)}{|1 - i\zeta|^2} \sin(2kx_0 + \varphi) \quad \left.\vphantom{\frac{\Re(\zeta)}{|1 - i\zeta|^2}}\right\} \text{ stimulated processes}$$

$$- 2\sqrt{I_0 I_1} \; \frac{\Im(\zeta)}{|1 - i\zeta|^2} \cos(2kx_0 + \varphi) \quad \left.\vphantom{\frac{\Im(\zeta)}{|1 - i\zeta|^2}}\right\} \text{ scattering out of beam.}$$

A similar expression holds for I_D, with I_0 and I_1 exchanged and opposite sign on the third line.

We can again take the limits $|\zeta| \ll 1$ and $\Gamma/|\Delta| \ll 1$ and find that the dominant terms are

$$I_A \simeq I_1 - 2\sqrt{I_0 I_1} \; \Re(\zeta) \sin(2kx_0 + \varphi) = I_1 + \frac{c}{2S} F_{\text{dip}}. \tag{9.42}$$

Consequently, the transmitted beam is power modulated by $\delta P_A = \frac{c}{2} F_{\text{dip}}$, and similarly $\delta P_D = -\frac{c}{2} F_{\text{dip}}$. This back-action of the atoms onto the light field has been observed in a number of experiments (Kozuma *et al.*, 1996; Görlitz *et al.*, 1997; Raithel *et al.*, 1998). For large atom number and relatively near-resonant lattices, the power modulation can be quite substantial, on the order of a few percent.

9.4.4   Generalization to Multiple Clouds and Multi-level Atoms

So far we considered only a single disc of atoms in one well of the optical lattice potential. In general, many neighbouring wells will be filled with atoms. The transfer matrix formalism based on Eq. (9.35) is ideally suited to analyse this situation. In the regime $|\zeta| \gtrsim 1$ where the atoms strongly perturb the light field, new phenomena emerge which can be understood in terms of the light-mediated interactions between atoms in different potential wells (Asboth *et al.*, 2008). On the other hand, many experiments operate in the regime $|\zeta| \ll 1$. In this perturbative regime, the effects on the light field of atoms in different wells of the lattice simply add up, leading to the same results as if all atoms were placed in the same well.

Finally, we point out that the formalism described here has been generalized to the case of atoms with multiple internal states interacting with light fields of general polarization (Xuereb *et al.*, 2010). This allows one e.g. to describe polarization gradient laser cooling of atoms in counterpropagating laser beams. In the context of atom optomechanics experiments, it allows description of spin-optomechanics experiments where the atomic hyperfine spin state couples to the polarization state of the light field.

9.5   Cavity Optomechanics with Atoms

Ultracold atoms trapped in the light field of an optical cavity are a powerful and flexible system to study cavity optomechanics in the quantum regime. The optomechanical interaction of the atoms is analogous to that of thin dielectric membranes or levitated nanoparticles in optical cavities. We will make this analogy evident by describing the atoms as small polarizable objects using the formalism presented in section 9.4. Because the atoms are microscopic objects, their mechanical quantum fluctuations are comparatively large, resulting in a very large optomechanical coupling strength. In combination with the ground-state cooling and quantum control techniques of atomic physics, this has allowed several experiments to enter the quantum regime of cavity optomechanics (Brennecke *et al.*, 2008; Murch *et al.*, 2008; Schleier-Smith *et al.*, 2011; Wolke *et al.*, 2012) and observe phenomena such as quantum measurement back-action (Murch *et al.*, 2008), ponderomotive squeezing (Brooks *et al.*, 2012), and novel many-body quantum phases (Baumann *et al.*, 2010). The field of cavity optomechanics with atoms has been reviewed in (Stamper-Kurn, 2014; Ritsch *et al.*, 2013).

9.5.1 'Atom-in-the-middle' Setup

As a simple model, we consider a cloud of N atoms in a trap of frequency Ω_a, which is placed in the standing-wave light field inside a Fabry–Perot optical cavity. This situation is closely analogous to cavity optomechanics experiments with thin dielectric membranes in optical cavities (Jayich *et al.*, 2008), see Fig. 9.3. Recall that the optomechanical single-photon single-phonon coupling constant in such a 'membrane-in-the-middle' setup is given by

$$g_0 = \frac{\partial \omega_{\text{cav}}}{\partial x} x_{\text{zpf}} = 2|r| \frac{\omega_{\text{cav}}}{L} x_{\text{zpf}} \tag{9.43}$$

where ω_{cav} is the resonance frequency and L the length of the cavity, and we have considered the case where the membrane of reflectivity r is placed on the slope of intracavity intensity standing wave, where g_0 is maximal.

To determine the optomechanical coupling for an 'atom-in-the-middle' system, we use the formalism of section 9.4 to calculate the reflectivity of the atoms. Note that in

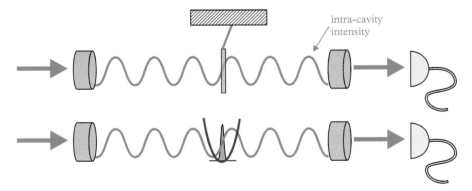

Fig. 9.3 *Analogy between a 'membrane-in-the-middle' setup (top) and an 'atom-in-the-middle' setup (bottom) for optomechanics. Figure courtesy of D. M. Stamper-Kurn.*

the regime where the atoms only weakly perturb the intracavity field ($|\zeta| \ll 1$), we have $r \simeq i\zeta$. For large detuning ($\Delta \gg \Gamma$), Eq. (9.36) gives a reflectivity of

$$|r| \simeq |\zeta| \simeq \frac{N\sigma}{2S} \cdot \frac{\Gamma}{2|\Delta|}. \qquad (9.44)$$

The first term is half the resonant optical depth of the atoms, the second term accounts for the reduction of the atom–light interaction due to the detuning. Furthermore, since the mechanical mode of interest is the centre-of-mass motion of N atoms in the trap of frequency Ω_a, the zero-point motion has an amplitude of $x_{\text{zpf}} = \sqrt{\hbar/2Nm\Omega_a}$, where m is the mass of a single atom. Hence, the atom-optomechanical coupling constant can be expressed as

$$g_0 = \sqrt{N} \cdot \frac{\sigma}{S} \cdot \frac{\Gamma}{2|\Delta|} \cdot \frac{\omega_{\text{cav}}}{L} \sqrt{\frac{\hbar}{2m\Omega_a}}. \qquad (9.45)$$

We find that the optomechanical coupling scales with $\sqrt{N}$, a characteristic feature of collective coupling. The same result can also be derived in the framework of cavity quantum electrodynamics in the dispersive limit, with N atoms collectively coupled to a single cavity mode (Stamper-Kurn, 2014). The interaction of the atomic centre-of-mass motion and the cavity mode is given by the generic optomechanical Hamiltonian (Aspelmeyer *et al.*, 2014)

$$H = \hbar\omega_{\text{cav}}c^\dagger c + \hbar\Omega_a a^\dagger a - \hbar g_0 c^\dagger c(a + a^\dagger), \qquad (9.46)$$

again in direct analogy with a 'membrane-in-the-middle' setup. Here, c ($c^\dagger$) and a ($a^\dagger$) are annihilation (creation) operators for the photons of cavity mode and the phonons of the atomic centre-of-mass mode, respectively.

It is interesting to compare the parameters of an 'atom-in-the-middle' setup with those of a 'membrane-in-the-middle' experiment, see Table 9.2. Although the reflectivity of the atomic ensemble is very small, this is more than compensated for by the large atomic zero-point motion. As a result, the optomechanical coupling of the atomic system is several

Table 9.2 *Comparison of optomechanical coupling parameters for a Fabry–Perot cavity of length L containing, respectively, a micromechanical membrane and an ensemble of N atoms as mechanical element. For the atoms, we consider $N = 5 \cdot 10^4$, $\Delta = 2\pi \cdot 100$ GHz, and $S = (30 \ \mu m)^2$.*

	$|r|$	x_{zpf}	$\Omega/2\pi$	L	$g_0/2\pi$
membrane	0.4	$7 \cdot 10^{-16}$ m	270 kHz	1 mm	0.22 kHz
N atoms	$2 \cdot 10^{-4}$	$2 \cdot 10^{-10}$ m	50 kHz	200 μm	150 kHz

orders of magnitude bigger than that of a typical membrane-in-the-middle system. This feature, in combination with the established ground-state cooling and quantum control techniques of atomic physics, makes atomic systems attractive for exploring cavity optomechanics deep in the quantum regime.

9.5.2   Atom Optomechanics in the Quantum Regime

Several experimental implementations of cavity optomechanics with ultracold atoms have been reported (Brennecke *et al.*, 2008; Murch *et al.*, 2008; Purdy *et al.*, 2010; Schleier-Smith *et al.*, 2011; Brahms *et al.*, 2012; Wolke *et al.*, 2012), for a review see (Stamper-Kurn, 2014; Ritsch *et al.*, 2013).

In the experiments of the Berkeley group (Gupta *et al.*, 2007; Murch *et al.*, 2008; Purdy *et al.*, 2010; Brahms *et al.*, 2012), an ensemble of ultracold thermal [87]Rb atoms is tightly trapped in an optical lattice potential inside a high-finesse Fabry–Perot cavity. The lattice is realized by driving a cavity mode with a laser at a wavelength of 850 nm, far detuned from the atomic transition. The vibrations of the atoms in this far-detuned lattice interact with a second cavity mode at 780 nm, much closer to atomic resonance, see Fig. 9.4. Because the periodicities of the trapping mode and the probe mode are different, each lattice well experiences a different optomechanical coupling $g_0(z)$, where z is the position along the cavity axis. In contrast to the simplified picture presented above, the mechanical oscillator thus corresponds to a collective mode of the ensemble where each atomic cloud contributes with a weight given by the local optomechanical coupling. More recently, experiments have been performed with individual clouds (Purdy *et al.*, 2010), in direct analogy to the model of section 9.5.1. The mechanical mode is ground-state cooled using a combination of laser and evaporative cooling. This system has been used to observe optomechanical effects such as quantum measurement back-action (Murch *et al.*, 2008) and ponderomotive squeezing (Brooks *et al.*, 2012).

The large optomechanical coupling strength that can be achieved in atomic systems (see Table 9.2) opens the way to investigations of the 'granular' regime of optomechanics, where the interaction of phonons and photons is significant on the level of single quanta

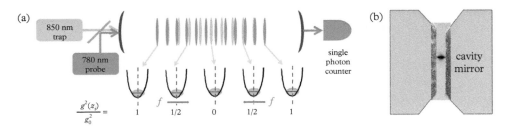

Fig. 9.4 *Optomechanics with atoms tightly confined in an intracavity lattice. (a) Schematic showing the intra-cavity optical lattice trap at 850 nm and the probe light at 780 nm. The mechanical mode is a collective oscillation of the array of atomic clouds. (b) Absorption image of the atoms between the mirrors of the optical cavity. Figure courtesy of D. M. Stamper-Kurn.*

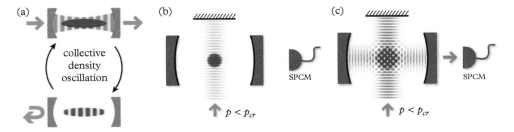

Fig. 9.5 *(a) Optomechanics with a weakly confined Bose–Einstein condensate. A collective density wave of the BEC acts as mechanical oscillator coupled to the cavity mode. (b,c) Setup used for the observation of the Dicke phase transition. (b) The gas is pumped from the side with light near-resonant with a cavity mode. (c) Above a certain threshold pump power, the atoms self-organize into a periodic pattern, maximizing the collective light scattering into the cavity mode. Figure courtesy of T. Esslinger.*

(Stamper-Kurn, 2014). The figure of merit is the ratio g_0/κ, where κ is the cavity linewidth. If $g_0/\kappa > 1$, a single phonon of vibration is sufficient to detune the cavity by more than its linewidth. Conversely, a single cavity photon gives a momentum kick to the mechanical oscillator that is larger than its zero-point momentum uncertainty. Experiments with atom-optomechanical systems are approaching the granular regime with $g_0/\kappa \sim 1$ (Brennecke *et al.*, 2008; Murch *et al.*, 2008), giving access to studies of nonlinear quantum optomechanics.

In the experiments of the Zürich group (Brennecke *et al.*, 2008), a Bose–Einstein condensate (BEC) is weakly trapped inside the cavity, distributed over many periods of the standing-wave cavity mode, see Fig. 9.5. In this system, the cavity mode is coupled to a collective density wave on the BEC, representing a mechanical oscillator of frequency $4\omega_r$, where $\omega_r = E_r/\hbar$ is the recoil frequency.

This experiment was used to explore optomechanical self-organization phenomena in many-body systems, such as the Dicke phase transition (Baumann *et al.*, 2010). When driven from the side with a standing-wave laser beam whose frequency is close to a cavity mode, the atoms can emit light into the cavity. Above a certain threshold driving strength, the atoms self-organize into a periodic pattern inside the cavity where they maximize their collective emission into the cavity mode and are simultaneously trapped by the combined cavity and driving laser field. These experiments open the path to studies of light–matter interactions at the intersection of optomechanics and many-body physics (Ritsch *et al.*, 2013).

9.6 Hybrid Mechanical-atomic Systems: Coupling Mechanisms

Hybrid mechanical systems are a promising way to achieve quantum control over the vibrations of mechanical oscillators (Treutlein *et al.*, 2014). In these systems, a mechanical oscillator is coupled to a microscopic quantum system for which a well-developed toolbox for quantum control is available. Through the coupling, this toolbox can be harnessed for enhanced cooling of mechanical structures, for detection of their quantum motion, for the preparation of non-classical states of vibration, and to implement new protocols for quantum measurement. In fact, the first experiment that reported a nanomechanical oscillator in the quantum regime used a hybrid approach (O'Connell *et al.*, 2010). More generally, since mechanical oscillators can be functionalized with electrodes, magnets and mirrors, they can couple to a variety of different quantum systems and serve as transducers in hybrid quantum information processing (Rabl *et al.*, 2010).

Different hybrid mechanical systems are currently being investigated, involving mechanical oscillators coupled to atoms (Wang *et al.*, 2006; Jöckel *et al.*, 2015; Camerer *et al.*, 2011; Hunger *et al.*, 2010; Montoya *et al.*, 2015), solid-state spin systems (Degen *et al.*, 2009; Arcizet *et al.*, 2011; Kolkowitz *et al.*, 2012; Teissier *et al.*, 2014; Ovartchaiyapong *et al.*, 2014), semiconductor quantum dots (Yeo *et al.*, 2013; Montinaro *et al.*, 2014), and superconducting qubits (O'Connell *et al.*, 2010; Pirkkalainen *et al.*, 2013; Lecocq *et al.*, 2015). Ultracold atoms are attractive in this context because all degrees of freedom, the internal states and the motion in a trap, can be controlled on the quantum level with long coherence times. In hybrid systems, the atoms can thus act as a microscopic mechanical oscillator deep in the quantum regime (Stamper-Kurn, 2014), or they can provide discrete internal levels that give access to new features difficult to realize in purely mechanical systems such as spin oscillators with negative effective mass (Hammerer *et al.*, 2009*a*; Polzik and Hammerer, 2015) and techniques for single-phonon control (Carmele *et al.*, 2014). Figure 9.6 shows an overview of different coupling

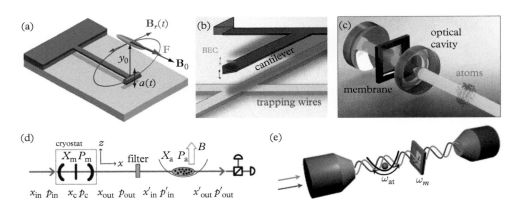

Fig. 9.6 *Coupling mechanisms for hybrid mechanical-atomic systems. (a) Cantilever with a magnetic tip coupled to the spin of atoms (Wang et al., 2006; Treutlein et al., 2007; Montoya et al., 2015). (b) Cantilever coupled by atom-surface forces to the vibrations of a Bose–Einstein condensate in a trap (Hunger et al., 2010). (c) Light-mediated coupling of a membrane oscillator to the vibrations of atoms in an optical lattice (Vogell et al., 2013; Jöckel et al., 2015). (d) Scheme for measurement-based entanglement generation between a membrane and the spin of an atomic ensemble (Hammerer et al., 2009a). (e) Coupling of a membrane to the motion of a single atom through intracavity light fields (Hammerer et al., 2009b).*

mechanisms for atoms and mechanical oscillators; see also the reviews in (Hunger *et al.*, 2011; Treutlein *et al.*, 2014)

First experimental implementations of hybrid mechanical-atomic systems in the classical regime were reported in (Wang *et al.*, 2006; Jöckel *et al.*, 2015; Camerer *et al.*, 2011; Hunger *et al.*, 2010; Montoya *et al.*, 2015). In (Wang *et al.*, 2006), a piezo-driven cantilever with a magnetic tip was used to excite magnetic resonance in a hot atomic vapour. In (Montoya *et al.*, 2015) a similar experiment was performed with trapped, ultracold atoms. In (Hunger *et al.*, 2010), a Bose–Einstein condensate was placed a few hundred nanometres from a classically driven cantilever, so that atom-surface forces led to the excitation of collective modes of the condensate. In all of these experiments, the atoms were used to detect classically driven mechanical vibrations at room temperature. However, the back-action of the atoms onto the mechanical oscillator, which is essential for controlling the oscillator with the atoms, could not be observed. More recently, hybrid mechanical-atomic systems coupled by light were implemented (Jöckel *et al.*, 2015; Camerer *et al.*, 2011). In these experiments, the back-action of the atoms onto the mechanical vibrations (Camerer *et al.*, 2011) as well as strong sympathetic cooling of the mechanical oscillator with the atoms (Jöckel *et al.*, 2015) were observed for the first time. Although the mechanical system still resided in the classical regime because of technical noise and its room-temperature environment, the experimental results showed good agreement with theory (Hammerer *et al.*, 2010*b*; Vogell *et al.*, 2013), which predicts that the quantum regime is accessible for realistic parameters.

In the following, we consider hybrid mechanical-atomic systems where the coupling is mediated by laser light. This results in a modular system, where the mechanical oscillator and the atoms reside in different experimental setups. The light can be routed from one setup to the other via a free-space link or an optical fibre. Such a modular setup circumvents the technological challenge of combining high-power lasers for atom trapping with a cryostat for pre-cooling of the mechanical device. Moreover, it is flexible, providing a great variety of coupling schemes, which can be implemented by simply changing the coupling laser configuration, without the need to open up the optomechanics setup or the cold atom machine. In particular, the light field can be arranged to couple either to the motion of the atoms (Hammerer *et al.*, 2010*b*; Vogell *et al.*, 2013) or to their internal state (Hammerer *et al.*, 2009*b*; Vogell *et al.*, 2015).

9.7   Optical Lattice with Vibrating Mirror

We consider a hybrid mechanical system as proposed in (Hammerer *et al.*, 2010*b*; Vogell *et al.*, 2013), in which the vibrations of a membrane oscillator are coupled via light to the centre-of-mass (c.o.m.) motion of ultracold atoms in an optical lattice. This system was experimentally realized in (Camerer *et al.*, 2011), where the coupling was studied and compared with theory. In a subsequent experiment, the coupling strength was strongly enhanced by placing the membrane in an optical cavity and sympathetic cooling of the membrane vibrations through their coupling to laser-cooled atoms was observed (Jöckel *et al.*, 2015).

9.7.1 Light-mediated Coupling

The coupled atom–membrane system is schematically shown in Fig. 9.7. Ultracold atoms are trapped in an optical lattice generated by reflecting a laser beam from an optomechanical system, essentially realizing an optical lattice with a vibrating mirror.

The optomechanical system is a Si_3N_4 membrane oscillator in an optical cavity in the 'membrane-in-the-middle' configuration (Thompson *et al.*, 2008; Jayich *et al.*, 2008). The Si_3N_4 film has a thickness of typically 50 nm and is supported by a Si frame, realizing a 'square drum' mechanical oscillator. For typical lateral dimensions of several hundred μm to a few mm such membranes feature vibrational modes with frequencies $\Omega_m/2\pi$ in the hundreds of kHz to few MHz range and very high mechanical quality factors of up to $Q = 5 \times 10^7$ (Wilson *et al.*, 2009; Jöckel *et al.*, 2011; Chakram *et al.*, 2014). In addition to their outstanding mechanical properties, these dielectric membranes also have very low optical absorption in the near infrared of order $10^{-5} - 10^{-6}$ in a single pass and a decent field reflectivity of typically $r_m = 0.4$.

The membrane is placed on the slope of the intracavity intensity standing wave. For concreteness, we consider the fundamental mode of the membrane, but coupling to higher-order modes can be realized in a similar way. As the membrane vibrates, it moves in and out of the intracavity field, periodically detuning the cavity resonance frequency ω_{cav}. This leads to an optomechanical coupling between the membrane vibrations and the intracavity field with single-photon single-phonon coupling constant

$$g_0 = G x_{m,zpf}, \tag{9.47}$$

where

$$G = -\frac{\partial \omega_{cav}}{\partial x} = 2|r_m|\frac{\omega_{cav}}{L} \tag{9.48}$$

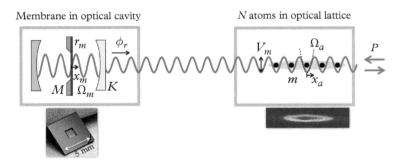

Fig. 9.7 *Atom-membrane coupling mediated by light. The vibrations x_m of a membrane in an optical cavity are coupled to the centre-of-mass motion x_a of an ensemble of N atoms in an optical lattice. The membrane cavity is single-sided and operates in the non-resolved sideband regime $\kappa \gg \Omega_m$. Atoms and membrane are placed in different experimental setups and coupled by light over a macroscopic distance.*

is the cavity frequency shift per membrane displacement, L is the cavity length, $x_{m,\mathrm{zpf}} = \sqrt{\hbar/2M\Omega_m}$ the zero-point amplitude and M the effective mass of the membrane mode.

The optical cavity is single-sided so that the light leaves the cavity again through the input port. Moreover, the cavity intensity decay rate $\kappa \gg \Omega_m$ so that the light leaving the cavity carries instantaneous information about the membrane displacement x_m. The cavity is resonantly driven by a laser of frequency $\omega_L = \omega_{\mathrm{cav}}$ and power P to a mean intracavity photon number

$$\bar{n}_c = \frac{4}{\kappa} \frac{P}{\hbar \omega_{\mathrm{cav}}}. \tag{9.49}$$

Under these conditions, the main effect of the membrane vibrations is to modulate the phase shift $\phi_r = \pi + \delta\phi_r$ of the beam reflected from the cavity by

$$\delta\phi_r = \frac{4}{\kappa} G x_m. \tag{9.50}$$

The interference of the driving laser beam and the light reflected from the cavity creates an optical standing wave outside the cavity. If the driving laser is detuned from an atomic transition, this creates an optical lattice potential for the atoms as described in section 9.3. An ensemble of N atoms is trapped in this lattice and cooled to the ground state along the lattice direction with additional laser cooling beams. Each atom $i = 1 \ldots N$ experiences an optical lattice potential

$$U_i = V_m \cos^2 \left(k x_{a,i} + \frac{\phi_r}{2} \right) \tag{9.51}$$

with modulation depth V_m given by Eq. (9.26). The lattice depends on the membrane position through ϕ_r. Expanding to second order in $k x_{a,i} \ll 1$ and $\delta\phi_r \ll 1$ around the potential minima we obtain

$$U_i \simeq V_m \left(k x_{a,i} + \frac{\delta\phi_r}{2} \right)^2 = V_m k^2 x_{a,i}^2 + V_m k x_{a,i} \delta\phi_r + V_m \left(\frac{\delta\phi_r}{2} \right)^2. \tag{9.52}$$

Using Eq. (9.50) and Eq. (9.28) for the trap frequency along the lattice we find

$$U_i \simeq \underbrace{\frac{1}{2} m \Omega_a^2 x_{a,i}^2}_{\text{atom trap}} + \underbrace{\frac{4}{\kappa} G V_m k x_{a,i} x_m}_{\text{coupling}} + \underbrace{V_m \left(\frac{2G}{\kappa} \right)^2 x_m^2}_{\text{freq. shift of membrane}}. \tag{9.53}$$

The first term is the atomic trapping potential in harmonic approximation, the second term describes a linear coupling of atomic and membrane motion, while the third term is a small correction to the membrane frequency that can be absorbed in the definition of Ω_m.

For N atoms trapped in the lattice, each atom experiences the optical potential and the resulting interaction Hamiltonian is $H_{\text{int}} = \sum_i H_i \sim x_m \sum_i x_{a,i}$ with H_i corresponding to the coupling term in Eq. (9.53). As a result, the membrane is coupled to the atomic c.o.m. coordinate $x_a = \frac{1}{N} \sum_i x_{a,i}$ with a Hamiltonian

$$H_{\text{int}} = N \frac{4}{\kappa} G V_m k x_a x_m. \tag{9.54}$$

Rewriting the position quadratures in terms of bosonic field operators

$$x_m = x_{\text{m,zpf}}(b + b^\dagger), \quad x_{\text{m,zpf}} = \sqrt{\frac{\hbar}{2M\Omega_m}}, \quad [b, b^\dagger] = 1 \tag{9.55}$$

for the membrane and

$$x_a = \frac{x_{\text{a,zpf}}}{\sqrt{N}}(a + a^\dagger), \quad x_{\text{a,zpf}} = \sqrt{\frac{\hbar}{2m\Omega_a}}, \quad [a, a^\dagger] = 1 \tag{9.56}$$

for the atomic c.o.m., we obtain the coupling Hamiltonian

$$H_{\text{int}} = \hbar g(b + b^\dagger)(a + a^\dagger) \tag{9.57}$$

with an atom-membrane single-phonon coupling constant

$$g = \sqrt{N} \frac{4g_0}{\kappa} \frac{V_m}{\hbar} k x_{\text{a,zpf}} \simeq |r_m| \Omega_a \sqrt{\frac{Nm}{M}} \frac{2\mathcal{F}}{\pi}, \tag{9.58}$$

where $\mathcal{F}$ is the cavity finesse and we have assumed near-resonant coupling $\Omega_a \approx \Omega_m$. Including the free evolution of membrane and atomic c.o.m., the full Hamiltonian is

$$H = \hbar\Omega_m b^\dagger b + \hbar\Omega_a a^\dagger a + \hbar g(b + b^\dagger)(a + a^\dagger). \tag{9.59}$$

We thus find that the light field acts like an optical 'spring' that couples the two mechanical oscillators, membrane and atoms. The coupling constant g in Eq. (9.58) contains a term $\sqrt{Nm/M}$, which is very small as the mass ratio of the atomic ensemble and the membrane is tiny (in the experiments of (Jöckel *et al.*, 2015), $Nm/M \approx 10^{-10}$). This can be understood as an impedance mismatch between the two mechanical oscillators. However, this small factor is compensated by the cavity finesse, which can be large ($\mathcal{F} = 10^3$–10^5). The cavity can be thought of as a 'lever' that impedance-matches the two mechanical oscillators.

The same interaction Hamiltonian can also be derived by considering the back-action of the atomic motion on the light field as in section 9.4. According to Eq. (9.42), the N atoms oscillating in the optical lattice potential modulate the power of the beam driving the optomechanical system by $\delta P = \frac{c}{2}F_{\mathrm{dip}} = -cNV_m k^2 x_a$. This leads to a modulation of the intracavity photon number by $\delta \bar{n}_c = \frac{4}{\kappa}\frac{\delta P}{\hbar \omega_{\mathrm{cav}}}$ and consequently to a modulation of the radiation pressure force acting on the membrane by $\delta F_{\mathrm{rad}} = \hbar G \delta \bar{n}_c = -N\frac{4}{\kappa}GV_m kx_a$. This interaction gives rise to a Hamiltonian $H_{\mathrm{int}} = N\frac{4}{\kappa}GV_m kx_a x_m$, in agreement with Eq. (9.54).

The simple semiclassical derivation of the atom-membrane coupling presented here is confirmed by a rigorous calculation using a fully quantum-mechanical description of atoms, membrane and light field (Vogell *et al.*, 2013). We point out that this coupling scheme can be implemented with a variety of different optomechanical systems featuring a single-sided cavity, such as the photonic crystal zipper cavities described in (Eichenfield *et al.*, 2009; Cohen *et al.*, 2013). A finite coupling efficiency to the membrane-cavity setup and losses in the optical beam path can be taken into account as in (Hammerer *et al.*, 2010b; Jöckel *et al.*, 2015).

9.7.2 Dissipation Mechanisms

In addition to the coherent coupling of atoms and membrane described by the Hamiltonian H in Eq. (9.59), there are a number of dissipation mechanisms that affect the atomic and membrane vibrations. As a result, the coupled system is described by a Lindblad master equation

$$\dot{\rho} = -\frac{i}{\hbar}[H,\rho] + \mathcal{L}_{\mathrm{th}}\rho + \mathcal{L}_{\mathrm{rp}}\rho + \mathcal{L}_{\mathrm{a}}\rho + \mathcal{L}_{\mathrm{a,cool}}\rho. \qquad (9.60)$$

The dissipation mechanisms were derived in (Hammerer *et al.*, 2010b; Vogell *et al.*, 2013) and include the following contributions:

- **thermal heating of the membrane**

$$\mathcal{L}_{\mathrm{th}}\rho = \frac{\Gamma_m}{2}(\bar{n}_{\mathrm{th}}+1)D[b]\rho + \frac{\Gamma_m}{2}\bar{n}_{\mathrm{th}}D[b^\dagger]\rho. \qquad (9.61)$$

Here, $D[b]\rho = 2b\rho b^\dagger - b^\dagger b\rho - \rho b^\dagger b$ and similar for the other operators, $\bar{n}_{\mathrm{th}} \simeq k_B T_{\mathrm{env}}/\hbar \Omega_m$ is the membrane's thermal phonon occupation at environment temperature T_{env} and $\Gamma_m = \Omega_m/Q$ is the mechanical energy decay rate. Thermal heating results in a decoherence rate of the membrane vibrational ground state of $\Gamma_m \bar{n}_{\mathrm{th}} = k_B T_{\mathrm{env}}/\hbar Q$. Heating of the membrane due to absorption of the coupling light can be accounted for by an increase of T_{env}.

- **radiation-pressure noise on membrane**

$$\mathcal{L}_{\mathrm{rp}}\rho = \frac{\Gamma_{\mathrm{rp}}}{2}D[b]\rho + \frac{\Gamma_{\mathrm{rp}}}{2}D[b^\dagger]\rho \qquad (9.62)$$

Radiation-pressure shot noise leads to diffusion of the membrane motion with rate $\Gamma_{\text{rp}} = 4g_0^2 \bar{n}_c/\kappa$. This is a fundamental limitation for cooling of the membrane with the atoms. For small coupling laser power, $\Gamma_{\text{rp}} \ll \Gamma_m \bar{n}_{\text{th}}$ and thermal heating dominates.

- **recoil heating of atoms**

$$\mathcal{L}_a \rho = \frac{\Gamma_{\text{heat}}}{2} D[a]\rho + \frac{\Gamma_{\text{heat}}}{2} D[a^\dagger]\rho \tag{9.63}$$

On the atomic side, spontaneous photon scattering and fluctuations of the dipole force lead to diffusion of the atomic motion in the lattice, as discussed in section 9.3.3, with a rate Γ_{heat} given in Eq. (9.30).

- **laser cooling of atoms**

$$\mathcal{L}_{a,\text{cool}} \rho = \frac{\Gamma_{a,\text{cool}}}{2} (\bar{n}_a + 1) D[a]\rho + \frac{\Gamma_{a,\text{cool}}}{2} \bar{n}_a D[a^\dagger]\rho. \tag{9.64}$$

On the atomic system, additional laser cooling beams can be applied to continuously cool the atomic motion. This can be described by a cooling rate $\Gamma_{a,\text{cool}}$ and a steady-state phonon occupation $\bar{n}_a$. Techniques such as Raman sideband cooling (Kerman *et al.*, 2000; Treutlein *et al.*, 2001) reach $\bar{n}_a < 1$. If furthermore $\Gamma_{a,\text{cool}} \gg \Gamma_{\text{heat}}$, the laser cooling is stronger than the additional heating due to the coupling lattice and the atoms remain in the ground state.

Depending on the strength of the coherent coupling g compared to the various dissipation rates, the coupled atom–membrane system shows different dynamics. In the following section 9.8, we first discuss how the atoms can be used for sympathetic cooling of the membrane vibrations. In section 9.9 we will discuss the regime of strong coherent coupling.

9.8 Sympathetic Cooling of a Membrane with Ultracold Atoms

In a recent experiment (Jöckel *et al.*, 2015), the hybrid atom–membrane system discussed in the previous section was implemented and the atoms were used for sympathetic cooling of the membrane.

To act as an efficient coolant, the atoms must be laser cooled by external cooling beams, see Fig. 9.8, with a rate $\Gamma_{a,\text{cool}} \gg (\Gamma_m \bar{n}_{\text{th}}, \Gamma_{\text{rp}}, \Gamma_{\text{heat}})$ that exceeds all other dissipation rates. For the parameters of the experiment, the atomic cooling rate was also larger than the coherent coupling, $\Gamma_{a,\text{cool}} \gg g$. In this regime, the atom–membrane coherence is quickly damped and can be eliminated from the equations of motion. For $\Omega_a \approx \Omega_m$ we can furthermore make the rotating-wave approximation. Under these

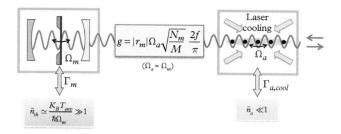

Fig. 9.8 *Sympathetic cooling of the membrane by coupling to laser-cooled atoms. Phonons enter the membrane with rate $\Gamma_m \bar{n}_{th}$ due to the coupling to the frame at temperature T_{env}. The light-mediated coupling g continuously transfers phonons between membrane and atoms. Phonons are removed from the coupled system by laser cooling the atoms with a set of laser cooling beams, corresponding to a coupling of the atoms with cooling rate $\Gamma_{a,cool}$ to a low-temperature bath ($\bar{n}_a \ll 1$).*

conditions, the master equation (9.60) yields the following coupled rate equations for the phonon occupations of the atomic c.o.m. and the membrane mode:

$$\frac{d}{dt}\langle a^\dagger a\rangle = -\left(\Gamma_{a,cool} + \Gamma_{sym}\right)\langle a^\dagger a\rangle + \Gamma_{sym}\langle b^\dagger b\rangle + \Gamma_{heat} + \Gamma_{a,cool}\bar{n}_a,$$

$$\frac{d}{dt}\langle b^\dagger b\rangle = -\left(\Gamma_m + \Gamma_{sym}\right)\langle b^\dagger b\rangle + \Gamma_{sym}\langle a^\dagger a\rangle + \Gamma_{rp} + \Gamma_m\bar{n}_{th}. \tag{9.65}$$

The rate at which the two systems exchange phonons is

$$\Gamma_{sym} = \frac{4g^2}{\Gamma_{a,cool}}, \tag{9.66}$$

which can be understood as the incoherent coupling of the membrane with rate g to the ultracold atomic reservoir of bandwidth $\Gamma_{a,cool} \gg g$. The membrane is sympathetically cooled at rate Γ_{sym}, while the atoms remain at low temperature due to the strong atomic laser cooling.

From Eqs. (9.65) we obtain the steady-state phonon occupation of the membrane

$$\bar{n}_{ss} = \langle b^\dagger b\rangle_{ss} \simeq \frac{\Gamma_m\bar{n}_{th} + \Gamma_{rp}}{\Gamma_{sym} + \Gamma_m} + \frac{\Gamma_{sym}}{\Gamma_{sym} + \Gamma_m}\cdot\frac{\Gamma_{heat}}{\Gamma_{a,cool}} + \frac{\Gamma_{sym}}{\Gamma_{sym} + \Gamma_m}\cdot\bar{n}_a. \tag{9.67}$$

The first term in this expression is the ratio of overall heating and cooling rates of the membrane, while the second and third term take into account the finite temperature of the atoms. Ground-state cooling of the membrane ($\bar{n}_{ss} \ll 1$) requires that

1. sympathetic cooling exceeds membrane heating, $\Gamma_{sym} \gg \Gamma_m\bar{n}_{th} + \Gamma_{rp}$,
2. the atoms are ground-state cooled, requiring $\Gamma_{a,cool} \gg \Gamma_{heat}$ and $\bar{n}_a \ll 1$.

We stress that unlike in standard cavity optomechanical cooling, the membrane-cavity system does *not* have to be in the resolved-sideband regime, i.e. sympathetic cooling with the atoms can reach the ground state for $\kappa \gg \Omega_m$. This can be understood by noting that the atomic oscillator provides the sideband resolution,[1] while the cavity simply enhances the optomechanical interaction. In section 9.9 we will see more generally that sympathetic cooling with atoms can reach the ground state in regimes where optomechanical techniques such as cavity cooling as well as feedback cooling (cold damping) fail to reach the ground state.

The sympathetic cooling scheme was implemented in a recent experiment (Jöckel *et al.*, 2015), where the membrane-cavity system was placed in a room-temperature environment. Figure 9.9 shows measurements of the membrane temperature as a function of time under different experimental conditions (Faber, 2016). If the coupling beam is turned on to sufficiently high power so that the atoms can resonantly couple to the membrane ($\Omega_a \approx \Omega_m$), the membrane temperature drops from $T_{env} = 300$ K to about $T_{min} = 600$ mK. The observed cooling rate extracted from the initial slope of the curve is $\Gamma_{sym} = 1.4 \times 10^3$ s^{-1}. For comparison, if the coupling beam is turned on without any atoms in the lattice, the membrane temperature drops only to about 10 K. This is due to cavity optomechanical cooling, because the driving laser was slightly red-detuned from the cavity resonance to avoid the optomechanical instability on the blue-detuned side. If the coupling beam is turned off, the membrane stays at T_{env}.

The fact that the atoms cool the membrane by a factor of about $T_{env}/T_{min} = 500$ is rather remarkable if one recalls that the mass of the membrane is ten orders of magnitude larger than the mass of the entire atomic ensemble, $Nm/M \approx 10^{-10}$. Sympathetic cooling with laser-cooled atoms and atomic ions is frequently used to cool molecules or other atomic species that cannot be directly laser-cooled. In these experiments, the target and the coolant species thermalize through collisions, and a large mass ratio reduces the cooling efficiency. The largest particles that were cooled in this way are protein molecules in an ion trap, using laser-cooled Ba ions as the coolant (Offenberg *et al.*, 2008). These experiments involved mass ratios of up to ≈ 90 and achieved similar cooling factors and final temperatures as in the atom-membrane sympathetic cooling experiment.

In the proof-of-principle experiment of (Jöckel *et al.*, 2015), the membrane oscillator was placed in a room-temperature environment, resulting in a large thermal heating rate $\Gamma_m \bar{n}_{th} \gg \Gamma_{sym}$. The cooling performance was further limited by technical laser noise in the coupling beam, which was generated by a diode laser. To reach the quantum regime, the membrane can be pre-cooled in a cryostat, as in many other cavity-optomechanics experiments. Moreover, the use of higher-frequency membranes and low-noise lasers will mitigate the effects of technical laser noise. In the following section it is shown that with such an improved system, ground-state cooling and strong coupling of atoms and membrane are within reach.

---

[1] If the atoms are cooled by Raman sideband cooling we have $\bar{n}_a \approx (\Gamma_{a,cool}/4\Omega_a)^2$. The condition $\bar{n}_a \ll 1$ thus limits $\Gamma_{a,cool} \ll \Omega_a$, i.e. the atomic laser cooling must resolve the vibrational sidebands.

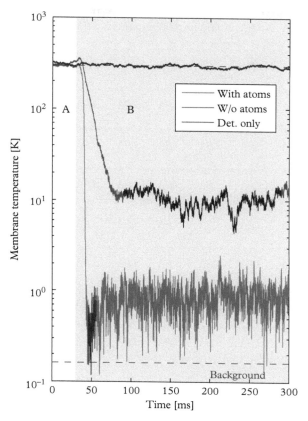

Fig. 9.9 *Sympathetic cooling results. The membrane temperature is shown as a function of time. In the grey-shaded region, the lattice laser beam is turned on so that the atoms can resonantly couple to the membrane, $\Omega_a \approx \Omega_m$. Red data: measurement with atoms in the lattice, showing strong sympathetic cooling. Blue: measurement without atoms, showing cavity optomechanical cooling. Black: lattice laser turned off. Figure courtesy of A. Faber.*

9.9 Ground-state Cooling, Strong Coupling, Cooperativity

In this section we discuss the conditions for observing quantum effects in the coupled atom–membrane system. We find that the atom–membrane cooperativity is an important figure of merit of the coupled system. We conclude by providing a set of parameters for which the system reaches the regime of large cooperativity.

9.9.1 Atom-Membrane Cooperativity

Figure 9.10 shows a schematic of the coupled atom–membrane system and its interaction with the environment. The strength of the coherent coupling is quantified by the rate

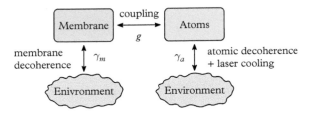

Fig. 9.10 *Coherent coupling and decoherence rates in the coupled atom–membrane system.*

g given in Eq. (9.58). On the membrane side, we define the decoherence rate of the vibrational ground state

$$\gamma_m = \Gamma_m \bar{n}_{th} + \Gamma_{rp}. \tag{9.68}$$

For the atoms, the ground-state decoherence rate depends on whether laser cooling is applied (since typically $\Gamma_{a,cool} \gg \Gamma_{heat}$),

$$\gamma_a = \Gamma_{heat} + \Gamma_{a,cool} \simeq \begin{cases} \Gamma_{a,cool} & \text{if laser cooling is on} \\ \Gamma_{heat} & \text{otherwise.} \end{cases} \tag{9.69}$$

We now define the atom–membrane cooperativity

$$C = \frac{4g^2}{\gamma_m \gamma_a}, \tag{9.70}$$

which compares the strength of the coherent interaction with the product of the decoherence rates of the two systems. In analogy with the corresponding parameter in cavity quantum electrodynamics (Tanji-Suzuki *et al.*, 2011) or cavity optomechanics (Aspelmeyer *et al.*, 2014), we expect the cooperativity to be the relevant figure of merit when analysing the ability of the system to show certain quantum effects.

Ground-state cooling

As a first example, we consider again ground-state cooling of the membrane by coupling to laser-cooled atoms as described in the previous section. Assuming that the atoms are ground-state cooled and focusing on the experimentally relevant regime of $\Gamma_{sym} \gg \Gamma_m$, we can express the steady-state phonon occupation of the membrane Eq. (9.67) as

$$\bar{n}_{ss} \simeq \frac{\gamma_m}{\Gamma_{sym}} = \frac{\gamma_m \Gamma_{a,cool}}{4g^2} = \frac{1}{C}. \tag{9.71}$$

Ground-state cooling of the membrane ($\bar{n}_{ss} < 1$) thus requires a cooperativity $C > 1$.

Strong coupling

For balanced decoherence rates $\gamma_a \approx \gamma_m$, the condition $C > 1$ implies $g > (\gamma_a/2, \gamma_m/2)$. This is the strong-coupling regime as defined e.g. in cavity quantum electrodynamics (Tanji-Suzuki *et al.*, 2011), where single-phonon Rabi oscillations can be observed and a quantum state swap between the two systems is possible.

Effects analogous to electromagnetically induced transparency

If $C > 1$ but we either have $g < \gamma_a/2$ or $g < \gamma_m/2$, a full single-phonon Rabi oscillation between the two systems cannot be observed in the time domain, but interference phenomena analogous to electromagnetically induced transparency are still observable and can be exploited for coherent control.

9.9.2 Connection to Optomechanical Cooperativity and Optical Depth

We can express the atom–membrane coupling Eq. (9.58) as

$$g = \frac{4\sqrt{N}\bar{n}_c g_0 g_a}{\kappa}, \tag{9.72}$$

where g_0 is the membrane–light and $g_a = V_1 k x_{\text{a,zpf}}/\hbar$ the atom–light coupling strength for a single photon in the cavity. Here, $V_1 = V_m/\bar{n}_c$ is the lattice potential experienced by the atoms for a single photon in the cavity.[2] We furthermore assume $\gamma_m \simeq \Gamma_m \bar{n}_{\text{th}}$ and $\gamma_a \simeq \Gamma_{\text{heat}}$. With this we can express the atom–membrane cooperativity as

$$C = \frac{4g^2}{\gamma_m \gamma_a} = 4 \cdot \frac{4g_0^2 \bar{n}_c}{\kappa \Gamma_m \bar{n}_{\text{th}}} \cdot \frac{4g_a^2 \bar{n}_c N}{\kappa \Gamma_{\text{heat}}} = 4 C_m C_a, \tag{9.73}$$

where the optomechanical (quantum) cooperativity referring to the membrane–light interaction is defined in the usual way (Aspelmeyer *et al.*, 2014),

$$C_m = \frac{4g_0^2 \bar{n}_c}{\kappa \Gamma_m \bar{n}_{\text{th}}}. \tag{9.74}$$

The atomic cooperativity referring to the atom–light interaction is

$$C_a = \frac{4g_a^2 \bar{n}_c N}{\kappa \Gamma_{\text{heat}}} = N\frac{\sigma}{S}, \tag{9.75}$$

---

[2] Note that the atoms are trapped in a lattice outside the cavity, but the power of the lattice beam and the intracavity photon number are related by Eq. (9.49).

which is simply the resonant optical depth of the atomic ensemble, as expected for a free-space atom–light interface (Tanji-Suzuki *et al.*, 2011; Hammerer *et al.*, 2010*a*). To derive the second equality in Eq. (9.75) we have used Eqs. (9.26), (9.29), (9.30), and (9.49).

Equation (9.73) implies that the coupled atom–membrane system can operate in a regime of large cooperativity $C > 1$ even if the optomechanical system has a cooperativity $C_m < 1$, because this can be compensated for by a large resonant optical depth C_a, which can reach hundreds or thousandths in state-of-the-art experiments. This has interesting consequences, as it implies that ground-state cooling and quantum control of the membrane by coupling to atoms is possible in the regime $C_m < 1$, where standard optomechanical techniques applied to the membrane–cavity system alone, i.e. without atoms in the beam path, cannot achieve these tasks.

For example, as discussed in (Bennett *et al.*, 2014), optomechanical ground-state cooling (without atoms) requires $C_m > \frac{1}{8}$ (feedback cooling/cold damping) or $C_m > 1$ (cavity feedback cooling). If neither of these conditions is fulfilled, the membrane can still be ground-state cooled by coupling it to an ensemble of laser-cooled atoms with $C_a \gg 1$ such that the overall cooperativity $C = 4C_m C_a > 1$. This can be understood in terms of the concept of quantum feedback (Bennett *et al.*, 2014): the atoms can be considered a coherent controller that can potentially outperform a classical controller that uses detection of light from the cavity and classical feedback. The hybrid atom–membrane setup thus allows one to experimentally study intriguing conceptual questions on the remote control of a quantum system (the mechanical oscillator) with the help of another quantum system (the atoms).

9.9.3   Experimental Parameters

In this section we give a set of experimental parameters of the coupled atom–membrane system for which the regime of large cooperativity and strong coupling is reached.

We consider the $(3,3)$-mode of a standard Si_3N_4 membrane of dimensions 1.5 mm $\times$ 1.5 mm $\times$ 50 nm with $r_m = 0.4$ and mechanical frequency $\Omega_m/2\pi = 0.8$ MHz, effective mass $M = 80$ ng, and quality factor $Q = 1 \times 10^7$, mounted in a cryostat at $T_{env} = 4$ K. The membrane is placed in an optical cavity of length $L = 1$ mm and finesse $\mathcal{F} = 2 \times 10^3$. The cavity is driven with a laser of power $P = 190$ μW, cross-sectional area $S = (25 \ \mu m)^2$, and detuning $\Delta/2\pi = 5$ GHz from the D_2 line of ^{87}Rb atoms. This generates an optical lattice potential with $\Omega_a = \Omega_m$ where we assume $N = 1 \times 10^6$ atoms are trapped.

For this system we find a coupling strength of $g = 1.2 \times 10^5$ s^{-1}, to be compared with decoherence rates $\Gamma_m \bar{n}_{th} = 8 \times 10^4$ s^{-1} and $\Gamma_a = 1.6 \times 10^3$ s^{-1}, placing the system in the strong-coupling regime. For the cooperativities we find $C_m = 0.5, C_a = 310$, and $C = 643$.

9.10   Coupling to the Atomic Internal State

By coupling the mechanical vibrations to the atomic internal state, the full toolbox of quantum control in atomic ensembles becomes accessible (Hammerer *et al.*, 2010*a*).

Such a coupling can be achieved in a setup very similar to Fig. 9.7, as shown in (Vogell *et al.*, 2015). The basic idea is to transduce the mechanical vibrations into a polarization rotation of the coupling light that couples to the atomic hyperfine spin. Conversely, changes in the atomic hyperfine state change the light polarization and modulate the radiation pressure on the mechanics. This again gives rise to a Hamiltonian of the form (Vogell *et al.*, 2015)

$$H = \hbar\Omega_m b^\dagger b + \hbar\omega_L a^\dagger a + \hbar g_{\text{int}}(b + b^\dagger)(a + a^\dagger), \qquad (9.76)$$

similar to Eq. (9.59), except that a and $a^\dagger$ now refer to excitations of the atomic spin state and ω_L is the Larmor frequency. We have assumed that the collective spin describing the internal state of the N atoms is polarized along a magnetic field so that it can be mapped to a harmonic oscillator, $S_z \simeq a^\dagger a - N/2$ and $S_x \simeq \sqrt{N/2}\,(a + a^\dagger)$ in the Holstein–Primakoff approximation (Hammerer *et al.*, 2010a). Several schemes have been analysed that give rise to such interactions (Hammerer *et al.*, 2009a; Bariani *et al.*, 2014; Vogell *et al.*, 2015) and strong coupling has been predicted for realistic parameters (Vogell *et al.*, 2015). Since the Larmor frequency ω_L can be tuned with magnetic fields over a wide range, the atoms can be coupled to mechanical oscillators in the MHz regime, where laser and other technical noise is much reduced.

9.10.1 Beam Splitter and Two-mode Squeezing Hamiltonians

Different types of interactions can be realized with this Hamiltonian: by magnetic-field tuning of the Larmor frequency to $\omega_L = \Omega_m$, the resonant interactions take the form of a 'beam-splitter' Hamiltonian $H_{\text{int}} = \hbar g_{\text{int}} \left(ba^\dagger + b^\dagger a \right)$, giving rise to normal mode splitting and Rabi oscillations between the mechanical and atomic system. This can be used to swap spin-squeezed and other non-classical states of the atoms to the mechanical system (Riedel *et al.*, 2010). Alternatively, by inverting the magnetic field orientation one can set $\omega_L = -\Omega_m$, so that $H_{\text{int}} = \hbar g_{\text{int}} \left(ba + b^\dagger a^\dagger \right)$. This 'two-mode squeezing' Hamiltonian produces entanglement between the atomic and mechanical modes. Interestingly, in this case the atomic ensemble can be thought of as realizing a spin oscillator with a 'negative effective mass' (Polzik and Hammerer, 2015), since $\omega_L < 0$ and creating atomic excitations thus reduces the energy. This illustrates one of the new features that atoms can provide in such hybrid systems.

9.10.2 Vibration Sensing beyond the Standard Quantum Limit

The two-mode squeezing Hamiltonian can be used to generate Einstein–Podolsky–Rosen entanglement between the mechanical oscillator and the atoms. As pointed out in (Hammerer *et al.*, 2009a), this can be used for sensing of mechanical vibrations beyond the standard quantum limit, which limits the precision of weak continuous position measurements (Aspelmeyer *et al.*, 2014). Entanglement is present if the variances of position and moment measurements on the two systems satisfy (Polzik and Hammerer, 2015)

$$\mathrm{Var}(X_m - X_a) + \mathrm{Var}(P_m + P_a) < 2, \tag{9.77}$$

where X_m (X_a) and P_m (P_a) here refer to dimensionless position and momentum quadratures of the mechanical oscillator (atoms), defined such that they satisfy the commutation relation $[X_m, P_m] = i$ ($[X_a, P_a] = i$). In the entangled state, a position (momentum) measurement on the atomic system can predict the outcome of a position (momentum) measurement on the mechanical system with a precision better than the SQL. Remarkably, this holds for both position and momentum quadratures. This distinguishes the approach from other schemes of back-action evasion in mechanical systems, which are limited to a single quadrature. It was pointed out (Polzik and Hammerer, 2015) that this allows one to follow trajectories of the mechanical system in the reference frame provided by the atoms in principle without quantum uncertainty. A further exciting aspect is that this is possible in a remote way, with atoms placed at a macroscopic distance from the mechanical oscillator.

9.10.3   Single-phonon Control

Exploiting the coupling of the mechanical oscillator to the atomic internal state can also be used to obtain control over single phonons. In the atomic ensemble, the effect of Rydberg blockade can be used to restrict the internal-state dynamics to an effective two-level system (Saffman *et al.*, 2010; Weber *et al.*, 2015). Creating single excitations of the atomic ensemble in this way and swapping them to the mechanical system provides a challenging, but in principle feasible, route to controlling single phonon excitations in hybrid mechanical–atomic systems. Indeed, a recent proposal suggests making use of Rydberg excitations in small atomic ensembles to achieve the desired nonlinearities and identifies a parameter regime for such experiments (Carmele *et al.*, 2014).

Bibliography

Adams, C. S. and Riis, E. (1997). Laser cooling and trapping of neutral atoms. *Prog. Quant. Electr.*, **21**(1), 1–79.

Arcizet, O., Jacques, V., Siria, A., Poncharal, P., Vincent, P., and Seidelin, S. (2011, September). A single nitrogen-vacancy defect coupled to a nanomechanical oscillator. *Nature Physics*, 7(11), 879–83.

Asboth, J. K., Ritsch, H., and Domokos, P. (2008). Optomechanical coupling in a one-dimensional optical lattice. *Physical Review A*, 77, 063424.

Aspelmeyer, M., Kippenberg, T. J., and Marquardt, F. (2014, December). Cavity optomechanics. *Rev. Mod. Phys.*, **86**, 1391.

Aspelmeyer, M., Meystre, P., and Schwab, K. (2012). Quantum optomechanics. *Physics Today*, **65**(7), 29.

Bariani, F., Singh, S., Buchmann, L. F., Vengalattore, M., and Meystre, P. (2014, September). Hybrid optomechanical cooling by atomic lambda systems. *Physical Review A*, **90**(3), 033838.

Baumann, K., Guerlin, C., Brennecke, F., and Esslinger, T. (2010, April). Dicke quantum phase transition with a superfluid gas in an optical cavity. *Nature*, **464**(7293), 1301–6.

Bennett, J. S., Madsen, L. S., Baker, M., Rubinsztein-Dunlop, H., and Bowen, W. P. (2014). Coherent control and feedback cooling in a remotely-coupled hybrid atom-optomechanical system. *New Journal of Physics*, **16**(8), 083036.

Bloch, I. (2005). Ultracold quantum gases in optical lattices. *Nature Physics*, **1**(1), 23.

Brahms, N., Botter, T., Schreppler, S., Brooks, D. W. C., and Stamper-Kurn, D. M. (2012, March). Optical detection of the quantization of collective atomic motion. *Physical Review Letters*, **108**(1), 133601.

Brennecke, F., Ritter, S., Donner, T., and Esslinger, T. (2008). Cavity optomechanics with a Bose–Einstein condensate. *Science*, **322**(5899), 235.

Brooks, D. W. C., Botter, T., Schreppler, S., Purdy, T. P., Brahms, N., and Stamper-Kurn, D. M. (2012, August). Non-classical light generated by quantum-noise-driven cavity optomechanics. *Nature*, **488**(7412), 476–80.

Camerer, S., Korppi, M., Jöckel, A., Hunger, D., Hänsch, T. W., and Treutlein, P. (2011, November). Realization of an optomechanical interface between ultracold atoms and a membrane. *Physical Review Letters*, **107**(22), 223001.

Carmele, A., Vogell, B., Stannigel, K., and Zoller, P. (2014, June). Opto-nanomechanics strongly coupled to a Rydberg superatom: coherent versus incoherent dynamics. *New Journal of Physics*, **16**(6), 063042.

Chakram, S., Patil, Y. S., Chang, L., and Vengalattore, M. (2014, March). Dissipation in ultrahigh quality factor SiN membrane resonators. *Physical Review Letters*, **112**(12), 127201.

Chu, S. (1991). Laser manipulation of atoms and particles. *Science*, **253**, 861.

Chu, S. (2002, March). Cold atoms and quantum control. *Nature*, **416**(6), 206–10.

Cohen, J. D., Meenehan, S. M., and Painter, O. (2013). Optical coupling to nanoscale optomechanical cavities for near quantum-limited motion transduction. *Optics Express*, **21**(9), 11227–36.

Degen, C. L., Poggio, M., Mamin, H. J., Rettner, C. T., and Rugar, D. (2009, February). Nanoscale magnetic resonance imaging. *Proceedings of the National Academy of Sciences*, **106**(5), 1313–17.

Eichenfield, M., Camacho, R., Chan, J., Vahala, K. J., and Painter, O. (2009, May). A picogram- and nanometre-scale photonic-crystal optomechanical cavity. *Nature*, **459**(7246), 550–5.

Faber, A. (2016). *Sympathetic cooling and self-oscillations in a hybrid atom-membrane system*. Ph.D. thesis, University of Basel.

Feynman, R. P., Leighton, R. B., and Sands, M. (1964). *The Feynman Lectures on Physics*. Volume 1. Addison-Wesley.

Genes, C., Ritsch, H., Drewsen, M., and Dantan, A. (2011, November). Atom-membrane cooling and entanglement using cavity electromagnetically induced transparency. *Physical Review A*, **84**(5), 051801.

Gordon, J. P. and Ashkin, A. (1980). Motion of atoms in a radiation trap. *Physical Review A*, **21**(5), 1606–17.

Görlitz, A., Weidemüller, M., Hänsch, T. W., and Hemmerich, A. (1997). Observing the position spread of atomic wave packets. *Physical Review Letters*, **78**(11), 2096–9.

Grimm, R., Weidemüller, M., and Ovchinnikov, Y. B. (2000). Optical dipole traps for neutral atoms. *Adv. At. Mol. Opt. Phys*, **42**, 95.

Gross, C., Zibold, T., Nicklas, E., Estève, J., and Oberthaler, M. K. (2010, March). Nonlinear atom interferometer surpasses classical precision limit. *Nature*, **464**(7292), 1165.

Gupta, S., Moore, K. L., Murch, K. W., and Stamper-Kurn, D. M. (2007, November). Cavity nonlinear optics at low photon numbers from collective atomic motion. *Physical Review Letters*, **99**(21), 213601.

Hammerer, K., Aspelmeyer, M., Polzik, E., and Zoller, P. (2009*a*, January). Establishing Einstein–Poldosky–Rosen channels between nanomechanics and atomic ensembles. *Physical Review Letters*, **102**(2), 020501.

Hammerer, K., Sørensen, A. S., and Polzik, E. S. (2010*a*). Quantum interface between light and atomic ensembles. *Reviews of Modern Physics*, **82**(2), 1041–93.

Hammerer, K., Stannigel, K., Genes, C., Zoller, P., Treutlein, P., Camerer, S., Hunger, D., and Hänsch, T. W. (2010*b*, August). Optical lattices with micromechanical mirrors. *Physical Review A*, **82**(2), 021803.

Hammerer, K., Wallquist, M., Genes, C., Ludwig, M., Marquardt, F., Treutlein, P., Zoller, P., Ye, J., and Kimble, H. J. (2009*b*, August). Strong coupling of a mechanical oscillator and a single atom. *Physical Review Letters*, **103**(6), 63005.

Hunger, D., Camerer, S., Hänsch, T. W., König, D., Kotthaus, J. P., Reichel, J., and Treutlein, P. (2010, April). Resonant coupling of a Bose–Einstein condensate to a micromechanical oscillator. *Physical Review Letters*, **104**(1), 143002.

Hunger, D., Camerer, S., Korppi, M., Jöckel, A., Hänsch, T. W., and Treutlein, P. (2011, August). Coupling ultracold atoms to mechanical oscillators. *Comptes Rendus Physique*, **12**, 871.

Jayich, A. M., Sankey, J. C., Zwickl, B. M., Yang, C., Thompson, J. D., Girvin, S. M., Clerk, A. A., Marquardt, F., and Harris, J. G. E. (2008, September). Dispersive optomechanics: a membrane inside a cavity. *New Journal of Physics*, **10**(9), 095008.

Jöckel, A., Faber, A., Kampschulte, T., Korppi, M., Rakher, M. T., and Treutlein, P. (2015, January). Sympathetic cooling of a membrane oscillator in a hybrid mechanical–atomic system. *Nature Nanotechnology*, **10**(1), 55–9.

Jöckel, A., Rakher, M. T., Korppi, M., Camerer, S., Hunger, D., Mader, M., and Treutlein, P. (2011). Spectroscopy of mechanical dissipation in micro-mechanical membranes. *Appl. Phys. Lett.*, **99**(14), 143109.

Kerman, A. J., Vuletic, V., Chin, C., and Chu, S. (2000). Beyond optical molasses: 3D Raman sideband cooling of atomic cesium to high phase-space density. *Physical Review Letters*, **84**(3), 439–42.

Kolkowitz, S., Bleszynski Jayich, A. C., Unterreithmeier, Q. P., Bennett, S. D., Rabl, P., Harris, J. G. E., and Lukin, M. D. (2012, March). Coherent sensing of a mechanical resonator with a single-spin qubit. *Science*, **335**(6076), 1603–6.

Kozuma, M., Nakagawa, K., Jhe, W., and Ohtsu, M. (1996). Observation of temporal behavior of an atomic wave packet localized in an optical potential. *Physical Review Letters*, **76**(14), 2428–31.

Lecocq, F., Teufel, J. D., Aumentado, J., and Simmonds, R. W. (2015, June). Resolving the vacuum fluctuations of an optomechanical system using an artificial atom. *Nature Physics*, **11**(8), 635–639.

McConnell, R., Zhang, H., Hu, J., Ćuk, S., and Vuletic, V. (2015, March). Entanglement with negative Wigner function of almost 3,000 atoms heralded by one photon. *Nature*, **519**(7544), 439–42.

Miller, R., Northup, T. E., Birnbaum, K. M., Boca, A., Boozer, A. D., and Kimble, H. J. (2005, February). Trapped atoms in cavity QED: coupling quantized light and matter. *Journal of Physics B Atomic Molecular and Optical Physics*, **38**, S551.

Montinaro, M., Wust, G., Munsch, M., Fontana, Y., Russo-Averchi, E., Heiss, M., Fontcuberta i Morral, A., Warburton, R. J., and Poggio, M. (2014, August). Quantum dot opto-mechanics in a fully self-assembled nanowire. *Nano Letters*, 14(8), 4454–60.

Montoya, C., Valencia, J., Geraci, A. A., Eardley, M., Moreland, J., Hollberg, L., and Kitching, J. (2015). Resonant interaction of trapped cold atoms with a magnetic cantilever tip. *Physical Review A*, 91(6), 063835.

Murch, K., Moore, K., Gupta, S., and Stamper-Kurn, D. M. (2008). Observation of quantum-measurement backaction with an ultracold atomic gas. *Nature Physics*, 4, 561.

O'Connell, A. D., Hofheinz, M., Ansmann, M., Bialczak, R. C., Lenander, M., Lucero, E., Neeley, M., Sank, D., Wang, H., Weides, M., Wenner, J., Martinis, J. M., and Cleland, A. N. (2010, March). Quantum ground state and single-phonon control of a mechanical resonator. *Nature*, 464, 697.

Offenberg, D., Zhang, C., Wellers, Ch., Roth, B., and Schiller, S. (2008, December). Translational cooling and storage of protonated proteins in an ion trap at subkelvin temperatures. *Physical Review A*, 78(6), 061401.

Ovartchaiyapong, P., Lee, K. W., Myers, B. A., and Bleszynski Jayich, A. C. (2014, July). Dynamic strain-mediated coupling of a single diamond spin to a mechanical resonator. *Nature Communications*, 5, 4429.

Pirkkalainen, J. M., Cho, S. U., Li, J., Paraoanu, G. S., Hakonen, P. J., and Sillanpää, M. A. (2013, February). Hybrid circuit cavity quantum electrodynamics with a micromechanical resonator. *Nature*, 494(7436), 211–15.

Polzik, E. S. and Hammerer, K. (2015, January). Trajectories without quantum uncertainties. *Annalen der Physik*, 527(1–2), A15–A20.

Purdy, T. P., Brooks, D. W. C., Botter, T., Brahms, N., Ma, Z. Y., and Stamper-Kurn, D. M. (2010). Tunable cavity optomechanics with ultracold atoms. *Physical Review Letters*, 105(13), 133602.

Rabl, P., Harris, J. G. E., Zoller, P., and Lukin, M. D. (2010, May). A quantum spin transducer based on nanoelectromechanical resonator arrays. *Nature Physics*, 6(8), 602.

Raithel, G., Phillips, W. D., and Rolston, S. L. (1998). Collapse and revivals of wave packets in optical lattices. *Physical Review Letters*, 81(17), 3615.

Riedel, M. F., Böhi, P. A., Li, Y., Hänsch, T. W., Sinatra, A., and Treutlein, P. (2010, March). Atom-chip-based generation of entanglement for quantum metrology. *Nature*, 464(7292), 1170.

Ritsch, H., Domokos, P., Brennecke, F., and Esslinger, T. (2013, April). Cold atoms in cavity-generated dynamical optical potentials. *Reviews of Modern Physics*, 85(2), 553–601.

Romero-Isart, O., Pflanzer, A. C., Juan, M. L., Quidant, R., Kiesel, N., Aspelmeyer, M., and Cirac, J. I. (2011, January). Optically levitating dielectrics in the quantum regime: Theory and protocols. *Physical Review A*, 83(1), 013803.

Saffman, M., Walker, T.G., and Mølmer, K. (2010). Quantum information with Rydberg atoms. *Reviews of Modern Physics*, 82(3), 2313.

Schleier-Smith, M. H., Leroux, I. D., Zhang, H., Van Camp, M. A., and Vuletic, V. (2011, September). Optomechanical cavity cooling of an atomic ensemble. *Physical Review Letters*, 107(14), 143005.

Stamper-Kurn, D. M. (2014). Cavity optomechanics with cold atoms. In *Cavity Optomechanics* (ed. M. Aspelmeyer, T. Kippenberg, and F. Marquardt), pp. 283–325. Springer, Berlin, Heidelberg.

Tanji-Suzuki, H., Leroux, I. D., Schleier-Smith, M. H, Cetina, M., Grier, A. T., Simon, J., and Vuletic, V. (2011). Interaction between atomic ensembles and optical resonators: Classical Description. *Advances in Atomic, Molecular and Optical Physics*, 60, 201–37.

Teissier, J., Barfuss, A., Appel, P., Neu, E., and Maletinsky, P. (2014, July). Strain coupling of a nitrogen-vacancy center spin to a diamond mechanical oscillator. *Physical Review Letters*, **113**(2), 020503.

Thompson, J. D., Zwickl, B. M., Jayich, A. M., Marquardt, F., Girvin, S. M., and Harris, J. G. E. (2008, March). Strong dispersive coupling of a high-finesse cavity to a micromechanical membrane. *Nature*, **452**(7183), 72–5.

Treutlein, P., Chung, K. Y., and Chu, S. (2001, May). High-brightness atom source for atomic fountains. *Physical Review A*, **63**(5), 51401.

Treutlein, P., Genes, C., Hammerer, K., Poggio, M., and Rabl, P. (2014). Hybrid mechanical systems. In *Cavity-Optomechanics* (ed. M. Aspelmeyer, T. Kippenberg, and F. Marquardt), pp. 327–351. Springer-Verlag, Berlin.

Treutlein, P., Hunger, D., Camerer, S., Hänsch, T. W., and Reichel, J. (2007). Bose–Einstein condensate coupled to a nanomechanical resonator on an atom chip. *Phys. Rev. Lett.*, **99**, 140403.

Vasilakis, G., Shen, H., Jensen, K., Balabas, M., Salart, D., Chen, B., and Polzik, E. S. (2015, March). Generation of a squeezed state of an oscillator by stroboscopic back-action-evading measurement. *Nature Physics*, **11**(5), 389–92.

Vogell, B., Kampschulte, T., Rakher, M. T., Faber, A., Treutlein, P., Hammerer, K., and Zoller, P. (2015, April). Long distance coupling of a quantum mechanical oscillator to the internal states of an atomic ensemble. *New Journal of Physics*, **17**(4), 043044.

Vogell, B., Stannigel, K., Zoller, P., Hammerer, K., Rakher, M. T., Korppi, M., Jöckel, A., and Treutlein, P. (2013, February). Cavity-enhanced long-distance coupling of an atomic ensemble to a micromechanical membrane. *Physical Review A*, **87**(2), 023816.

Wang, Y.-J., Eardley, M., Knappe, S., Moreland, J., Hollberg, L., and Kitching, J. (2006, December). Magnetic resonance in an atomic vapor excited by a mechanical resonator. *Physical Review Letters*, **97**(22), 227602.

Weber, T. M., Höning, M., Niederprüm, T., Manthey, T., Thomas, O., Guarrera, V., Fleischhauer, M., Barontini, G., and Ott, H. (2015, January). Mesoscopic Rydberg-blockaded ensembles in the superatom regime and beyond. *Nature Physics*, **11**, 157.

Weidemüller, M. and Zimmermann, C. (eds.) (2009). *Cold Atoms and Molecules*. Wiley-VCH, Berlin.

Wilson, D. J., Regal, C. A., Papp, S. B., and Kimble, H. J. (2009, November). Cavity optomechanics with stoichiometric SiN films. *Physical Review Letters*, **103**(2), 207204.

Winoto, S. L., DePue, M. T., Bramall, N. E., and Weiss, D. S. (1999). Laser cooling at high density in deep far-detuned optical lattices. *Physical Review A*, **59**(1), 19–22.

Wolke, M., Klinner, J., Kessler, H., and Hemmerich, A. (2012, July). Cavity cooling below the recoil limit. *Science*, **337**(6090), 75–8.

Xuereb, A., Domokos, P., Horak, P., and Freegarde, T. (2010, September). Scattering theory of multilevel atoms interacting with arbitrary radiation fields. *Physica Scripta*, **T140**(T140), 014010.

Yeo, I., de Assis, P.-L., Gloppe, A., Dupont-Ferrier, E., Verlot, P., Malik, N. S., Dupuy, E., Claudon, J., Gérard, J.-M., Auffeves, A., Nogues, G., Seidelin, S., Poizat, J-Ph., Arcizet, O., and Richard, M. (2013, December). Strain-mediated coupling in a quantum dot–mechanical oscillator hybrid system. *Nature Nanotechnology*, **9**(2), 106–10.

10

Optically Levitated Nanospheres for Cavity Quantum Optomechanics

Oriol Romero-Isart

Institute for Quantum Optics and Quantum Information of the
Austrian Academy of Sciences, and
Institute for Theoretical Physics, University of Innsbruck,
Austria

Oriol Romero-Isart

Oriol Romero-Isart, *Optically Levitated Nanospheres for Cavity Quantum Optomechanics* In: *Quantum Optomechanics and Nanomechanics.* Edited by: Pierre-Francois Cohadon, Jack Harris, Florian Marquardt, Leticia F. Cugliandolo, Oxford University Press (2020). © Oxford University Press. DOI: 10.1093/oso/9780198828143.003.0010

Chapter Contents

This is a transcript[1] of the lectures (3×90 minutes) that took place in Les Houches during the *Summer School: Quantum Optomechanics and Nanomechanics* (August, 2015). I was asked to give lectures on cavity quantum optomechanics with levitated nanospheres with some emphasis on preparing mesoscopic quantum superpositions and testing collapse models. I divided the lectures into three parts:

1. **Levitated quantum optomechanics: atoms vs. sphere**. It is shown how the master equation describing the dynamics of a polarizable object in a cavity along the cavity axis and that of the cavity mode is derived. The master equation describes the optomechanical coupling as well as the recoil heating due to light scattering. Optical levitation is also discussed. While some details are skipped, this provides an overview of how this derivation can be done starting from first principles for a general polarizable object with a given polarizability. This general derivation is used to compare the case of a subwavelength dielectric nanosphere with a far-detuned two-level atom.

2. **Decoherence in levitated nanospheres**. Most of the decoherence sources in levitated nanospheres can be cast into a relatively simple master equation describing position localization type of decoherence. Such decoherence tends to suppress the centre-of-mass position coherences. After a general discussion of position localization decoherence, several specific physical mechanisms that produce such decoherence are detailed: scattering of light, emission, scattering and absorption of black-body radiation, scattering of air molecules, and stochastic classical linear forces as induced by fluctuations of the trap centre. Most of the known collapse models can also be described as a position localization type of decoherence and hence one can easily analyse the requirements to falsify them in the presence of standard sources of decoherence. Throughout the discussion some typical numbers are given such that one can get a feeling for how strong these effects are.

3. **Wave-packet dynamics: coherence vs. decoherence**. With the motivation of using levitated nanospheres for matter-wave interferometry, that is, to create macroscopic quantum superpositions for testing quantum mechanics in unprecedented parameter regimes, a discussion of wave-packet dynamics is given. In particular, it is analysed how to describe the dynamics triggered by releasing a cold nanoparticle from the trap by taking into account standard sources of decoherence. This is a good application of the discussion given in the second part of the lectures. This can also be used to get a good feeling about the strength of standard sources of decoherence and how challenging but exciting it is to prepare a large quantum delocalization of a nanosphere.

---

[1] I am very grateful to Cosimo C. Rusconi, Kanupriya Sinha, and B. Prasanna Venkatesh for carefully reading and improving these lecture notes.

As a disclaimer I would like to emphasize that these lectures should not be considered a review article of levitated nanospheres for quantum nanomechanics. Also, I didn't put emphasis on rigorously supporting all the statements given and providing all the formulas with the exact pre-factors; instead, my goal was to give the students an overview of levitating nanospheres with a grasp on how things can be calculated and analysed carefully. Also, I'm not using references as in regular articles, that is references in theses notes are not aimed at giving scientific credit but instead are used to help students find places where more details are given.

Before starting let me thank very much the organizers, Pierre-François, Florian and Jack, for giving me the opportunity to give these lectures in such a wonderful summer school. I also want to warmly thank the students attending the school for their enthusiasm, motivation and great spirit. It was really a pleasure!

10.1 Levitated Quantum Optomechanics: Atom vs. Sphere

10.1.1 Introduction

Let us start with a brief motivation of levitating nano- and microspheres[2] in high vacuum. In my opinion levitating spheres have applications in different contexts:

- In **cavity quantum optomechanics**, the topic of this school. By optically levitating a dielectric subwavelength nanosphere in a high-finesse optical cavity, a cavity frequency shift is induced depending on the centre-of-mass position of the sphere (as we will discuss later). Thus an optomechanical coupling is obtained. Such an optomechanical system has some distinctive features:

  - Due to levitation, there are **no clamping losses** in the mechanical mode.
  - The mechanical centre-of-mass motion is **decoupled from internal defects**. The intuition is that internal defects couple to mechanical modes via strain, however, these internal mechanical modes are decoupled from the centre-of-mass mode due to a huge mismatch of mechanical frequencies. The internal mechanical modes have a frequency of the order of the speed of sound divided by the size of the nanosphere, which results in the 10^{11} Hz regime, while the centre-of-mass motion will typically have a frequency of 10^5 Hz.
  - The above two features should allow for **ground state cooling from room temperature** at high vacuum. This has not been experimentally demonstrated but some research groups might hopefully achieve this goal relatively soon.

---

[2] Let me remark that there is always a bit of confusion and arbitrariness in either using the prefix nano- or micro-. Typically I will use the term nanosphere for particles with a radius smaller than ~ 500 nanometres, and microsphere for radius larger than ~ 500 nm.

- The **mechanical frequency can be modulated**. This apparently irrelevant feature turns out to be quite useful. In particular, after cooling the mechanical mode one can reduce the mechanical frequency to expand the wave function, either dynamically (free expansion) or adiabatically. This makes a nice link between quantum nanomechanics and matter-wave interferometry, as we discuss below. One could also use a periodic modulation of the frequency to create mechanical squeezing.

- In **matter-wave interferometry**, namely in the possibility to observe quantum interference of a massive object, as in the beautiful and seminal experiments of the group of Markus Arndt in Vienna where interference of organic molecules up to 10^4 amu have been reported. We think that quantum nanomechanics with levitated nano- and microspheres can be of interest in this context because of the following two reasons:

  - Cavity cooling the centre-of-mass motion allows preparation of **initial pure states** for the matter-wave interferometer. Purification of the centre-of-mass thermal state of the objects to be interfered is typically done by strongly collimating the beam of molecules. For more massive objects, such as nano- and microspheres, actively cooling the centre-of-mass motion would be very useful as a pre-stage before entering into the matter-wave interferometer.
  - In quantum nanomechanics we have learned how to coherently control the quantum mechanical state of massive objects. For matter-wave interferometry these tools could be exploited for **quantum control of matter waves**. For instance, after expanding the wave function one requires a non-Gaussian operation to create a superposition state (which has a negative Wigner function); this could be done by using cavity quantum nanomechanics, for instance, by performing a continuous-time quantum measurement of the squared position. Such a *quantum* double slit could provide a better control of decoherence in this crucial step.

- In **sensing**, which is a typical application of quantum nanomechanics. We find it particularly interesting to exploit the distinctive feature of nanomechanics, the large mass, to sense gravity. In our opinion, levitating nanospheres can offer the following advantageous features in this context:

  - A nanosphere has a mass comparable to a gas of ultracold atoms; however, there is a very important difference between a cold nanosphere and a BEC of neutral atoms: the nanosphere has a **high mass density**. This can be particularly interesting for short-distance force detectors, such as measurements of Newton's law at short distances, where theories beyond the standard model predict corrections, or to measure Casimir forces. The size of a nanosphere is of a few tens of nanometres whereas the same mass in a BEC occupies several tens of micrometres.

- In experiments aimed at measuring G (the gravitational constant), inhomogeneities in the mass density of the test masses, where for instance little holes are present, are very relevant. Nanospheres should more easily have an **homogeneous mass density**, which could be useful for measuring G. Actually in this context the nanosphere can often be approximated by a point-particle of mass $m = \rho V$.

- In quantum metrology it is well known that highly entangled states, for instance NOON states $|0\rangle^{\otimes N} + |1\rangle^{\otimes N}$, can enhance the sensitivity of the measurement. A quantum superposition of the centre-of-mass position of a massive object containing N atoms can be considered as a **NOON state for measuring gravity**. One has to be nevertheless very careful with this since the analysis of sensitivity has to account for decoherence, which typically increases for NOON states and thus the overall gain might be compromised.

- In **nanophysics**, since having a nanosphere in high vacuum so well isolated from the environment that its centre-of-mass can be brought to the quantum regime seems an ideal lab for studying the nanophysics of such an object. In particular, optically levitated nanospheres are already being used as a nice tool to study thermodynamics and statistical physics at the nanoscale. One could also consider levitating other type of nanospheres (nanodiamonds, nanomagnets, superconducting microspheres, etc.) and study their properties in free space, without the noise coming from the substrate.

- In **quantum foundations**, in particular in testing quantum mechanics at unprecedented large length and mass scales. In our opinion levitated nanospheres offer an ideal platform to prepare 'macroscopic' quantum superpositions of massive objects, where by 'macroscopic' I mean the centre-of-mass being delocalized over distances comparable to the radius of the nanosphere. Levitated nanospheres offer a bottom-up approach where with the same tools applied for preparing superpositions of atoms and molecules, one can increase the scales to test quantum mechanics at unprecedented length scales where collapse models conjecture a breakdown of the quantum superposition principle.

Apart from these applications one should also have the motivation given by the broader perspective attained when looking back at the history of quantum optics. Namely, the main goal should always be to first fully understand and control the system in consideration, in our case levitated nanospheres, in the quantum regime. Once this is achieved, many more unforeseen applications should arise, as has always happened with the plethora of systems in nature that we have learned how to control in the quantum regime. In the lectures I will not discuss anymore the applications of levitating nanospheres but rather I will aim at providing an introduction to the underlying physics with particular emphasis on cavity optomechanics with optically levitated dielectric nanospheres and sources of decoherence.

10.1.2 Optical Dipole Force

Let us start with considering the optical dipole force acting on a **polarizable** object due to a spatial dependence of the light intensity. Consider an electromagnetic field of frequency ω such that the time-averaged squared electric field is position-dependent, namely $E^2(\mathbf{r}) \equiv 2\langle |\mathbf{E}(\mathbf{r}, t)|^2 \rangle$, where $\langle \cdot \rangle$ denotes a time average. If such an electromagnetic field interacts with an object, assumed much smaller than the relevant wavelength of the electromagnetic field, then it exerts an **optical dipole force** given by

$$\mathbf{F} \equiv \frac{\alpha}{4} \nabla \left[E^2(\mathbf{r}) \right] \tag{10.1}$$

where $\alpha \in \mathbb{R}$ is the real part of the **polarizability**. The polarizability is a parameter that will be crucial for cavity optomechanics with levitated polarizable objects.

For a dielectric nanosphere of radius R in a sufficiently low light intensity such that its response is linear one can show that (see for instance section 4.4 of Jackson's book on Classical Electrodynamics) the polarizability, as defined here, is given by

$$\alpha_s = 3\epsilon_0 V \frac{n^2 - 1}{n^2 + 2} \tag{10.2}$$

where $V = 4\pi R^3 / 3$ and n is the refractive index at the light frequency ω. Note that the polarizability **scales with the volume** of the nanosphere provided $R \ll 2\pi c / \omega$ and is **always positive**. Thus the force always points in the direction in which the light intensity grows. This force is used to trap dielectric objects with a highly focused laser light, as in the so-called optical tweezers, see Fig. 10.1.

A neutral atom is also a polarizable object and thus one should obtain an expression for the dipole force Eq. (10.1.2) for a given polarizability α_q. In particular, consider a two-level system $\{|g\rangle, |e\rangle\}$ with transition frequency ω_0, hence detuning $\Delta = \omega - \omega_0$, and spontaneous emission rate given by

$$\Gamma = \frac{d^2 \omega_0^3}{3\epsilon_0 \pi \hbar c^3}. \tag{10.3}$$

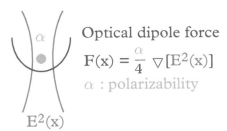

Optical dipole force

$$F(x) = \frac{\alpha}{4} \nabla [E^2(x)]$$

α : polarizability

$E^2(x)$

Fig. 10.1 *Optical dipole potential for a polarizable object.*

We assume here an isotropic atom with a dipole moment d. Recall that in the optical Bloch equations (see Chapter 5 of (Steck)), our definition of Γ corresponds to $\partial_t \langle |e\rangle\langle e|\rangle = -\Gamma\langle |e\rangle\langle e|\rangle + \dots$. The dephasing is given by $\partial_t \langle |e\rangle\langle g|\rangle = -\gamma\langle |e\rangle\langle g|\rangle + \dots$ with $\gamma = \Gamma/2 + \gamma_c$, where γ_c accounts for any inhomogeneous broadening. We also define the Rabi frequency $\Omega = 2\langle|\vec{d}\cdot\mathbf{E}(\mathbf{r},t)|\rangle/\hbar = \sqrt{2/3}dE(\mathbf{r})/\hbar$ that quantifies the coupling between the light and the neutral atom. We can now define a very important parameter in quantum optics, the saturation parameter given by

$$s \equiv \frac{\Omega^2}{\gamma\Gamma}\frac{1}{1+\Delta^2/\gamma^2}. \tag{10.4}$$

Recall that for instance, if we drive a two-level atom the population of the excited state in the steady state is given by $\langle |e\rangle\langle e|\rangle(t\to\infty) = s/[2(s+1)] \leq 1/2$. The expression of the saturation parameter Eq. (10.4) and the free-space spontaneous emission rate Eq. (10.3) are very much worth remembering! Using the rotating wave approximation, valid provided $|\Delta| \ll \omega_0$, and the Born–Oppenheimer approximation, valid provided the motion of the atom is much slower than the timescale required for the internal state to reach the steady state at a given position, then one can show that the optical dipole force is given by Eq. (10.1.2) with the polarizability given by

$$\alpha_q = -\frac{2\Delta d^2\Gamma}{3\hbar\Omega^2\gamma}\frac{s}{1+s}. \tag{10.5}$$

For details on how to obtain this expression I recommend reading Section 5.8 of (Steck). Note that the polarizability α_q depends on the light intensity via Ω which is also present in the saturation parameter. The polarizability can be **either positive or negative** depending on the sign of the detuning Δ. For $\omega_0 > \omega$ (red detuned), $\Delta < 0$, and thus the force points to the direction in which the intensity grows, as for a dielectric nanosphere. However the opposite case $\omega_0 < \omega$ (blue detuned) pushes the atom in the direction where the intensity decreases. As said before, in the general case the polarizability depends on the intensity and thus the polarization does not linearly depend on the external field. The linear case (polarizability independent of the light intensity) is obtained in the far off-resonance case $|\Delta| \gg \gamma$. To simplify the expression we will further assume $\gamma_c = 0$ such that $\gamma = \Gamma/2$. In this regime the saturation parameter is given by $s \approx \Omega^2/(2\Delta^2) \ll 1$, and thus

$$\alpha_q \approx -\frac{2d^2}{3\hbar\Delta} = -\frac{\epsilon_0}{(2\pi)^2}\frac{\Gamma}{\Delta}\lambda_0^3 \tag{10.6}$$

The second equality is obtained using Eq. (10.3) and defining $\lambda_0 = 2\pi c/\omega_0$. Note that indeed the polarizability does not depend on the light intensity, as in the case of the dielectric nanosphere. The formula Eq. (10.6) is also very much worth remembering! It tells us how a neutral atom responds to far off-resonant light, a response that is in some sense classical since the polarizability does not depend on $\hbar$.

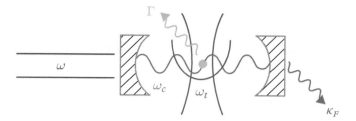

Fig. 10.2 *Cavity quantum optomechanics with a levitated polarizable object. The relevant degrees of freedom are: the mechanical mode along the cavity axis, achieved by optically confining the polarizable object in a harmonic trap of frequency ω_t, and the cavity optical mode of frequency ω_c. The mechanical mode can decohere with a rate Γ and the optical photons can leak out of the cavity via the imperfect mirrors with a rate κ_F. The cavity can be driven with a laser light of frequency ω.*

It is actually very interesting to calculate the ratio between the polarizability of a far off-resonance two-level neutral atom and a dielectric nanosphere, namely

$$\frac{|\alpha_q|}{\alpha_s} = \left(\frac{\lambda_0}{R}\right)^3 \frac{\Gamma}{|\Delta|} \frac{3}{(4\pi)^3} \frac{n^2 + 1}{n^2 - 1}. \tag{10.7}$$

This ratio is surprisingly large considering that a nanosphere has many atoms! In particular, considering $n = 2$, $|\Delta| = 100\Gamma$, $\lambda_0 = 20R$, then $|\alpha_q|/\alpha_s \sim 0.2$! The reason is that the polarizability per atom in a dielectric nanosphere without any optical resonance is much smaller than the polarizability of a two-level atom with an optical transition. Roughly speaking, a two-level optical transition interacting with optical light can be considered as a sphere of a radius of the order of λ_0! This is actually one of the reasons why you have learned in Philipp Treutlein's lectures that neutral atoms can be strongly coupled to optomechanical systems.

To conclude this subsection we note that in the case when the polarizability does not depend on the light intensity, the optical dipole force $\mathbf{F} = \alpha \nabla E^2/4$ can be easily integrated and the interaction energy is given by

$$H_{\text{int}} = -\frac{\alpha}{4} [E(\mathbf{r})]^2. \tag{10.8}$$

10.1.3   Dispersive Cavity Quantum Optomechanics

Let us now consider an optical cavity mode with frequency ω_c and creation operator $\hat{a}$. We assume that a polarizable object of mass m is placed inside the cavity and can move along the optical axis, assumed to be the x-axis. By tracing out the output cavity modes, the dynamics of the cavity mode, the position along the x-axis and the free (i.e. non-cavity) electromagnetic modes is given by

$$\dot{\rho} = \frac{1}{i\hbar}[\hat{H}, \hat{\rho}] + \kappa_F(\hat{a}\hat{\rho}\hat{a}^\dagger - \frac{1}{2}[\hat{a}^\dagger\hat{a}, \hat{\rho}]_+), \tag{10.9}$$

where κ_F is the decay rate of the cavity due to a finite finesse of the mirrors, and the Hamiltonian is given by

$$\hat{H} = \hbar\omega_c\hat{a}^\dagger\hat{a} + \frac{\hat{p}^2}{2m} + \hat{H}_{\text{free}} + \hat{H}_{\text{int}}. \tag{10.10}$$

The first term is the energy of the cavity mode, the second the kinetic energy of the polarizable object of mass m along the optical axis, and

$$\hat{H}_{\text{free}} = \hbar \int d^3k \sum_\epsilon \omega_k \hat{a}^\dagger_\epsilon(\mathbf{k})\hat{a}_\epsilon(\mathbf{k}) \tag{10.11}$$

accounts for the energy of the electromagnetic field modes in free space. The term $\hat{H}_{\text{int}}$ describes the dipole interaction between the polarizable object and the electromagnetic field modes and is given by

$$\hat{H}_{\text{int}} = -\frac{\alpha}{4}[\hat{\mathbf{E}}(\hat{x})]^2. \tag{10.12}$$

This expression can be derived in a more rigorous way; here, we just use the discussion of the previous subsection, see Eq. (10.8). Note that $\hat{\mathbf{E}}(\hat{x})$ has, in general, the following contributions

$$\hat{\mathbf{E}}(\hat{x}) = \hat{\mathbf{E}}_{\text{cav}}(\hat{x}) + \hat{\mathbf{E}}_{\text{free}}(\hat{x}) + \mathbf{E}_{\text{class}}(\hat{x}). \tag{10.13}$$

The electromagnetic field operator of the cavity mode is given by

$$\hat{\mathbf{E}}_{\text{cav}}(\hat{x}) = iE_0\vec{\epsilon}\sin\left(\frac{\omega_c}{c}\hat{x} + \phi\right)\left(\hat{a}^\dagger - \hat{a}\right), \tag{10.14}$$

where the zero-point electric field is given by $E_0 \equiv \sqrt{\hbar\omega_c/(V_c\epsilon_0)}$. Here V_c is the cavity mode volume, $\vec{\epsilon}$ the polarization of the cavity mode, and ϕ parameterizes the position of the sphere with respect to the cavity mode. The Gaussian spatial dependence of the cavity mode along the y- and z-axes is neglected. The electromagnetic field operator of the free modes is given by

$$\hat{\mathbf{E}}_{\text{free}}(\hat{x}) = \int d^3k \sum_\epsilon i\sqrt{\frac{\hbar\omega_k}{2\epsilon_0(2\pi)^3}}\left[\hat{a}_\epsilon(\mathbf{k})e^{-i\mathbf{k}\cdot\hat{\mathbf{r}}} - \hat{a}^\dagger_\epsilon(\mathbf{k})e^{i\mathbf{k}\cdot\hat{\mathbf{r}}}\right]. \tag{10.15}$$

Here $\hat{\mathbf{r}} = (\hat{x}, y, z)$ (we only consider x to be the quantum degree of freedom). Note that by considering all the free modes we are double counting the cavity mode, which is a linear combination of free modes. This has a negligible effect since a single mode has zero measure. Finally $\mathbf{E}_{\text{class}}(\hat{x})$ accounts for any external classical field, for instance it could be the optical tweezers used for optical trapping which would be given by

$$\mathbf{E}_{\text{class}}(\hat{x}) = \vec{\epsilon}_{\text{tw}} E_{\text{tw}} \exp\left[-\frac{\hat{x}^2 + y^2}{2W^2}\right], \tag{10.16}$$

where W is the laser waist. Here we neglected the z dependence of the optical tweezers. By plugging Eq. (10.13) into Eq. (10.12) we obtain the following terms in the Hamiltonian:

- A term of the form $[\mathbf{E}_{\text{class}}(\hat{x})]^2$ which leads to **optical trapping**. In particular by Taylor expanding around $\hat{x} = 0$ at $y = 0$ and $z = 0$, one obtains that

$$-\frac{\alpha}{4}\left[\mathbf{E}_{\text{class}}(\hat{x})\right]^2 = -\frac{\alpha}{4}E_{\text{tw}}^2 e^{-\hat{x}^2/W^2} \approx -\frac{\alpha}{4}E_{\text{tw}}^2 + \frac{\alpha}{4}\frac{E_{\text{tw}}^2}{W^2}\hat{x}^2. \tag{10.17}$$

The second term is a harmonic potential $m\omega_t^2\hat{x}^2/2$ with trap frequency given by

$$\omega_t \equiv \sqrt{\frac{\alpha}{2m}\frac{E_{\text{tw}}^2}{W^2}} \tag{10.18}$$

Note that since for a nanosphere $\alpha \propto V$, the trap frequency does not depend on the volume of the sphere provided it is smaller than the optical wavelength. The trap frequency is proportional to the square root of the intensity of the optical tweezers and it is larger the smaller the waist, since then larger intensity field gradients are achieved. The motion along the x-axis is then confined in a harmonic potential where we define the mechanical phonon mode of the centre-of-mass motion via $\hat{x} = x_{\text{zp}}(\hat{b} + \hat{b}^\dagger)$, with $x_{\text{zp}} = [\hbar/(2m\omega_t)]^{1/2}$.

- A term of the form $[\hat{\mathbf{E}}_{\text{cav}}(\hat{x})]^2$ which leads to the **optomechanical coupling**. In particular, for $\phi = \pi/4$ such that the polarizable object is placed at the maximum of the cavity-field-mode intensity gradient, one obtains

$$-\frac{\alpha}{4}\left[\hat{\mathbf{E}}_{\text{cav}}(\hat{x})\right]^2 = \frac{\alpha E_0^2}{4}\sin^2\left(\frac{\omega_c}{c}\hat{x} + \frac{\pi}{4}\right)\left(\hat{a} - \hat{a}^\dagger\right)^2. \tag{10.19}$$

Expanding around $\hat{x} = 0$ and dropping the fast rotating terms, one has

$$-\frac{\alpha E_0^2}{2}\sin^2\left(\frac{\omega_c}{c}\hat{x} + \frac{\pi}{4}\right)\hat{a}^\dagger\hat{a} \approx -\frac{\alpha E_0^2}{4}\hat{a}^\dagger\hat{a} - \frac{\alpha E_0^2}{2}\frac{\omega_c}{c}x_{\text{zp}}\hat{a}^\dagger\hat{a}\left(\hat{b} + \hat{b}^\dagger\right). \tag{10.20}$$

Thus, one can identify the single-photon cavity optomechanical coupling $\hbar g_0 \hat{a}^\dagger \hat{a} (\hat{b}^\dagger + \hat{b})$ with

$$g_0 \equiv \frac{\alpha E_0^2}{2\hbar} \frac{\omega_c}{c} x_{zp} \qquad (10.21)$$

Note that g_0 is proportional to the polarizability α and is larger the smaller the cavity since $E_0^2 \propto 1/V_c$. This coupling is called dispersive as the presence of the polarizable object in the cavity shifts the resonance frequency. This shift depends on the position of the object, since the cavity mode intensity does, and hence leads to a dispersive optomechanical cavity. The expression Eq. (10.21) can be used for either a dielectric nanosphere or for a far-detuned two-level atom.

- A term of the form $\hat{\mathbf{E}}_{\text{free}}(\hat{x}) \cdot \left[\hat{\mathbf{E}}_{\text{cav}}(\hat{x}) + \mathbf{E}_{\text{class}}(\hat{x}) \right]$ which leads to **light scattering** effects, in particular photon losses and recoil heating. The free electromagnetic field modes can be traced out and a Born–Markov master equation can be derived. In this derivation the Lamb–Dicke approximation $x_{zp}\omega_c/c \ll 1$ is assumed. For details on this derivation, see for instance (Romero-Isart, Pflanzer, Juan, Quidant, Kiesel, Aspelmeyer and Cirac, 2011). The term $\hat{\mathbf{E}}_{\text{free}}(\hat{x}) \cdot \hat{\mathbf{E}}_{\text{cav}}(\hat{x})$ (at $\hat{x} = 0$) couples the free modes with the cavity mode due to the presence of the polarizable object, and will lead to **photon losses** given by

$$R_{\text{sc}} \left(\hat{a}\hat{\rho}\hat{a}^\dagger - \frac{1}{2} \left[\hat{a}^\dagger \hat{a}, \hat{\rho} \right]_+ \right). \qquad (10.22)$$

The term $\hat{\mathbf{E}}_{\text{free}}(\hat{x}) \cdot \mathbf{E}_{\text{class}}(\hat{x})$ couples the motion with the free modes and leads to **recoil heating** given by

$$\Gamma_{\text{sc}} \left\{ \left(\hat{b} + \hat{b}^\dagger \right) \hat{\rho} \left(\hat{b} + \hat{b}^\dagger \right) - \frac{1}{2} \left[\left(\hat{b} + \hat{b}^\dagger \right)^2, \hat{\rho} \right]_+ \right\}. \qquad (10.23)$$

One can show that the scattering rate is given by

$$R_{\text{sc}} \sim \alpha^2 E_0^2 \left(\frac{\omega_c}{c} \right)^3 \frac{1}{\epsilon_0 \hbar} \qquad (10.24)$$

and the recoil heating rate is given by

$$\Gamma_{\text{sc}} \sim \left(\frac{\omega_c}{c} x_{zp} \right)^2 R_{\text{sc}} \left(\frac{E_{\text{tw}}}{E_0} \right)^2 \sim \alpha^2 \left(\frac{\omega_c}{c} \right)^5 \frac{x_{zp}^2 E_{\text{tw}}^2}{\epsilon_0 \hbar}. \qquad (10.25)$$

Note that both the scattering rate and the recoil heating rate depend on α^2, as opposed to the optomechanical coupling that depends on α.

As a consistency check recall that the scattering rate of a two-level system is given by $R_{\text{sc}} = \Gamma \langle |e\rangle \langle e| \rangle_{\text{ss}} = \Gamma s/[2(s+1)]$. For a far-detuned driving one has that

$s \ll 1$ and $s \approx \Omega^2/(2\Delta^2)$. By recalling that $\Omega^2 \sim d^2 E_0^2/\hbar^2$ and $\Gamma \sim d^2\omega_0^3/(\epsilon_0\hbar c^3)$, one obtains

$$R_{sc} \sim \frac{\Gamma\Omega^2}{4\Delta^2} \sim \underbrace{\left(\frac{d^2}{\hbar\Delta}\right)^2}_{\alpha^2} E_0^2 \left(\frac{\omega_0}{c}\right)^3 \frac{1}{\epsilon_0\hbar}, \tag{10.26}$$

in agreement with Eq. (10.24).

- A term of the form $[\hat{\mathbf{E}}_{\text{free}}(\hat{x})]^2$ which induces a coupling between the free electromagnetic field modes due to the presence of the polarizable object. This would lead to corrections to g_0 which are negligible for small polarizable objects. This term is however important as soon as a nanosphere is comparable to λ, see (Pflanzer, Romero-Isart and Cirac, 2012) for details.

- A term of the form $\hat{\mathbf{E}}_{\text{cav}}(\hat{x}) \cdot \mathbf{E}_{\text{class}}(\hat{x})$, which leads to a shift in the trapping frequency and the equilibrium position of the polarizable object.

Note that starting from the Hamiltonian describing the interaction between a polarizable object and an electromagnetic field, given by Eq. (10.12), one can obtain the trapping frequency ω_t, the single-photon optomechanical coupling g_0, the scattering rate R_{sc}, and the recoil heating rate Γ_{sc} as a function of the polarizability α. These expressions can be used either for a dielectric nanosphere or for a far-detuned two-level neutral atom.

Cooperativity

In cavity quantum optomechanics it is useful to consider the single photon cooperativity defined by

$$\mathcal{C}_0 = \frac{g_0^2}{\kappa\Gamma}, \tag{10.27}$$

where κ is the cavity mode decay rate and Γ the decoherence rate of the mechanical mode. The cooperativity quantifies how much coherent the single-photon optomechanical coupling is. For a levitated polarizable object we have that $\kappa = \kappa_F + R_{sc}$, where κ_F is the contribution from the imperfect reflectivity of the mirrors and R_{sc} corresponds to the scattering of photons off the cavity due to the presence of the polarizable object, and $\Gamma = \Gamma_{sc} + \Gamma_{\text{other}}$, where apart from the recoil heating, Γ_{other} accounts for other potential sources of decoherence in the mechanical mode, as for instance those discussed in the next section. In the particular case of having mechanical decoherence only due to recoil heating, namely for $\Gamma_{\text{other}} = 0$, there are two important regimes:

- When the cavity mode decay rate is dominated by scattering of photons, namely the quality of the cavity mirrors is so good that $R_{sc} \gg \kappa_F$. Then one has that $\kappa \approx R_{sc}$ and thus

$$\mathcal{C}_0 = \frac{g_0^2}{R_{sc}\Gamma_{sc}} \sim \frac{1}{\alpha^2}\epsilon_0^2\lambda_c^6\left(\frac{E_0}{E_{tw}}\right)^2. \tag{10.28}$$

Note that $\mathcal{C}_0$ decreases with the polarizability of the particle as $\mathcal{C}_0 \sim \alpha^{-2}$. For a dielectric nanosphere sphere and a neutral two-level atom one has

$$\mathcal{C}_0^{sphere} \sim \left(\frac{\lambda_c}{R}\right)^6 \quad \mathcal{C}_0^{atom} \sim \left(\frac{\Delta}{\Gamma}\right)^2, \tag{10.29}$$

where $\lambda_c = 2\pi c/\omega_c$. In this regime the smaller the sphere (or the more far-detuned the neutral atom), the smaller the polarizability, and thus the larger the optomechanical cooperativity.

- When the cavity mode decay rate is dominated by photon losses at the mirrors, namely when $R_{sc} \ll \kappa_F$ such that $\kappa \approx \kappa_F$. Then,

$$\mathcal{C}_0 = \frac{g_0^2}{\kappa_F\Gamma_{sc}} \sim \frac{E_0^2}{E_{tw}^4}\frac{\epsilon_0}{\hbar}\frac{\lambda_c^3}{\kappa_F} \tag{10.30}$$

Note that $\mathcal{C}_0$ is **independent of the polarizability**! Thus the optomechanical cooperativity is the same for a dielectric nanosphere and for a two-level neutral atom. This regime is actually the typical one in current experiments with optical levitation of dielectric nanospheres in high-finesse optical cavities.

In cavity optomechanics we typically drive the cavity and linearize the optomechanical coupling such that we have an enhanced linear coupling of the form $\sqrt{n_{ph}}g_0(\hat{a}^\dagger + \hat{a})(\hat{b}^\dagger + \hat{b})$, namely the optomechanical coupling is enhanced by the square root of the steady-state number of photons in the cavity, $\sqrt{n_{ph}}$. While the probability of losing a cavity photon is still the same, that is the R_{sc} contribution to κ is not modified, the recoil heating due to the scattering of the driving field is increased as $n_{ph}\Gamma_{sc}$. Assuming that the recoil heating due to the driving field is larger than the recoil heating due to the trapping field, one can note that the cooperativity with the linearized optomechanical coupling does not depend on n_{ph}! This result might seem quite surprising when compared with other standard optomechanical systems. Note, however, that in the regime $\Gamma_{other} \gg n_{ph}\Gamma_{sc}$, that is when mechanical decoherence is not dominated by recoil heating, then the optomechanical cooperativity with the linearized coupling does increase linearly with n_{ph}, as one typically has in standard optomechanical systems. But we emphasize that this happens only while $\Gamma_{other} \gg n_{ph}\Gamma_{sc}$ is fulfilled, that is for sufficiently low driving intensity.

10.2 Decoherence in Levitated Nanospheres

As mentioned in the introduction, levitated nanospheres for cavity quantum nanomechanics have the advantage of not having clamping losses. However, this does not mean that there is no decoherence. Actually we have already seen in the previous section that recoil heating does create decoherence. For matter-wave interferometry it is also very important to assess decoherence effects in free expansion of the centre-of-mass motion. Let us then in this section discuss decoherence in the context of levitated nanopsheres. We will first discuss a relatively simple model to describe the so-called position localization type of decoherence. This type of decoherence captures most of the unavoidable decoherence mechanisms present in levitating nanospheres. We will also put some numbers for some of these unavoidable sources of decoherence such that we can get a feeling and intuition for how strong these effects can be. We will also compare these standard sources of decoherence with the exotic decoherence predicted by collapse models, whose effects can also be modelled by position localization type of decoherence.

10.2.1 Position Localization Type of Decoherence

In levitating nanospheres standard sources of decoherence cause the off-diagonal terms of the density matrix written in the position basis to exponentially decay in time, namely

$$\langle x|\hat{\rho}(t)|x'\rangle \sim e^{-\Gamma t}\langle x|\hat{\rho}(0)|x'\rangle. \tag{10.31}$$

The decoherence rate Γ usually depends on $|x - x'|$. This type of decoherence is usually called **position localization**. This effect is also predicted by **collape models**, as we will discuss later.

This source of decoherence is very well modelled by the following master equation

$$\langle x|\dot{\hat{\rho}}(t)|x'\rangle = \frac{1}{i\hbar}\langle x|[\hat{H},\hat{\rho}(t)]|x'\rangle - \Gamma(x - x')\langle x|\hat{\rho}(t)|x'\rangle, \tag{10.32}$$

with

$$\Gamma(x) = \gamma\left[1 - \exp\left(-\frac{x^2}{4a^2}\right)\right]. \tag{10.33}$$

Here γ, which has units of frequency, and a, which has units of distance, are two parameters modelling the particular sources of decoherence. Note that $\Gamma(x)$ grows quadratically in x and then **saturates** to the value of γ for $x \gg 2a$. This master equation has two important limits:

- The **long wavelength limit,** obtained when $2a$ is much larger than the relevant correlation length scale of the density matrix. In this limit the decay rate is given by:

$$\Gamma(x \ll 2a) \approx \frac{\gamma}{4a^2}x^2 \equiv \Lambda x^2. \tag{10.34}$$

We have introduced the **localization parameter** $\Lambda \equiv \gamma/(4a^2)$, which has dimensions of frequency divided by a squared distance. In this limit, the position localization type of decoherence only depends on a single parameter, the localization parameter Λ, which will be used very often hereafter. It is a very important parameter! Note that in this limit, the decoherence rate of the off-diagonal terms of the density matrices scales quadratically with the distance from the diagonal. Later we provide more details of this important limit.

- The **short wavelength limit**, obtained when $2a$ is much smaller than the relevant correlation length scale of the density matrix. In this limit the decay rate is given by:

$$\Gamma(x \gg 2a) \approx \gamma. \qquad (10.35)$$

In this regime, all the off-diagonal terms of the density matrix in the position basis decay at the same rate.

This type of decoherence arises in a very typical physical situation which provides an intuitive **physical interpretation** to the position localization type of decoherence. Consider particles of wavelength $\lambda \sim 2a$ scattering with the nanosphere with a rate γ, see Fig. 10.3. Each particle scattering off the nanoparticle carries away information about the position of the nanosphere, it gets entangled, and therefore when tracing them out it produces a loss of coherence in the position degree of freedom of the nanosphere. For $2a \gg |x - x'|$ a single scattering event does not completely resolve the position, which implies that the decoherence rate grows with $|x - x'|$; the larger the separation, the easier to resolve it. For $2a \ll |x - x'|$ a single scattering event resolves the position and thus the decoherence rate is constant and given by the scattering rate; each scattering event provides which-path information.

Long wavelength limit: diffusion without friction

Let us now focus again on the **long wavelength limit** Eq. (10.34) of the position localization type of decoherence. The master equation in this limit is of the Lindbland form with the position operator being the jump operator and it can be written as

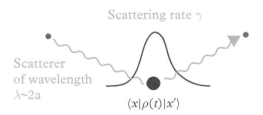

Fig. 10.3 *Physical interpretation of the position localization type of decoherence.*

$$\dot{\rho} = \frac{1}{i\hbar}[\hat{H}, \hat{\rho}] - \Lambda\left[\hat{x}, [\hat{x}, \hat{\rho}(t)]\right] = \frac{1}{i\hbar}[\hat{H}, \hat{\rho}] - 2\Lambda\left(\hat{x}\hat{\rho}\hat{x} - \frac{1}{2}[\hat{x}^2, \hat{\rho}]_+\right) \qquad (10.36)$$

As we said before, this equation implies that the off-diagonal terms decay exponentially in time with a rate that scales quadratically with the distances, namely

$$\langle x|\hat{\rho}(t)|x'\rangle \sim \exp\left[-\Lambda(x - x')^2 t\right]\langle x|\hat{\rho}(0)|x'\rangle. \qquad (10.37)$$

Recall however that decoherence rate saturates to a constant value for $|x - x'| \gg 2a$.

While we will further discuss the dynamics that this master equation produces in section 10.3, let us here comment the following. Let us define $v_x(t) \equiv \langle\hat{x}^2\rangle = \mathrm{tr}\,[\hat{x}^2\hat{\rho}(t)]$, $v_p \equiv \langle\hat{p}^2\rangle = \mathrm{tr}\,[\hat{p}^2\hat{\rho}(t)]$ and $c \equiv \langle\hat{x}\hat{p} + \hat{p}\hat{x}\rangle/2 = \mathrm{tr}\,[(\hat{x}\hat{p} + \hat{p}\hat{x})\hat{\rho}(t)]/2$. Considering the free dynamics case $\hat{H} = \hat{p}^2/(2m)$, one can show that

$$v_x(t) = v_x^{\Lambda=0}(t) + \underbrace{\frac{2\Lambda\hbar^2}{3m^2}t^3}_{\sim t^3!}, \qquad (10.38)$$

$$v_p(t) = v_p^{\Lambda=0}(t) + 2\Lambda\hbar^2 t, \qquad (10.39)$$

$$c(t) = c^{\Lambda=0}(t) + \frac{\Lambda\hbar^2}{m}t^2. \qquad (10.40)$$

We have denoted $v_x^{\Lambda=0}(t)$ (the same for v_p and c) as the evolution due to purely coherent dynamics; see section 10.3 for more details. Note that the position fluctuations diffuse with a rate of $t^{3/2}$ which is a signature of **Brownian motion without friction**. Actually, one can show that $\langle\hat{x}(t)\rangle = \langle\hat{x}(t)\rangle^{\Lambda=0}$ and $\langle\hat{p}(t)\rangle = \langle\hat{p}(t)\rangle^{\Lambda=0}$, that is, there is no friction!

Let us now consider the harmonic potential case $\hat{H} = \hbar\omega_t\hat{b}^\dagger\hat{b}$ with $\hat{x} = x_{\mathrm{zp}}(\hat{b} + \hat{b}^\dagger)$. Assuming that the rotating wave approximation can be performed, namely $\omega_t \gg \Lambda x_{\mathrm{zp}}^2 \equiv \Gamma/2$, then Eq. (10.36) reads

$$\dot{\rho}(t) = -i\omega_t[\hat{b}^\dagger\hat{b}, \hat{\rho}(t)] + \Gamma\left(\hat{b}\hat{\rho}\hat{b}^\dagger - \frac{1}{2}[\hat{b}^\dagger\hat{b}, \hat{\rho}]_+\right) + \Gamma\left(\hat{b}^\dagger\hat{\rho}\hat{b} - \frac{1}{2}[\hat{b}\hat{b}^\dagger, \hat{\rho}]_+\right). \qquad (10.41)$$

Compare this equation with the standard master equation describing the coupling, with rate γ, to an infinite set of harmonic oscillators in a thermal state with $\bar{n} = K_b T/(\hbar\omega)$, that is

$$\dot{\rho} = -i\omega_t[\hat{b}^\dagger\hat{b}, \hat{\rho}] + \gamma(\bar{n} + 1)\left(\hat{b}^\dagger\hat{\rho}\hat{b} - \frac{1}{2}[\hat{b}\hat{b}^\dagger, \hat{\rho}]_+\right) + \gamma\bar{n}\left(\hat{b}^\dagger\hat{\rho}\hat{b} - \frac{1}{2}[\hat{b}\hat{b}^\dagger, \hat{\rho}]_+\right). \qquad (10.42)$$

This is the typical master equation describing clamping losses in an optomechanical system. See the difference between Eq. (10.41) and Eq. (10.42). Note that we recover Eq. (10.41) by making the limit $\bar{n} \to \infty$, $\gamma \to 0$ with $\gamma\bar{n} \to \Gamma$, namely the infinitely small coupling limit to a bath of infinite temperature. The mechanical quality factor associated to Eq. (10.42) is $Q = \omega/\gamma$ and the decoherence rate is $\bar{n}\gamma = K_b T/(\hbar Q)$, as in standard clamped optomechanics. For Eq. (10.41), which is the common scenario with levitated nanospheres, it does not make sense to talk about the Q factor since $\gamma = 0$, instead we should always talk about decoherence rate. One has to be careful about this point! Levitated spheres are in this sense similar to ions where one usually does not mention the mechanical quality factor of a trapped ion but instead the decoherence rate.

10.2.2 Standard Sources of Decoherence in Levitation

Let us now discuss some unavoidable sources of decoherence in levitated nanospheres. In some cases we will not derive the exact expressions (with the particular numerical factors). The exact pre-factors are of course relevant for a quantitative analysis. Instead we will provide some intuition on how to obtain the scalings and therefore how to obtain an order of magnitude of their strength.

Light scattering

When the dielectric nanosphere is illuminated with light, for instance, due to optical tweezers, we have already seen in the previous section that light scattering will create decoherence (recoil heating) in the form of Eq. (10.23). Note that the form of this decoherence is exactly of the position localization type of decoherence in the long wavelength limit with a localization parameter given by

$$\Lambda_L \sim \alpha^2 E_{\text{tw}}^2 \left(\frac{\omega_c}{c}\right)^5 \frac{1}{\epsilon_0 \hbar} \tag{10.43}$$

The long wavelength limit is in agreement with the Lamb–Dicke approximation, namely than the zero point motion x_{zp}, which is the length scale of the position coherence, is much smaller than the wavelength of the laser light.

Indeed, given light of frequency ω, the localization parameter will always scale as $\Lambda \sim R_{\text{sc}}(\omega/c)^2$, where R_{sc} is the scattering rate given by $R_{\text{sc}} \sim \alpha^2 E^2 (\omega/c)^3/(\epsilon_0 \hbar)$ for some given light intensity E^2.

Black-body radiation.

Even if laser light is absent, there is always electromagnetic radiation at a given environmental temperature T_e with typical wavelength

$$\lambda_T \sim \frac{\hbar c}{K_b T_e} \sim \frac{1\text{mm}}{T_e[1\text{K}]}. \tag{10.44}$$

Thus for typical temperatures, black-body radiation decoherence can be modelled by the long wavelength limit of position localization type of decoherence. Black-body radiation induces decoherence via three different processes: scattering, absorption and emission of black-body radiation.

Regarding **scattering of black-body radiation,** one can estimate the localization parameter for scattering of black-body radiation similarly to the previous subsection. It will be given by $\Lambda_{BB} \sim (\omega_T/c)^2 R_{sc} \sim \alpha^2 (\omega_T/c)^5 E_T^2/(\epsilon_0 \hbar)$. The relevant thermal frequency is given by $\omega_T \sim K_b T_e/\hbar$ and the relevant thermal field intensity is obtained via $K_b T_e \sim \epsilon_0 E_T^2 (c/\omega_T)^3$. Using this, one already obtains that for a dielectric nanosphere

$$\Lambda_{BB} \sim cR^6 \left(\frac{K_b T_e}{\hbar c}\right)^9 \left(\frac{n^2-1}{n^2+2}\right)^2 \tag{10.45}$$

This expression that we have readily obtained differs from the exact formula actually only with the prefactor $8!8\zeta(9)/(9\pi) \sim 10^4$; see for instance arXiv:1110.4495 (Romero-Isart, 2011) and references therein. Note that the localization parameter strongly depends on the environmental temperature as T_e^9! Note also the strong dependence on the size of the nanosphere R^6, which is a trace of the Rayleigh scattering cross-section. When the nanosphere is trapped, the decoherence rate due to black-body scattering is given by $\Gamma_{BB} \sim x_{zp}^2 \Lambda_{BB}$. We can now plug in some numbers by recalling that in SI units, $x_{zp} \sim 10^{-12}$, $c \sim 10^8$, $R \sim 10^{-7}$, $K_b \sim 10^{-23}$, $\hbar \sim 10^{-34}$, and $n \sim 2$, then $\Gamma_{BB} \sim 10^{-27}(T_e[\mathrm{K}])^9 \mathrm{Hz}$. Thus, for a sphere of the order of 100 nm, at room temperature, and confined in a trap, decoherence due to scattering of black-body radiation is negligible. However, note that this is not the case if either the sphere is much larger and/or its wavefunction is much more expanded than x_{zp} as happens in free expansion for matter-wave interferometry purposes; see section 10.3 for further discussion.

Absorption and emission of black-body radiation also creates decoherence that can be modelled by the long wavelength limit of the position localization type of decoherence. However, this is related to the imaginary part of the dielectric constant ϵ at the relevant frequency which accounts for inelastic processes. Here we directly provide expression for the localization parameter

$$\Lambda_{e(a)} = \frac{16\pi^5}{189} cR^3 \left(\frac{K_b T_{i(e)}}{\hbar c}\right)^6 \mathrm{Im}\left[\frac{\epsilon-1}{\epsilon+2}\right] \tag{10.46}$$

The subindex e (a) stands for emission (absorption) and $T_{i(e)}$ for the internal (environmental) temperature of the nanosphere. This expression can be readily obtained for the emission and abosprtion rate of a dielectric nanosphere; see for instance the supplementary material of arXiv:1312.0500 (Bateman, Nimmrichter, Hornberger and Ulbricht, 2014). Since the internal temperature of the nanosphere is typically larger than that of the environment due to light absorption, decoherence due to emission of black-body radiation dominates over absorption. Note that the decoherence rate will strongly depend on the imaginary part of the dielectric constant as well as the internal temperature of the nanosphere, with a large dependence T_i^6. Note that the decoherence due to emission of black-body radiation depends on the volume of the sphere. This is due to the fact the sphere is smaller than the relevant thermal wavelengths, otherwise one would expect the standard dependence on the surface area.

Scattering of air molecules

In thermal equilibrium, the kinetic energy of air molecules has an energy of the order of $p^2/(2m_a) \sim K_b T_e$. Thus the de Broglie wavelength of air molecules is of the order of

$$\lambda_{\text{air}} \sim \frac{h}{p} \sim \frac{h}{\sqrt{2m_a K_b T_e}} \underset{m_a \sim 30\text{amu}}{\sim} \frac{0.15\,\text{nm}}{\sqrt{T_e[1\text{K}]}}. \tag{10.47}$$

Hence, in matter-wave interferometry scattering of air molecules will rapidly resolve the position and thus one is typically within the short wavelength limit where $\langle x|\hat{\rho}(t)|x'\rangle \sim e^{-\gamma_{\text{air}}t}\langle x|\hat{\rho}(0)|x'\rangle$. The saturated decoherence rate is given by

$$\gamma_{\text{air}} = \frac{16\sqrt{2\pi}}{\sqrt{3}} \frac{PR^2}{m_a \bar{v}} \tag{10.48}$$

where $\bar{v} \sim \sqrt{K_b T_e 2m_a}/m_a$. This rate (up to numerical factors) can be easily derived by realizing that the number of scattering events per second is (in the reference frame where the sphere is moving with velocity $\bar{v}$) the volume density of air molecules ρ_N times the volume covered by the sphere in one second, namely $R^2\bar{v}$, therefore

$$\gamma_{\text{air}} \sim \rho_N R^2 \bar{v} \underset{P=\rho_N K_b T_e}{\sim} \frac{P}{K_b T_e} \bar{v} R^2 \underset{K_b T_e \sim m_a \bar{v}^2}{\sim} \frac{PR^2}{m_a \bar{v}} \tag{10.49}$$

Let's put in some numbers, recalling that $1\,\text{mbar} = 10^2\,\text{Pa}$. In SI units, using $R \sim 10^{-7}$, $m_a \sim 10^{-26}$, $T \sim 10^2$, then $\gamma_{\text{air}} \sim 10^{12} P[\text{mbar}]\,\text{Hz}$. This is a significant decoherence rate, even for ultrahigh vacuum. Recall that in a trap, x_{zp} is still smaller than λ_{air} and thus the long wavelength limit should be used to calculate the decoherence rate, which would lead to a smaller decoherence rate. The decoherence rate calculated here applies when the correlation distance is larger than λ_{air}.

At this point one should not confuse γ_{air} with the damping due to air friction which we call η. The scattering of air molecules can be modelled with the Langevin equation

$$\ddot{x} + 2\eta\dot{x} + \omega^2 x = \xi(t), \tag{10.50}$$

with $\langle \xi(t)\xi(t')\rangle = 4K_b T m \eta \delta(t-t')$ for white noise. η can be estimated by computing the variation of linear momentum due to forward and backward scattering events. In the reference frame of the sphere, which is moving with velocity v, air molecules scattering backwards have a velocity $\bar{v} - v$ and those scattering forwards have a velocity $\bar{v} + v$. Thus the variation of momentum is given by

$$\dot{p} \sim m_a[(\bar{v}-v)R^2\rho_N(\bar{v}-v) - (\bar{v}+v)R^2\rho_N(\bar{v}+v)] \sim -m_a R^2 \rho_N \bar{v}v. \tag{10.51}$$

By comparing this result with $-2\eta mv$, we obtain $\eta \sim m_a R^2 \rho_N \bar{v}/m$. Recalling that $\rho_N = P/(K_b T)$ and that $K_b T \sim m_a \bar{v}^2$, then

$$\eta \sim \gamma_{\text{air}} \frac{m_a}{m} \ll \gamma_{\text{air}}. \tag{10.52}$$

Note that in levitated nanospheres, one usually discusses the mechanical quality factor due to air damping, namely $Q = \omega_t/\eta$. However, this should not be confused with the ratio $\omega_t/\gamma_{\text{air}}$ which is the number of coherent oscillations.

Trap fluctuations: classically fluctuating force

Consider a levitated nanosphere whose trap centre can fluctuate, namely $V(\hat{x}) = m\omega_t^2[\hat{x} - \xi(t)]^2/2$. The fluctuating variable $\xi(t)$, with dimensions of distance, fulfills that $\langle \xi(t) \rangle = 0$ while $\langle \xi(t)\xi(t') \rangle \neq 0$. The fluctuation of the trap centre leads to a fluctuating term in the total Hamiltonian given by

$$\hat{H}(t) = \hat{H} + f(t)\hat{x}, \tag{10.53}$$

where $f(t) = -m\omega_t^2 \xi(t)$. Written in this way, Eq. (10.53) also describes the effect of any fluctuating classical force $f(t)$, not necessarily coming from the fluctuations of the trap centre.

One can show that, in the **free dynamics case** $\hat{H} = \hat{p}^2/(2m)$, this classical force leads to a position localization type of decoherence in the long wavelength limit with localization parameter given by

$$\Lambda_f = \frac{1}{2\hbar^2} S_f(0) \tag{10.54}$$

where $S_f(\omega) = \int d\tau \langle f(t+\tau)f(\tau) \rangle e^{i\omega\tau}$ is the force spectral correlation function. This can be readily obtained from the quantum Brownian motion master equation in the limit of infinitely small coupling to a thermal bath of infinite temperature; see for instance section 3.6 of (Breuer and Petruccione, 2002).

In the case of the **particle in the trap**, namely $\hat{H} = \hat{p}^2/(2m) + m\omega_t^2 \hat{x}^2/2$, considering that the rotating wave approximation can be performed, and that the classical force is such that $S_f(\omega_t) = S_f(-\omega_t)$, then the localization parameter is given by

$$\Lambda_f = \frac{1}{2\hbar^2} S_f(\omega_t) \tag{10.55}$$

For more details see (Henkel, Pötting and Wilkens, 1999). For the case of a fluctuating trap centre, it is useful to write the fluctuating position amplitude in units of x_{zp}, namely

$\xi(t) = x_{zp}\tilde{\xi}(t)$ ($\tilde{\xi}(t)$ is now dimensionless), and define $\tilde{S}(\omega) = \int d\tau \langle \tilde{\xi}(t+\tau)\tilde{\xi}(t)\rangle e^{i\omega\tau}$. Then one has that

$$\Lambda_f = \frac{m^2\omega_t^4 x_{zp}^2}{2\hbar^2}\tilde{S}(\omega_t) = \frac{\omega_t^2}{8x_{zp}^2}\tilde{S}(\omega_t). \tag{10.56}$$

The decoherence rate in the trap is given by $\Gamma_f = \Lambda_f x_{zp}^2 = \omega_t^2\tilde{S}(\omega_t)/8$, which is only negligible if the position fluctuation at the trap frequency of the trap centre is several orders of magnitude smaller than the zero point motion x_{zp}! This decoherence effect scales linearly with the mass of the nanosphere, and thus the decoherence rate due to fluctuations of the trap centre for a nanosphere is as many orders of magnitude larger than an atom as the number of atoms in the sphere! This is not the case for the noise arising from the trap frequency fluctuations which does not scale with the mass of the trapped particle, see (Henkel, Pötting and Wilkens, 1999).

Collapse models

Collapse models refer to a class of models that predict a breakdown of the quantum superposition principle for sufficiently large masses and large delocalization distances. While I will not enter here into the details of these models, I just want to mention that most of them predict, at the level of density matrices, a position localization type of decoherence, see (Romero-Isart, 2011). Actually, the master equation Eq. (10.32) with the decay rate function Eq. (10.33) was discussed in the original Ghirardi–Rimini–Weber collapse model in the 1980s; see equation 3.4 in (Ghirardi, Rimini and Weber, 1986).

Let me start by discussing the strength of one of the first collapse models, **gravitationally induced decoherence** or the Penrose–Diósi collapse model. In terms of the density matrix, the collapse model can be well cast into the long wavelength limit of the position localization type of decoherence with the following localization parameter

$$\Lambda_G = \frac{Gm^2}{2R^3\hbar} \tag{10.57}$$

Here G is Newton's gravity constant. Note that this collapse model is **parameter-free**, namely does not require any phenomenological parameter. The saturation of the decoherence happens at $|x - x'| \sim R$, such that when the sphere is delocalized over distances comparable to its size, then the decoherence rate is given by $\Gamma_G = \Lambda_G R^2 = Gm^2/(2R\hbar)$. Note that Γ_G is the gravitational interaction energy of two point masses of mass m separated by a distance $2R$ divided by $\hbar$. From the aesthetical point of view, this is quite nice, namely it gives a very natural timescale involving G, $\hbar$ and the radius of a sphere given some mass density. I want to emphasize that, commonly in the literature, the gravitationally induced decoherence is discussed assuming a fine structure beyond the constant average mass as a conglomerate of identical small balls of mass m_0 and radius r_0 such that saturation is at $|x - x'| = r_0$. This has an impact in the value of the localization parameter, which then is given by

$$\tilde{\Lambda}_G = \left(\frac{R}{r_0}\right)^3 \Lambda_G \gg \Lambda_G. \tag{10.58}$$

Note that this makes the model **parameter dependent**, since the value of r_0 is somehow arbitrary. Typically, one chooses $10^{-14}\text{m} \lesssim r_0 \lesssim 10^{-7}\text{m}$, and hence the gravitationally induced decoherence is greatly boosted. One should be very careful when reading discussions of gravitationally induced decoherence and pay attention whether they consider an homogeneous mass density or not, and in the latter, what is the value of r_0 that they use. By the way, recall that this point is very nicely discussed in Yanbei's lectures where a value for $r_0 \sim 10^{-14}\text{m}$ is argued to be a very natural choice.

Another famous collapse model is the **continuous spontaneous localization (CSL) model**; for details see the review (Bassi, Lochan, Satin, Singh and Ulbricht, 2013). This model introduces a nonlinear stochastic term into the Schrödinger equation that typically depends on two phenomonelogical parameters that can be bounded by experiments. One can show that at the level of density matrix dynamics for the centre-of-mass of a nanosphere, it leads to a position localization type of decoherence with localization parameter given by

$$\Lambda_{\text{CSL}} = \gamma_{\text{CSL}}^0 \frac{m^2}{4a_{\text{CSL}}(\text{amu})^2} f(R/a_{\text{CSL}}) \tag{10.59}$$

Here $f(x) = 6x^{-4}[1 - 2x^{-2} + (1 + x^{-2})e^{-x^2}]$, $a_{\text{CSL}} = 100\,\text{nm}$, and a_{CSL} and γ_{CSL}^0 are the two phenomenological parameters of the model. While a_{CSL} is typically chosen to be around 100 nm, there is much more discrepancy about the value of γ_{CSL}^0 which ranges from the originally (by Ghirardi–Rimini–Weber) suggested value of 10^{-16} Hz to the values suggested more recently by Stephen Adler of 10^{-8} Hz. a_{CSL} is also the saturation distance in the language of the position localization type of decoherence. It is interesting to compare the CSL model with the gravitationally induced collapse model with homogeneous mass density (parameter free); one typically has

$$\frac{\Lambda_G}{\Lambda_{\text{CSL}}} = 10^{-6} \frac{10^{-16}\,\text{Hz}}{\gamma_{\text{CSL}}^0}. \tag{10.60}$$

Thus, even for the weakest choice of the CSL model with $\gamma_{\text{CSL}}^0 \sim 10^{-16}\text{Hz}$, the gravitationally induced decoherence model with homogeneous mass density is six orders of magnitude weaker.

Since we have provided the localization parameter for standard sources of decoherence, I recommend you to compare their strength for feasible experimental scenarios with the collapse models and analyse when could they be falsified. This, in essence, is what is done in arXiv:1110.4495 (Romero-Isart, 2011).

10.3   Wave-Packet Dynamics: Coherence vs. Decoherence

In this final section we discuss how the quantum dynamics of a nanosphere evolving in free space can be calculated. After a brief introduction based on basic but usually forgotten quantum mechanics of a free evolving wave packet, we will discuss the effect of position localization type of decoherence. The tools here are very useful to quantitatively analyse the strength of the sources of decoherence in the goal of coherently expanding the centre-of-mass wavefunction of a nanosphere. The latter is a resource for preparing macroscopic quantum superposition in the spirit of matter-wave interferometry.

10.3.1   Gaussian States

Consider a single mode $[\hat{x}, \hat{p}] = i\hbar$ to describe the position of a massive particle in a one dimension. There is a class of states $\hat{\rho}$ for this mode, called **Gaussian states,** for which the density matrix $\hat{\rho}$ of a state with tr $[\hat{x}\hat{\rho}]$ = tr $[\hat{p}\hat{\rho}]$ = 0 is completely determined by the following three real parameters

$$v_x = \text{tr}\left[\hat{x}^2\hat{\rho}\right],\tag{10.61}$$

$$v_p = \text{tr}\left[\hat{p}^2\hat{\rho}\right],\tag{10.62}$$

$$c = \frac{1}{2}\text{tr}\left[(\hat{x}\hat{p}+\hat{p}\hat{x})\hat{\rho}\right].\tag{10.63}$$

Indeed, the density matrix of any Gaussian state in the position basis reads

$$\langle x|\hat{\rho}|x'\rangle = \sqrt{\frac{a_1 + a_1^* + a_2}{\pi}}e^{-a_1 x^2 - a_1^* x'^2 - a_2 xx'},\tag{10.64}$$

where

$$a_1 = \frac{4v_x v_p - (2c + i\hbar)^2}{8\hbar^2 v_x},\tag{10.65}$$

$$a_2 = \frac{4c^2 + \hbar^2 - 4v_x v_p}{4\hbar^2 v_x}.\tag{10.66}$$

For those familiar with the Wigner function formalism, Gaussian states have a positive Gaussian Wigner function. The following properties of Gaussian states will be very useful for us:

- The Heisenberg uncertainty condition reads

$$v_x v_p - c^2 \geq \frac{\hbar^2}{4}.\tag{10.67}$$

- The purity of the state $\mathcal{P} = \mathrm{tr}\left[\hat{\rho}^2\right]$ reads

$$\mathcal{P} = \frac{1}{2}\frac{\hbar}{\sqrt{v_x v_p - c^2}}. \tag{10.68}$$

- The **coherence length** ξ, which is defined by

$$\langle x/2|\hat{\rho}|-x/2\rangle = \frac{1}{\sqrt{2\pi v_x}}\exp\left[-\frac{x^2}{\xi^2}\right], \tag{10.69}$$

reads

$$\xi = \mathcal{P}\sqrt{8v_x}. \tag{10.70}$$

The coherence length is a very important parameter to describe the coherent expansion of a wave packet. Note that, for instance, if the purity is kept constant during time evolution, the coherence length grows as the square root of v_x.

Finally let us recall that the **thermal state of a harmonic oscillator** described by the Hamiltonian $\hat{H} = \hat{p}^2/(2m) + m\omega^2\hat{x}^2/2$, namely $\hat{\rho}_T = \hat{\mathcal{Z}}/\mathrm{tr}\left[\hat{\mathcal{Z}}\right]$, where $\hat{\mathcal{Z}} = \exp[-\beta\hat{H}]$ with $\beta^{-1} = k_b T$, is a Gaussian state defined by

$$v_x = x_{\mathrm{zp}}^2(2\bar{n}+1), \tag{10.71}$$

$$v_p = \frac{\hbar^2}{4x_{\mathrm{zp}}^2}(2\bar{n}+1), \tag{10.72}$$

$$c = 0, \tag{10.73}$$

where $\bar{n} = (\exp[\beta\hbar\omega] - 1)^{-1}$, and $x_{\mathrm{zp}} \equiv \sqrt{\hbar/(2m\omega)}$. Note that the purity of a thermal state is then given by $\mathcal{P} = 1/(2\bar{n}+1)$, and thus the hotter the thermal state, the more mixed.

10.3.2 Coherent Evolution

Consider an initial Gaussian state $\hat{\rho}$ with $\mathrm{tr}\left[\hat{x}\hat{\rho}\right] = \mathrm{tr}\left[\hat{p}\hat{\rho}\right] = 0$ evolving in free space without decoherence such that $\hat{H} = \hat{p}^2/2m$ and $\dot{\rho} = -i[\hat{H},\hat{\rho}]/\hbar$. One can readily show that that the three parameters of the Gaussian state fulfill the following set of differential equations

$$\dot{v}_x(t) = \frac{2}{m}c(t),$$

$$\dot{v}_p(t) = -2m\omega^2 c(t),$$

$$\dot{c}(t) = -m\omega^2 v_x(t) + \frac{1}{m}v_p(t). \tag{10.74}$$

This can be readily solved and the solution is given by

$$v_x(t) = v_x(0) + \frac{2c(0)}{m}t + \frac{v_p(0)}{m^2}t^2, \tag{10.75}$$

$$v_p(t) = v_p(0), \tag{10.76}$$

$$c(t) = c(0) + \frac{v_p(0)}{m}t. \tag{10.77}$$

This solution carries all the properties of the free propagation of a wave packet. For instance, note that if $c(0) = 0$, then the wave packet expands linearly in time, namely $\sqrt{v_x(t)} \propto t$. See also that if $c(0) < 0$, the wave packet initially contracts at a rate $\propto t^{1/2}$ until the linear expansion dominates. In free space the momentum distribution is kept constant. Finally, note that the coherence length grows linearly in t as $\xi(t) = \mathcal{P}\sqrt{8v_x(t)}$ with a speed in the long time limit given by

$$\lim_{t\to\infty} \dot{\xi}(t) = \mathcal{P}\frac{\sqrt{8v_p(0)}}{m}. \tag{10.78}$$

This is quite an interesting result worth remembering. For the case in which the initial state was a thermal state, then

$$\lim_{t\to\infty} \dot{\xi}(t) = \frac{2}{\sqrt{2\bar{n}+1}}\sqrt{\frac{\hbar\omega}{m}} \tag{10.79}$$

Considering that the mass of the nanosphere is $m = N \times 1$ *amu* and that it is initially in the ground state $\bar{n} = 0$ of a harmonic trap of frequency $\omega = 2\pi \times 10^5$ Hz, then

$$\lim_{t\to\infty} \dot{\xi}(t) \approx \frac{0.4}{\sqrt{N}}\frac{\text{m}}{\text{s}}. \tag{10.80}$$

This is quite a speed, isn't it? An atom of mass 1 amu trapped in a harmonic potential of 10^5 Hz coherently expands to 40 cm in 1 second! This is actually one of the reasons why matter-wave interferometers of ultracold gases, as those in the experiments of the group of Mark Kasevich at Stanford, can measure the small gravity constant with such a high precision. A nanosphere can easily have a mass of the order of several millions of amu's and thus, the coherent speed expansion reduces by at least three orders of magnitude.

Nonetheless, this allows the possibility of expanding the centre-of-mass position of a cold nanosphere to the micrometre scale after one second of free expansion.

Let us now discuss what is the influence of the position localization type of decoherence in the free propagation of a wave packet.

10.3.3 The Effect of Decoherence

Let us now consider that the initial Gaussian state evolves in free expansion but with position localization type of decoherence in the long wavelength limit, namely

$$\dot{\rho} = \frac{1}{i\hbar}[\hat{p}^2/(2m), \hat{\rho}] - \Lambda[\hat{x},[\hat{x},\hat{\rho}]]. \tag{10.81}$$

One can readily show that in this case the differential equations for v_x, v_p, and c are given by

$$\dot{v}_x(t) = \frac{2}{m}c(t),$$
$$\dot{v}_p(t) = -2m\omega^2 c(t) + 2\hbar^2\Lambda, \tag{10.82}$$
$$\dot{c}(t) = -m\omega^2 v_x(t) + \frac{1}{m}v_p(t).$$

Note that Λ only appears in the equation of $\dot{v}_p$. This can also be solved analytically, one finds

$$v_x(t) = v_x(0) + \frac{2c(0)}{m}t + \frac{v_p(0)}{m^2}t^2 + \frac{2\Lambda\hbar^2}{3m^2}t^3, \tag{10.83}$$

$$v_p(t) = v_p(0) + 2\Lambda\hbar^2 t, \tag{10.84}$$

$$c(t) = c(0) + \frac{v_p(0)}{m}t + \frac{\Lambda\hbar^2 t^2}{m}. \tag{10.85}$$

By recalling the expression of the purity $\mathcal{P}$ Eq. (10.68) and the coherence length ξ Eq. (10.69), one can make the following observations:

- For $t \ll 3v_p(0)/(2\Lambda\hbar)$ the evolution is purely unitary, due to the Hamiltonian part, and $\sqrt{v_x} \propto t$ grows linearly in time. In this regime the purity $\mathcal{P}$ is nearly constant and thus the coherence length $\xi(t)$ grows linearly in time.
- For $t \gg 3v_p(0)/(2\Lambda\hbar)$ decoherence due to the Λ term dominates and then $\sqrt{v_x} \sim t^{3/2}$. In this regime the purity decreases as $\mathcal{P} \sim t^{-2}$ and thus the coherence length also decreases as $\xi \sim t^{-1/2}$.
- Thus the coherence length grows linearly in time, has a maximum, and decays as $t^{-1/2}$. See Fig. 10.4 for the typical time dependence of the coherence length for the particular case in which $c(0) = 0$ (as in a thermal state).

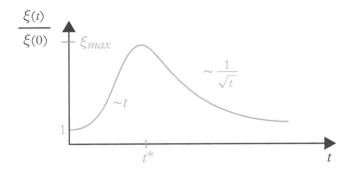

Fig. 10.4 *Typical time dependence of the coherence length in the presence of position localization type of decoherence in the long wavelength limit.*

You can show that for an initial thermal state $(c(0) = 0)$, the coherence length $\xi(t)$ has a maximum at

$$t^\star = \left[\frac{3m^2 v_x(0)}{\Lambda \hbar^2} \right]^{1/3}, \tag{10.86}$$

where the coherence length is very well approximated by

$$\xi_{\max} = \xi(t^\star) \approx \sqrt{2} \left[\frac{\hbar^2}{3 v_x(0) m^2 \Lambda^2} \right]^{1/6}. \tag{10.87}$$

Note that the maximum of the coherence length is a strict upper bound of the maximum quantum superposition size that could be created by using a non-Gaussian operation splitting the coherently expanded wave function. I encourage you to play around with the expressions given in section 10.2 for the different sources of decoherence. Note that quite a rich and general analysis can be done on how challenging testing collapse models are to study under typical experimental conditions. This is done in detail in arXiv:1110.4495 (Romero-Isart, 2011). In particular you should consider the following points:

- $t^\star$ should always be smaller than $1/\gamma_{\text{air}}$; recall Eq. (10.48), namely the timescale in which decoherence due to scattering of air molecules starts to exponentially reduce the off-diagonal terms of the density matrix in the position basis via the short wavelength limit of position localization type of decoherence.
- Note that $t^\star$ for the emission of black-body radiation, recall Eq. (10.46), does not depend on the radius of the sphere and thus $t^\star$ can be plotted as a function of the internal temperature. It gives a good intuition of the maximum coherence time allowed for a given internal temperature. What is the order of magnitude of $t^\star$ at room temperature?

- To see how challenging it is to test collapse models, plot the coherence length evolution for some given source of decoherence Λ and the same by adding the presence of a collapse model parameterized by Λ_{CM}, namely for $\Lambda + \Lambda_{CM}$. In which parameter regimes can the two cases be distinguishable?

Thank you very much for reading these notes. Should you have questions and or comments, please do not hesitate to contact me at oriol.romero-isart@uibk.ac.at

Bibliography

Bassi, A., Lochan, K., Satin, S., Singh, T. P., and Ulbricht, H. (2013). Models of wave-function collapse, underlying theories, and experimental tests. *Rev. Mod. Phys.*, **85**, 471.

Bateman, J., Nimmrichter, S., Hornberger, K., and Ulbricht, H. (2014). Near-field interferometry of a free-falling nanoparticle from a point-like source. *Nat. Commun.*, **5**, 4788.

Breuer, H.-P., Petruccione, F. (2002). The Theory of Open Quantum Systems. Oxford University Press.

Ghirardi, G. C., Rimini, A., and Weber, T. (1986). Unified dynamics for microscopic and macroscopic systems. *Phys. Rev. D*, **34**, 470.

Henkel, C., Pötting, S., and Wilkens, M. (1999). Loss and heating of particles in small and noisy traps. *App. Phys. B*, **69**, 379.

Pflanzer, A. C., Romero-Isart, O., and Cirac, J. I. (2012). Master-equation approach to optomechanics with arbitrary dielectrics. *PRA*, **86**, 013802.

Romero-Isart, O. (2011). Quantum superposition of massive objects and collapse models. *Phys. Rev. A*, **84**, 052121.

Romero-Isart, O., Pflanzer, A. C., Juan, M. L., Quidant, R., Kiesel, N., Aspelmeyer, M., and Cirac, J. I. (2011). Optically levitating dielectrics in the quantum regime: Theory and protocols. *PRA*, **83**, 013803.

Steck, D. A., Quantum and Atom Optics. http://steck.us/teaching.

11

Quantum Optomechanics Thermodynamics and Heat Engines

Pierre Meystre

Department of Physics and College of Optical Sciences, University of Arizona, Tucson, Arizona, USA

Pierre Meystre

Pierre Meystre, *Quantum Optomechanics, Thermodynamics, and Heat Engines* In: *Quantum Optomechanics and Nanomechanics.* Edited by: Pierre-Francois Cohadon, Jack Harris, Florian Marquardt, Leticia F. Cugliandolo, Oxford University Press (2020). © Oxford University Press. DOI: 10.1093/oso/9780198828143.003.0011

Chapter Contents

11.1   Introduction

Cavity optomechanics (Aspelmeyer, Meystre and Schwab, 2012), (Meystre, 2013), (Aspelmeyer, Kippenberg and Marquardt, 2014) provides a universal tool to achieve quantum optical control over mechanical motion—or conversely mechanical control over optical or microwave fields. It covers a wide spectrum of systems from nanometre-sized devices, comprising as little as 10^7 atoms, to nano- and micromechanical structures and to the centimetre-sized mirrors used in gravitational wave detectors, weighing up to several kilograms. Because their mechanical elements can be functionalized by coupling their motion to other physical systems, including for example photons, spins, charges, atomic systems, or qubits, they offer a rich potential both in fundamental science and for applications. From the latter point of view optomechanical techniques in both the optical and microwave regimes readily provide motion and force detection near the standard quantum limit, with the expectation of soon overcoming that limit. The force sensitivity of these systems now exceeds $10^{-18} \text{N}/\sqrt{\text{Hz}}$, an astoundingly weak force that corresponds to the gravitational attraction between a person in Los Angeles and another in New York.

Micromechanical oscillators also offer a route to the development of new tests of of the foundations of physics, including tests of quantum theory at unprecedented size and mass scales. For instance, future spatial quantum superpositions of massive objects could be used to probe various theories of decoherence, i.e. the transition from quantum behaviour to classical behaviour. It is widely accepted that quantum mechanics is the fundamental theory of nature, yet in everyday life we don't observe the remarkable quantum effects that we can achieve with small ensembles of atoms or with photons under exquisitely controlled conditions. We cannot make a car be 'in two places at the same time', or, in the famous example of Schrödinger, we cannot have a cat that is both alive and dead at the same time. Even large quantum condensates such as a cup of superfluid helium do not display macroscopic superposition states and entanglement. The everyday world seems most definitely governed by the laws of classical physics. This is disconcerting: if the quantum mechanical description of nature is more fundamental than its classical description, then quantum mechanics should govern not just the microscopic world, but the macroscopic world as well.

Romero-Isart and colleagues (Romero-Isart, Pflanzer, Blaser, Kaltenbaek, Kiesel, Aspelmeyer and Cirac, 2011), (Romero-Isart, 2011) have recently considered a method to prepare and probe the quantum superposition of a nanometre-sized object with a spatial separation of its two partial wave functions comparable to its size. It is hoped that such systems will eventually be realized in a parameter regime where it will be possible to test various mechanisms that have been proposed beyond traditional quantum mechanics to explain the washing out of quantum coherence in macroscopic objects. It will be exciting indeed to see these experiments be realized and start answering questions that have surrounded quantum mechanics and its interpretation since its early days, nearly 100 years ago.

These lecture notes address selected topics in quantum thermodynamics, another area where optomechanics is in a position to contribute an attractive set of experimental tests and additional understanding. Quantum thermodynamics can be loosely defined as the study of thermodynamics in regimes where quantum mechanical noise coexists with thermal noise and has a significant impact on the dynamics of the system. In some ways it is an extension of stochastic thermodynamics, which deals with the consistency of macroscopic thermodynamic quantities such as work, heat and entropy with the random, erratic motion of small systems. These systems are not within the scope of conventional macroscopic thermodynamics, because they are dominated by fluctuations. Examples include colloidal particles immersed in a viscous fluid or a laser trap, and biomolecules such as RNA or DNA manipulated by optical tweezers. Considerable progress has recently been made toward understanding how work, heat and entropy can be defined along the individual stochastic trajectories that characterize the dynamics of these systems, including a first-law like energy balance equation, introduced by Sekimoto (Sekimoto, 2007), and a definition of entropy along single trajectories (Seifert, 2005).

Quantum thermodynamics presents, however, a number of additional challenges, open questions and opportunities. They include proper definitions of work and heat, especially for open quantum systems, and the influence of the back-action of quantum measurements on thermodynamic quantities, one example being the work delivered by heat engines. Further fascinating questions include whether the unique attributes of quantum coherence and fluctuations can be exploited to provide advantages over classical systems (Scully, 2002), (Scully, Zubairy, Agarwal and Walther, 2003), with promising developments including increases in engine efficiency by 'quantum afterburners'. Another possibility that we discuss at some length in these notes is the use of the controllable nature of quantum quasiparticles as a working fluid.

The definition of thermodynamic quantities in the quantum context presents conceptual difficulties (Allahverdyan and Nieuwenhuizen, 2005), (Esposito and Mukamel, 2006), (Boukobza and Tannor, 2006), (Talkner, Lutz and Hänggi, 2007) and much attention has been devoted to the proper definition and the quantum statistical properties of heat, work and entropy (Fusco, Pigeon, Apollaro, Xuereb, Mazzola, Campisi, Ferraro, Paternostro and De Chiara, 2014), (Dorner, Goold, Cormick, Paternostro and Vedral, 2012), (Mascarenhas, Braganca, Dorner, Franca Santos, Vedral, Modi and Goold, 2014), (Apollaro, Francica, Paternostro and Campisi, 2015), (Campisi, Blattmann, Kohler, Zueco and Hänggi, 2013), (Joshi and Campisi, 2013), (Smacchia and Silva, 2013), (Saira, Yoon, Tanttu, Möttönen, Averin and Pekola, 2012), (Batalhão, Souza, Mazzola, Auccaise, Sarthour, Oliveira, Goold, De Chiara, Paternostro and Serra, 2014), (Jarzynski, 2015), (Hänggi and Talkner, 2015), (An, Zhang, Um, Lv, Lu, Zhang, Yin, Quan and Kim, 2015), (Suomela, Salmilehto, Savenko, Ala-Nissila and Möttönen, 2015), (Salmilehto, Solinas and Möttönen, 2014), (Frenzel, Jennings and Rudolph, 2014). In closed quantum systems work may be defined in terms of a two-time measurement scheme (Esposito, Harbola and Mukamel, 2009), (Campisi,

Hänggi and Talkner, 2011), (Talkner, Hänggi and Morillo, 2008), (Mukamel, 2003) or, in a recently proposed alternative approach, of a single projective measurement (Roncaglia, Cerisola and Paz, 2014). However, the situation is less clear for open quantum systems, where there are still unsettled questions regarding the definition and the experimental implementation of measurements of work and heat (Esposito, Harbola and Mukamel, 2009), (Campisi, Talkner and Hänggi, 2009), (Campisi, Talkner and Hänggi, 2010), (Campisi, Hänggi and Talkner, 2011), (Ritort, 2009) due to the lack of energy conservation in the reservoir(s). In this context quantum stochastic thermodynamics (Horowitz, 2012), (Esposito, Harbola and Mukamel, 2007), (Esposito and Van den Broeck, 2010), like its classical counterpart (Seifert, 2005), (Schmiedl, Speck and Seifert, 2007), (Schmiedl and Seifert, 2007), (Seifert, 2008), offers an interesting framework to discuss thermodynamic properties and also to simulate numerically the system behaviour.

These lecture notes focus on the specific example of an optomechanical quantum heat engine (QHE) as a test system to discuss several of these issues. Section 11.2 sets the stage by briefly reviewing some of the outstanding questions associated with the characterization of quantum work. Not surprisingly, quantum measurements are a cornerstone of that discussion. Section 11.3 then outlines the key steps leading to the formulation of continuous measurements in terms of stochastic Schrödinger equations. Turning more specifically to QHE, section 11.4 reviews their main characteristics, comparing and contrasting thermodynamic processes and engine cycles in the classical and the quantum regimes.

The working substance of the optomechanical heat engine that we consider is a photon–phonon polariton fluid. Polaritons are quasiparticles formed by the quantum superposition of two (or more) types of particles coupled by some interaction, in the present example optomechanical coupling. A remarkable property of quasiparticles is that their physical nature can be changed continuously by varying the relative weight of their constituents. This can be done by controlling the details of their interaction. For instance, photon–phonon polaritons can become predominantly photon-like or phonon-like, and as a result behave for all practical purposes as either photons or phonons. Because these elementary constituents may be coupled to vastly different environments, for instance to thermal reservoirs at different temperatures, this allows one to operate heat engines in ways that are classically unavailable, via the interplay of quantum coherence and dissipation.

The additional opportunities offered by quasiparticles in the operation of QHE justify reviewing their main properties in some detail. This is done in Section 11.5, before introducing the optomechanical QHE system itself in section 11.6. Section 11.7 discusses the quantum properties of the engine. The quantum fluctuations of the work are discussed in the context of two continuous measurement schemes that can be used to monitor them. Section 11.8 briefly expands the discussion to the related system of a polariton based quantum heat pump, illustrating again the potential of polaritonic working substances. Finally, section 11.9 is a conclusion and outlook.

11.2 Quantum Thermodynamics: Work and Heat

11.2.1 Average Work and Heat

We mentioned that the measurement of work and of its fluctuations, as well as the role of measurement back-action, are important questions in quantum thermodynamics. One difficulty is that work is not a quantum observable in the usual sense; its value depends on the protocol followed to go from the initial point to the final point in state space. That is, it is characterized by a process. This leads to subtle issues, especially in the case of open systems.

In classical thermodynamics the expression of the First Law is

$$dU = dQ + dW, \tag{11.1}$$

where U, Q, and W are energy, heat and work, respectively. This law states that the energy exchanged by a system in a transformation is divided between work W and heat Q.

To obtain a quantum version of this expression, we express the average energy U in terms of the eigenstates $|i\rangle$ of the system Hamiltonian $\hat{H}_s = \sum_i E_i |i\rangle\langle i|$ as

$$U = \langle \hat{H}_s \rangle = \sum_i p_i E_i, \tag{11.2}$$

where E_i is the energy of the eigenstate $|i\rangle$, with corresponding occupation probability p_i. An infinitesimal change in average energy is given by

$$dU = \sum_i dp_i E_i + \sum_i p_i dE_i \tag{11.3}$$

and one can identify the first term on the right-hand side of this equation as the infinitesimal heat transferred to or from the quantum system, and the second as the infinitesimal work,

$$dQ = \sum_i dp_i E_i, \tag{11.4}$$

$$dW = \sum_i p_i dE_i. \tag{11.5}$$

These quantum expressions for the infinitesimal average heat and work variations are consistent with their definitions in classical thermodynamics and statistical physics. The heat dQ corresponds to a change in the populations p_i without change of the energy eigenvalues, while the work dW done on or by the system corresponds to a redistribution of the energy eigenvalues. That is, the heat exchange results in a change in the statistical distribution of the microstates of different energies while the work is a change in the energy structure of the system.[1]

But how do we measure the average work? To address that question Solinas and coworkers (Solinas, Averin and Pekola, 2013) consider work in analogy to the classical case as an integral over the injected power during the evolution of the system. The work performed on the system is then shown to be the internal energy difference of the system between its initial and final states, minus the heat released to the heat bath. The determination of the average work requires therefore two measurements, one at the beginning and one at the end of the process.[2]

More concretely, consider a system described by a Hamiltonian $\hat{H}_s(\lambda)$ that can be varied by a control parameter $\lambda(t)$ and is in addition coupled to a thermal reservoir by some Liouvillian $\mathcal{L}(\hat{\rho})$, $\hat{\rho}$ being the system reduced density operator. The injected power operator and average power are given by

$$\hat{P} = \frac{\partial \hat{H}_s}{\partial \lambda} \dot{\lambda} = \frac{d\hat{H}_s}{dt} \tag{11.6}$$

and

$$\langle \hat{P} \rangle = \left\langle \frac{d\hat{H}_s}{dt} \right\rangle, \tag{11.7}$$

so that

$$\langle W \rangle = \int_{t_i}^{t_f} dt' \langle \hat{P}(t') \rangle. \tag{11.8}$$

As noted by Solinas (Solinas, Averin and Pekola, 2013) this shows that the work depends explicitly on the full evolution of the system through its density operator $\hat{\rho}(t)$, which in turn includes contributions from both its unitary and its non-unitary evolution. To make this latter point clearer, consider the average energy $\langle \hat{H}_s \rangle = \mathrm{Tr}(\hat{\rho}\hat{H}_s)$ of the system. It is governed by the equation of motion

---

[1] The sign conventions used in thermodynamics are sometimes a source of confusion. Equation (11.5) shows that positive work increments dW result in an increase in the internal energy of the system. That is, in the convention used in these notes positive work corresponds to work done *on* the system, and negative work to work done *by* the system.

[2] Note however the recent paper by Roncaglia (Roncaglia, Cerisola and Paz, 2014), who considered work measurement as a generalized quantum measurement.

$$\frac{d\langle \hat{H}_s \rangle}{dt} = \text{Tr}\left(\frac{d\hat{\rho}}{dt}\hat{H}_s + \hat{\rho}\frac{d\hat{H}_s}{dt}\right) = \text{Tr}\left(\frac{d\hat{\rho}}{dt}\hat{H}_s\right) + \langle \hat{P} \rangle,$$

where we have used Eq. (11.7). The master equation for the reduced density operator $\hat{\rho}(t)$ of a dissipative system has the familiar form

$$\frac{d\hat{\rho}}{dt} = -\frac{i}{\hbar}[\hat{H}_s, \hat{\rho}] + \mathcal{L}(\hat{\rho}) \tag{11.9}$$

so that

$$\text{Tr}\left(\frac{d\hat{\rho}}{dt}\hat{H}_s\right) = \text{Tr}\left(\frac{i}{\hbar}[\hat{H}_s, \hat{\rho}]\hat{H}_s + \mathcal{L}(\hat{\rho})\hat{H}_s\right) = \text{Tr}\left(\mathcal{L}(\hat{\rho})\hat{H}_s\right). \tag{11.10}$$

Since $\text{Tr}\{[\hat{H}_s, \hat{\rho}]\hat{H}_s\} = 0$ it follows from Eq. (11.9) that

$$\frac{d\langle \hat{H}_s \rangle}{dt} = \langle \hat{P} \rangle + \text{Tr}(\mathcal{L}(\hat{\rho})\hat{H}_s) \tag{11.11}$$

and

$$\langle W \rangle = \int_{t_i}^{t_f} dt' \left[\left\langle\frac{d\hat{H}_s}{dt'}\right\rangle + \text{Tr}[\mathcal{L}(\hat{\rho})\hat{H}_s]\right] = \langle \hat{H}_s(t_f) \rangle$$
$$- \langle \hat{H}_s(t_i) \rangle - \int_{t_i}^{t_f} dt' \text{Tr}\left(\mathcal{L}(\hat{\rho})\hat{H}_s\right). \tag{11.12}$$

Recognizing that

$$\langle U \rangle \equiv \langle \hat{H}_s(t_f) \rangle - \langle \hat{H}_s(t_i) \rangle \tag{11.13}$$

is the change in average internal energy U, we can therefore identify the average heat $\langle Q \rangle$ as

$$\langle Q \rangle \equiv \int_{t_i}^{t_f} dt' \text{Tr}\left(\mathcal{L}(\hat{\rho})\hat{H}_s\right). \tag{11.14}$$

and in that way recover the First Law of thermodynamics.

Equation (11.14) shows that the average heat $\langle Q \rangle$ is associated with the exchange of energy of the system with the environment during its evolution. Like the average work $\langle W \rangle$ it depends on the specific protocol followed by the control parameter $\lambda(t)$. Importantly, this result implies that in isolated quantum systems the average work can be determined unambiguously via a two-time measurement process (Esposito, Harbola and Mukamel, 2009), (Campisi, Hänggi and Talkner, 2011) of the internal energy of the system. But this approach becomes problematic for open quantum systems, in which case measurements on both the system and its reservoir(s) are in general required.

In analogy with stochastic thermodynamics, it is possible in principle to use quantum trajectory methods and two-time measurement schemes to evaluate the quantum fluctuations of these quantities as well. Section 11.3.2 will summarize two examples of that approach, a driven harmonic oscillator coupled to a reservoir of two-state systems (Horowitz, 2012), and a two-state system coupled to a reservoir of harmonic oscillators (Hekking and Pekola, 2013). These examples rely for their analysis on the so-called Quantum Monte Carlo wave functions formalism, which we first review before turning to a summary of these results.

11.3   Stochastic Quantum Trajectories

In quantum mechanical systems—and in all stochastic systems for that matter—the mean value of observables provides only limited information about their properties. Characterizing these systems beyond the mean value evolution requires considering higher moments of the relevant observables. We mentioned in the introduction that small classical systems whose evolution is dominated by (thermal) noise are not within the scope of macroscopic thermodynamics. We also hinted at the progress recently achieved in understanding how thermodynamic quantities can be understood along individual stochastic trajectories, with the introduction of a first-law-like energy balance equation (Sekimoto, 2007) and the introduction of entropy along single trajectories (Seifert, 2005).

The extension of these ideas to quantum systems is a somewhat more recent development. Horowitz (Horowitz, 2012) formulated an approach to quantum stochastic thermodynamics based on the idea of quantum trajectories, and introduced trajectory-dependent definitions of work, heat and entropy. Section 11.3.2 will briefly review some aspects of a specific example that he considers, a forced harmonic oscillator coupled to a thermal reservoir, as well as the somewhat complementary example of a driven damped two-state system analysed by Hekking and Pekola (Hekking and Pekola, 2013).

There is additional merit in focusing on single trajectories: while the knowledge of the system density operator permits in principle evaluation of any observable of interest, this approach does not provide much intuition about what happens for a single realization of its stochastic evolution, what an experimentalist would typically observe before starting to average over measurements. Stochastic quantum trajectories can provide such intuition.

11.3.1   Quantum Monte Carlo wave functions

The quantum Monte Carlo wave functions method provides an attractive and powerful way to investigate the evolution of an open quantum system coupled to a (thermal) reservoir by generating a large ensemble of time-dependent wave functions representative of single realizations of the system. In contrast to the situation for closed systems, these wave functions are intrinsically stochastic, resulting from the combination of a non-Hermitian Schrödinger-like evolution and random 'quantum jumps'. Notably, such stochastic quantum trajectories are also central to the discussion of measurements that concerns us, as will become clear in the next section. This section outlines the main ingredients of this approach, see e.g. Meystre and Sargent (Meystre and Sargent, 2007) or Carmichael (Carmichael, 2008) for more details.

In the master equation (11.9) describing the evolution of the reduced density operator for a system coupled to a general Markovian reservoir the Liouvillian $\mathcal{L}[\hat{\rho}]$ that accounts for dissipation is of the general Lindblad form

$$\mathcal{L}[\hat{\rho}] = -\frac{1}{2}\sum_i(\hat{C}_i^\dagger\hat{C}_i\hat{\rho} + \hat{\rho}\hat{C}_i^\dagger\hat{C}_i) + \sum_i\hat{C}_i\hat{\rho}\hat{C}_i^\dagger, \tag{11.15}$$

where the operators $\hat{C}_i$ and $\hat{C}_i^\dagger$ are system operators whose form depends on the explicit form of the coupling of the system to its reservoir(s). For a two-state system with upper to lower level decay coupled linearly to a broadband bath of harmonic oscillators, for example, they would be $\hat{C}_1 = \sqrt{\Gamma(\bar{n}+1)}\hat{\sigma}_-$ and $\hat{C}_2 = \sqrt{\Gamma\bar{n}}\hat{\sigma}_+$, where Γ is the decay constant and $\bar{n}$ the thermal excitation of the reservoir at the transition frequency of the two-state system (Meystre and Sargent, 2007).

The starting point of the Monte Carlo wave functions method is to decompose the master equation (11.15) as

$$\frac{d\hat{\rho}}{dt} = -\frac{i}{\hbar}\left(\hat{H}_{\text{eff}}\hat{\rho} - \hat{\rho}\hat{H}_{\text{eff}}^\dagger\right) + \mathcal{L}_{\text{jump}}\hat{\rho}, \tag{11.16}$$

where we introduced the effective non-Hermitian Hamiltonian

$$\hat{H}_{\text{eff}} \equiv \hat{H}_s - \frac{i\hbar}{2}\sum_i\hat{C}_i^\dagger\hat{C}_i \tag{11.17}$$

and the 'jump' Liouvillian

$$\mathcal{L}_{\text{jump}}[\hat{\rho}] \equiv \sum_i\hat{C}_i\hat{\rho}\hat{C}_i^\dagger. \tag{11.18}$$

One can justify associating $\mathcal{L}_{\text{jump}}[\hat{\rho}]$ with 'quantum jumps' by returning to a state vectors rather than a density operator description of the problem. Expressing the density operator $\hat{\rho}$ as a statistical mixture of state vectors $|\psi\rangle$ as

$$\hat{\rho} = \sum_\psi P_\psi|\psi\rangle\langle\psi|, \tag{11.19}$$

where the summation over ψ results from a classical average over the various states that the system can occupy with probability P_ψ, and substituting that form of $\hat{\rho}$ into the master equation (11.16) gives

$$\sum_\psi P_\psi \left[|\dot{\psi}\,\rangle\langle\psi| + |\psi\rangle\langle\dot{\psi}| = -\frac{i}{\hbar}\left(\hat{H}_{\text{eff}}|\psi\rangle\langle\psi| - |\psi\rangle\langle\psi|\hat{H}_{\text{eff}}^\dagger \right) + \sum_i \hat{C}_i|\psi\rangle\langle\psi|\hat{C}_i^\dagger \right].$$

$$\text{(11.20)}$$

Focusing temporarily on a single representative state vector $|\psi\rangle$ of the mixture (11.19) we recognize that the first term on the right-hand side of Eq. (11.20) can be interpreted as resulting from its Schrödinger-like evolution under the influence of the non-Hermitian Hamiltonian $\hat{H}_{\text{eff}}$,

$$|\dot{\psi}\rangle = -\frac{i}{\hbar}\hat{H}_{\text{eff}}|\psi\rangle. \tag{11.21}$$

Things are different for the second term. It is clearly not a Schrödinger-like term. Rather, it can be interpreted as resulting from a discontinuous evolution whereby the state $|\psi\rangle$ is projected, or 'jumps', onto one of the possible states $|\psi\rangle_i$ via the transformation

$$|\psi\rangle \to |\psi\rangle_i = \hat{C}_i|\psi\rangle. \tag{11.22}$$

This is the observation that leads to the identification of $\mathcal{L}_{\text{jump}}[\hat{\rho}]$ as a 'quantum jump' Liouvillian.

The evolution of the master equation (11.16) can be obtained by carrying out an ensemble average over the evolution of a large number of state vectors instead. It proceeds numerically by selecting an arbitrary state vector $|\psi\rangle$ out of an initial ensemble that reproduces the initial density operator when averaged over, and first evolving it for a short time δt under the influence of $\hat{H}_{\text{eff}}$ only. For sufficiently small time intervals this gives

$$|\tilde{\psi}(t+\delta t)\rangle = \left(1 - \frac{i\hat{H}_{\text{eff}}\delta t}{\hbar} \right)|\psi(t)\rangle. \tag{11.23}$$

Because of the non-Hermitian nature of $\hat{H}_{\text{eff}}$, though, the norm of the state vector is not conserved during that evolution. The square of the norm of $|\tilde{\psi}(t+\delta t)\rangle$ is

$$\langle\tilde{\psi}(t+\delta t)|\tilde{\psi}(t+\delta t)\rangle = \langle\psi(t)| \left(1 + \frac{i\hat{H}_{\text{eff}}^\dagger\delta t}{\hbar} \right)\left(1 - \frac{i\hat{H}_{\text{eff}}\delta t}{\hbar} \right)|\psi(t)\rangle = 1 - \delta p, \quad \text{(11.24)}$$

where to lowest order in δt

$$\delta p = \frac{i}{\hbar}\delta t\langle\psi(t)|\hat{H}_{\text{eff}} - \hat{H}_{\text{eff}}^\dagger|\psi(t)\rangle = \delta t \sum_i \langle\psi(t)|\hat{C}_i^\dagger\hat{C}_i|\psi(t)\rangle \equiv \sum_i \delta p_i. \tag{11.25}$$

Of course the full master equation evolution does preserve the norm. The lack of norm conservation that we now encounter results from the fact that we have so far ignored the effects of $\mathcal{L}_{\text{jump}}[\hat{\rho}]$. The 'missing norm' δp is contained in the states $|\psi\rangle_i = \hat{C}_i|\psi\rangle$ resulting from the jumps part of the evolution. We interpret this as a result of the fact that $\mathcal{L}_{\text{jump}}[\hat{\rho}]$ projects the system into the state $|\psi\rangle_i = \hat{C}_i|\psi\rangle$ with probability δp_i such that $\sum_i \delta p_i = \delta p$.

Hence the next step of a Monte Carlo simulation consists in deciding whether a jump occurred or not. Numerically, this is achieved by choosing a random variate from uniformly distributed pseudorandom numbers $0 \leq r \leq 1$. If its value is larger than δp, no jump is said to have occurred, and the next integration step proceeds from the *normalized* state vector

$$|\psi(t+\delta t)\rangle = \frac{|\tilde{\psi}(t+\delta t)\rangle}{\||\tilde{\psi}(t+\delta t)\rangle\|}. \tag{11.26}$$

If on the other hand $r \leq \delta p$, a jump is said to have occurred. The state vector $|\tilde{\psi}(t+\delta t)\rangle$ is then projected to the *normalized* new state

$$|\psi(t+\delta t)\rangle = \frac{\hat{C}_i|\psi(t)\rangle}{\|\hat{C}_i|\psi(t)\rangle\|} = \sqrt{\frac{\delta t}{\delta p_i}}\,\hat{C}_i|\psi(t)\rangle \tag{11.27}$$

with probability $\delta p_i/\delta p$, and this state is taken as the initial condition for the next integration step. The procedure is then repeated for as many iterations as desired, yielding one possible realization of the time evolution of the initial state vector $|\psi\rangle$ called a quantum trajectory. The stochastic nature of the jumps implies that repeating the simulation from the same initial state will result in a different trajectory. It is easy to verify that averaging over a large number of such realizations reproduces the results of the master equation (11.9), as it should.

In some situations it is possible to interpret the reservoir as a 'measurement apparatus', in which case each quantum trajectory can be thought of as representative of a sequence of measurements on that system. This is the idea underlying the approaches of Horowitz (Horowitz, 2012) and of Hekking and Pekola (Hekking and Pekola, 2013) to the quantum work measurement problem, to which we now turn.

11.3.2   Quantum Work Measurements: Examples

Damped forced harmonic oscillator.

In this example Horowitz (Horowitz, 2012) considers a simple harmonic oscillator driven by an externally controlled force

$$f = \sqrt{2/(m\omega)}\,vt \tag{11.28}$$

and described by the Hamiltonian

$$\hat{H}(f) = \frac{\hat{p}^2}{2m} + \frac{1}{2}m\omega^2\hat{x}^2 - m\omega^2 f\hat{x}$$

$$= \frac{\hat{p}^2}{2m} + \frac{1}{2}m\omega^2(\hat{x} - f)^2 - \frac{1}{2}m\omega^2 f^2 \qquad (11.29)$$

where m is its mass, ω its frequency, $\hat{x}$ its position and $\hat{p}$ its momentum. The second line of this equation shows that this Hamiltonian describes an oscillator that has been displaced from its original position by f, with its energy lowered by $(1/2)m\omega^2 f^2$. Introducing the creation operator

$$\hat{a}_t^\dagger = \sqrt{\frac{\hbar m\omega}{2}}\left(\hat{x} - f_t + \frac{i\hat{p}}{m\omega}\right) \qquad (11.30)$$

with a similar expression for the annihilation operator $\hat{a}_t$, $\hat{H}(f)$ may be further reexpressed as

$$\hat{H}(f) = \hbar\omega(\hat{a}_t^\dagger\hat{a}_t + 1/2) - \frac{1}{2}m\omega^2 f^2. \qquad (11.31)$$

where the index t labels the operators at the time t when $f(t)$ has the value f_t.

In addition the oscillator is also coupled to a thermal reservoir of two-level atoms with ground state $|g\rangle$ and excited state $|e\rangle$ that act as a measuring apparatus. The idea is to determine the average work on the oscillator and the heat exchanged with the reservoir through projective measurements of the state of the atoms at the beginning and the end of the protocol from a large number of Monte Carlo wave function simulations. This is possible in this system because the knowledge of whether the atoms have undergone a transition during their interaction with the oscillator permits to directly infer how much energy was exchanged with the reservoir. That energy is interpreted as heat, consistently with the discussion of section 11.2.1. Since it depends on the stochastic outcomes of the individual Monte Carlo trajectories it is itself also a stochastic variable. At the same time, the time-dependence of the Hamiltonian also results in a change in the internal energy of the system. Combining that information with the stochastic transfer of energy to the reservoir allows one to determine the change in internal energy and work for each quantum trajectory.

Each two-state atom is described by a Hamiltonian of the form

$$\hat{H}_A = \hbar\omega\hat{\sigma}_+\hat{\sigma}_- \qquad (11.32)$$

where $\hat{\sigma}_- = |g\rangle\langle e|$ and $\hat{\sigma}_+ = |e\rangle\langle g|$. As shown by Horowitz (Horowitz, 2012), for weak coupling $\lambda \ll \omega$ the interaction between these atoms and the forced harmonic oscillator is described by the Hamiltonian

$$V(f_t) = \hbar\lambda(\bar{a}_t^\dagger\hat{\sigma}_- + \bar{a}_t\hat{\sigma}_+) \qquad (11.33)$$

where

$$\hat{\bar{a}}_t^\dagger = \hat{a}_t^\dagger + iv/\omega,$$
$$\bar{a}_t = \hat{a}_t - iv/\omega. \tag{11.34}$$

Before interacting with the oscillator the atoms are prepared either in their ground state $|g\rangle$ or their excited state $|e\rangle$ with probabilities

$$r_g = \frac{1}{1 + e^{-\beta\hbar\omega}}, \tag{11.35}$$

$$r_e = \frac{e^{-\beta\hbar\omega}}{1 + e^{-\beta\hbar\omega}}, \tag{11.36}$$

where $\beta = k_B T$. The interaction between individual atoms and the oscillator is assumed to be weak and of a duration δt short enough that it can be described to first order in perturbation theory. The probability for any one of the atoms in the reservoir to undergo a transition from $|g\rangle$ to $|e\rangle$ during a short interval $dt \gg \delta t$ about t is then given by

$$p_{ge} = (\lambda^2 \delta t) r_g \langle \bar{a}_t^\dagger \bar{a}_t \rangle_t dt \equiv g r_g \langle \bar{a}_t^\dagger \bar{a}_t \rangle_t dt \tag{11.37}$$

where the subscript t outside the brackets indicates that the expectation value is taken at time t and

$$g = \lambda^2 \delta t \tag{11.38}$$

can be interpreted as the jump rate. Similarly the probability to observe a jump from state $|e\rangle$ to $|g\rangle$ is

$$p_{eg} = g r_e \langle \hat{a}(t) \hat{a}^\dagger(t) \rangle dt, \tag{11.39}$$

so that the probability of observing no transition is $1 - p_{ge} - p_{eg}$.

Since the probability of observing a jump scales with dt the process is Poissonian. One can therefore introduce two Poisson increments dN^+ and dN^- corresponding to an atom jumping up or down, and the oscillator jumping correspondingly down or up. The increments are a sequence of random numbers that are either 1 or 0, so that

$$(dN^+)^2 = dN^+ \quad ; \quad (dN^-)^2 = dN^- \tag{11.40}$$

with ensemble expectation values $E[dN^+] = g r_g \langle \hat{a}_t^\dagger \hat{a}_t \rangle_t$ and $E[dN^-] = g r_e \langle \hat{a}_t \hat{a}_t^\dagger \rangle_t$. This permits description of the evolution of the system in terms of the stochastic Monte Carlo Schrödinger approach outlined in section 11.3.1.

Each time an atom jumps from its ground state $|g\rangle$ to the excited state $|e\rangle$ along the stochastic trajectory $|\psi\rangle$ it absorbs $\hbar\omega$ of energy, and in the reverse process it releases that same amount of energy. Since heat is the energy exchanged with the thermal reservoir, the energy released or absorbed by the reservoir atoms can therefore be identified as the heat absorbed or emitted by the oscillator. For a given trajectory the heat increment absorbed by the oscillator between t and $t + dt$ is therefore

$$dQ_t[\psi] = \hbar\omega[dN_t^- - dN_t^+], \tag{11.41}$$

and

$$Q_t[\psi] = \int_0^t \hbar\omega[dN_t^- - dN_t^+]. \tag{11.42}$$

Similarly, the change in internal energy $U_t[\psi] = \langle\psi_t|\hat{H}(f_t)|\psi_t\rangle$, which is due as we have seen both to the time dependence of the Hamiltonian and the stochastic evolution of the state vector along the Monte Carlo trajectory is found to be (Horowitz, 2012)

$$
\begin{aligned}
dU_t[\psi] =\ & \dot{f}_t\langle\psi_t|\partial_f\hat{H}(f_t)|\psi_t\rangle\,dt \\
& - gr_g\langle\psi_t|\left[\frac{1}{2}\{\hat{H}(f_t),\bar{a}_t^\dagger\bar{a}_t\} - \hat{H}(f_t)\langle\bar{a}_t^\dagger\bar{a}_t\,\rangle_t\right]|\psi_t\rangle\,dt \\
& - gr_e\langle\psi_t|\left[\frac{1}{2}\{\hat{H}(f_t),\bar{a}_t\bar{a}_t^\dagger\} - \hat{H}(f_t)\langle\bar{a}_t\bar{a}_t^\dagger\,\rangle_t\right]|\psi_t\rangle\,dt \\
& + \left[\frac{\langle\bar{a}_t\hat{H}(f_t)\bar{a}_t^\dagger\rangle_t}{\langle\bar{a}_t\bar{a}_t^\dagger\rangle_t} - \langle\hat{H}(f_t)\rangle_t\right]dN_t^- + \left[\frac{\langle\bar{a}_t^\dagger\hat{H}(f_t)\bar{a}_t\rangle_t}{\langle\bar{a}_t^\dagger\bar{a}_t\rangle_t} - \langle\hat{H}(f_t)\rangle_t\right]dN_t^+,
\end{aligned}
\tag{11.43}
$$

where $\{\hat{A},\hat{B}\} = \hat{A}\hat{B} + \hat{B}\hat{A}$ is the anticommutator of $\hat{A}$ and $\hat{B}$. Finally, the work performed during the interval for t to $t + dt$ is the change in energy not accounted by the heat

$$dW_t[\psi] = dU_t[\psi] - dQ_t[\psi]. \tag{11.44}$$

Averaging over a large number of quantum Monte Carlo trajectories permits determination of the mean work and heat, as well as their higher moments.

Driven damped two-state system

In this second example, due to Hekking and Pekola (Hekking and Pekola, 2013), a driven two-state system described by the Hamiltonian

$$\hat{H}_s = -\hbar\omega_0\hat{\sigma}_z/2 + \lambda(t)(\hat{\sigma}_+ + \hat{\sigma}_-), \tag{11.45}$$

where $\hat{\sigma}_z = |e\rangle\langle e| - |g\rangle\langle g|$ and $\lambda(t)$ is a control parameter, is linearly coupled to a bath of harmonic oscillators by the interaction Hamiltonian

$$\hat{H}_c = \sum_\mu (c_\mu \hat{\sigma}_+ \hat{b}_\mu + \text{h.c.}). \tag{11.46}$$

The annihilation and creation operators $\hat{b}_\mu$ and $\hat{b}_\mu^\dagger$ of the mode of frequency ω_μ satisfy the familiar bosonic commutation relations $[\hat{b}_\mu, \hat{b}_\mu^\dagger] = \delta_{\mu,\mu'}$, and the c_μ are coupling constants.

Like Horowitz (Horowitz, 2012), Hekking and Pekola evaluate the work on this system by making two measurements, one before and one after the driving period and use Monte Carlo wave function simulations to generate a number of stochastic realizations of the protocol, an approach that allows them to obtain the quantum statistics of the work distribution in the system.

Assuming that at time t the two-state system is in the superposition

$$|\psi(t)\rangle = [a(t)|g\rangle + b(t)|e\rangle]|0\rangle \tag{11.47}$$

the probabilities $\Gamma_\uparrow$ and $\Gamma_\downarrow$ that it either absorbs or emits a photon in a time interval Δt short enough that at most one photon is exchanged with the bath are

$$\Gamma_\downarrow = \frac{2\pi}{\hbar} \sum_\mu (\bar{n}_\mu + 1)|c_\mu|^2 \delta(\omega_0 - \omega_\mu),$$

$$\Gamma_\uparrow = \frac{2\pi}{\hbar} \sum_\mu \bar{n}_\mu |c_\mu|^2 \delta(\omega_0 - \omega_\mu), \tag{11.48}$$

where $\bar{n}_\mu = [\exp \hbar\omega_\mu / k_B T) - 1]^{-1}$. From the discussion of section 11.3.1, if the two-state system does not perform a jump then its evolution is then governed by the effective non-Hermitian Hamiltonian

$$\hat{H}_{\text{eff}} = \hat{H}_s - i\hbar\Gamma_\downarrow |e\rangle\langle e| - i\hbar\Gamma_\uparrow |g\rangle\langle g|. \tag{11.49}$$

The probability that it undergoes a quantum jump instead during that same interval Δt is

$$\Delta p = \left[|a(t)|^2 \Gamma_\uparrow + |b(t)|^2 \Gamma_\downarrow \right] \Delta t. \tag{11.50}$$

with the atom winding up in its excited state if the jump corresponds to the absorption of a photon and in its ground state $|g\rangle$ if the atom emits a photon.

The dynamics of the system can be simulated by the Monte Carlo wave functions approach of section 11.3.1, picking a random number ϵ between 0 and 1 for each time increment Δt. If that number is larger than Δp then the system follows the dynamics given by $\hat{H}_{\text{eff}}$, and the wave function is properly renormalized at the end of the step. And if ϵ is less that Δp then the system will have either emitted a photon with probability $|b(t)|^2 \Gamma_{\downarrow}/[|a(t)|^2 \gamma_{\uparrow} + |b(t)|^2 \gamma_{\downarrow}]$ or absorbed it with probability $|a(t)|^2 \Gamma_{\uparrow}/[|a(t)|^2 \gamma_{\uparrow} + |b(t)|^2 \gamma_{\downarrow}]$. By averaging the Monte Carlo simulations over a distribution of initial states whose distribution corresponds to the initial system density operator one reproduces the full master equation dynamics, with the substantial benefit of generating a set of quantum trajectories typical of what would be observed in single runs of the experiment.

Hekking and Pekola evaluated the work for each quantum trajectory by making one measurement before and one after the driving period. These measurements were realized by the detection (projective measurement) of the last photon emitted to the environment, or absorbed by the two-state system. If the results of both the first and the second measurement indicate that the atom is in the ground state $|g\rangle$, then the internal energy of the system was not changed by its coupling to the bath for that particular realization. The other two possible outcomes are $\Delta U = \pm \hbar \omega_0$, in which case heat is taken from or given to the environment during the quantum jump events. The quantum statistics of the work can then be directly extracted from a statistical analysis of the results of a large number of stochastic Monte Carlo trajectories, and taking into account the time dependence of the system Hamiltonian.

11.4   Continuous Measurements

The two examples of the previous section exploit projective measurements on the bath to extract information on the heat exchange with the environment and the work being performed. In essence, one can think of the reservoir of two-level atoms considered by Horowitz (Horowitz, 2012) as a probe beam on which projective measurements are carried out to extract information about the system. In practice this is not easy to do, though. Thermal reservoirs are normally assumed to be large enough that they remain essentially unchanged as a result of their interaction with the system, so measuring a small change in the state of one of their constituents is no small challenge. Also, the use of projective measurements is not without its own limitations, as they produce in general a significant back-action on the system. This back-action impacts its later evolution and hence the outcome of subsequent measurements, and presumably the work and heat extracted from or delivered to the system.

Still, there are obvious benefits and useful information to be gained by monitoring a single system repeatedly without having to reset it. One way to minimize the effects of measurement back-action would be to probe the systems gently so as to not significantly perturb its subsequent evolution. Quantum non-demolition measurements can achieve this goal in principle, but they are not necessarily easy to realize. An alternative approach is offered by non-projective weak measurements. This section discusses how to do that, and outlines the main steps in the derivation of a stochastic formalism that leads to the description of weak continuous quantum measurements in terms of stochastic master equations (or alternatively stochastic Schrödinger equations). This is taken directly from the excellent tutorial presentation of Jacobs and Steck (Jacobs and Steck, 2006), to which the reader is referred for more details.

Projective measurements use a set of projection operators $\{\hat{P}_n = |n\rangle\langle n|\}$ that describe what happens as one of the possible outcomes of the measurement. If the pre-measurement density operator is $\hat{\rho} = |\psi\rangle\langle\psi|$ with $|\psi\rangle = \sum_n c_n|n\rangle$ then the nth possible final state is

$$\hat{\rho}_f = |n\rangle\langle n| = \frac{\hat{P}_n\hat{\rho}\hat{P}_n}{\text{Tr}(\hat{P}_n\hat{\rho}\hat{P}_n)} \qquad (11.51)$$

with probability $P(n) = \text{Tr}(\hat{P}_n\hat{\rho}\hat{P}_n) = |c_n|^2$.

As discussed by Jacobs and Steck (Jacobs and Steck, 2006) every measurement, not just projective measurements, can be described in a similar fashion by generalizing the set of projection operators $\hat{P}_n$. If we pick any set of m_{max} operators $\hat{\Omega}_m$ with the restriction

$$\sum_{m=1}^{m_{\text{max}}} \hat{\Omega}_m^\dagger\hat{\Omega}_m = I, \qquad (11.52)$$

where I is the identity operator, then it is possible to design a measurement with possible outcomes

$$\hat{\rho}_f = \frac{\hat{\Omega}_m\hat{\rho}\hat{\Omega}_m^\dagger}{\text{Tr}[\hat{\Omega}_m\hat{\rho}\hat{\Omega}_m^\dagger]} \qquad (11.53)$$

occurring with probabilities

$$P(m) = \text{Tr}[\hat{\Omega}_m\hat{\rho}\hat{\Omega}_m^\dagger] \qquad (11.54)$$

and with the total probability of obtaining a result in the range $[a, b]$ given by

$$P(m \in [a,b]) = \sum_{m=a}^{b} \text{Tr}[\hat{\Omega}_m\hat{\rho}\hat{\Omega}_m^\dagger] = \text{Tr}\left[\sum_{m=a}^{b} \hat{\Omega}_m^\dagger\hat{\Omega}_m\hat{\rho}\right]. \qquad (11.55)$$

These generalized measurements are referred to as Positive Operator-Valued Measures, or POVM. They can be implemented by performing a unitary interaction between the system to be characterized and an auxiliary system and then performing a von Neumann measurement on that system.

As is clear from their name, continuous measurements extract information from the system continuously. To construct such measurement schemes time is divided into a sequence of small intervals Δt and a weak measurement is performed during each of them. We denote an observable to be monitored in this way by the Hermitian operator $\hat{X}$ and assume for simplicity that it has a continuous spectrum of eigenvalues $\{x\}$ with corresponding eigenstates $|x\rangle$, so that $\langle x|x'\rangle = \delta(x - x')$. We are not interested in making projective measurements that would leave $\hat{X}$ in one of its eigenstates. Rather, we consider 'weaker measurements' characterized by the POVM

$$\hat{A}(\alpha) = \left(\frac{4k\Delta t}{\pi}\right)^{1/4} \int_{-\infty}^{+\infty} e^{-2k\Delta t(x-\alpha)^2} |x\rangle \langle x| dx. \tag{11.56}$$

This is a Gaussian-weighted sum of projectors onto the eigenstates of $\hat{X}$ that provides only partial information about the observable. Here α is a continuous index, such that the spectrum of measurement results is a continuum labelled by it. As we shall see, the parameter k can be understood as a measure of the measurement strength. Continuous measurements result from taking the limit $\Delta t \to 0$.

In practice the measurements are realized by coupling the system to a measuring apparatus through an interaction that is proportional to $\hat{X}$ and to some observable of the measuring device that is then determined by a projective measurement. In the quantum heat engine that we consider in the following sections the 'measuring apparatus' will consist of a low density beam of two-level atoms that can be either resonant (absorptive case) or off-resonant (dispersive case) with the intracavity field to be measured. This use of two-state systems as a measuring apparatus is reminiscent to the situation considered by Horowitz (Horowitz, 2012) and discussed in the previous section, with the important difference that the atoms are not treated as a reservoir and that the interaction with the monitored system is weak and not akin to projective measurements.

For the initial state $|\psi\rangle = \int \psi(x)|x\rangle dx$ we have

$$P(\alpha) = \text{Tr}[\hat{A}(\alpha)|\psi\rangle\langle\psi|\hat{A}(\alpha)^\dagger], \tag{11.57}$$

compare with Eq. (11.54), and

$$\langle \alpha \rangle = \int_{-\infty}^{\infty} \alpha P(\alpha) d\alpha = \int_{-\infty}^{\infty} \alpha \text{Tr}[\hat{A}(\alpha)^\dagger \hat{A}(\alpha)|\psi\rangle\langle\psi|] = \int_{-\infty}^{\infty} x|\psi(x)|^2 dx = \langle \hat{X} \rangle. \tag{11.58}$$

This shows that the expectation value of α is equal to the expectation value of $\hat{X}$, as should be the case. Furthermore the probability density to obtain the measurement outcome α is (Jacobs and Steck 2006)

$$P(\alpha) = \text{Tr}[\hat{A}(\alpha)|\psi\rangle\langle\psi|\hat{A}^\dagger(\alpha)]$$
$$= \sqrt{\frac{4k\Delta t}{\pi}} \int_{-\infty}^{+\infty} |\psi(x)|^2 e^{-4k\Delta t(\alpha-x)^2} dx.$$

For Δt sufficiently small the Gaussian under the integral is much broader than $\psi(x)$, so that one can approximate $|\psi(x)|^2$ by a delta function centred at the expectation value $\langle \hat{X} \rangle$. We then have that

$$P(\alpha) \simeq \sqrt{\frac{4k\Delta t}{\pi}} e^{-4k\Delta t(\alpha - \langle \hat{X} \rangle)^2}, \tag{11.59}$$

which is a Gaussian of variance $\sigma^2 = 1/(8k\Delta t)$ centred at the expectation value $\langle \hat{X} \rangle$. We can therefore think of α as the stochastic quantity

$$\alpha = \langle \hat{X} \rangle + \sigma \frac{\Delta w}{\sqrt{\Delta t}} = \langle \hat{X} \rangle + \frac{\Delta w}{\sqrt{8k\Delta t}}. \tag{11.60}$$

Here Δw is a zero-mean Gaussian random variable of variance Δt. That is, its root mean square scales as $(\Delta t)^{1/2}$ and its variance scales as Δt. It is that stochastic nature of α that accounts for the random nature of the quantum successive measurements. The larger k, the smaller the fluctuations in the measurement outcomes. This justifies associating it with the measurement strength.

In the infinitesimal limit $\Delta t \to dt$ and $\Delta w \to dw$ the stochastic variable $w(t)$ is referred to as a Wiener process, a random walk with arbitrary small, independent steps taken arbitrarily often. Importantly, one needs to keep in mind that the Wiener differential dw satisfies the Itô rule $dw^2 = dt$. This might appear surprising since dw is a stochastic quantity but dt is not, and so neither is dw^2, a subtle point discussed in detail in Jacobs and Steck (Jacobs and Steck, 2006).

These results permit us to numerically determine at each time step the evolution of the wave function $|\psi(t)\rangle$ subject to measurements characterized by the POVM $\hat{A}(\alpha)$. The infinitesimal, stochastic change in the quantum state following a single measurement is given by

$$|\psi(t + \Delta t)\rangle \propto \hat{A}(\alpha)|\psi(t)\rangle \propto e^{-2k\Delta t(\alpha - \hat{X})^2}|\psi(t)\rangle. \tag{11.61}$$

Expanding the exponential to first order in $\Delta t \to dt$ and to second order in the Wiener process dw (necessary since $dw^2 = dt$) and normalizing $|\psi(t + dt)\rangle$ finally gives the stochastic Schrödinger equation

$$d|\psi\rangle = [-k(\hat{X} - \langle \hat{X} \rangle)^2 dt + \sqrt{2k}(\hat{X} - \langle \hat{X} \rangle)dw]|\psi(t)\rangle. \tag{11.62}$$

For example, for measurements on a single-mode field amplitude, $\hat{X} = (\hat{a} + \hat{a}^\dagger)$, one finds (Imoto, Ueda, and Ogawa, 1990), (Jacobs and Steck, 2006)

$$d|\psi(t)\rangle = \left\{ \left[-\frac{i}{\hbar}\hat{H}_{ab} - \frac{1}{2}\lambda_a \left(\hat{a}^\dagger \hat{a} - \langle \hat{a} + \hat{a}^\dagger \rangle \hat{a} + \frac{\langle \hat{a} + \hat{a}^\dagger \rangle^2}{4} \right) \right] dt \right.$$
$$\left. + \sqrt{\lambda_a} \left(\hat{a} - \frac{\langle \hat{a} + \hat{a}^\dagger \rangle}{2} \right) dw \right\} |\psi(t)\rangle, \tag{11.63}$$

where λ_a measures the strength of the measurement. Likewise, for measurements of the field intensity, $\hat{X} = \hat{a}^\dagger \hat{a}$, with measurement strength λ_d the stochastic Schrödinger equation reads (Ueda, Imoto, Nagaoka and Ogawa, 1992)

$$d|\psi(t)\rangle = \left\{ \left[-\frac{i}{\hbar}\hat{H}_{ab} - \frac{1}{2}\lambda_d \left(\hat{n}_a - \langle \hat{n}_a \rangle \right)^2 \right] dt + \sqrt{\lambda_d}(\hat{n}_a - \langle \hat{n}_a \rangle) dw \right\} |\psi(t)\rangle. \qquad (11.64)$$

These two equations will reappear in section 11.8 in the discussion of continuous measurement schemes of the work and its quantum fluctuations of a quantum optomechanical quantum heat engine.

11.5   Quantum Heat Engines

Quantum heat engines (QHE) are heat engines that use quantum matter rather than classical matter as their working substance. This is not a new topic: as early as 1959 Scovil and Schultz-Dubois (Scovil and Schultz-Dubois, 1959), see also Geusic (Geusic, Scovil and Schultz-Dubois, 1967), discussed the three-level maser as an example of a QHE and introduced the concept of negative temperatures in these systems. The recent resurgence of interest in QHE was triggered in part by advances in ultracold science, single atom and ion manipulation, and nanofabrication. A number of new systems have been proposed and some have already been demonstrated (Roßnagel, Dawkins, Tolazzi, Abah, Lutz, Schmidt-Kaler and Singer, 2016) or are close to experimental realization. This section, which follows the excellent introduction by Quan (Quan, Liu, Sun and Nori, 2007), briefly summarizes the most important thermodynamic cycles used in these engines.

11.5.1   Quantum thermodynamic processes

Section 11.2.1 showed how the First Law of thermodynamics permits expression of differential changes in average work and heat in terms of the eigenenergies of a quantum working substance and the probabilities of occupation of the corresponding eigenstates. This decomposition allows one to easily extend the basic isothermal, isochoric and adiabatic thermodynamic processes familiar from classical thermodynamics to the quantum regime.

Quantum isothermal processes are similar to classical isothermal processes: the working substance is in contact with a heat bath at constant temperature and can perform work to the outside by absorbing heat from that bath. In that situation both the eigenenergies E_i and their probabilities of occupation p_i in Eq. (11.2)

$$U = \langle \hat{H}_s \rangle = \sum_i p_i E_i, \qquad (11.65)$$

need to change simultaneously to account for the work performed and heat exchange with the thermal bath.

Quantum isochoric processes share with their classical counterparts the property that no work is performed, and heat is exchanged with the thermal bath. Classically, this corresponds to keeping the volume of the working substance constant. In their quantum version, isochoric processes are characterized by the fact that the occupation probabilities p_i change, but the eigenenergies E_i remain constant.

Quantum adiabatic processes are somewhat more subtle, because of the difference between classical and quantum adiabaticity. In classical adiabaticity there is no heat exchange with the reservoir(s), $dQ = 0$, but work can be performed. Quantum adiabaticity requires in addition that the occupation probabilities p_i remain constant. That is, quantum adiabatic processes are also adiabatic in the classical sense, but classical adiabatic processes need not satisfy quantum adiabaticity.

Because of these differences in the characterization of quantum versus classical thermodynamic processes their common nature is best visualized in entropy–temperature S-T diagrams, as shown in Quan (Quan, Liu, Sun and Nori, 2007). We now discuss how they enter into two of the most familiar cycles used in QHE, Carnot cycles and Otto cycles.

11.5.2   Carnot and Otto engine cycles

Like their classical counterparts, *quantum Carnot engines* consist of two isothermal processes and two adiabatic processes. During the isothermal expansion processes the quantum working substance is kept in contact with a heat bath, with its energy levels changing slowly enough that it remains in constant thermal equilibrium with that bath. Quantum adiabatic processes keep the populations p_i of the energy levels E_i constant, but change the values of these energies. As mentioned by Quan (Quan, Liu, Sun and Nori, 2007) ideally all energy differences $E_i - E_j$ should change by the same ratio in these processes, and this ratio must be equal to the ratio of the temperatures of the heat baths. If that is not the case then a thermalization of the working substance is inevitable before the subsequent isothermal process. In that step the combined system plus bath entropy increases, indicating that it is not reversible.

Quantum Otto cycles consist of two isochoric and two quantum adiabatic processes. In the isochoric strokes no work is done, but heat is absorbed. Importantly the heat bath and the working substance are not always in thermal equilibrium, so that the process is not thermodynamically reversible. In contrast to quantum Carnot engines, there is no constraint on maintaining the ratio of the energy separations during the adiabatic processes.

Experimental advances in single atom and ion manipulation and in nanofabrication have led to an increased interest in QHE. Abah and coworkers Abah *et al.* (2012) considered a scheme to realize a nanoscale heat engine with a single ion, the idea being to confine that ion in a linear Paul trap with tapered geometry and coupling it to engineered laser reservoirs to realize an Otto cycle. Fialko and Hallwood (Fialko and Hallwood, 2012) analysed theoretically an isolated system using cold bosonic atoms confined to a double

		isothermal	isochoric	adiabatic
classical	invariant	T	V	
	variable	P, V	P, T	P, T, V
	heat absorbed or released	✓	✓	
	work done	✓		✓
quantum	invariant	T	E_n	P_n
	variable	U, E_n, P_n	P_n, T_{eff}	E_n, T_{eff}
	heat absorbed or released	✓	✓	
	work done	✓		✓

Fig. 11.1 *Comparison of the classical and quantum properties of basic thermodynamic processes. Adapted with permission from (Quan, Liu, Sun and Nori, 2007). Copyright (2007) by the American Physical Society.*

well potential created by splitting a harmonic trap with a focused laser. The system shows thermalization and is predicted to be able to operate as a heat engine with a finite quantum heat bath, thereby addressing fundamental questions related to the thermalization of isolated quantum systems and the so-called eigenstate thermalization hypothesis. More recently, Bergenfeldt (Bergenfeldt, Samuelsson, Sothmann, Flindt and Büttiker, 2014) analysed the use of hybrid microwave cavities as QHE, a possible realization consisting of two macroscopically separated quantum-dot conductors coupled capacitively to the fundamental mode of a microwave cavity. A QHE consisting of a photon gas inside an optical cavity as the working fluid and multi-atom coherent atomic systems as the fuel was also recently analysed by Hardal and Müstecaplıoğlu (Hardal and Müstecaplıoğlu, 2015), and we have already mentioned the recent single-ion experimental realization of a QHE by Roßnagel (Roßnagel, Dawkins, Tolazzi, Abah, Lutz, Schmidt-Kaler and Singer, 2016).

11.6 The Optomechanical Interaction–Polariton Picture

In the following sections we consider an optomechanical QHE and an optomechanical heat pump whose working fluid consists of photon–photon polaritons, the normal modes of optomechanically coupled photon and phonon fields. We mentioned in the introduction that the physical nature of these quasiparticles can be changed continuously from photon-like to phonon-like by varying some control parameter in the photon–phonon interaction. Because both photons and phonons are typically coupled to thermal reservoirs at different temperatures, this allows one to exploit the interplay of the quantum coherence inherent to the polaritons and dissipation mechanisms to operate heat engines. This is the principle on which the optomechanical QHE is based. Since polaritons are central to its understanding we proceed by first reviewing their main properties in the context of a generic two-mode optomechanical system.

That system consists of a Fabry–Pérot resonator with a compliant end mirror of effective mass m and oscillation frequency ω_m. We assume that the cavity modes are sufficiently well separated that a single mode of frequency ω_{cav} needs to be considered. It is driven by a classical field of amplitude α_{in} and frequency ω_p. The mean-field steady state is characterized by an intracavity field amplitude α and normalized mirror displacement $x/x_{\text{zpt}} = \beta$, where $x_{\text{zpt}} = (\hbar/2m\omega_m)^{1/2}$ is the zero-point mirror displacement. For small optical damping rates $\kappa \ll |\omega_p - \omega_{\text{cav}}|$ we have $\alpha \approx \alpha_{\text{in}}/\Delta$ and $\beta \approx -g_0\alpha^2/\omega_m$, where

$$\Delta = \omega_p - \omega_{\text{cav}} - 2g_0\beta \tag{11.66}$$

is the effective detuning between the optical pump and the cavity mode.

Linearizing the optomechanical interaction about α the effective optomechanical Hamiltonian simplifies to (see e.g. (Aspelmeyer, Kippenberg and Marquardt, 2014))

$$\hat{H}_0 \equiv \hat{H}_{(ab)} = -\hbar\Delta\hat{a}^\dagger\hat{a} + \hbar\omega_m\hat{b}^\dagger\hat{b} + \hbar G(\hat{b} + \hat{b}^\dagger)(\hat{a} + \hat{a}^\dagger). \tag{11.67}$$

Here $G = g_0\alpha$ is taken for simplicity to be real and positive,

$$g_0 = x_{\text{zpt}}\frac{\partial\omega_{\text{cav}}}{\partial x} \tag{11.68}$$

is the single-photon optomechanical coupling constant, $\hat{a}$ is the photon annihilation operator for the quantum fluctuations of the optical mode, and $\hat{b}$ describes the quantum fluctuations of the mechanics.[3]

11.6.1   Polariton spectrum

The optomechanical Hamiltonian (11.67) can be diagonalized in terms of two bosonic normal modes, or polaritons, with annihilation operators $\hat{A}$ and $\hat{B}$ as

$$\hat{H}_0 \equiv \hat{H}_{(AB)} = \hbar\omega_A\hat{A}^\dagger\hat{A} + \hbar\omega_B\hat{B}^\dagger\hat{B} + \text{const.} \tag{11.69}$$

---

[3] We use the notation $\hat{H}_{(ab)}$ when working in the bare modes basis and $\hat{H}_{(AB)}$ when working in the normal modes (polariton) basis if there is a possible ambiguity.

Briefly, the diagonalization proceeds by expressing the polariton annihilation and creation operators $\hat{A}$ and $\hat{B}$ in terms of the bare modes via the Bogoliubov transformation

$$\begin{pmatrix} \hat{A} \\ \hat{B} \\ \hat{A}^\dagger \\ \hat{B}^\dagger \end{pmatrix} = \begin{pmatrix} U^\dagger & -V^\dagger \\ -V^T & U^T \end{pmatrix} \begin{pmatrix} \hat{a} \\ \hat{b} \\ \hat{a}^\dagger \\ \hat{b}^\dagger \end{pmatrix}, \tag{11.70}$$

where U and V are 2×2 submatrices that satisfy the relationships

$$U^\dagger U - V^\dagger V = I, \tag{11.71}$$
$$U^T V - V^T U = 0, \tag{11.72}$$

with the inverse transformation

$$\begin{pmatrix} \hat{a} \\ \hat{b} \\ \hat{a}^\dagger \\ \hat{b}^\dagger \end{pmatrix} = \begin{pmatrix} U & V^* \\ V & U^* \end{pmatrix} \begin{pmatrix} \hat{A} \\ \hat{B} \\ \hat{A}^\dagger \\ \hat{B}^\dagger \end{pmatrix}. \tag{11.73}$$

One finds that the polariton mode frequencies are (Zhang, Bariani and Meystre, 2014b)

$$\omega_A(\Delta) = \frac{1}{\sqrt{2}} \sqrt{\Delta^2 + \omega_m^2 + \sqrt{(\Delta^2 - \omega_m^2)^2 - 16G^2 \Delta \omega_m}}, \tag{11.74}$$

$$\omega_B(\Delta) = \frac{1}{\sqrt{2}} \sqrt{\Delta^2 + \omega_m^2 - \sqrt{(\Delta^2 - \omega_m^2)^2 - 16G^2 \Delta \omega_m}}, \tag{11.75}$$

see Fig. 11.2. At the avoided crossing $\Delta = -\omega_m$ they simplify to

$$\omega_{A,B}(-\omega_m) = \omega_m \sqrt{1 \pm \frac{2G}{\omega_m}}, \tag{11.76}$$

indicating that the minimum frequency difference between the polariton branches 'A' and 'B' is proportional to the dimensionless optomechanical coupling $g \equiv G/\omega_m$. Importantly, for the dimensionless detuning $\delta \equiv \Delta/\omega_m \to -\infty$ we have $\omega_A(-\infty) \to -\Delta$ and $\omega_B(-\infty) \to \omega_m$. In that limit the branch 'A' describes a photon-like excitation and the branch 'B' a phonon-like excitation. In contrast, for $-\omega_m < \Delta < 0$ we have

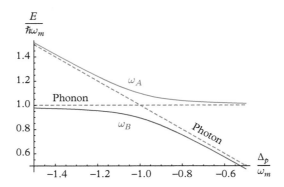

Fig. 11.2 *Frequencies of the two polariton branches (normal modes) of the optomechanical system for* $G/\omega_m = 0.05$ *in the red-detuned case* $\Delta < 0$. *The dashed curves correspond to the frequencies of the bare photon and phonon modes.*

$$\omega_A(\Delta) \approx \omega_m \left(1 + \frac{2G^2\Delta}{(\Delta^2 - \omega_m^2)\omega_m} \right),$$

$$\omega_B(\Delta) \approx -\Delta \left(1 - \frac{2G^2\omega_m}{(\Delta^2 - \omega_m^2)\Delta} \right). \tag{11.77}$$

so that or $\Delta \to 0^{(-)}$ the polariton 'A' is phonon-like and 'B' is photon-like.

For small dimensionless optomechanical couplings $g \ll 1$ and detunings $\delta \ll -1$ the normal mode operators are given in terms of the bare mode operators by

$$\hat{A} = \left[1 + \frac{2\delta g^2}{(\delta-1)^2} \right] \hat{a} - \frac{g}{1+\delta}\hat{b} + \frac{g^2}{\delta(1-\delta^2)}\hat{a}^\dagger + \frac{g}{1-\delta}\hat{b}^\dagger, \tag{11.78}$$

$$\hat{B} = \frac{g}{1+\delta}\hat{a} + \left[1 + \frac{2\delta g^2}{(\delta-1)^2} \right] \hat{b} + \frac{g}{1-\delta}\hat{a}^\dagger + \frac{g^2\delta}{\delta^2-1}\hat{b}^\dagger, \tag{11.79}$$

and the polariton number operators $\hat{n}_A \equiv \hat{A}^\dagger\hat{A}$ and $\hat{n}_B \equiv \hat{B}^\dagger\hat{B}$ are then

$$\hat{n}_A = \left[1 + \frac{4\delta g^2}{(\delta-1)^2} \right]\hat{a}^\dagger\hat{a} + \frac{2(1+\delta^2)g^2}{(\delta-1)^2}\hat{b}^\dagger\hat{b} + \left(\frac{g}{1-\delta} \right)^2$$

$$- \frac{g}{1+\delta}(\hat{a}^\dagger\hat{b} + \hat{b}^\dagger\hat{a}) + \frac{g^2}{\delta(1-\delta^2)}(\hat{a}^2 + \hat{a}^{\dagger 2}) + \frac{g}{1-\delta}(\hat{a}\hat{b} + \hat{b}^\dagger\hat{a}^\dagger) - \frac{g^2}{1-\delta^2}(\hat{b}^2 + \hat{b}^{\dagger 2}), \tag{11.80}$$

$$\hat{n}_B = \left[1 + \frac{4\delta g^2}{(\delta-1)^2} \right]\hat{b}^\dagger\hat{b} + \frac{2(1+\delta^2)g^2}{(\delta-1)^2}\hat{a}^\dagger\hat{a} + \left(\frac{g}{1-\delta} \right)^2$$

$$+ \frac{g}{1+\delta}(\hat{a}^\dagger\hat{b} + \hat{b}^\dagger\hat{a}) + \frac{g^2}{1-\delta^2}(\hat{a}^2 + \hat{a}^{\dagger 2}) + \frac{g}{1-\delta}(\hat{a}\hat{b} + \hat{b}^\dagger\hat{a}^\dagger) + \frac{g^2\delta}{\delta^2-1}(\hat{b}^2 + \hat{b}^{\dagger 2}). \tag{11.81}$$

To lowest order one can neglect the intermode correlations and squeezing terms appearing in these expressions. This approximates the mean polariton numbers by the mean thermal occupations of the optical reservoir $\bar{n}_a$ and of the mechanical reservoir $\bar{n}_b$, respectively. When the optomechanical coupling is small but finite, though, all terms in Eqs. (11.80) and (11.81) contribute, and the steady-state polariton populations deviate from thermal equilibrium. For $-1 < \delta < 0$, the expressions for $\hat{A}$ and $\hat{B}$ are simply interchanged.

11.6.2 Dissipation

In addition to the unitary evolution governed by the Hamiltonian (11.67), the optical mode and the mechanical mode also suffer from dissipation at rates κ and γ, respectively. Since for high-Q mechanical oscillators the effect of Brownian thermal motion can be described by a familiar Lindblad form (see e.g. Gröblacher (Gröblacher, Trubarov, Prigge, Cole, Aspelmeyer and Eisert, 2015)), the master equation describing these two dissipation channels is

$$\frac{d\hat{\rho}}{dt} = -\frac{i}{\hbar}[\hat{H}_0, \hat{\rho}] + \kappa(\bar{n}_a + 1)\mathcal{L}_{\hat{a}}[\hat{\rho}] + \kappa\bar{n}_a\mathcal{L}_{\hat{a}^\dagger}[\hat{\rho}]$$
$$+ \gamma(\bar{n}_b + 1)\mathcal{L}_{\hat{b}}[\hat{\rho}] + \gamma\bar{n}_b\mathcal{L}_{\hat{b}^\dagger}[\hat{\rho}],$$

where

$$\mathcal{L}_{\hat{x}}[\hat{\rho}] = \hat{x}\hat{\rho}\hat{x}^\dagger - \frac{1}{2}\hat{x}^\dagger\hat{x}\hat{\rho} - \frac{1}{2}\hat{\rho}\hat{x}^\dagger\hat{x}. \qquad (11.82)$$

The Lindblad superoperators $\mathcal{L}_{\hat{a}}[\hat{\rho}]$ and $\mathcal{L}_{\hat{b}}[\hat{\rho}]$ couple the two polariton modes, so that it is not possible to describe their evolution in terms of two uncoupled master equations. As a consequence we shall see that the dissipative coupling of the polariton modes can have a negative impact on the work extracted from the QHE that we consider in the next sections.

To illustrate more concretely how dissipation impacts the polariton dynamics we consider the simple case where the normal mode 'A' remains in the vacuum state at all times, $\hat{\rho} \approx \hat{\rho}_B \otimes |0\rangle \langle 0|_A$, so that $\hat{H}_{(AB)} \to \hat{H}_B = \hbar\omega_B(\Delta)\hat{B}^\dagger\hat{B}$. The dissipation of mode 'B' is then governed by the master equation (Zhang, Bariani and Meystre, 2014b)

$$\frac{d\hat{\rho}_B}{dt} = -\frac{i}{\hbar}[\hat{H}_B, \hat{\rho}_B] + \Gamma_B(\bar{n}_B + 1)\mathcal{L}_{\hat{B}}[\hat{\rho}_B] + \Gamma_B\bar{n}_B\mathcal{L}_{\hat{B}^\dagger}[\hat{\rho}_B]$$
$$+ \Gamma_B\bar{m}_B\mathcal{J}_{\hat{B}}[\hat{\rho}_B] + \Gamma_B\bar{m}_B^*\mathcal{J}_{\hat{B}^\dagger}[\hat{\rho}_B] \qquad (11.83)$$

where

$$\bar{n}_B = \frac{\kappa(\bar{n}_a + 1)|V_{12}|^2 + \kappa\bar{n}_a|U_{12}|^2 + \gamma(\bar{n}_b + 1)|V_{22}|^2 + \gamma\bar{n}_b|U_{22}|^2}{\kappa(|U_{12}|^2 - |V_{12}|^2) + \gamma(|U_{22}|^2 - |V_{22}|^2)}, \qquad (11.84)$$

$$\bar{m}_B = \frac{\kappa(2\bar{n}_a + 1)V_{12}U_{12} + \gamma(2\bar{n}_b + 1)V_{22}U_{22}}{\kappa(|U_{12}|^2 - |V_{12}|^2) + \gamma(|U_{22}|^2 - |V_{22}|^2)}, \tag{11.85}$$

and $\Gamma_B = \kappa(|U_{12}|^2 - |V_{12}|^2) + \gamma(|U_{22}|^2 - |V_{22}|^2)$. Here U_{ij} and V_{ij} are the elements of the sub-matrices U and V of the Bogoliubov transformation. The master equation (11.83) differs from the bare mode master equation through the appearance of the additional Lindblad operators

$$\mathcal{J}_{\hat{x}}[\hat{\rho}] = \hat{x}\hat{\rho}\hat{x} - \frac{1}{2}\hat{x}\hat{x}\hat{\rho} - \frac{1}{2}\hat{\rho}\hat{x}\hat{x}, \tag{11.86}$$

with $\hat{x} = \hat{B}$ and $\hat{B}^\dagger$. This shows that the polariton mode is effectively coupled to a squeezed thermal reservoir (Dum *et al.*, 1992). Hence the steady state of mode 'B' is not thermal. Its steady-state population is

$$\bar{n}_B = \langle \hat{B}^\dagger \hat{B} \rangle_s = \frac{\langle \hat{X}^2 \rangle_s + \langle \hat{Y}^2 \rangle_s - 1}{2}, \tag{11.87}$$

with

$$\langle \hat{X}^2 \rangle_s = \bar{n}_B - \bar{m}_B + \frac{1}{2}, \tag{11.88}$$

$$\langle \hat{Y}^2 \rangle_s = \bar{n}_B + \bar{m}_B + \frac{1}{2}, \tag{11.89}$$

$\hat{X}$ and $\hat{Y}$ being its quadrature operators. For the case $\delta_i < -1$ and $\bar{n}_a = 0$ we find to second order in g

$$\bar{n}_B = \left[1 + \frac{4\delta_i g^2 \kappa}{\gamma(\delta_i^2 - 1)^2}\right]\bar{n}_b + \frac{\kappa}{\gamma}\left(\frac{g}{1 - \delta_i}\right)^2, \tag{11.90}$$

$$\Gamma_B = \gamma + (\gamma - \kappa)\frac{4g^2\delta_i}{(\delta_i^2 - 1)^2}. \tag{11.91}$$

Both the steady-state population and the effective decay rate of the polariton 'B' are therefore close to the phonon case for $\delta_i \to -\infty$. On the other side of the avoided crossing, $-1 < \delta_f < 0$, one finds likewise that they approach the values of the bare phonon mode for $\delta_f \to 0^{(-)}$.

11.7   A Quantum Optomechanical Heat Engine

Armed with the results of the previous sections we now discuss a specific example of optomechanical QHE (Zhang, Bariani and Meystre, 2014*a*) whose working substance is a polariton fluid. The basic idea is that by adiabatically switching the nature of the polaritons from phonon-like to photon-like, and enabling their thermalization with the associated 'hot' and 'cold' reservoirs, it is possible to operate an optomechanical QHE on a quantum Otto cycle. Experimentally this could be realized by varying the detuning Δ while keeping the intracavity optical field α constant. Provided that non-adiabatic transitions between the two polariton branches 'A' and 'B' can be avoided, or at least significantly reduced, each of them is associated with its own cycle. In the following we focus on the Otto cycle of the 'B' branch, ignoring for now the complications associated with the fact that in practice it is not possible to fully avoid non-adiabatic transitions and the dissipative coupling between the two branches.

11.7.1   The Four Strokes

We assume that the optomechanical system is initially in thermal equilibrium at large red-detuning Δ_i, in dimensionless units $\delta_i \equiv \Delta_i/\omega_m \sim -\infty$, so that the phonon-like lower polariton branch 'B' is in thermal equilibrium with a reservoir at effective temperature T_{Bi}—for all practical purposes the temperature of the phonon heat bath. For that same detuning the photon-like upper polariton branch 'A' is in thermal equilibrium with a reservoir at temperature T_{Ai}, with $T_{Ai} \approx 0K$ at optical frequencies for systems operating at room temperature or below, so that the initial polariton populations are related by $\langle \hat{n}_B \rangle_i \sim \bar{n}_b \gg \langle \hat{n}_A \rangle_i \sim \bar{n}_a = 0$.

The first stroke of the cycle is an adiabatic change of Δ from Δ_i to a small negative final value Δ_f, in dimensionless units $\delta_f \equiv \Delta_f/\omega_m \to 0^{(-)}$. This step must be fast enough that the interaction of the system with the thermal reservoirs can be largely neglected, yet slow enough that non-adiabatic transitions between the two polariton branches remain negligible. At the end of that stroke the lower branch polariton 'B' has become photon-like. In the second stroke it is then allowed to reach thermal equilibrium at the temperature $T_{Bf} \approx 0K$ of the photon reservoir. The third stroke involves switching the detuning back to its initial large negative value Δ_i. Again, this step must be fast compared to κ^{-1} to avoid thermalization, yet slow enough to avoid non-adiabatic 'A' $\leftrightarrow$ 'B' transitions. The fourth and final stroke is the re-thermalization at the detuning Δ_i to the temperature T_{Bi} for the lower polariton branch and to T_{Ai} for the upper branch.

During the first stroke the phonon-like thermal excitations, which are initially large due to the contact of the 'B' polariton mode with a thermal reservoir essentially at the temperature of the mechanics, become photon-like. The amplitude of vibrations of the mechanics decreases, with the excess energy transferred to the intracavity field. As a result of the increased radiation pressure the resonator length increases slightly. It is at this point that mechanical work is produced by the engine, but this work is quite small due to the disproportion between the steady amplitudes and the quantum fluctuations of the photon and phonon fields. During the thermalization of stroke 2 the population of photon-like excitations decays at rate κ, with the cavity length unchanged. In stroke 3 the remaining photon-like excitations are then turned back to phonon-like. This results in a small contraction of the cavity length and costs a small amount of work. In the final stroke 4 the polariton population, now back to phonon-like, grows back to its initial value via thermal contact with the mechanical reservoir.

11.7.2 Average Work and Efficiency

Classically, and assuming no significant coupling between the 'A' and 'B' polariton branches, the heat exchanged with the reservoirs and the work performed by the engine during the various strokes are given by $W_{1,B} = E_{2,B} - E_{1,B}$, $W_{3,B} = E_{4,B} - E_{3,B}$, $Q_{2,B} = E_{3,B} - E_{2,B}$ and $Q_{4,B} = E_{1,B} - E_{4,B}$, with $W_{1,B} + W_{3,B} + Q_{2,B} + Q_{4,B} = 0$. Here $E_{i,B}, i = 1, \ldots, 4$ are the energies of the 'B' polariton at the nodes of the four strokes. From Eq. (11.2), the work is positive if the change in internal energy of the system is positive, that is, if work is performed *on* the engine, and negative if it is performed *by* the engine. In the absence of heat exchange this corresponds to work performed *on* the system.[4]

The efficiency of cycle 'B' is the ratio between the total work and the input heat (Abah, Roßnagel, Jacob, Deffner, Schmidt-Kaler, Singer and Lutz, 2012)

$$\eta_B = \frac{-W_{B,\text{tot}}}{Q_{4,B}} = 1 - \frac{\omega_{Bf}}{\omega_{Bi}}. \tag{11.92}$$

Figure 11.3 shows η_B and the absolute value $|W_{B,\text{tot}}|$ of the work as functions of δ_f and the dimensionless interaction strength $g = G/\omega_m$, see Zhang (Zhang, Bariani and

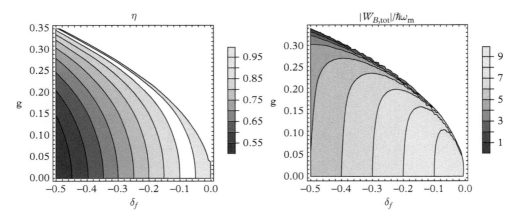

Fig. 11.3 *Contour maps of the total efficiency, η, and the absolute value of the total work, $|W_{B,\text{tot}}|$, in units of $\hbar\omega_m$, for the 'B'-branch Otto cycle, as a function of the dimensional optomechanical coupling $g = G/\omega_m$ and the dimensionless final detuning $\delta_f = \Delta_f/\hbar\omega_m$. Here $\delta_i = -3$, and the thermal mean photon and phonon number are $\bar{n}_a = 0$ and $\bar{n}_b = 10$, respectively. The white region is mechanically unstable. Reprinted with permission from (Zhang, Bariani and Meystre, 2014a). Copyright (2014) by the American Physical Society.*

---

[4] The opposite convention, where work done *by* the engine is defined as positive, is also frequently used. In an attempt to limit confusion we denote work defined in that way as W_{out}, with obviously $W_{\text{out}} = -W$.

Meystre, 2014b) for details. ($W_{B,\text{tot}} < 0$ for our choice of parameters.) To second order in g^2, and with δ_i large and negative and $\delta_f = 0^{(-)}$ one finds $\eta \approx 1 - (-\delta_f - 2g^2)$, which is maximum for $g^2 = -\delta_f/2$. The total average work is also readily found to be $W_{B,\text{tot}} \approx \hbar\omega_m(-\delta_f - 2g^2 - 1)[(1 - 2g^2)\bar{n}_b - g^2]$. It is minimum for $g^2 = -\delta_f/4 - \hbar\omega_m/(8k_BT_b)$, assuming that the phonon temperature T_b is high enough that $\bar{n}_b \approx (k_BT_b/(\hbar\omega_m)) - 1/2$. This yields the efficiency at maximum power

$$\eta_P = 1 - \left(\frac{-\Delta_f}{2\omega_m} + \frac{\hbar\omega_m}{4k_BT_b} \right), \tag{11.93}$$

which with the help of the triangle inequality gives

$$\eta_P < 1 - \sqrt{\frac{-\hbar\Delta_f}{2k_BT_b}}. \tag{11.94}$$

This is the quantum version of the Curzon–Ahlborn efficiency limit $1 - \sqrt{T_{\text{low}}/T_{\text{high}}}$ (Curzon and Ahlborn, 1975).

11.7.3   Bare modes physics

We saw in sections 11.2 and 11.5 that the average work and heat are defined in terms of the energy eigenstates of the system, in the present case the polariton modes. However, physical measurements are not performed on these quasiparticles—there are no polariton detectors—but rather on the bare modes, most easily on the optical field. It is therefore important to also understand the physics of the QHE in terms of the photon and phonon fields. We concentrate in the following on the first stroke of the QHE, where work is produced by the engine (Zhang, Bariani and Meystre, 2014b). In this discussion we ignore for simplicity the polariton 'A', whose population remains negligible throughout the cycle if non-adiabatic transitions and dissipative coupling between the two branches remain unimportant.

Since under these simplifying assumptions the first stroke of the engine is adiabatic, we have in the polariton picture that $d\hat{\rho}_B = 0$, where $\hat{\rho}_B$ is the density operator of the normal mode 'B'. From section 11.2.1 it follows that $dQ_B = 0$ so that

$$dU = d\langle \hat{H}_B \rangle = dW_B = \text{Tr}[\hat{\rho}_B d\hat{H}_B] \tag{11.95}$$

where

$$\hat{H}_B = \hbar\omega_B(\Delta)\hat{B}^\dagger\hat{B}. \tag{11.96}$$

In the bare modes representation the description of the physics is somewhat more complex: the system obviously still doesn't exchange any heat with its reservoirs, so that we have again that $dU = dW_B$, but now its explicit form is

$$dU = d\langle \hat{H}_{(ab)} \rangle = \mathrm{Tr}[d\hat{\rho}_a \hat{H}_a] + \mathrm{Tr}[\hat{\rho}_a d\hat{H}_a] + \mathrm{Tr}[d\hat{\rho}_b \hat{H}_b] + \mathrm{Tr}[\hat{\rho}_b d\hat{H}_b]$$
$$+ \mathrm{Tr}[d\hat{\rho}_{ab} \hat{V}] + \mathrm{Tr}[\hat{\rho}_{ab} d\hat{V}]$$

where $\hat{\rho}_{ab}$ is the density operator of the two-mode system and $\hat{\rho}_a$ and $\hat{\rho}_b$ are the reduced density operators of the photon and phonon mode, respectively. Since the external control only acts on the detuning $\Delta(t)$ it follows that $d\hat{H}_b = d\hat{V} = 0$ and we have

$$dU = dW_B = d\mathcal{Q}_a + dW_a + d\mathcal{Q}_b + \mathrm{Tr}[d\hat{\rho}_{ab} \hat{V}]. \tag{11.97}$$

where $d\mathcal{Q}_a \equiv \mathrm{Tr}[d\hat{\rho}_a \hat{H}_a]$, $d\mathcal{Q}_b \equiv \mathrm{Tr}[d\hat{\rho}_b \hat{H}_b]$ and $dW_a \equiv \mathrm{Tr}[\hat{\rho}_a d\hat{H}_a]$. If the initial population of the photon mode is zero, we have $dW_a = 0$, and, accounting for the changes in photon and phonon populations during the first stroke, we conclude that $d\mathcal{Q}_a > 0$ and $d\mathcal{Q}_b < 0$, so that

$$dW_B = d\mathcal{Q}_a + d\mathcal{Q}_b + \mathrm{Tr}[d\hat{\rho}_{ab} \hat{V}_{ab}], \tag{11.98}$$

where the last term is the change in quantum correlations between the photon and phonon fields. It is initially zero for a product of thermal states, but becomes finite as a result of the optomechanical coupling. This term, which does not have a classical thermodynamical correspondent, is much smaller than $d\mathcal{Q}_a$ and $d\mathcal{Q}_b$ for the weak optomechanical couplings considered here.

If one were to blindly associate terms of the form $\mathrm{Tr}[d\hat{\rho}\hat{H}]$ with infinitesimal changes in average heat dQ, then Eq. (11.98) could be interpreted as follows: the phonon mode releases 'heat', part of it being absorbed by the photon field, a small amount contributing to the generation of quantum correlations, and the rest being absorbed by the external control field. However, it is not correct to interpret the terms $d\mathcal{Q}_a$ and $d\mathcal{Q}_b$ as true heat exchange processes. The form $dQ = \mathrm{Tr}[d\hat{\rho}\hat{H}]$ only holds in the energy basis of the *full* system Hamiltonian, not for its subsystems taken independently. In Eq. (11.98) these terms are *not* associated with the irreversible exchange of energy with a reservoir, in the fashion discussed in section 11.2.1. Their origin is instead the unitary Hamiltonian evolution of the coupled photon–phonon system. This is why we label them with the symbols $d\mathcal{Q}_a$ and $d\mathcal{Q}_b$ rather than dQ_a and dQ_b.

11.7.4   Numerical Simulations, Mean Values

Figure 11.4 summarizes results of a numerical simulation of a full cycle of the heat engine that fully accounts for non-adiabatic transitions and of dissipative coupling between the lower and upper polariton branches. The four curves on the figure, obtained by numerically solving the master equation (11.82), show the time dependence of the average populations of the bare modes 'a' and 'b' and of the polariton modes 'A' and 'B' for a full cycle of the engine. This example assumes a small thermal mean phonon number $\bar{n}_b = 4$. For a mechanical resonator of frequency $\omega_m = 2\pi \times 200$ MHz this corresponds to a phonon reservoir temperature $T_b = 45$ mK, which is in the parameter

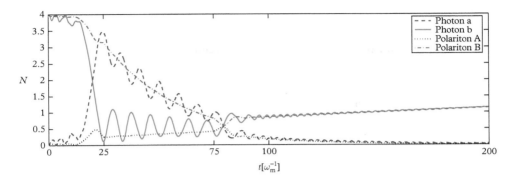

Fig. 11.4 *Time evolution of the mean numbers of photons (blue dashed line), phonons (red solid line), polariton 'A' (black dotted line) and 'B' (pink dot-dashed line) during a full cycle of the Otto engine. Here* $G = 0.2\omega_m$, $\kappa = 0.03\omega_m$, *and* $\gamma = 10^{-3}\omega_m$, *all normalized to* $\omega_m = 2\pi \times 200$ MHz. *The stroke times are* $\tau_1 = \tau_3 = 25\omega_m^{-1}$, $\tau_2 = 50\omega_m^{-1}$, *and* $\tau_4 = 10^4\omega_m^{-1}$. *Reprinted with permission from (Zhang, Bariani and Meystre, 2014b). Copyright (2014) by the American Physical Society.*

range of current optomechanical experiments. The initial mean photon number is $\bar{n}_a = 0$, an excellent approximation for visible frequencies and an optomechanical device operating at room temperature or below. The initial detuning is $\Delta_i = -3\omega_m$ and we choose the final detuning $\Delta_f = -0.4\omega_m$ to avoid the unstable region near the cavity resonance. For the small optomechanical coupling considered here it is appropriate to start from the factorized thermal state $\hat{\rho}_{\text{sys}}(0) = \hat{\rho}_{a,\text{th}} \otimes \hat{\rho}_{b,\text{th}}$, where $\hat{\rho}_{a,\text{th}}$ and $\hat{\rho}_{b,\text{th}}$ are the thermal states of the photon and the phonon mode. All other parameters are given in the figure caption.

During the first stroke the populations of the polariton modes 'A' and 'B' initially coincide to an excellent approximation with the mean photon and phonon number, respectively. In the ideal case, as Δ is varied from Δ_i to Δ_f the photon and phonon mode would simply exchange populations, while keeping the mean polariton numbers constant. In practice, the variation of the detuning is not quite slow enough to avoid non-adiabatic transitions between the two polariton branches. In addition, dissipative coupling between them results in a small decrease in the polariton 'B' population and corresponding increase in the polariton 'A' population.

During the second stroke the photon-like polariton 'B' decays fast due to the cavity decay, while the thermalization of the phonon-like polariton 'A' at the much slower rate γ remains negligible. This step is also characterized by Rabi oscillations between photon and phonon populations due to the optomechanical coupling. The polariton 'B' then recovers its phonon-like properties in the third stroke and finally thermalizes back to its initial population at the end of the last stroke (not shown, at $t > 10^4\omega_m^{-1}$). The total cycle takes a time of the order 10^{-4} s. Note that as the population of polariton 'A' remains small throughout the whole cycle, its effect on the engine can be safely neglected as we did in the considerations of the previous section. More details on these simulations can be found in Zhang (Zhang, Bariani and Meystre, 2014a).

11.8   Quantum Fluctuations

We mentioned earlier that it is not possible to directly monitor the occupation of the polariton modes. What is experimentally accessible instead is the intracavity optical field. Fortunately the average work (11.102) performed by the engine can be expressed in terms of changes in the mean intracavity photon number, as we show below. Building on that observation we then discuss two protocols that permit one to extract information on both the work performed by the QHE and its fluctuations through continuous quantum measurements of that field following the protocols discussed in section 11.4.

11.8.1   Average Work Measurement

Since the effective optical reservoir temperature is $T_a = 0$ for visible frequencies and systems operating at room temperature or below, it follows that $\bar{n}_B = n_a = 0$ after thermalization of the 'B' polariton in its photon-like form at the end of the second stroke. It follows that no work is produced during the third stroke of the Otto cycle if non-adiabatic transitions and dissipative coupling between the polariton branches can be neglected. We can then restrict the determination of the work to the first adiabatic stroke as we did in section 11.7.2 so that

$$dW_B = \mathrm{Tr}[\hat{\rho}_B(d\hat{H}_B)] = \hbar \bar{n}_B d\omega_B \qquad (11.99)$$

and the average work is simply (Dong, Zhang, Bariani and Meystre, 2015*a*)

$$W_B = \int_{\Delta_i}^{\Delta_f} dW_B = \hbar \bar{n}_B [\omega_B(\Delta_f) - \omega_B(\Delta_i)]. \qquad (11.100)$$

Alternatively one can also work in the bare modes representation, where the photon and phonon distributions are time-dependent, as we have seen. In that picture

$$dW_B = \mathrm{Tr}[\hat{\rho}_{ab}(d\hat{H}_{(ab)})] = -\hbar \bar{n}_a(\Delta) d\Delta \qquad (11.101)$$

where $\bar{n}_a(\Delta) = \langle \hat{a}^\dagger \hat{a} \rangle (\Delta)$ is the average number of photons at detuning Δ and we have exploited the fact that the only term in the Hamiltonian (11.67) with a dependence on $\Delta(t)$ is the free field part $-\hbar\Delta(t)\hat{a}^\dagger \hat{a}$. In that representation the average work produced in the first stroke of the engine is therefore

$$W_B = -\hbar \int_{\Delta_i}^{\Delta_f} \bar{n}_a(\Delta) d\Delta. \qquad (11.102)$$

If the change in detuning is perfectly adiabatic the values of the average work obtained from the expressions (11.100) and (11.102) are identical. However if either the optical or the mechanical damping is significant on the timescale of the adiabatic stroke, or

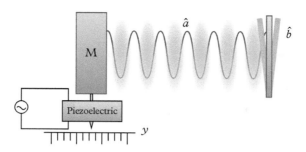

Fig. 11.5 *Schematic setup for a classical measurement of the output work, whereby the work performed by radiation pressure acting on the mirror of large mass M can be stored in the control system. Reprinted with permission from (Dong, Zhang, Bariani and Meystre, 2015a). Copyright (2015) by the American Physical Society.*

if the variation of the optical detuning is sufficiently fast that it induces non-adiabatic transitions, then Eq. (11.100) is no longer exact, but Eq. (11.102) remains valid.

This suggests using the idealized setup sketched in Fig. 11.5 to measure the average work of the QHE. The main difference with the 'canonical' optomechanical arrangement is that the position y of the input mirror, which is normally assumed to be fixed, is now controlled externally by a potential $V(y)$ provided, for example, by a piezoelectric element. Its mass is taken to be extremely large compared to that of the oscillating end mirror driven by radiation pressure, so that its motion can be described classically.

To understand how this arrangement permits one to measure W_B we first note that the classical control of the detuning $\Delta(t)$ can be achieved through a change in cavity length. With this mapping of detunings onto positions the average work of Eq. (11.102) can be expressed in terms of the spatial integral of the position-dependent radiation pressure force

$$F_{\rm rp}(y) = \hbar g_M \bar{n}_a(y), \tag{11.103}$$

where g_M is the optomechanical coupling normalized to the mirror zero-point motion y_M, as

$$W_B = \int_{y_i}^{y_f} F_{\rm rp}(y) \, dy. \tag{11.104}$$

Accounting for the dynamics of the massive control mirror the full optomechanical system is described by the Hamiltonian

$$\hat{H}_m = \hat{H}_{(ab)} + H_M, \tag{11.105}$$

where the *classical* Hamiltonian for the input mirror

$$H_M = \frac{p^2}{2M} + V(y) \qquad (11.106)$$

includes the potential $V(y)$ that controls $\Delta(y)$ in the presence of the radiation force F_{rp}. The classical equations of motion for that mirror are

$$\frac{dy}{dt} = \frac{\partial H_m}{\partial p} = \frac{p}{M}, \qquad (11.107)$$

$$\frac{dp}{dt} = -\frac{\partial H_m}{\partial y} = -\frac{\partial V(y)}{\partial y} - F_{rp}, \qquad (11.108)$$

where H_m is the classical limit of $\hat{H}_m$. If the mass M is large enough that it can be considered as essentially infinite compared to all other optomechanical elements we have $dy/dt \approx dp/dt \approx 0$. That is, the force exerted by the control system balances the expectation value of the radiation pressure force,

$$-\frac{\partial V(y)}{\partial y} = F_{rp}. \qquad (11.109)$$

This demonstrates that, provided the kinetic energy of the large mirror remains essentially zero, all work performed by the photons is converted into the control potential energy and can be measured in that way. This confirms the intuitive result that the work can be measured by sensing the radiation pressure force, which is proportional to the mean number of intracavity photons.

11.8.2 Continuous Quantum Measurements

Building on this classical result we now consider two continuous quantum measurement protocols to monitor both the mean intracavity photon number and its fluctuations, thereby inferring the statistical properties of the work generated by the QHE and the measurement back-action (Dong, Zhang, Bariani and Meystre, 2015a). Both are realized by propagating a beam of two-state atoms of transition frequency ω_0 across the resonator, with a flux low enough that at most one atom at a time interacts with the cavity mode for a short time τ. The atom–field interaction is resonant in the first scheme, and dispersive in the second. Measuring the state of the successive atoms as they exit the resonator provides information on the field, following well-established methods of cavity QED.

Absorptive measurements

We first consider the resonant situation $\omega_0 = \omega_c$ and assume that the atoms are prepared in their ground state $|g\rangle$ before being injected inside the cavity, see Fig. 11.6(a). The atom-field coupling is then given in the rotating wave approximation by

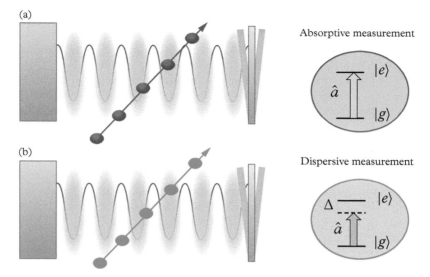

Fig. 11.6 *Schematic setup for the continuous quantum measurement of the output work of the QHE by a beam of two-level atoms. (a) Absorptive measurements: the cavity field is resonant with the atomic transition and the coupling induces real oscillations in the atomic population, resulting in the loss of intracavity photons. (b) Dispersive measurements: the cavity mode frequency is far off-resonant from the two-level atom transition frequency, resulting in a dispersive interaction that modifies the phase of the atomic ground state wave function. Reprinted with permission from (Dong, Zhang, Bariani and Meystre, 2015a). Copyright (2015) by the American Physical Society.*

$$\hat{V}_a = \hbar g_a(\hat{a}^\dagger \hat{\sigma}_{ge} + \hat{a}\hat{\sigma}_{eg}) \tag{11.110}$$

where g_a is the single-photon Rabi frequency of the transition and $\hat{\sigma}_{ij} = |i\rangle \langle j|$. The optical field induces Rabi oscillations in the atomic population, which can then be monitored by projective measurements on the atoms after they exit the cavity. Because of the atom–field entanglement resulting from the interaction (11.110) this typically results in the loss of cavity photons, with important consequences for the measurement back-action as we shall see.

Each cycle of the QHE is characterized by a stochastic measurement sequence, or quantum trajectory 'j' described by a stochastic wave function $|\psi_j(t)\rangle$, as discussed in section 11.4. Since the interaction (11.110) corresponds to measurements of the field amplitude, these wave functions are described by the stochastic Schrödinger equation (11.63)

$$d|\psi_j(t)\rangle = \left\{ \left[-\frac{i}{\hbar}\hat{H}_{(ab)} - \frac{1}{2}\lambda_a \left(\hat{a}^\dagger \hat{a} - \langle \hat{a} + \hat{a}^\dagger \rangle \hat{a} + \frac{\langle \hat{a} + \hat{a}^\dagger \rangle^2}{4} \right) \right] dt \right.$$
$$\left. + \sqrt{\lambda_a} \left(\hat{a} - \frac{\langle \hat{a} + \hat{a}^\dagger \rangle}{2} \right) dw \right\} |\psi_j(t)\rangle \tag{11.111}$$

where dw is a Wiener process and $\lambda_a = g_a^2 \tau$ is a measure of the strength of the measurement (Imoto, Ueda, and Ogawa, 1990), (Jacobs and Steck, 2006).

The term proportional to λ_a on the right-hand side of Eq. (11.111) accounts for the measurement-induced dissipation of the intracavity field, a consequence of the absorption of photons by the successive probe atoms previously mentioned. The stochastic term proportional to $\sqrt{\lambda_a}$ and to the Wiener process dw describes the stochastic changes of the intracavity field about its expected value $\langle \hat{a} + \hat{a}^\dagger \rangle$ resulting from the measurement outcomes.

Dispersive measurements

In this second measurement scheme the atoms are far off-resonant from the optical field and prepared in the coherent superposition $|+\rangle = (|e\rangle + |g\rangle)/\sqrt{2}$ of the ground and excited states before entering the resonator. Information on the intracavity field is then inferred from a change in the phase of the atomic state.

For far off-resonant interactions between the two-level atoms and the intracavity field mode the upper electronic state can be adiabatically eliminated in the usual fashion, resulting in the effective Hamiltonian

$$\hat{V}_d = \hbar g_d \hat{a}^\dagger \hat{a} (\hat{\sigma}_{ee} - \hat{\sigma}_{gg}) = \hbar g_d \hat{a}^\dagger \hat{a} (\hat{\sigma}_{+-} + \hat{\sigma}_{-+}), \tag{11.112}$$

where $|\pm\rangle = (|e\rangle \pm |g\rangle)/\sqrt{2}$ and $g_d = g_a^2/2\delta$ is the off-resonant effective Rabi frequency coupling, with $\delta = \omega_c - \omega_0$. The effect of the measurements on the optical field is now described by the stochastic Schrödinger equation (11.64)

$$d|\psi_j(t)\rangle = \left\{ \left[-\frac{i}{\hbar}\hat{H}_{(ab)} - \frac{1}{2}\lambda_d \left(\hat{n}_a - \langle \hat{n}_a \rangle \right)^2 \right] dt + \sqrt{\lambda_d}(\hat{n}_a - \langle \hat{n}_a \rangle) dw \right\} |\psi_j(t)\rangle. \tag{11.113}$$

where $\lambda_d = g_d^2 \tau$ (Ueda, Imoto, Nagaoka and Ogawa, 1992).

Although like Eq. (11.111) this equation comprises two contributions, the underlying physics that they describe accounts for the important differences in the back-action of the two measurement schemes on the QHE. Specifically, because the non-resonant atom-field coupling $\hat{V}_d$ of Eq. (11.112) is a quantum non-demolition interaction for the photon number $\hat{n}_a = \hat{a}^\dagger \hat{a}$, the dissipative channel of Eq. (11.111) is replaced by a number-conserving term that results in additional damping of the phase of the optical field. Importantly $\hat{V}_d$ also couples the polariton branches 'A' and 'B' and transfers excitations between them. This is in contrast to the resonant interaction (11.110), which does not significantly couple the two normal modes. Still, both measurement schemes result in the appearance of additional photon loss channels that limit the amount of extractable work.

Fluctuations

Solving the stochastic Schrödinger equations (11.111) and (11.113) repeatedly for a large number N of cycles of the QHE generates a set of trajectories $|\psi_j(t)\rangle$ that can be used to evaluate the statistics of any field observable. Just like in the examples of section 11.3.2, we use the state vector $|\psi_j(t)\rangle$ (or the density operator $\hat{\rho}_j(t)$) associated with each trajectory to operationally define a stochastic variable associated with the work along that trajectory for the time interval t_i to t_f as (Horowitz, 2012), (Solinas, Averin and Pekola, 2013)

$$W_j = \text{Tr}[\hat{\rho}_j \hat{H}_{(ab)}(t_f)] - \text{Tr}[\hat{\rho}_j \hat{H}_{(ab)}(t_i)] - \int_{t_i}^{t_f} \text{Tr}[(\partial_t \hat{\rho}_j)\hat{H}_{(ab)}(t)]dt \qquad (11.114)$$

where $\hat{\rho}_j(t) = |\psi_j(t)\rangle\langle\psi_j(t)|$ is the stochastic density operator associated with the jth trajectory, see Eq. (11.12). In terms of the state vectors $|\psi_j(t)\rangle$, which can conveniently be used instead of the density operators $\hat{\rho}_j(t)$ to trade memory for time requirements in the numerical simulations, this expression reduces to

$$W_j = \int_{t_i}^{t_f} \langle\psi_j(t)| \frac{\partial \hat{H}_{(ab)}}{\partial t} |\psi_j(t)\rangle dt. \qquad (11.115)$$

The difference between the first and second terms on the right side of Eq. (11.114) gives the internal energy difference of the system, and we identify the third term as the heat exchanged with the reservoirs along the trajectory. While admittedly leaving open some fundamental questions about the definition of work in open quantum systems, operationally this is how an experimentalist would extract the work performed by radiation pressure on the QHE.

The average and the variance of the work are then obtained as the first and second moments of this distribution,

$$W \equiv \sum_{j=1}^{N} \frac{W_j}{N}, \qquad (11.116)$$

and

$$\Delta W^2 \equiv \sum_{j=1}^{N} \frac{(W_j - W)^2}{N}. \qquad (11.117)$$

In the limit $N \to \infty$, the statistics resulting from the quantum trajectories approach the correct result. The numerical simulations presented in the following section are based on this approach.

11.8.3 Simulations

This section presents results from numerical simulations of the continuous measurement of the work output and its fluctuations.For each set of parameters they were obtained by averaging over typically 20,000 trajectories obtained from the stochastic Schrödinger

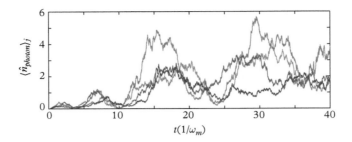

Fig. 11.7 *Typical trajectories during the first stroke of the QHE, illustrating both the stochastic character of the intracavity mean photon number $\langle \hat{n}_a \rangle$ inferred from continuous dispersive measurements as well as the variations from cycle to cycle. In these examples $\bar{n}_a = 0$, $\bar{n}_b = 4$, $g = 0.2$, $\kappa = 5 \cdot 10^{-3}$, $\gamma = 10^{-4}$, $G = 0.2$, and $\lambda_d = 0.04$. The initial state of the intracavity field is taken to be a Fock state $|4\rangle$. Time in units of $1/\omega_m$.*

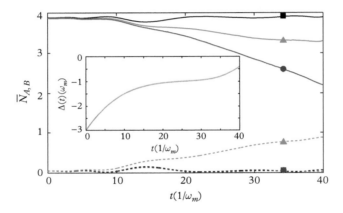

Fig. 11.8 *Time evolution of the mean excitations of the polariton modes 'B' (solid lines) and 'A' (dashed lines) during the first stroke of the heat engine, averaged over 20,000 trajectories of the stochastic Schrödinger equation. Black lines marked by squares: no measurement. Red lines marked by triangles: dispersive measurements with $\lambda_d = 0.04\omega_m$. Blue lines with circles: absorptive measurements with $\lambda_a = 0.04\omega_m$. The other parameters are $G = 0.2\omega_m$, $\Delta_i = -3\omega_m$, $\Delta_f = -0.4\omega_m$, $\kappa = 5 \times 10^{-3}\omega_m$ and $\gamma = 10^{-4}\omega_m$. Inset: time dependence of the pump-cavity detuning $\Delta(t)$. Time in units of $1/\omega_m$. Reprinted with permission from (Dong, Zhang, Bariani and Meystre, 2015a). Copyright (2015) by the American Physical Society.*

equations (11.111) or (11.113). A few of these trajectories are shown in Fig. 11.7. Details and additional results can be found in Dong (Dong, Zhang, Bariani and Meystre, 2015a).

Figure 11.8 shows the average populations $\langle n_A \rangle (t)$ and $\langle n_B \rangle (t)$ of the 'A' and 'B' polaritons during the first stroke of the engine. The insert shows the time variation of the detuning $\Delta(t)$ (in units of ω_m) used in these simulations, with $\Delta(t)$ changing rapidly away from the avoided crossing and slowly in its vicinity to reduce the amount of non-adiabatic coupling between the polariton modes. As a result non-adiabatic transitions remain largely negligible, as illustrated by the black curves labelled by a square in Fig. 11.8. In the absence of measurements the population of the 'B' polariton mode is essentially constant and the 'A' polariton population remains nearly zero. The additional curves show $\bar{n}_A(t)$ and $\bar{n}_B(t)$ during the first stroke of the heat engine, again neglecting mechanical and optical damping, for both dispersive (red lines with triangles) and absorptive (blue lines with circles) measurements.

Comparing the mean polariton populations for absorptive and dispersive measurements illustrates the difference in their back-action on the operation of the QHE. For absorptive measurements the 'B' polariton population decreases significantly during what would otherwise be an adiabatic, population-conserving stroke. Since that stroke occurs rapidly compared to κ^{-1}, the observed damping results solely from the additional photon dissipation at rate λ_a due to the measurements, the term proportional to λ_a in Eq. (11.111). On the other hand these measurements do not result in any significant transfer of population to the 'A' polariton.

The situation is qualitatively different for dispersive measurements, with no significant loss in *total* polariton population, but with a significant transfer of population from mode 'B' to mode 'A'. The stochastic phase fluctuations associated with dispersive measurements change the frequency of the photons randomly, as seen by the term proportional to $\hat{a}^\dagger \hat{a}$ in Eq. (11.113), thereby changing the structure of the polaritons. The population transfer between polariton modes results in a reduction in the work performed by the engine that can become quite dramatic since the Otto cycle associated with the polariton branch 'A' is reversed and produces positive work, that is, negative *output* work (Zhang, Bariani and Meystre, 2014*b*).

Figure 11.9 shows the *output* work probability distribution $P(W_{\text{out}})$ in the absence of measurements as well as for two measurement strengths, for both dispersive and absorptive measurements. (Remember that as discussed in section 11.7.2 we have that $W_{\text{out}} = -W$ with our sign convention for the work W.) Without measurements $P(W_{\text{out}})$ exhibits a series of peaks that correspond to the initial thermal Fock states phonon distribution of the mechanics. Each of these is converted to a photonic Fock state during the first adiabatic stroke and produces a specific amount of work. The width of the peaks is due to residual non-adiabatic effects.

Due to the stochastic nature of the detection process, continuous measurements result in a broadening of these peaks. In the case of absorptive measurements one also observes a decrease in amplitude of all peaks except for the one corresponding to the vacuum field. This is a consequence of the additional photon decay channel associated with those measurements, which also results in a clear reduction in variance ΔW^2 as the measurement strength is increased (Dong, Zhang, Bariani and Meystre, 2015*b*). In contrast, for dispersive measurements the distribution $P(W_{\text{out}})$ shifts toward negative values of W_{out} as a consequence of the coupling with the 'A'-polariton engine cycle which, as we have seen, tends to be characterized by negative output work W_{out}. Because absorptive measurements don't couple the polariton modes in any significant way, this effect is almost completely absent in that case.

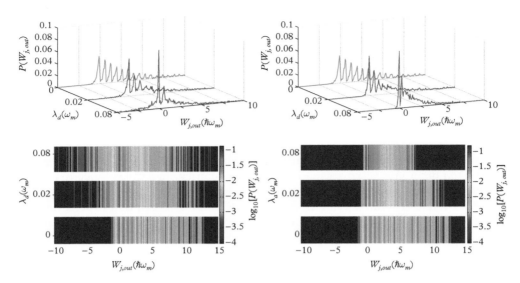

Fig. 11.9 *Linear (upper panel) and logarithmic (lower panel) plots of the probability distribution of the output work, W_{out} (in units of ω_m) for dispersive (left column) and absorptive (right column) measurement respectively in the absence of measurement and for two measurement strength values which label the axes. All parameters are as in Fig. 11.8. Reprinted with permission from (Dong, Zhang, Bariani and Meystre, 2015a). Copyright (2015) by the American Physical Society.*

11.9   Polaritonic Heat Pump

The operation of the optomechanical QHE relies on polariton dispersion management to change the nature of the 'B' polaritons from photon-like to phonon-like, combined with the fact that photons and phonons are coupled to thermal baths at different temperatures. Other applications of this general idea can also be considered. One possibility is to exploit the properties of the 'A' mode, which we have largely ignored so far and which is photon-like at large negative detunings and phonon-like at small negative detunings. We now illustrate how this property can be exploited in a polaritonic heat pump that can be used to cool one or more mechanical modes to their quantum ground state (Dong, Zhang, Bariani and Meystre, 2015b) in a way that might prove useful for modes that can't readily be cooled by conventional sideband cooling.

Conventional cavity-assisted optomechanical sideband cooling relies on the use of a driving laser field tuned to the red side of a cavity resonance (see e.g. (Marquardt, Chen, Clerk and Girvin, 2007), (Wilson-Rae, Nooshi, Zwerger and Kippenberg, 2007), (Aspelmeyer, Kippenberg and Marquardt, 2014)) and is typically applied to the cooling of a single mechanical mode. However in the emerging area of nonlinear quantum phononics ((Abdi, Bahrampour and Vitali, 2012), (Seok, Buchmann, Singh and Meystre, 2012), (Xuereb, Genes and Dantan, 2012),(Xuereb, Genes, Pupillo, Paternostro and Dantan, 2014), (Kipf and Agarwal, 2014), (Mahboob, Okamoto, Onomitsu and Yamaguchi, 2014)) there is much interest in simultaneously cooling two or more phonon modes of arbitrary frequencies deep into the quantum regime (Metzger, Favero, Ortlieb and Karrai, 2008). This can be achieved by the proposed heat pump, which combines the use of the 'A' mode as a cooling agent with a sequential external modulation of the mechanical membrane to parametrically couple two or more of its modes (Patil, Chakram, Chang and Vengalattore, 2015).

The basic idea is to change the nature of polariton 'A' from photon-like for large negative detunings Δ, in which case it is coupled to a thermal reservoir at $T_a = 0$, to phonon-like for small negative detunings. It is then parametrically coupled to the phonon mode(s) to be cooled just long enough to realize an approximate quantum state transfer between them. The detuning is then adiabatically returned to its large negative value, where the polariton is again photon-like and left to thermalize by dumping its excitations to the reservoir at $T_a = 0$.

This cooling scheme is reminiscent of a conventional heat pump operating in the cooling regime, with, however, a subtle difference. Refrigerators do not require reservoirs colder than their target temperatures. They satisfy the Second Law of thermodynamics, which states that one can only transfer heat from hot to cold, by expanding or compressing a cooling fluid. When the fluid is in the refrigerator it is colder than the target temperature, and when it exchanges the heat that it has accumulated to the outside it is warmer than the external environment. In contrast, the polariton heat pump satisfies the Second Law by changing the *nature* of a fluid of quasiparticles rather than its *temperature* to make it first colder than the mechanical mode to be cooled, and then warmer than the environment into which it irreversibly dumps its energy. Instead of moving a cooling fluid spatially between a compressor and an expansion valve to change its temperature, the role of the expansion phase is now realized by changing the polariton fluid from photon-like to phonon-like, and the role of the compression stage by its return to photon-like. The heat exchange between the mode(s) to be cooled and the fluid is achieved by phonon population transfer, and heat dumping is achieved by coupling the polariton fluid in its photon-like form to a thermal reservoir at $T \approx 0$.

To describe the operation of the polaritonic heat pump we consider again the generic optomechanics arrangement where a single mode of an electromagnetic cavity is driven by a laser at frequency ω_p detuned from the cavity mode frequency by $\Delta = \omega_p - \omega_{\mathrm{cav}}$ and optomechanically coupled to a single vibration mode of a compliant mechanical resonator of frequency ω_m and annihilation and creation operators $\hat{b}$ and $\hat{b}^\dagger$. This mode can in turn be coupled to a second mechanical oscillation mode via external actuation. The total Hamiltonian of the system is

$$\hat{H} = \hat{H}_{\mathrm{om}} + \hat{H}_{\mathrm{st}} + \hat{H}_{\mathrm{dis}}. \qquad (11.118)$$

Here $\hat{H}_{\mathrm{om}}$ accounts for the optomechanical coupling between the cavity mode and the mechanical mode at frequency ω_m, where we have changed the notation of Eq. (11.67) from $\hat{H}_0$ to $\hat{H}_{\mathrm{om}}$ for notational clarity, $\hat{H}_{\mathrm{st}}$ describes the parametric interaction between the mechanical modes, and $\hat{H}_{\mathrm{dis}}$ accounts for the intracavity field and mechanical oscillators damping of rates κ and γ. The coupling $\hat{H}_{\mathrm{st}}$ between the phonon mode 'b' at frequency ω_m and the mechanical mode 'c' at frequency ω_c to be cooled can be realized by actuating the substrate at a frequency ω_s close to their frequency difference, $\omega_s \approx \omega_c - \omega_m$ (Patil, Chakram, Chang and Vengalattore, 2015). In the rotating wave approximation and a rotating frame at the modulation frequency ω_s it is described by the 'state transfer' Hamiltonian

$$\hat{H}_{\mathrm{st}} = \hbar\delta\hat{c}^\dagger\hat{c} + \hbar\Omega_0(t)(\hat{b}^\dagger\hat{c} + \hat{c}^\dagger\hat{b}) \tag{11.119}$$

where $\hat{c}$ and $\hat{c}^\dagger$ are the annihilation and creation operators for mode 'c',

$$\delta = \omega_c - \omega_s \simeq \omega_m, \tag{11.120}$$

and Ω_0 is the parametric coupling frequency, proportional to the amplitude of oscillations of the substrate.

11.9.1   Cooling Cycle

The first step of the cooling cycle changes the polariton fluid from photon-like to phonon-like. The effect of this transformation is to leave it in a state where the mode to be cooled can transfer heat to it. Since its purpose is the same as the expansion of the cooling fluid in conventional heat pumps, we loosely refer to it as the 'expansion' step in the following. The second step is the 'heat exchange' interaction, at the end of which the thermal excitations of mechanical mode 'c' have been transferred to the polariton and it is effectively cooled. The third step is the adiabatic transformation of the polariton back to its photon-like state. This is the analogue of the compression stage in conventional heat pumps, following which its energy can be dumped into the photon thermal reservoir at $T \approx 0$.

The 'expansion' is realized by adiabatically changing the detuning $\Delta(t)$ from a large negative value $\Delta_i \ll -\omega_m$ to a small negative value $-\omega_m \ll \Delta_f < 0$ close to 0. The parametric mechanical coupling is turned off during this step, $\Omega_0 = 0$. If perfectly adiabatic this transformation conserves the initial thermal populations of the two polariton modes: the polariton 'A' remains unpopulated since $T_a = 0$ while the polariton 'B' stays at the temperature of mode 'b'. Much like in the QHE discussed earlier the duration τ_1 of this step must satisfy two conflicting conditions: it should be slow enough to guarantee the adiabaticity of the evolution and avoid transitions between the two polariton branches, $\tau_1 \gg 1/(2g)$, and at the same time it should be fast compared to the photon and phonon damping times, $\tau_1 \ll 1/\kappa, 1/\gamma$. This is because it is desirable to avoid as much as possible the contamination of the polariton fluid 'A' by thermal polaritons from branch 'B'. Furthermore, the intracavity field amplitude α should ideally be kept constant while varying the detuning Δ so that the resulting optomechanical coupling strength g remains fixed, although this is not a strong requirement.

Once the detuning has reached the value Δ_f the state transfer coupling $\Omega_0(t)$ between the mechanical modes 'b' and 'c' is switched on for a duration τ_2. At this point the polariton 'A' is phonon-like and well approximated by the phonon mode 'b'. In the absence of other interactions and for a properly adjusted interaction time the Hamiltonian (11.119) accomplishes a perfect quantum state transfer, which for thermal states can be regarded as 'heat exchange'. Since the previous step removed the thermal occupation of the mechanical mode 'b' by transferring it into photons, this ensures maximal efficiency to the process. In practice both optomechanical coupling and dissipation are still effective during τ_2, but this does not change the dynamics significantly for large detunings between the photon and phonon modes and short enough τ_2.

Neglecting these effects, one can estimate the mean excitation transfer between the two modes from $\hat{H}_{st}$ only. For Ω_0 constant during τ_2, the Heisenberg equations of motion $d\hat{b}/dt = -i(\omega_b\hat{b} + \Omega_0\hat{c})$ and $d\hat{c}/dt = -i(\delta\hat{c} + \Omega_0\hat{b})$ yield readily

$$\langle\hat{n}_c\rangle(t) = \langle\hat{n}_c\rangle(0) + \left[\langle\hat{n}_b\rangle(0) - \langle\hat{n}_c\rangle(0)\right]\left(\Omega_0^2/\Omega^2\right)\sin^2(\Omega t), \qquad (11.121)$$

where $\Omega = [(\omega_b - \delta)^2/4 + \Omega_0^2]^{1/2}$ plays the role of an effective Rabi frequency. The maximum exchange of excitation between the two modes occurs after the interaction time $\tau_2 = \pi/(2\Omega)$, but we have verified numerically that it is not necessary to achieve perfect state transfer to achieve essentially ground state cooling of mode 'c'. It is sufficient that at the end of that stage the average excitation $\langle\hat{n}_c\rangle$ of mode 'c' is lower than $\langle\hat{n}_b\rangle$. The result of imperfect state transfer is merely the need for an increased number of cooling cycles.

Following the heat exchange process the polariton mode 'A' is returned to its photon-like nature by adiabatically changing Δ_f back to Δ_i. The final step is the thermalization of the now photon-like polariton with its reservoir at $T_a = 0$ over a time $\tau_4 \gg \kappa^{-1}$. Repeating the full cycle allows one in principle to achieve ground state cooling of mode 'c' from room temperature.

This intuitive description is confirmed by full numerical simulations of a master equation that includes all three modes (Dong, Zhang, Bariani and Meystre, 2015b). Results of such a simulation are summarized in Fig. 11.10. Here the initial phonon number in mode 'c' is a low $\langle\hat{n}_c\rangle(0) = 12$ for numerical convenience, while the polariton mode 'A' is photon-like and has a very small thermal population. (It has the unrealistically large value $\langle\hat{n}_A\rangle(0) = 0.5$ in that example to illustrate the role of thermal photons.) In contrast the polariton mode 'B' is phonon-like, with $\langle\hat{n}_B\rangle(0) = 2$.

For short times $t \leq \tau_1$ both $\langle\hat{n}_A\rangle$ and $\langle\hat{n}_c\rangle$ remain essentially constant. During that time the detuning Δ is adiabatically changed from Δ_i to Δ_f, and the polariton and phonon modes are uncoupled. The adiabaticity of the 'expansion' step is confirmed by the fact that $\langle\hat{n}_A\rangle$ remains approximately constant, with modest non-adiabatic transitions between the two polariton modes. In the short interval between $\tau_1 < t \leq \tau_1 + \tau_2$ the interaction between modes 'b' and 'c' is switched on. Since $\hat{A} \approx \hat{b}$ at Δ_f, we observe an almost perfect mean population transfer between polariton 'A' and mode 'c'. Meanwhile the mean population of polariton 'B' remains essentially unchanged, which confirms the validity of a simple two-mode description (11.121) of that step.

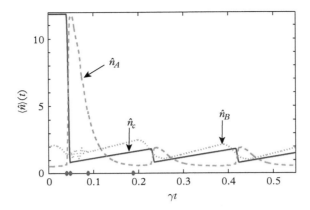

Fig. 11.10 *Mean populations $\langle \hat{n}_A \rangle(t)$ (green dashed line), $\langle \hat{n}_B \rangle(t)$ (dotted dotted line) and $\langle \hat{n}_c \rangle(t)$ (solid blue line) for a few cooling cycles deep in the quantum regime. Here $\omega_b = \delta = 2 \times 10^3$, $\kappa = 40$, $\Delta_i = -6 \times 10^3$, $\Delta_f = -6 \times 10^2$, $g = \Omega_0 = 2 \times 10^2$, $\tau_1 = \tau_3 = 4 \times 10^{-2}$, $\tau_2 = 8 \times 10^{-3}$, and $\tau_4 = 0.1$, with times in units of γ^{-1}. The four dots mark the beginning and end of the 'heat exchange' step, the beginning of thermalization, and the end of the first cooling cycle. Here $\langle \hat{n}_A \rangle(0) = 0.5$, $\langle \hat{n}_B \rangle(0) = 2$ and $\langle \hat{n}_c \rangle(0) = 12$. Reprinted with permission from (Dong, Zhang, Bariani and Meystre, 2015b). Copyright (2015) by the American Physical Society.*

The third stroke, of duration τ_3, is a near-adiabatic change of the detuning Δ back from Δ_f to Δ_i. Here non-adiabatic effects are most apparent in oscillations of $\langle \hat{n}_B \rangle(t)$. This step lasts until $\gamma t \approx 0.08$ in Fig. 11.10. Finally, heat dissipation into the environment results in the decay of the mean population $\langle \hat{n}_A \rangle(t)$ of mode 'A', which is now photon-like, to its thermal equilibrium value n_a. At the same time, though, the phonon mode 'c' is also coupled to a thermal reservoir and consequently $\langle \hat{n}_c \rangle(t)$ slowly increases. Subsequent cooling cycles permit to keep it at a value near the quantum ground state, as illustrated by the next two cooling cycles in Fig. 11.10.

11.9.2   Cooling Limit

To evaluate the cooling limit we first consider the 'heat exchange' step. At the small negative detuning Δ_f where it occurs the polariton mode 'A' consists almost entirely of the phonon mode 'b', with a small photonic component that plays essentially no role, and mode 'B' is far detuned and remains uncoupled. Under these conditions we can approximate the Hamiltonian (11.119) by

$$\hat{H}_{\text{st}} \approx \hbar \delta \hat{c}^\dagger \hat{c} + \hbar \Omega_0'(t)(\hat{A}^\dagger \hat{c} + \hat{c}^\dagger \hat{A}), \tag{11.122}$$

a form that results from carrying out the Bogoliubov transformation that diagonalizes $\hat{H}_{\text{om}}$, dropping the counter-rotating terms in the resulting interaction between the polariton 'A' and mode 'c', and omitting the coupling to polariton 'B' altogether. The effective coupling strength is $\Omega'_0 = u\Omega_0$, where u is the Bogoliubov transformation coefficient between modes 'A' and 'b', with $u \approx 1$ for $\Delta_f \to 0^{(-)}$.

The expression for the mode dynamics obtained from Eq. (11.122) has the same analytical form as Eq. (11.121), with $\Omega_0 \to \Omega'_0$. For the jth cooling cycle and the optimal choice $\Omega' t = \pi/2$, with $\Omega' = [\omega_A(\Delta_f) - \delta]^2/4 + \Omega'^2_0]^{1/2}$, the population $\langle \hat{n}_{c,\text{out}}\rangle_j$ at the end of the heat exchange step is related to its beginning value $\langle \hat{n}_{c,\text{in}}\rangle_j$ by

$$\langle \hat{n}_{c,\text{out}}\rangle_j = (1 - \eta)\langle \hat{n}_{c,\text{in}}\rangle_j + \eta \langle \hat{n}_{A,\text{in}}\rangle_j, \qquad (11.123)$$

where $\eta = (\Omega'_0/\Omega')^2$ and $\langle \hat{n}_{A,\text{in}}\rangle_j$ is the mean number of 'A' polaritons, with $\langle \hat{n}_{A,\text{in}}\rangle_j = \bar{n}_a$ if that mode properly thermalized at the temperature of the photon reservoir at the end of the previous cycle.

Between the heat exchange steps the mode 'c' is decoupled from the polaritons and only subject to thermalization, so that

$$\langle \hat{n}_{c,\text{in}}\rangle_j = \bar{n}_c + r(\langle \hat{n}_{c,\text{out}}\rangle_{j-1} - \bar{n}_c), \qquad (11.124)$$

where $r = \exp[-\gamma(\tau_1 + \tau_2 + \tau_3 + \tau_4)]$ and $\bar{n}_c$ is the mean phonon number at its reservoir temperature. Substituting Eq. (11.124) into Eq. (11.123) and taking the asymptotic limit $\langle \hat{n}_{c,\text{out}}\rangle_{j-1} = \langle \hat{n}_{c,\text{out}}\rangle_j$ gives then the cooling limit

$$\langle \hat{n}_{c,\text{out}}\rangle_\infty = \frac{1}{1 - r(1 - \eta)}\left[\eta \bar{n}_a + (1 - r)(1 - \eta)\bar{n}_c\right]. \qquad (11.125)$$

The contribution of the phonon thermal noise is fully suppressed for $\eta = 1$, which corresponds to the resonance condition $\omega_A(\Delta_f) = \delta$. With $\omega_A(\Delta_f) \approx \omega_b$ and Eq. (11.120) this gives $\omega_s = \omega_c - \omega_b$, which is precisely the resonance condition for the substrate modulation frequency to establish the state transfer coupling between the phonon modes 'b' and 'c'. This ideal case yields the fundamental limit $\langle \hat{n}_{c,\text{out}}\rangle_\infty = \bar{n}_a$. Ideally the phonon mode 'c' can therefore be cooled to the temperature of the electromagnetic field, $T_a \approx 0$ for visible radiation. The polaritonic heat pump also opens up the possibility of simultaneously cooling multiple modes by using an appropriate sequence of parametric couplings to address different modes (Dong, Zhang, Bariani and Meystre, 2015b).

11.10 Outlook

The pace of development of quantum optomechanics is breathtaking. The numerous designs that achieve optomechanical control via radiation pressure effects in high-quality resonators range from nanometre-sized devices with as little as 10^7 atoms and a mass of 10^{-20} kg to micromechanical structures of 10^{14} atoms and 10^{-11} kg, and to the centimetre-sized mirrors used in gravitational wave detectors comprising more than 10^{20} atoms and weighing up to several kilos. The mechanical elements can serve as transducers that enable the coupling between otherwise incompatible systems.

The examples considered in this chapter involved only photons and phonons, but the motional degree(s) of freedom can be coupled to a broad range of other physical systems, including spins, electric charges, atoms, molecules and 'artificial atoms.' This opens up a rich spectrum of possibilities that are only starting to be explored, both for fundamental studies and for device applications.

The main goal of this chapter was to illustrate how quantum optomechanics also provides a powerful testbed to investigate fundamental questions in quantum thermodynamics and to explore the boundary between the classical and the quantum worlds in situations where thermal and quantum noise coexist. In both the heat engine and heat pump examples that we discussed a central aspect was the use of a polariton fluid as a working substance. In the quantum regime, these normal modes are comprised of quasiparticles whose physical nature can be controlled by changing the details of the coupling between their constituents. By exploiting this remarkable 'quantum werewolf' capability to change the nature of the polaritons from photon-like to phonon-like, and in doing so by controlling the environment to which their constituents are coupled, including its effective temperature, one can manipulate the interplay between quantum coherence and dissipation channels in a way that permits one to realize thermodynamic systems that are not possible with classical systems, thereby opening up powerful new approaches to the realization of QHE and heat pumps that we have discussed.

Although we focused on situations where the phonon modes are coupled to a reservoir at a higher temperature than the optical field, this need not be the case. The assumption that the temperature of the optical bath is essentially $T = 0$, so that the phonon bath is by default the 'hot' reservoir, is usually an excellent approximation in the optical regime, but this need not be so in general. For example, in the microwave regime the 2.7 K cosmic black-body background results in significant photon occupation numbers for frequencies in the 10^2 GHz range. By the same token, it is also possible to realize quantum mechanical oscillators that operate essentially at $T = 0$, for instance in ultracold atomic gases (see e.g. (Brahms, Botter, Schreppler, Brooks and Stamper-Kurn, 2012), (Brennecke, Ritter, Donner and Esslinger, 2008). This suggests that it might be possible to exchange the roles of photons and phonons in optomechanical heat engines, provided that the mechanical oscillator is cold enough. A key requirement is that the temperature of the atomic system must be low enough that thermal motion does not wash out the momentum recoil $2\hbar k$ of the atoms due to their interaction with photons of wave vector k, the condition under which ultracold atomic systems can behave as coherent mechanical oscillators. As an example, for a condensate of lithium atoms this results in the requirement that the temperature of the sample does not exceed a pK or so for $2\pi \times 300$ GHz microwave photons. While challenging, this does not seem completely impossible. If realized, a quantum heat engine operating on the upper polariton branch 'A' would therefore be able to extract heat energy from the cosmic microwave background! While such ideas are highly speculative, one cannot rule out that future technological advances will some day change that situation.

More generally one can think of a number of other systems of quasiparticles that can be controlled and manipulated coherently so that they morph into essentially just one or the other of their constituents, which are coupled to environments at very different temperatures and/or to reservoirs with distinct quantum features. This may open up intriguing new avenues for the development of heat engines and other quantum thermodynamical systems with no direct classical counterparts, and expand the playground on which to explore basic questions at the interface between classical and quantum physics.

Acknowledgements

I am pleased to acknowledge my colleagues and collaborators Francesco Bariani, Ying Dong, Keye Zhang, and Weiping Zhang, with whom much of the work on quantum heat engines and heat pumps was carried out. I have also benefited from numerous discussions with Aash Clerk, Keith Schwab, Swati Singh, Mukund Vengalattore, and Ewan Wright. This work was supported by NSF, ARO and the DARPA QuASAR and ORCHID programs through grants from AFOSR and ARO.

Bibliography

Abah, O., Roßnagel, G., Jacob, J., Deffner, S., Schmidt-Kaler, F., Singer, K., and Lutz, E. (2012). Single-ion heat engine at maximum power. *Phys. Rev. Lett.*, **109**, 203006.

Abdi, M., Bahrampour, A. R., and Vitali, D. (2012). Quantum optomechanics of a multimode system coupled via a photothermal and a radiation pressure force. *Phys. Rev. A*, **86**, 043803.

Allahverdyan, A. E. and Nieuwenhuizen, Th. M. (2005). Fluctuations of work from quantum subensembles: The case against quantum work-fluctuation theorems. *Phys. Rev. E*, **71**, 066102.

An, S., Zhang, J. N., Um, M., Lv, D., Lu, Y., Zhang, J., Yin, Z. Q., Quan, H. T., and Kim, K. (2015). Experimental test of the quantum Jarzynski equality with a trapped-ion system. *Nat. Phys.*, **11**, 193.

Apollaro, T. J. G., Francica, G., Paternostro, M., and Campisi, M. (2015). Work statistics, irreversible heat and correlations build-up in joining two spin chains. *Physica Scripta*, **T165**, 014023.

Aspelmeyer, M., Kippenberg, T., and Marquardt, F. (2014). Cavity optomechanics. *Rev. Mod. Phys.*, **86**, 1391.

Aspelmeyer, M., Meystre, P., and Schwab, K. (2012). Quantum optomechanics. *Physics Today (7)*, **65**, 29.

Batalhão, T. B., Souza, A. M., Mazzola, L., Auccaise, R., Sarthour, R. S., Oliveira, I. S., Goold, J., De Chiara, G., Paternostro, M., and Serra, R. M. (2014). Experimental reconstruction of work distribution and study of fluctuation relations in a closed quantum system. *Phys. Rev. Lett.*, **113**, 140601.

Bergenfeldt, C., Samuelsson, P., Sothmann, B., Flindt, C., and Büttiker, M. (2014). Hybrid microwave-cavity heat engine. *Phys. Rev. Lett.*, **112**, 076803.

Boukobza, E. and Tannor, D. J. (2006). Thermodynamics of bipartite systems: Application to light-matter interactions. *Phys. Rev. A*, **74**, 063823.

Brahms, N., Botter, T., Schreppler, S., Brooks, D. W. C., and Stamper-Kurn, D. M. (2012). Optical detection of the quantization of collective atomic motion. *Phys. Rev. Lett.*, **108**, 133601.

Brennecke, F., Ritter, S., Donner, T., and Esslinger, T. (2008). Cavity optomechanics with a Bose–Einstein condensate. *Science*, **322**, 235.

Campisi, M., Blattmann, R., Kohler, S., Zueco, D., and Hänggi, P. (2013). Employing circuit QED to measure non-equilibrium work fluctuations. *New J. Phys.*, **15**, 105028.

Campisi, M., Hänggi, P., and Talkner, P. (2011). Quantum fluctuation relations: Foundations and applications. *Rev. Mod. Phys.*, **83**, 771.

Campisi, M., Talkner, P., and Hänggi, P. (2009). Fluctuation theorem for arbitrary open quantum systems. *Phys. Rev. Lett.*, **102**, 210401.

Campisi, M., Talkner, P., and Hänggi, P. (2010). Fluctuation theorems for continuously monitored quantum fluxes. *Phys. Rev. Lett.*, **105**, 140601.

Carmichael, H. J. (2008). *Statistical Methods in Quantum Optics, Vol. 2*. Springer-Verlag, Berlin Heidelberg.

Curzon, F. L. and Ahlborn, B. (1975). Efficiency of a Carnot engine at maximum power output. *Am. J. Phys.*, **43**, 22.

Dong, Y., Zhang, K., Bariani, F., and Meystre, P. (2015a). Work measurement in an optomechanical quantum heat engine. *Phys. Rev. A*, **92**, 033854.

Dong, Y., Zhang, K., Bariani, F., and Meystre, P. (2015b). Phonon cooling by an optomechanical heat pump. *Phys. Rev. Lett.*, **115**, 223602.

Dorner, R., Goold, J., Cormick, C., Paternostro, M., and Vedral, V. (2012). Emergent thermodynamics in a quenched quantum many-body system. *Phys. Rev. Lett.*, **109**, 160601.

Dum, R., Parkins, A. S., Zoller, P., and Gardiner, C. W. (1992). Monte Carlo simulation of master equations in quantum optics for vacuum, thermal, and squeezed reservoirs. *Phys. Rev. A*, **46**, 4382.

Esposito, M., Harbola, U., and Mukamel, S. (2007). Entropy fluctuation theorems in driven open systems: Application to electron counting statistics. *Phys. Rev. E*, **76**, 031132.

Esposito, M., Harbola, U., and Mukamel, S. (2009). Nonequilibrium fluctuations, fluctuation theorems, and counting statistics in quantum systems. *Rev. Mod. Phys.*, **81**, 1665.

Esposito, M. and Mukamel, S. (2006). Fluctuation theorems for quantum master equations. *Phys. Rev. E*, **73**, 046129.

Esposito, M. and Van den Broeck, C. (2010). Three detailed fluctuation theorems. *Phys. Rev. Lett.*, **104**, 090601.

Fialko, O. and Hallwood, D. W. (2012). Isolated quantum heat engine. *Phys. Rev. Lett.*, **108**, 085303.

Frenzel, M. F., Jennings, D., and Rudolph, T. (2014). Reexamination of pure qubit work extraction. *Phys. Rev. E*, **90**, 052136.

Fusco, L., Pigeon, S., Apollaro, T. J. G., Xuereb, A., Mazzola, L., Campisi, M., Ferraro, A., Paternostro, M., and De Chiara, G. (2014). Assessing the nonequilibrium thermodynamics in a quenched quantum many-body system via single projective measurements. *Phys. Rev. X*, **4**, 031029.

Geusic, J. E., Scovil, H. E. D., and Schultz-Dubois, E. O. (1967). Quantum equivalent of the Carnot cycle. *Phys. Rev.*, **156**, 343.

Gröblacher, S., Trubarov, A., Prigge, N., Cole, G. D., Aspelmeyer, M., and Eisert, J. (2015). Observation of non-Markovian micromechanical Brownian motion. *Nature Commun.*, **6**, 7606.

Hänggi, P. and Talkner, P. (2015). The other QFT. *Nature Phys.*, **11**, 108.

Hardal, Ü. C. and Müstecaplıoğlu, Ö. E. (2015). Superradiant quantum heat engine. *Scientific Reports*, **5**, 12953.

Hekking, F. W. J. and Pekola, J. P. (2013). Quantum jump approach for work and dissipation in a two-level system. *Phys. Rev. Lett.*, **111**, 093602.

Horowitz, J. M. (2012). Quantum-trajectory approach to the stochastic thermodynamics of a forced harmonic oscillator. *Phys. Rev. E*, **85**, 031110.

Imoto, N., Ueda, M., , and Ogawa, T. (1990). Microscopic theory of the continuous measurement of photon number. *Phys. Rev. A*, **41**, 4127.

Jacobs, K. and Steck, D. A. (2006). A straightforward introduction to continuous quantum measurement. *Contemporary Phys.*, **47**, 279.

Jarzynski, C. (2015). Diverse phenomena, common themes. *Nat. Phys.*, **11**, 105.

Joshi, D. G. and Campisi, M. (2013). Quantum hertz entropy increase in a quenched spin chain. *Eur. Phys. J. B*, **86**, 157.

Kipf, T. and Agarwal, G. S. (2014). Superradiance and collective gain in multimode optomechanics. *Phys. Rev. A*, **90**, 053808.

Mahboob, I., Okamoto, H., Onomitsu, K., and Yamaguchi, H. (2014). Two-mode thermal-noise squeezing in an electromechanical resonator. *Phys. Rev. Lett.*, **113**, 167203.

Marquardt, F., Chen, J. P., Clerk, A. A., and Girvin, S. M. (2007). Quantum theory of cavity-assisted sideband cooling of mechanical motion. *Phys. Rev. Lett.*, **99**, 093902.

Mascarenhas, E., Braganca, H., Dorner, R., Franca Santos, M., Vedral, V., Modi, K., and Goold, J. (2014). Work and quantum phase transitions: Quantum latency. *Phys. Rev. E*, **89**, 062103.

Metzger, C., Favero, I., Ortlieb, A., and Karrai, K. (2008). Optical self cooling of a deformable Fabry–Pérot cavity in the classical limit. *Phys. Rev. B*, **78**, 035309.

Meystre, P. (2013). A short walk through quantum optomechanics. *Annalen der Physik*, **525**, 215.

Meystre, P. and Sargent III, M. (2007). *Elements of Quantum Optics, 4th Edition*. Springer-Verlag, Berlin Heidelberg.

Mukamel, S. (2003). Quantum extension of the Jarzynski relation: Analogy with stochastic dephasing. *Phys. Rev. Lett.*, **90**, 170604.

Patil, Y. S., Chakram, S., Chang, L., and Vengalattore, M. (2015). Thermomechanical two-mode squeezing in an ultrahigh-Q membrane resonator. *Phys. Rev. Lett.*, **115**, 017202.

Quan, H. T., Liu, Y. X., Sun, C. P., and Nori, F. (2007). Quantum thermodynamic cycles and quantum heat engines. *Phys. Rev. E*, **76**, 031105.

Ritort, F. (2009). Viewpoint: Fluctuations in open systems. *Physics*, **2**, 43.

Romero-Isart, O. (2011). Quantum superposition of massive objects and collapse models. *Phys. Rev. A*, **84**, 052121.

Romero-Isart, O., Pflanzer, A. C., Blaser, F., Kaltenbaek, R., Kiesel, N., Aspelmeyer, M., and Cirac, J. I. (2011). Large quantum superpositions and interference of massive nanometer-sized objects. *Phys. Rev. Lett*, **107**, 020405.

Roncaglia, A. J., Cerisola, F., and Paz, J. P. (2014). Work measurement as a generalized quantum measurement. *Phys. Rev. Lett.*, **113**, 250601.

Roßnagel, G., Dawkins, S. T., Tolazzi, K. N., Abah, O., Lutz, E., Schmidt-Kaler, F., and Singer, K. (2016). A single-atom heat engine. *Science*, **352**, 325.

Saira, O. P., Yoon, Y., Tanttu, T., Möttönen, M., Averin, D. V., and Pekola, J. (2012). Test of the Jarzynski and Crooks fluctuation relations in an electronic system. *Phys. Rev. Lett.*, **109**, 180601.

Salmilehto, J., Solinas, P., and Möttönen, M. (2014). Quantum driving and work. *Phys. Rev. E*, **89**, 052128.

Schmiedl, T. and Seifert, U. (2007). Stochastic thermodynamics of chemical reaction networks. *J. Chem. Phys.*, **126**, 044101.

Schmiedl, T., Speck, T., and Seifert, U. (2007). Entropy production for mechanically or chemically driven biomolecules. *J. Stat. Phys.*, **128**, 77.

Scovil, H. E. D. and Schultz-Dubois, E. O. (1959). Three-level masers as heat engines. *Phys. Rev. Lett.*, **2**, 262.

Scully, M. O. (2002). Quantum afterburner: Improving the efficiency of an ideal heat engine. *Phys. Rev. Lett.*, **88**, 050602.

Scully, M. O., Zubairy, M. S., Agarwal, G. S., and Walther, H. (2003). Extracting work from a single heat bath via vanishing quantum coherence. *Science*, **299**, 862.

Seifert, U. (2005). Entropy production along a stochastic trajectory and an integral fluctuation theorem. *Phys. Rev. Lett.*, **95**, 040602.

Seifert, U. (2008). Stochastic thermodynamics: Principles and perspectives. *Eur. Phys. J. B*, **64**, 423.

Sekimoto, K. (2007). Microscopic heat from the energetics of stochastic phenomena. *Phys. Rev. E*, **76**, 060103 (R).

Seok, H., Buchmann, L. F., Singh, S., and Meystre, P. (2012). Optically mediated nonlinear quantum optomechanics. *Phys. Rev. A*, **86**, 063829.

Smacchia, P. and Silva, A. (2013). Work distribution and edge singularities for generic time-dependent protocols in extended systems. *Phys. Rev. E*, **88**, 042109.

Solinas, P., Averin, D. V., and Pekola, J. P. (2013). Work and its fluctuations in a driven quantum system. *Phys. Rev. B*, **87**, 060508 (R).

Suomela, S., Salmilehto, J., Savenko, I. G., Ala-Nissila, T., and Möttönen, M. (2015). Fluctuations of work in nearly adiabatically driven open quantum systems. *Phys. Rev. E*, **91**, 022126.

Talkner, P., Hänggi, P., and Morillo, M. (2008). Microcanonical quantum fluctuation theorems. *Phys. Rev. E*, **77**, 051131.

Talkner, P., Lutz, E., and Hänggi, P. (2007). Fluctuation theorems: Work is not an observable. *Phys. Rev. E*, **75**, 050102 (R).

Ueda, M., Imoto, N., Nagaoka, H., and Ogawa, T. (1992). Continuous quantum-nondemolition measurement of photon number. *Phys. Rev. A*, **46**, 2859.

Wilson-Rae, I., Nooshi, N., Zwerger, W., and Kippenberg, T. J. (2007). Theory of ground state cooling of a mechanical oscillator using dynamical backaction. *Phys. Rev. Lett.*, **99**, 093901.

Xuereb, A., Genes, C., and Dantan, A. (2012). Strong coupling and long-range collective interactions in optomechanical arrays. *Phys. Rev. Lett.*, **109**, 223601.

Xuereb, A., Genes, C., Pupillo, G., Paternostro, M., and Dantan, A. (2014). Reconfigurable long-range phonon dynamics in optomechanical arrays. *Phys. Rev. Lett.*, **112**, 133604.

Zhang, K., Bariani, F., and Meystre, P. (2014*a*). Quantum optomechanical heat engine. *Phys. Rev. Lett.*, **112**, 150602.

Zhang, K., Bariani, F., and Meystre, P. (2014*b*). Theory of an optomechanical quantum heat engine. *Phys. Rev. A*, **90**, 023819.